12th High-Tech Plasma Processes Conference

(HTPP-12)

Journal of Physics: Conference Series Volume 406

Bologna, Italy
24-29 June 2012

ISBN: 978-1-62276-872-1
ISSN: 1742-6588

Printed from e-media with permission by:

Curran Associates, Inc.
57 Morehouse Lane
Red Hook, NY 12571

Some format issues inherent in the e-media version may also appear in this print version.

Printed by Curran Associates, Inc. (2013)

For permission requests, please contact the Institute of Physics
at the address below.

Institute of Physics
Dirac House, Temple Back
Bristol BS1 6BE UK

Phone: 44 1 17 929 7481
Fax: 44 1 17 920 0979

techtracking@iop.org

Additional copies of this publication are available from:

Curran Associates, Inc.
57 Morehouse Lane
Red Hook, NY 12571 USA
Phone: 845-758-0400
Fax: 845-758-2634
Email: curran@proceedings.com
Web: www.proceedings.com

TABLE OF CONTENTS

Journal of Physics: Conference Series **406** (2012) 011001 doi:10.1088/1742-6596/406/1/011001

PREFACE

The High-Tech Plasma Processes - 12th European Plasma Conference (HTPP-12) was held in Bologna (Italy) on June 24-29, 2012.

The conference series started in 1990 as a thermal plasma conference and gradually expanded to include other topic fields as well. Now the High-Tech Plasma Processes - European Plasma Conference (HTPP) is a bi-annual international conference based in Europe with topics encompassing the whole area of plasma processing science.
The aim of the conference is to bring different scientific communities together, facilitate the contacts between science, technology and industry and provide a platform for the exploration of both fundamental topics and new applications of plasmas.

Thanks to the efforts of the conference chairman, Prof. Vittorio Colombo and of the co-chair, Prof. Piero Favia, a well balanced participation from both the communities of thermal and non-thermal plasma researchers was achieved; this resulted in just about 196 attendees from 39 countries, with 8 plenary and 15 invited talks, plus 50 oral and 140 poster contributions.

The present volume of *Journal of Physics: Conference Series* gathers papers from regular contributions of HTPP-12; each contribution submitted for publication has been peer reviewed and the Editors are very grateful to the referees for their careful support in improving the original manuscripts.

In the end, 39manuscripts were accepted for publication, covering different topics of plasma processing science: from plasma fundamentals and modelling to source design and process diagnostics, from nanomaterial synthesis to surface modification, from waste treatment to plasma applications in a liquid environment.
It is an honour to present this volume of *Journal of Physics: Conference Series* and we deeply thank the authors for their enthusiastic and high-grade contribution.

Finally, we would like to thank the conference chairmen, the members of the steering committee, the international scientific committee, the local organizing committee, the organizing secretariat and the financial support from the conference sponsors that allowed the success of HTPP-12.

the Editors of the HTPP 12 Proceedings

Prof. Alain Gleizes (head of the ISC)
Dr. Emanuele Ghedini
Dr. Matteo Gherardi
Dr. Paolo Sanibondi
Dr. Giorgio Dilecce

Bologna, 30th October 2012

htpp

high-tech plasma processes 12

UNDER THE AUSPICES OF:

UNIVERSITÀ DEGLI STUDI DI BARI ALDO MORO

COMUNE DI BOLOGNA

SPONSORS:

ALMA MATER STUDIORUM UNIVERSITÀ DI BOLOGNA

Simba

SEVENTH FRAMEWORK PROGRAMME

CEBORA

Centro Sviluppo Materiali S.p.A.

TEKNA Plasma Systems inc.

umicore

AIR LIQUIDE

EMPA Materials Science & Technology

ewm

HIDEN ANALYTICAL

nac Image Technology

Plasma Sources Science and Technology

12th EUROPEAN PLASMA CONFERENCE

HIGH-TECH PLASMA PROCESSES

24-29 June 2012 - Bologna, Italy

Conference Center San Giovanni in Monte
Aula Giorgio Prodi

TOPICS

Plasma fundamentals, modeling and simulation
Plasma sources, design and characterization
Plasma processes, diagnostics and monitoring

AREAS AND APPLICATION FIELDS

Thermal plasmas
Low pressure and atmospheric non-equilibrium plasmas
Non-equilibrium phenomena
Plasma spraying and surface treatments
Plasma deposition and treatment of polymers
Plasma cutting and welding
Plasma electrodes interaction
Plasma medicine and biomedical applications
Light generation and radiation transport
Plasma synthesis (nanomaterials, fullerenes, polymers, ultrafine powders)
Plasmas and liquids
Material treatment and metallurgy
Plasma processing for microelectronics and micromechanics
Aeronautical and space industry
Plasma aided combustion
Environmental applications (waste abatment, VOC conversion, DeSOx, DeNOx, ...)
Plasmas for energy (fuel cells, solar cells, ...)
Other

PLENARY SPEAKERS ON THE SUBJECT...
"WHERE ARE WE COMING FROM, WHERE ARE WE GOING IN PLASMA SCIENCE?"

Maher I. Boulos *(Tekna Plasma Systems, Canada)*
Riccardo d'Agostino *(Università degli Studi di Bari, Italy)*
Alexander Fridman *(Drexel University, U.S.A.)*
Alix Gicquel *(Université Paris 13, France)*
Mark Kushner *(University of Michigan, U.S.A.)*
John Lowke *(CSIRO, Australia)*
Yuri P. Raizer *(Russian Academy of Science, Russia)*
Toyonobu Yoshida *(University of Tokyo, Japan)*
Michael R. Wertheimer *(École Polytechnique de Montréal, Canada)*

YOUNG SCIENTISTS' SESSION

HTPP12 is offering a Young Scientists' session open to Master students and PhD-students. It is an opportunity for promising young talents to present their work at an international conference. The time for each presentation will be limited to 15 minutes. All attendees to this session will elect their favorite contribution that will win the **HTPP12 Young scientist award**

REGISTRATION

The registration form is available on the website **www.htpp12.com** and has to be sent **no later than 10 June 2012.**

REGISTRATION FEES

- **EARLY REGISTRATION** *(before 15 May 2012)* .... € 480,00
- **LATE REGISTRATION** *(after 15 May 2012)* .... € 530,00
- **STUDENT REGISTRATION** .... € 230,00
- **ACCOMPANYING PERSON** .... € 120,00

OPTIONAL EVENTS

- **Tour of Ferrara** .... € 60,00
- **Tour of Bologna** .... € 30,00
- **Conference banquet** .... € 70,00
- **Conference tour in Ravenna** *(included in the registration fees)*
- **Post Conference tour in Siena and at its Palio**

KEY DATES - DEADLINES

Abstract submission	**15 March 2012**
Early registration	**15 May 2012**
Late registration	**10 June 2012**
Post Conference tour	**30 April 2012**
Hotel reservation	**30 April 2012**

COMMITTEES

STEERING COMMITTEE FOR HTPP CONFERENCES

Vittorio Colombo (Italy)
Alma Mater Studiorum-Università di Bologna, Department of Mechanical Engineering (DIEM) and Research Center for Applied Mathematics (CIRAM)
Alain Gleizes (France)
Université Paul Sabatier, LAPLACE
Daniel Morvan (France)
Université Pierre et Marie Curie - Paris 6, Laboratoire de Génie des Procédés Plasmas et Traitement de Surface
Dimitrios Rapakoulias (Greece)
University of Patras, Plasma Technology Laboratory, Department of Chemical Engineering
Armelle Vardelle (France)
University of Limoges, Laboratoire Sciences des Procédés Céramiques et de Traitements de Surface

LOCAL ORGANIZING COMMITTEE

Vittorio Colombo *(Alma Mater Studiorum - Università di Bologna, Italy)* **CHAIR**
Pietro Favia *(Università degli Studi di Bari, Italy)* **CO-CHAIR**
Emanuele Ghedini *(Alma Mater Studiorum - Università di Bologna, Italy)*
Matteo Gherardi *(Alma Mater Studiorum - Università di Bologna, Italy)*
Paolo Sanibondi *(Alma Mater Studiorum - Università di Bologna, Italy)*
Giorgio Dilecce *(CNR-IMIP, Bari, Italy)*
Luca Lusvarghi *(Università di Modena e Reggio Emilia, Italy)*
Enrico Malfa *(Centro Sviluppo Materiali - C.S.M., Dalmine, Italy)*
Cristina Paradisi *(Università di Padova, Italy)*
Vincenzo Palleschi *(CNR-ICCOM, Pisa, Italy)*
Mauro Vancini *(CEBORA S.p.A., Italy)*

INTERNATIONAL SCIENTIFIC COMMITTEE

Alain Gleizes *(LAPLACE, Université Paul Sabatier, France)* **Head of ISC**
Mikhail Benilov *(Universidade da Madeira, Portugal)*
Luc Bianchi *(CEA DAM Le Ripault, Monts, France)*
Annemie Bogaerts *(University of Antwerp, Belgium)*
Giovanni Bolelli *(Università di Modena e Reggio Emilia, Italy)*
Peter Bruggeman *(Eindhoven University of Technology, The Netherlands)*
David Cameron *(Lappeenranta University of Technology, Finland)*
Jean-Marie Cormier *(CNRS, GREMI, Orléans, France)*
Gérard Degrez *(Université Libre de Bruxelles, Belgium)*
Francesco Fracassi *(Università degli Studi di Bari, Italy)*
Laurent Fulcheri *(MINES Paris Tech, France)*
Vasco Guerra *(Instituto Superior Técnico, Lisboa, Portugal)*
Christoph Hollenstein *(École Polytechnique Fédérale de Lausanne, Switzerland)*
Milan Hrabowsky *(Academy of Science-IPP, Czech Republic)*
Marc Leparoux *(EMPA - Thun, Switzerland)*
Dimitrios Mataras *(University of Patras, Greece)*
Jerzy Mizeraczyk *(Polish Academy of Sciences, Gdansk, Poland)*
Daniel Morvan *(Université Pierre et Marie Curie - Paris 6, France)*
Per Nylen *(West University, Sweden)*
Christian Oehr *(Fraunhofer Institute for Interfacial Engineering and Biotechnology, Germany)*
Stijn Put *(Umicore Research, Belgium)*
Dimitrios Rapakoulias *(University of Patras, Greece)*
Thierry Renault *(Air Liquide-CTAS, France)*
Jochen Schein *(Universität der Bundeswehr München, Germany)*
Micheal Schnick *(Technische Universität Dresden, Germany)*
Frédéric Schuster *(CEA, Saclay, France)*
Andrea Scrivani *(Artec S.p.A., Italy)*
Janos Szlepovgli *(Hungarian Academy of Sciences, IMEC CRC, Hungary)*
Alexander Surov *(Russian Academy of Sciences, St. Petersburg, Russia)*
Michael Tatoulian *(Université Pierre et Marie Curie - Paris 6, France)*
Richard Van De Sanden *(FOM Institute for Plasma Physics Rijnhuizen, Nieuwegein, The Netherlands)*
Armelle Vardelle *(Université de Limoges, France)*
Klaus Dieter Weltmann *(University of Greifswald, Germany)*

For updates visit the website **www.htpp12.com**

SCIENTIFIC SECRETARIAT

Emanuele Ghedini, Paolo Sanibondi and **Matteo Gherardi**
Alma Mater Studiorum - Università di Bologna
Department of Mechanical Engineering (DIEM) and Research Center for Applied Mathematics (CIRAM)
Via Saragozza 8 - 40123 Bologna, Italy
Vittorio Colombo phone: +39 051 20 9 3978
Emanuele Ghedini phone: +39 051 20 9 3984
e-mail: scientific@htpp12.com

ORGANIZING SECRETARIAT

Avenue media®

Via Riva Reno 61 - 40122 Bologna, Italy
phone: +39 051 6564300
fax: +39 051 6564334
e-mail: congressi@avenuemedia.eu
website: www.avenuemedia.eu

Conference Center San Giovanni in Monte - University of Bologna

PLENARY SPEAKERS

"Where are we coming from, where are we going in plasma science?"

MAHER I. BOULOS
University of Sherbrooke - Tekna Plasma Systems and Tekna Advanced Materials

"The Induction Plasma, four decades of Research and development work"

RICCARDO D'AGOSTINO
University of Bari

"Cold Plasma Nano-Technology"

ALEXANDER FRIDMAN
Drexel University, Philadelphia

"Plasma Medicine: Fundamentals and Applications"

ALIX GICQUEL
Université Paris 13

"H2-CH4-(0-ε)B2H6 microwave plasmas for intrinsic and boron-doped diamond single crystal growth"

MARK J. KUSHNER
University of Michigan

"Model Based Design for Non-Equilibrium Plasmas: Reality, Expectation or Fantasy?"

JOHN J. LOWKE
CSIRO Materials Science and Engineering, Sidney

"Plasma Predictions: Past, Present and Future."

YURI P. RAIZER
Institute for Problems in Mechanics of the Russian Academy of Sciences, Moscow

"Self-organization phenomena and formation of the current patterns in the gas discharges"

TOYONOBU YOSHIDA
The University of Tokyo

"Innovation of plasma spray technology"

MICHAEL R. WERTHEIMER
École Polytechnique de Montréal

"Plasmas and polymers: a personal perspective"

Peer review statement

All papers published in this volume of *Journal of Physics: Conference Series* have been peer reviewed through processes administered by the proceedings Editors. Reviews were conducted by expert referees to the professional and scientific standards expected of a proceedings journal published by IOP Publishing.

Journal of Physics: Conference Series **406** (2012) 012001 doi:10.1088/1742-6596/406/1/012001

A large amount synthesis of nanopowder using modulated induction thermal plasmas synchronized with intermittent feeding of raw materials

Yasunori Tanaka, Tatsuya Tsuke, Weixuan Guo, Yoshihiko Uesugi, Tatsuo Ishijima, Shu Watanabe†, Keitaro Nakamura†

Faculty of Electrical and Computer Engineering, Kanazawa University, Kakuma, Kanazawa 920-1192, Japan

‡Research Center for Production & Technology, Nisshin Seifun Group Inc.,5-3-1 Tsurugaoka, Fujimino 356-8511, JAPAN

E-mail: tanaka@ec.t.kanazawa-u.ac.jp

Abstract.

A large amount synthesis method for titanium dioxide (TiO_2) nanopowder is proposed by direct evaporation of titanium powders using Ar-O_2 pulse-modulated induction thermal plasma (PMITP). To realize a large amount synthesis of nanopowder, the PMITP method was combined with the intermittent and heavy load feeding of raw material powder, as well as the quenching gas injection. The intermittent powder feeding was synchronized with the modulation of the coil current sustaining the PMITP for complete evaporation of the injected powder. Synthesized particles by the developed method were analyzed by FE-SEM and XRD. Results indicated that the synthesized particles by the 20-kW PMITP with a heavy loading rate of 12.3 g min^{-1} had a similar particle size distribution with the mean diameter about 40 nm to those with light loading of 4.2 g min^{-1}.

1. Introduction

The inductively coupled thermal plasmas (ICTPs) have been widely adopted as effective heat and chemical species sources for various materials processings such as syntheses of diamond films [1], thermal barrier coatings [2], fullerene synthesis [3], and surface modifications, etc [4]. The ICTP has some advantages like their high enthalpy and high radical density without any contamination. The ICTP is also useful to synthesize nanoparticles of various kinds [5]–[13]. It provides one-step direct processing including rapid evaporation of raw materials injected, and rapid cooling of evaporated materials. It can also offer nanoparticles in non-equilibrium or metastable phase [7, 8]. However, there are some issues about difficulty in controlling synthesized particle size and phase constituent. In addition, the production rate was relatively low compared to liquid phase synthesis for industrial application [14]. A large amount production method of nanopowder is thus strongly desired for future demand in nanotechnology fields.

On the other hand, the authors previously developed some series of modulated induction thermal plasma system including a pulse-modulated induction thermal plasmas (PMITP) system and an arbitrary-waveform-modulated induction thermal plasma (AMITP) system [15, 16]. The PMITP and AMITP are established by the coil current modulated into a rectangular or other waveforms. Such a millisecond modulation can perturb the thermal plasma markedly, and then it can change the temperature and radical densities as well as the gas flow field in the thermal plasma in time domain [15]. Through this modulation, the temperature and radical densities can be controlled intentionally. We found actually that the Ar-N_2 PMITP provides simultaneous control of the nitrogen radical flow and the enthalpy flow for surface modification [17].

We have also studied to adopt the Ar-O_2 PMITP to nanoparticle synthesis of titanium dioxide (TiO_2). The TiO_2 nanoparticles in anatase phase are continuously receiving attention for use as photocatalysts [18], photonic crystals [19], photovoltaic cells [20], gas sensors [21], and a strong deoxidation material used for producing hydrogen gas from water for fuel cells [22]. The rutile type of TiO_2 is also useful mainly to pigment, cosmetic and medical powder due to its UV cut properties [23] , and to electronics material because of its higher dielectric constant and electrical resistance [24]. In our previous works, we have found that the modulation of the coil current can control the size of synthesized nanoparticles, while keeping the weight fraction of anatase-TiO_2 in synthesized nanoparticles to be about 85%–90% almost independent of the modulation condition [25]. It was also found that the size of synthesized nanoparticles depends on the temperature decay rate in the cooling period in the PMITP [26]. Our another work has developed a system for intermittent feeding of raw materials and studied the influence of the intermittent feeding fundamentally [27].

In this paper, we newly propose a large amount synthesis method of nanopowder using the PMITP. For a large amount synthesis of nanopowder, the PMITP was combined with intermittent heavy rate feeding of raw material, and also the quenching gas injection. Raw materials were fed to the PMITP only during the on-time for their efficient evaporation, and were stopped during the off-time for efficient cooling of the vapor by the quenching gas. Synthesized particles were analyzed by the field emission scanning electron microscopy (FE-SEM) and X-Ray Diffraction (XRD). Results indicated that the synthesized particles by the 20-kW PMITP with a heavy loading rate of 12.3 g min^{-1} had a particle size distribution with the mean diameter about 40 nm and the coefficient of variation of 0.53. This distribution was similar to those with light loading of 4.2 g min^{-1}.

2. Concept of a large amount synthesis of nanoparticles using modulated induction plasmas

Our previous work has studied the influence of coil current modulation of PMITP on the synthesized nanoparticles [26]. The PMITP can repetitively produce a high-temperature field during the 'on-time' and a low-temperature field during the 'off-time' in thermal plasmas. Here, the on-time means the time period with the higher current level (HCL), while the off-time is the time period with the lower current level (LCL). It was found from the experiments that a larger modulation degree in the coil current, which means a larger difference between HCL and LCL, decreased the mean diameter of the synthesized nanoparticles [26]. This means that the size of synthesized nanoparticle can be controlled by the shimmer current level (SCL) which is defined as a ratio of LCL to HCL, that is SCL=LCL/HCL. Our next step was concerned with the intermittent powder feeding effect on the synthesized nanoparticles [27]. That result indicated that particle size depended on the timing of the powder intermittent feeding, and that the size was smallest when the powder feeding timing was synchronized with the coil current modulation of the PMITP.

In this paper, we newly propose a method combined with the above methods for a large amount synthesis of nanopowder: the PMITP with intermittent feeding of raw material, and also the quenching gas injection. Figure 1 illustrates a comparison between (a) the conventional method using a non-modulated ICTP with continuous powder feeding, and (b) the proposed method using the PMITP with intermittent powder feeding method for a large amount of nanopowder production. In the conventional method, the raw material powder is continuously

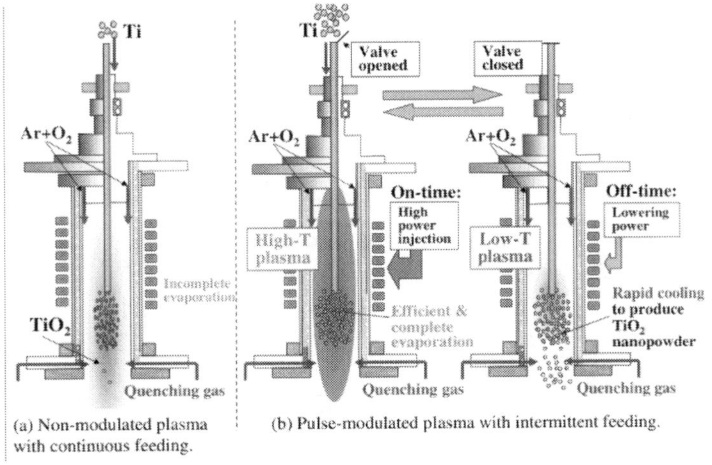

Figure 1. Concept of intermittent feeding of raw material.

fed directly to the ICTP. The ICTP evaporates the injected powder, and then the evaporated material is cooled down by the quenching gas downstream of the torch to create nanoparticles. However, the increasing powder feed rate causes a temperature drop around the center of the ICTP mainly for required latent heat of the material evaporation. Thus, the ICTP cannot completely evaporate the powders injected at heavy rate power feeding rate.

The proposed method uses the PMITP and the intermittent powder feeding technique. A high-speed solenoid valve is installed on the powder feeding tube for intermittent powder feeding. Setting the open and/or close timing of the valve controls the timing and duration of powder feeding. Thus, the developed method can make raw materials powder fed to the PMITP only during the on-time, and stopped during the off-time in the PMITP. In the on-time of the PMITP, the instantaneous input power to the plasma increases, and then it elevates the plasma temperature. Therefore, the injected powder can be evaporated efficiently and completely in the PMITP. On the other hand, in the off-time of the PMITP, the input power decreases to decay the plasma temperature in milliseconds. Meanwhile, the powder feeding is stopped by closing the valve. Thus, the above evaporated materials are efficiently cooled down during this time. In addition, this cool-down efficiency of the evaporated materials is promoted by the quenching gas injected into downstream of the plasma torch. The Ti atom and oxygen atom are cooled down to produce TiO molecules in gas phase, and then nucleation of TiO_2 may occur before nucleation of Ti [11]. After that, TiO_2 condensation, and agglomeration may take place until coagulation.

3. Experimental

3.1. Experimental arrangements

Figure 2 shows the whole nanoparticle synthesis system using the PMITP used in the present work. This whole system has three parts: the rf power source, intermittent powder feeding system, and the plasma torch and the chambers.

The rf power source is operated with a metal-oxide semiconductor field emission transistor (MOSFET) at rated power of 30 kW. This power source can modulate the electric current amplitude of several hundreds of amperes into a rectangular waveform in this work. The modulated coil current has modulation parameters such as the on-time, and the off-time as indicated in the previous section [15]. The shimmer current level (SCL) is defined as LCL/HCL,

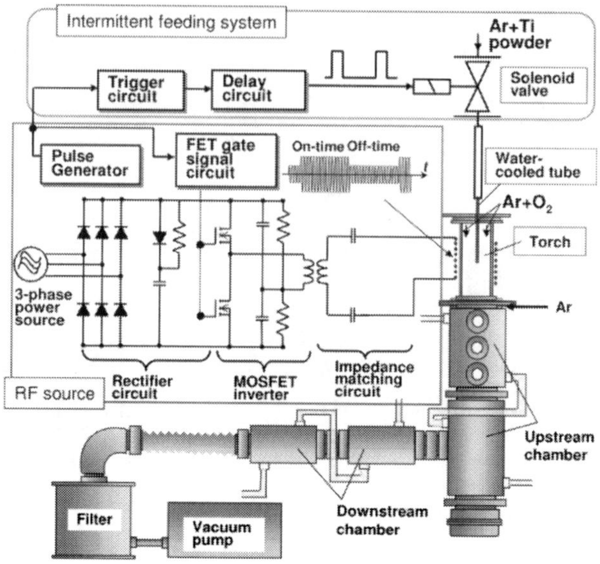

Figure 2. System for intermittent feeding of raw material.

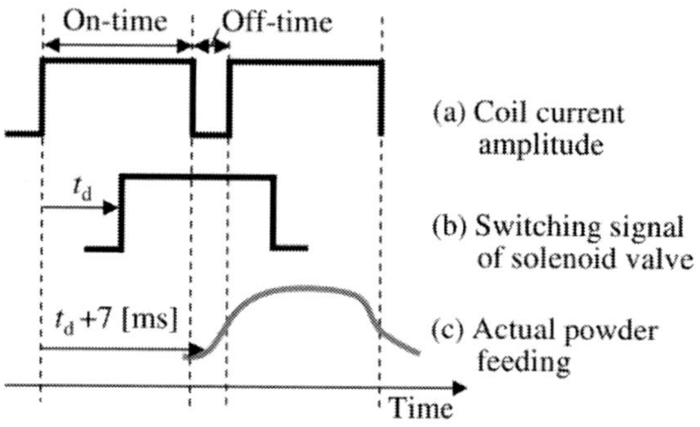

Figure 3. Timing chart in feeding of raw material.

which is often used in the experiment. In addition, the duty factor (DF) is set to a ratio of the on-time to one modulation cycle: DF= 'on-time' /('on-time' +'off-time').

The plasma torch and the chambers in Fig. 2 are configured identically to that used in our previous work [25, 26]. The plasma torch has two coaxial quartz tubes. The inner diameter of the interior quartz tube is 70 mm; its length is 370 mm. Into the plasma torch, a water-cooled tube for powder feeding is inserted from the plasma torch head. Downstream of the plasma torch, quenching gas can be injected in radial direction from eight holes installed on the flange. Further downstream has the water-cooled chambers, separately named as the upstream chamber and the downstream chamber, which are respectively installed in vertical and then in horizontal, as depicted in Fig. 2. Finally, a powder-collecting filter is set up and then a vacuum pump.

The upper side in Fig. 2 portrays the system for intermittent feeding of raw material to the PMITP. The system contains the trigger circuit, the delay circuit and a high speed solenoid valve. The pulse generator in the rf source provides the modulation control signal with 0–10 V to the FET gate signal circuit and also to the trigger circuit in the intermittent feeding system. The FET gate signal circuit controls the fire angle of the MOSFET elements for rf power supply and also for the amplitude modulation of the coil current. On the other hand, the trigger circuit and then the delay circuit controls the opening and closing of a high-speed solenoid valve. This

Table 1. Experimental condition.

Time-averaged input power	20 kW
Fundamental frequency of the coil current	450 kHz
Modulation cycle	15 ms
Shimmer current level, SCL	60%[a] or 80%[b]
Duty factor, DF	80% (On-time / Off-time=12 ms / 3 ms)
Pressure	300 torr (= 40 kPa)
Gas composition	Ar:90 L min^{-1},O_2:10 L min^{-1}
Carrier gas flow rate	Ar:4 L min^{-1}
Quenching gas flow rate	Ar:50 L min^{-1}
Delay time of solenoid valve open t_d	6 ms
Powder feed rate g_{pow}	12.3 g min^{-1}[a] or 4.2 g min^{-1}[b]
Raw powder and its maximum size	Ti, 45 μm

a. SCL=60% for g_{pow}=12.3 g min^{-1}
b. SCL=80% for g_{pow}=4.2 g min^{-1}

solenoid valve has a response time constant of less than 2 ms. Use of this system enables us to intermittently feed the raw material powder to the thermal plasma with controlling feed timing.

Figure 3 schematically illustrates the timing chart of (a) the coil current amplitude, (b) the switching signal of solenoid valve, and (c) the actual powder feeding. The coil current amplitude was assumed to change from LCL to HCL at a certain timing t=0 ms. By using the current modulation signal, the signal in the TTL level was created in the trigger circuit to the delay circuit. The delay circuit outputs a signal with the delay time t_d against the current modulation signal to open the valve. After opening the solenoid valve, the powder was supplied actually with another delay time t_{adt} because the distance between the valve and the tip of the water-cool tube is about 870 mm long. This delay time t_{adt} has been measured to be 6 – 8 ms by spectroscopic observation in our previous work [26] Thus, there is the actual total delay time of $t_d + t_{adt} \sim t_d + 7$ ms before the powder is actually injected from the tip of the water-cooled tube. Controlling t_d makes it possible to change the timing of powder injection intentionally. The powder can be controlled to be supplied only during the on-time of the PMITP.

3.2. Experimental conditions

Table 1 summarizes the experimental condition in the present work. The time-averaged input power was fixed at 20 kW to the inverter power supply. This inverter power supply in the experiment was a semiconductor power supply with a fundamental frequency of 450 kHz. This power supply was measured to have a high power conversion efficiency more than 95%. Thus, the more than 19 kW was outputed from the power supply. The instantaneous input power $P_{in}(t)$ was also periodically changes with time according to the coil current modulation. For 80%SCL-80%DF, the quantity $P_{in}(t)$ increased gradually with time constant about 1.6 ms during the on-time and reached to 28 kW, whereas it decreased with time constant of 2.0 ms in the off-time, reaching to 10 kW. The modulation cycle was fixed at 15 ms. The reason to select the modulation cycle of 15 ms is attributed to the fact that the residence time of the reactant vapor was estimated as 10–20 ms according to the gas flow velocity calculation by numerical thermofluid simulation [25]. It is expected to provide sufficient evaporation of the injected powders in the high-temperature region in the plasma during the on-time, and successive rapid cooling of the plasma tail during the off-time. The DF of the modulated coil current was fixed at 80%. The SCL was set to 60% or 80%.

The total flow rate of Ar-O_2 sheath gas was fixed at 100.0 L min^{-1} (liters per minute). The oxygen gas admixture ratio to Ar was 10% in the sheath gas in the gas flow rate. The gas flow rate of Ar carrier gas was 4 L min^{-1}. Titanium powder with maximum diameter of 45 μm and with mean diameter of 27 μm is fed as the raw material together with Ar carrier gas. The powder was injected through a water-cooled tube probe made of stainless steel. The powder feed rate $g_{\rm pow}$ of the Ti raw material was set at 4.2 g min^{-1} (=7.0 × 10^{-5} kg s^{-1}), and also at 12.3 g min^{-1} (=21 × 10^{-5} kg s^{-1}) as a heavy load feeding condition for nanopowder mass production. The magnitude of this powder feed rate was time-averaged value which is determined as follows: the weight w_1 [g] of Ti raw material powder in the powder feeder was measured before the experiment, and then the weight w_2 [g] of Ti raw material powder in the powder feeder was also measured after the experiment. The time-averaged value of the powder feed rate was calculated as $(w_1 - w_2)/t_{\rm exp}$ where $t_{\rm exp}$ is the experiment time. The definition of the time-averaged feeding rate is necessary because the actual instantaneous feeding rate changes with time by opening and closing the valve intermittently in the present work. Note that powder feeding rate of $g_{\rm pow}$=12.3 g min^{-1} at 20 kW that we used here is eminently a larger amount among the conventional ICTP nanopowder synthesis methods. The conventional methods usually use the rate of 1–4 g min^{-1} for 20–30 kW. Note that the powder feed rate of $g_{\rm pow}$=12.3 g min^{-1} requires 2.3 kW for heating Ti bulk from 300 K through complete melting to complete evaporation.

As described in the previous section, a high-speed solenoid valve was installed on the powder feeding tube for intermittent powder feeding. The response time of the solenoid valve was about 2 ms. The delay time $t_{\rm d}$ of the valve open signal was set to 6 ms against the signal. In this case, the powder is initiated to be supplied at about 13 ms after the current transition from LCL to HCL. This timing corresponds to the synchronized timing of powder feeding with the next on-time of the coil current [27].

Argon quenching gas was injected downstream of the torch in radial direction with a gas flow rate of 50 L min^{-1}. The pressure inside the chamber was controlled at 300 Torr (=40 kPa) with an automatic feedback control for pressure. Synthesized particles were collected in the upstream chamber and the downstream chamber, and in the collecting filter.

4. Results for a heavy rate production of nanoparticles

4.1. SEM images and particle size distribution

Figure 4 presents examples of FE-SEM micrographs of particles synthesized using the PMITP with intermittent powder feeding. The powder feed rate $g_{\rm pow}$ was 12.3 g min^{-1}, i.e., a heavy rate injection. For comparison, those for $g_{\rm pow}$=4.2 g min^{-1} are also indicated in Fig. 5. Panel (a) in these figures portrays the image for particles collected in the upstream chamber, whereas panel (b) shows the one in the collecting filter. The collecting yields were both about 0.5–0.6. As seen in these figures, particles for $g_{\rm pow}$=12.3 g min^{-1} generally have larger sizes than those $g_{\rm pow}$=4.2 g min^{-1}. Nevertheless, an important point is that even a heavy rate feeding condition $g_{\rm pow}$=12.3 g min^{-1} could provide many nanoparticles with diameter sizes less than 100 nm.

Figures 6 and 7 show the particle size distributions for $g_{\rm pow}$=12.3 g min^{-1} and $g_{\rm pow}$=4.2 g min^{-1}, respectively. Similarly, panel (a) in these figures portrays the result for particles collected in the upstream chamber, whereas panel (b) shows the one in the collecting filter. These size distributions were obtained from 200 randomly sampled particles in the micrographs in Figs. 4 and 5. These figures also include the cumulative fraction, the mean diameter $\bar{d}$, the median diameter d_{50} and the standard deviation σ.

(a) (b)

Figure 4. SEM picture of fabricated particle in large amount powder feeding. Powder feed rate was set to 12.3 g/min. (a), Upstream chamber ; (b), Filter.

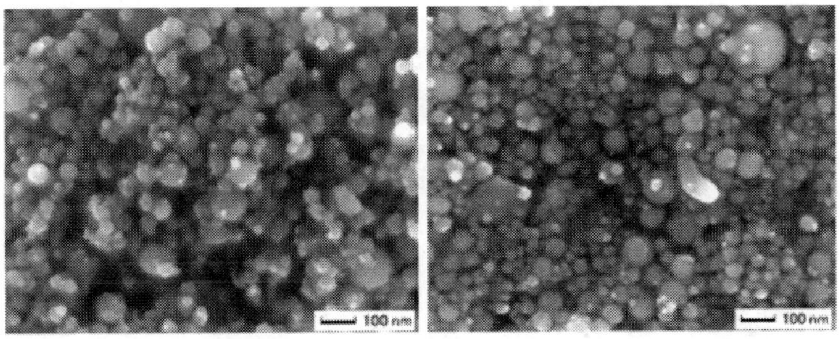

(a) (b)

Figure 5. SEM picture of fabricated particle in little amount powder feeding. Powder feed rate was set to 4.2 g/min.(a), Upstream chamber ; (b), Filter.

For $g_{\rm pow}$=4.2 g min^{-1}, the $\bar{d}$ is about 32−33 nm irrespective of the powder collection position. This independence of the collection position means that the cooling effect from the quenching gas was enough for creation of nanoparticles from a small amount of Ti vapor. The standard deviation σ of 16 − 17 nm is relatively low. On the other hand, for $g_{\rm pow}$=12.3 g min^{-1}, $\bar{d}$ and σ depend on the collection position. The particles collected in the filter have larger $\bar{d}$ of 42.5 nm and σ of 22.5 nm. The dependence of the collection position is considered to arise from the cooling history of the material vapor. The particles collected in the upstream chamber are regarded as the ones fabricated by rapid cooling of the mixture of Ti and oxygen atoms directly on the short distance to the chamber wall. The rapid cooling generally prevents the particles from growing up due to coagulation. From this reason, the smaller nanoparticles might be obtained in the upper chamber. In the meanwhile, the particles collected in the filter are regarded as the ones created around the axis of the chamber which is transferred to the filter on the long way. The cooling rate for the mixture of Ti and oxygen atoms is relatively low compared to the above situation, and then the particles grow up in size. Hence, the relatively large particles may be found in this case. Nevertheless, the most particles are limited in size smaller than 100 nm, i.e. the nanoparticle size even for the heavy rate injection of Ti raw material. The coefficient of variation $\sigma/\bar{d}$ is 0.52 − 0.53 for both cases $g_{\rm pow}$=4.2 g min^{-1} and 12.3 g min^{-1}.

4.2. XRD spectra and weight fraction of anatase and rutile TiO_2

The TiO_2 has mainly three phases in the structure: rutile, anatase and brookite. Such crystallographic structure and chemical composition of the synthesized nanoparticles were investigated using XRD analysis. Figures 8 and 9 represent the XRD spectra for $g_{\rm pow}$=12.3 g min^{-1} and $g_{\rm pow}$=4.2 g min^{-1}, respectively. Again, panel (a) portrays the image for particles collected in the upstream chamber, whereas panel (b) is the one in the collecting filter. The XRD spectra weakly depend on the powder collection position for both powder feed rate conditions. This implies that cooling rate hardly influenced the crystallographic structure but oxygen content did it in the present conditions as described later. What should be emphasized here is that there is no peak from Ti raw material even in case of the heavy rate injection. The fact of no Ti peak indicates that injected Ti powder was completely evaporated and oxidized in the PMITP for $g_{\rm pow}$=12.3 g min^{-1} as well as $g_{\rm pow}$=4.2 g min^{-1}. Another noticeable fact is that there is quite a difference in results between Fig. 8 and Fig. 9. The particles for $g_{\rm pow}$=12.3 g min^{-1} indicates many peaks from the rutile phase as designated by 'R'. On the other hand, the dominant peaks from the particles for $g_{\rm pow}$=4.2 g min^{-1} are due to the metastable anatase phase denoted by 'A'. This difference may be attributed to the oxygen amount against Ti vapor as described later.

Calculating the ratio of the XRD intensity for these spectral lines, we estimated the weight fraction $f_{\rm A}$ of TiO_2 in anatase form to whole TiO_2 particles using the method indicated in [13, 25]:

$$f_{\rm A} = \frac{1}{1 + 1.26\frac{I_{\rm R_{27.4^\circ}}}{I_{\rm A_{25.3^\circ}}}}, \tag{1}$$

where $I_{\rm R_{27.4^\circ}}$ and $I_{\rm A_{25.3^\circ}}$ respectively signify the XRD spectral intensities at 2θ=27.4° for rutile and 2θ=25.3° for anatase. This equation was obtained using the test line for standard TiO_2. Figure 10 presents the weight fraction of anatase TiO_2 to whole TiO_2 particles for different powder collection positions for two $g_{\rm pow}$ conditions. The weight fraction of anatase TiO_2 is a high value of 0.84 for $g_{\rm pow}$=4.2 g min^{-1}, almost irrespective of the powder collection position. On the other hand, the weight fraction is about 0.35 for $g_{\rm pow}$=12.3 g min^{-1}. In other words, 0.65 of the whole particles are in rutile phase. This difference may be explained by oxygen amount as described in literature [9]. They had showed experimentally that the phase composition of the TiO_2 nanoparticles from the solidification of vaporized species depends on the ambient oxygen concentration. They also deduced that the weight fraction of anatase-TiO_2 could be increased by increasing oxygen gas flow rate, because rutile TiO_2 tends to be solidified from oxygen-deficient clusters while anatase TiO_2 inclines to be obtained from oxygen-rich clusters [9]. According to their results, the oxygen amount in the present work was too low to obtain anatase-TiO_2 for a large amount of powder feed rate $g_{\rm pow}$=12.3 g min^{-1}, while it was enough for $g_{\rm pow}$=4.2 g min^{-1}. The weight fraction of anatase-TiO_2 could be expected to be improved by more oxygen gas injection.

5. Conclusions

This paper has proposed a new method for a large amount synthesis of TiO_2 nanopowder using the Ar-O_2 PMITP with an intermittent heavy rate powder feeding and quenching gas injection. Ti raw material powder was intermittently injected into the PMITP only during the on-time, synchronized with the coil current modulation. This synchronized intermittent powder feeding with a rate of 12.3 g min^{-1} offered complete evaporation of Ti powder in the 20-kW PMITP. The evaporated materials were rapidly cooled down to create nanoparticles by quenching gas injection. Synthesized particles were analyzed by SEM and XRD. Results indicated that almost all particles were TiO_2, no Ti remaining. The mean diameter and standard deviation were 40 nm and 21 nm, respectively, for a feed rate condition of 12.3 g min^{-1}.

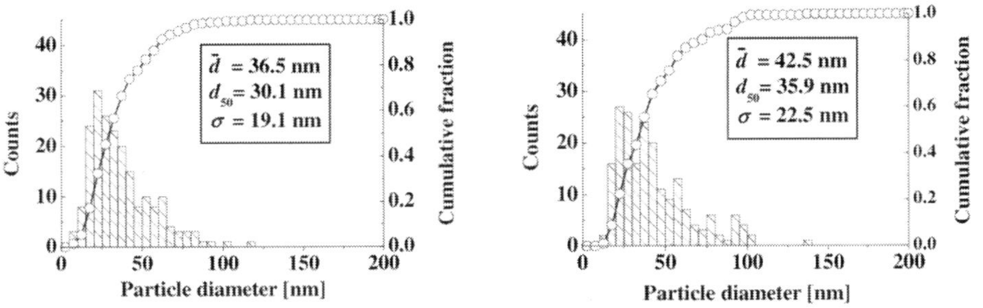

(a) (b)

Figure 6. Size distribution of fabricated particle in large amount powder feeding. Powder feed rate was set to 12.3 g/min. (a), Upstream chamber ; (b) Filter.

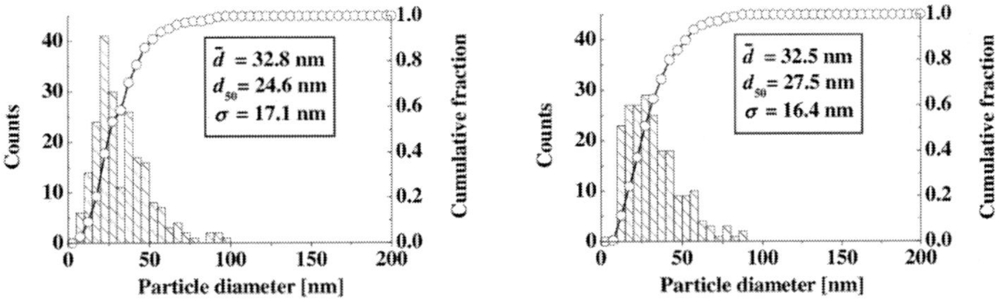

(a) (b)

Figure 7. Size distribution of fabricated particle in large amount powder feeding. Powder feed rate was set to 4.2 g/min. (a), Upstream chamber ; (b) Filter.

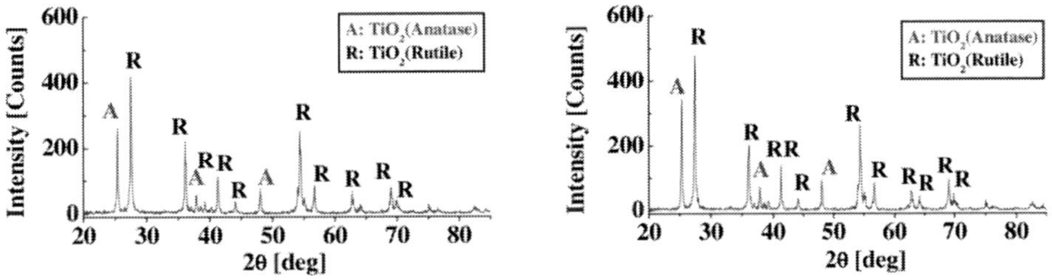

(a) (b)

Figure 8. XRD spectra of fabricated particle in large amount powder feeding. Powder feed rate was set to 12.3 g/min. (a), Upstream chamber ; (b) Filter.

References

[1] Berghaus J O, Meunier J L and Gitzhofer F 2004 *Meas. Sci. Technol.* **15** 161
[2] Chen W R, Wu X, Marple B R and Patnaik P C 2005 *Surface & Coatings Technol.* **197** 109
[3] Wang C, Inazaki A, Shirai T, Tanaka Y, Sakuta T, Takikawa H and Matsuo H 2003 *Thin Solid Films* **425** 41
[4] Tanaka H, Osawa T, Moriyoshi Y, Kurihara M, Maruyama S and Ishigaki T 2004 *Thin Solid Films* **457** 209
[5] Shigeta M and Watanabe T 2010 *J. Appl. Phys.* **108** 043306
[6] Chen L C 2010 *J.Alloys & Compounds,* **495** 476
[7] Li Y and Ishigaki T 2002 *J. Cryst. Growth* **242** 511
[8] Huang J Y, Ishigaki T, Tanaka T and Horiuchi S 1998 *J. Mater. Sci.* **33** 4141

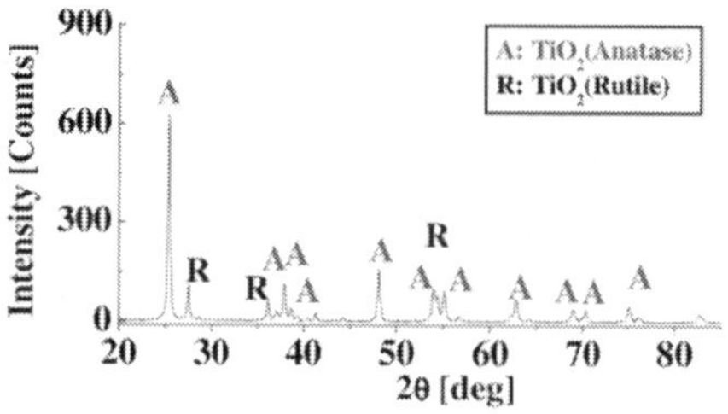

(a)

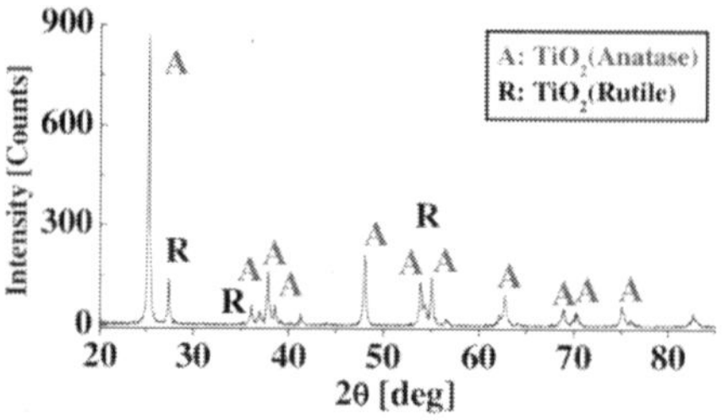

(b)

Figure 9. XRD spectra of fabricated particle in large amount powder feeding. Powder feed rate was set to 4.2 g/min. (a), Upstream chamber ; (b) Filter.

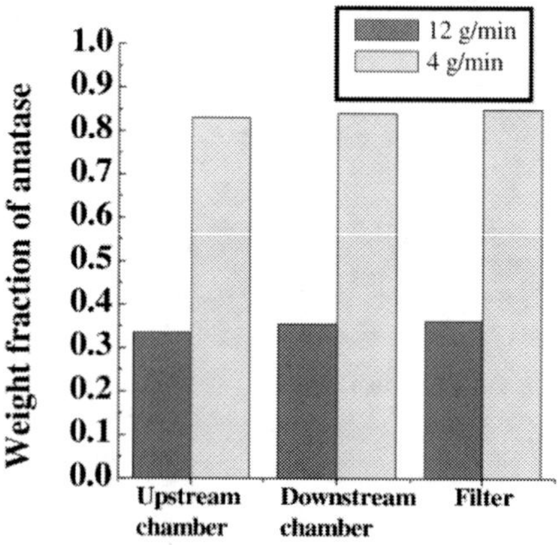

Figure 10. Weight fraction of TiO_2 in anatase phase in the fabricated nanoparticles.

[9] Oh S M and Ishigaki T 2004 *Thin Solid Films* **457** 186
[10] Lee J E, Oh S M and Park D W 2004 *Thin Solid Films* **457** 230
[11] Watanabe T and Fujiwara K 2004 *Chem. Eng. Comm.* **191** 1343
[12] Ishigaki T and Li J G 2007 *Sci. & Technol. Advanced Mater.* **8** 617
[13] Li J G, Ikeda M, Ye R, Moriyoshi Y and Ishigaki T 2007 *J. Phys. D: Appl. Phys.* **40** 2348
[14] Chen X and Mao S S 2007 *Chem. Rev.* **107** Issue 7, 2891
[15] Tanaka Y, Uesugi Y, Sakuta T 2007 *Plasma Sources Sci. & Technol.* **16** 281
[16] Tanaka Y, Morishita Y, Okunaga K, Fushie S and Uesugi Y 2007 *Appl. Phys. Lett.* **90** 071502
[17] Tanaka Y, Hayashi K, Nakamura T and Uesugi Y 2008 *J. Phys. D: Appl. Phys.* **41** 185203
[18] Malato S, Blanco J, Alarcon D C, Maldonado M I, Fernandez-Ibanez P and Gernjak W 2007 *Catalysis Today* **122** 137
[19] Li J, Zhao X, Wei H, Gu Z Z, Lu Z 2008 *Analytica Chemica Acta* **625** 63
[20] Gratzel M 2001 *Nature* **414** 338
[21] Biju K P, Jain M K 2008 *Thin Solid Films* **516** 2175
[22] Park J H, Park O O and Kim S 2006 *Appl. Phys. Lett.* **89** 163106
[23] Lee K N, Kim Y, Lee C W and Jai-Sung Lee J S 2011 *Mater. Sci. Eng.* **18** 082021
[24] Liu P S, Cai W P, Wan L X, Shi M D, Luo X D and Jing W P 2009 *Trans. Nonferrous Met. Soc. China* **19** s743
[25] Tanaka Y, Nagumo T, Sakai H, Uesugi Y, Sakai Y and Nakamura K 2010 *J. Phys. D: Appl. Phys.* **43** 265201
[26] Tanaka Y, Sakai H, Tsuke T, Uesugi Y, Sakai Y and Nakamura K 2011 *Thin Solid Films* **519** Issue 20, 7100
[27] Tanaka Y, Sakai H, Tsuke T, Guo W, Uesugi Y, Sakai Y and Nakamura K 2011 *Proc. 20th Int. Symp. Plasma Chem. ISPC-20*, 346, Philadelphia, USA

Journal of Physics: Conference Series **406** (2012) 012002 doi:10.1088/1742-6596/406/1/012002

Incineration of radioactive organic liquid wastes by underwater thermal plasma

M Mabrouk[1], F Lemont[2], J M Baronnet[1],

[1] University of Limoges, 123 Avenue Albert Thomas, 87 060 Limoges, France,
[2] CEA Marcoule, BP 17 171, 30207 Bagnols sur Cèze, France.

E-mail: majdi.mabrouk@etu.unilim.fr

Abstract. This work deals with incineration of radioactive organic liquid wastes using an oxygen thermal plasma jet, submerged under water. The results presented here are focused on incineration of three different wastes: a mixture of tributylphosphate (TBP) and dodecane, a perfluoropolyether oil (PFPE) and trichloroethylene (TCE). To evaluate the plutonium behavior in used TBP/dodecane incineration, zirconium is used as a surrogate of plutonium; the method to enrich TBP/dodecane mixture in zirconium is detailed. Experimental set-up is described. During a trial run, CO2 and CO contents in the exhaust gas are continuously measured; samples, periodically taken from the solution, are analyzed by appropriate chemical methods: contents in total organic carbon (COT), phosphorus, fluoride and nitrates are measured. Condensed residues are characterized by RX diffraction and SEM with EDS. Process efficiency, during tests with a few L/h of separated or mixed wastes, is given by mineralization rate which is better than 99.9 % for feed rate up to 4 L/h. Trapping rate is also better than 99 % for phosphorous as for fluorine and chlorine. Those trials, with long duration, have shown that there is no corrosion problems, also the hydrogen chloride and fluoride have been neutralized by an aqueous solution of potassium carbonate.

1. Introduction

Radioactive organic liquid waste is a particular category of radioactive waste precisely because of their organic nature. Therefore they are systematically separated from other liquid wastes, stored and processed separately.

They consist of:

- used solvents, implemented especially in combustible cycle installations. The main solvent is TBP (tributylphosphate) used diluted with a hydrocarbon, typically dodecane. TBP evolves under irradiation, giving rise to degradation products, mono and dibutylphosphates.
- solvents used in decontamination operations. These solvents are often halogenated (trichlorethylene, perchlorethylene, freon, ...).
- lubricating oils of various mechanical devices used in workshops or nuclear installations and which may be contaminated.
- liquids of scintillation.

Several techniques for conditioning organic radioactive waste have been developed like incineration and wet oxidation, but the combustion of solution containing waste produces corrosive products for the process, and leads to off-gas purification problems. Moreover, combustion generates relatively large volumes of radioactive gaseous, liquid and solid wastes. Wet oxidation also generates various undesirable secondary waste byproducts.

Also few works [1] are devoted to submerged quench incineration : combustion takes place in a chamber above a liquid surface, and the products of combustion are introduced through a downcomer into the liquid bath for a rapid quench. It is an effective means to exchange heat and to absorb components of combustion products and inorganic salts while cooling and scrubbing the flue gas. The hydrogen chloride and fluoride are soluble in water and can be easily scrubbed and removed in the form of weak acid and (or) neutralized by an aqueous solution of sodium or potassium hydroxide.

New processes based on plasma technologies [2] appeared and proved their efficiency to decontaminate aqueous effluents: intense UV radiation emission and very high concentrations of radicals allow to mineralize persistent organic pollutants.

Since a few years we are working about a new process of oxidation of organic compounds by an outflow of an oxygen thermal plasma jet into the bulk of a liquid [3, 4]. Under these conditions, the gas-liquid system has an intense turbulence resulting in a rapid degradation of the jet into small sized gaseous inclusions. This ensures high rates of mass transfer processes and has the possibility of transport of dissociated entities to the gas-liquid interface and their subsequent participation in chemical reactions with the liquid phase components. This kind of processes [5-7] have many advantages:

- high temperature but cold reactor in bulk,
- high concentrations of radicals,
- clean reactor without thermal inertia,
- high chemical reaction rates,
- mineral radioactive pollutants may be trapped in the solution.

2. The wastes of interest

2.1. TBP / dodecane

The first aim of this work was to incinerate a mixture of tributylphosphate (TBP)(Sigma Aldrich(SA) ref:158615) and dodecane(SA ref:44030).

TBP is a extraction solvent involved in nuclear fuel reprocessing, and particularly in the Purex process to recover uranium and plutonium from spent fuel. Additionally, it is listed among the highly radioactive liquid wastes which must be mineralized before conditioning. Different kind of processes are around the world under investigation to improve treatment of used TBP/dodecane, such as glidarc [8], steam-assisted pyrolysis system [9], submerged combustion technique [10], pebble bed pyrolysis [11] and a few others. Some remaining problems and the characteristics of this particular waste were an incitation to test our process with it.

To understand the behavior of radioactive elements, especially plutonium, in the process, zirconium was chosen as a surrogate. Nevertheless to enrich TBP/dodecane with zirconium needs a particular method. An 1M TBP/dodécane solution (with 30 % TBP in volume) is contacted in 1 : 1 phase ratio with nitric acid 4M (SA ref:438073) during 30 minutes with intensive agitation. In these conditions, if the initial HNO_3 concentration is 4M, after equilibration, the HNO_3 concentration in organic phase is 0.7 M [12]. Water is simultaneously extracted from aqueous phase : solvent adduct in form of a Lewis acid-base complex is formed :

$$TBP_{org} + x\ HNO_{3\,aq} + y\ H_2O_{aq} \rightarrow (TBP(HNO_3)_x(H_2O)_y)_{org}$$

$TBP(HNO_3)_{1.3}(H_2O)_{0.6}$ and $TBP(HNO_3)_{1.8}(H_2O)_{0.6}$ are proposed in [13].

Then a determined quantity of zirconium carbonate (SA ref:14616) (typically few g of zirconium) is dissolved in nitric acid 4M. Then this aqueous phase is contacted, in 1 : 1 phase ratio, with the pre-equilibrated organic phase, during 45 minutes with intensive agitation. Zirconium is supposed extracted in the form of di,tri, or tetra nitrate complex : $Zr(OH)_2(NO_3)_2(TBP)_2$, $ZrOH(NO_3)_3(TBP)_2$ and $Zr(NO_3)_4(TBP)_2$ [13]. Taking into account the aqueous phase molarity –4 M-, the distribution coefficient of Zr by TBP 30 % in dodecane may be estimated to 0.144 [13].

The basic principle of the process may be summarized by:

$$\text{TBP + dodecane + Zr} \xrightarrow[O_2]{\text{thermal plasma}} CO_2\uparrow + H_2O + H_3PO_4 + Zr(?)\downarrow \qquad (1)$$

Carbon is burnt into CO_2 which exits in the off-gas, hydrogen to H_2O vapor which is immediately condensed inside the solution, phosphorus is presumed trapped inside the solution in the form of H_3PO_4 since pH is rapidly less than 2. A question remains: is zirconium dissolved in the solution as zirconium nitrate or phosphate, or, if it is condensed, in what species: zirconium oxide or zirconium phosphate?

2.2. Perfluoropolyether oil

The second aim is to evaluate the efficiency of the process on a very particular waste: a pump oil, which is, unlike TBP/dodecane, non-flammable. It is a perfluoropolyether oil (PFPE). Their thermal stability and oxidation resistance are exceptional, as well as their resistance to fire: they are completely fireproof. They have a very high chemical inertia. These properties justify their use in the transfer pumps for UF6, by example.

The ratio of carbon, oxygen and fluorine in this oil may be approximated in a pseudo formula - $C_2O_{1.1}F_{4.2}$- used to strike the mass balance. Besides a pseudo formation enthalpy may be calculated, using bonds energies, giving -536 kJ for a pseudo mole $C_2O_{1.1}F_{4.2}$. So incineration of this oil may be schematized by :

$$(C_2O_{1.1}F_{4.2})_{liq} + 0.4\ O_2 + 2.1\ H_2O_{liq} \rightarrow 2\ CO_2 + 4.2\ HF \quad (2)$$

2.3. Trichloroethylene (TCE)

Trichlorethylene (C_2HCl_3 or TCE) is also known by a variety of chemical and trade names : Trimar, Trilene ...Due to its effectiveness, non corrosivity, non flammability and ease of recycling, few solvents matched the performance of TCE in cleaning and degreasing.

A wide variety of entities used TCE, including the electronics, defense, chemical, rail, automotive, boat, food processing, textile, and dry-cleaning industries. Although TCE works best on fats, grease, and waxes, it is also a powerful solvent for a large variety of natural and synthetic substances [14].

Moreover, this waste has been heavily used in the nuclear industry. But its separated incineration involves some difficulties in standard processes in particular for certain co-products. Today the production and use of this solvent are prohibited but stocks of spent solvents exist and must be destroyed.

We can simplify the decomposition reaction product in the reactor as:

$$(C_2HCl_3)_{liq} + 1.5\ O_2 + H_2O_{liq} \rightarrow 2\ CO_2 + 3\ HCl_{liq} \quad (3)$$

3. Experimental set-up

3.1. The process

The process is based on a non-transferred arc torch of an original design producing a plasma jet of oxygen.

The torch, with gas-vortex stabilization of the arc column is made of a tungsten cathode, a copper anode, between both a diaphragm. The cathode tungsten tip is protected by a stream of argon. Oxygen is the main gas, it is injected downstream the diaphragm. The change in diameter inside the nozzle anode creates a zone of the recirculation flow, colder thus more electrically resistive, fixing the anodic spot behind the ledge [15].

The liquid waste is injected, with a peristaltic pump, at the nozzle exit inside the jet; mean enthalpic temperature is about 3500 K. At this temperature the oxidation reactions are fast enough to reach immediately thermodynamic equilibrium.

The plasma torch is disposed vertically to the base of a water-jacket reactor filled with water (see figure 1). At the top, we find a condenser/mist eliminator to trap water, a heat exchanger to cold the gas flow and to decrease the steam partial pressure, a cyclone to trap residual water, then a demister to improve the water balance.

To control the solution temperature, we added a cooling loop which includes a pump and a heat exchanger. To trap any solid particles in the solution, we placed upstream of the heat exchanger, two filters, positioned in parallel.

A tube placed in the axis at the top of the reactor, can take a sample of gas. It crosses a pretreatment device for trapping dust or liquid droplet before reaching the analyzer. The principle of analysis is by infrared absorption which allows to give the gas content of CO and CO_2.

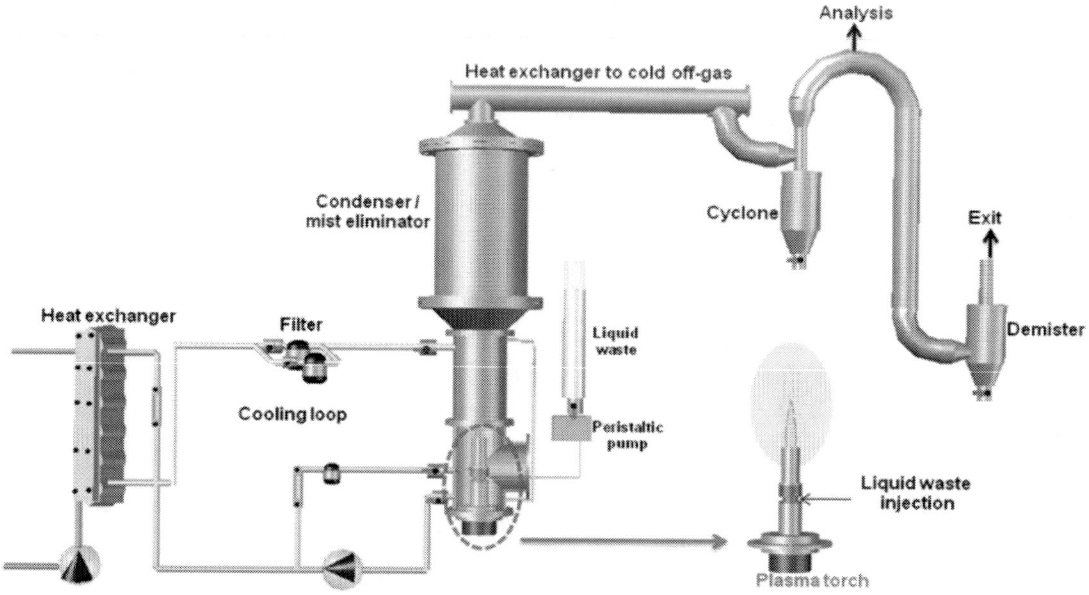

Figure 1. Experimental set-up

3.2. Operating parameters

The gas flow rates for a typical trial are:

- 30 NL / min of argon at the cathode
- 180 NL / min of oxygen at the anode
- 7 NL / min of argon to inject the waste
- 40 NL / min of oxygen in the stage of dilution.

The current intensity of the arc is 200 A, the tension is ~ 250 V; the thermal efficiency of the torch is about 62 %. Once the arc is established, the reactor and the loop are filled with water ~ 30 liters; the flow-rate in the loop is 2 m^3/h; after a few minutes the waste is introduced via a peristaltic pump.

The solution temperature is maintained below 40°C and the off gas temperature after the heat exchanger is about 20°C. Over the course of the trial, samples are taken at the bottom of the reactor for earlier analysis.

At the end of the test, the reactor is evacuated before the arc is switched off.

4. Results

4.1. TBP/dodecane

The simulated waste was prepared with 30% in mass of TBP and the zirconium concentration is estimated to 1.2 g/L. During the trial described here, waste was fed at ~ 1.2 kg/h (1.5 L/h). In other studies, it can reach 2.4 kg/h (3 L/h).

At high temperatures, the oxidation reactions are sufficiently fast so it can be assumed that thermodynamic equilibrium is reached. So a predictive model [16] shows that we find argon, oxygen in excess and CO_2 ; water in excess and phosphoric acid are supposed to be completely condensed in the reactor. Thus the CO_2 content in the outlet gas should be around 11 %.

The combustion of TBP/dodecane produces 15 kW besides 30 kW effectively delivered by the plasma torch. So a total power of 45 kW is delivered inside only 10 cm^3. 6 kW are transferred in the cooling stage, 10 kW in the reactor walls and condenser, the rest is found in the cooling circuit of the loop.

Measurements during waste injection give 11.2 % for CO_2 and 0.07 % for CO. The combustion efficiency defined by: [CO2]/([CO2]+[CO]) is ≈ 99%; pH falls rapidly to below 2.

At the end of the trial, the amount of TOC content in the whole solution was ~ 0.2 g; while 778 g of carbon were introduced: So we can inferred a destruction rate better than 99.9 %, within measurement accuracy.

The simulated waste introduced contains 36.15 g of phosphorus while the amount of phosphorus content in the final solution is ~ 35.97 g. Thus, the trapping rate of phosphorus is evaluated at 99.5%, within measurement accuracy. Nitrate content in the solution varies like phosphorus content (figure 2 where nitrate content is given in mg of nitrogen/L). Nitrogen mass balance shows nitrogen loss which may be explained by NO formation in the plasma: NO might escape in the off-gas.

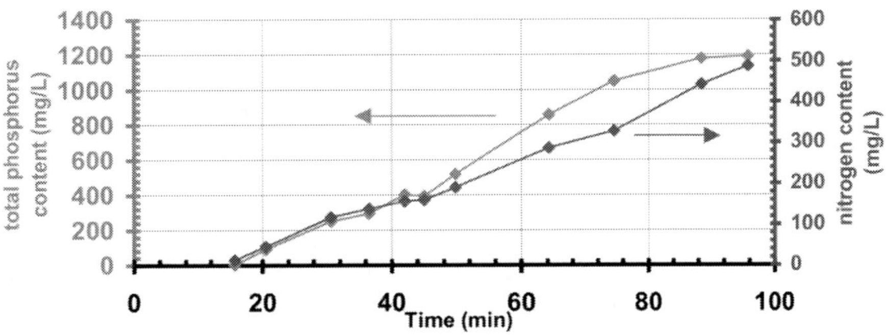

Figure 2. Phosphorus and nitrogen contents

No zirconium trace in solid residues has been detected by RX diffraction, neither by SEM with EDS. In the liquid phase, we detected by ICP/MS only 0.32 g of Zr, while the Zr introduced mass is estimated at 1.56 g! Besides, analyzes of Zr content, on the one hand in the raw solution and in the other hand in the filtered solution, showed a difference of almost one order of magnitude. We therefore plan to filter the remaining solution and analyze the obtained powder by plasma emission spectrometry. In all instances, further trials must be carried out with a larger amount of Zr and especially we must check the real mass of Zr in the commercial product (zirconium carbonate) whose purity does not exceed 40%.

4.2. PFPE

Waste was fed at ~ 3.5 kg/h (1.8 L/h), 0.350 L are processed during the test run.

The combustion of PFPE produces 8.6 kW besides 31 kW effectively delivered by the plasma torch; so the total power is ~ 40 kW. 3.2 kW are transferred in the cooling stage, 7.4 kW in the reactor walls and condenser, the rest is found in the cooling circuit of the loop.

Supposing complete thermodynamic equilibrium at each step and all the water and fluoride acid condensed inside the reactor, CO_2 content in the off-gas is calculated to be around 7.9 %. Measurements give 7.6 % for CO_2 and 0.3 % for CO: The combustion efficiency is ~ 96%.

At the end of the trial, the amount of TOC content in the whole solution is measured at about 0.3 g; while 131 g of carbon were introduced: So we can inferred a destruction rate better than 99.75 %, within measurement accuracy.

The amount of fluorine content in the solution is ~ 411 g while the mass of fluorine introduced was 436 g. Thus, the trapping rate of fluorine is evaluated at 94.1 %, within measurement accuracy.

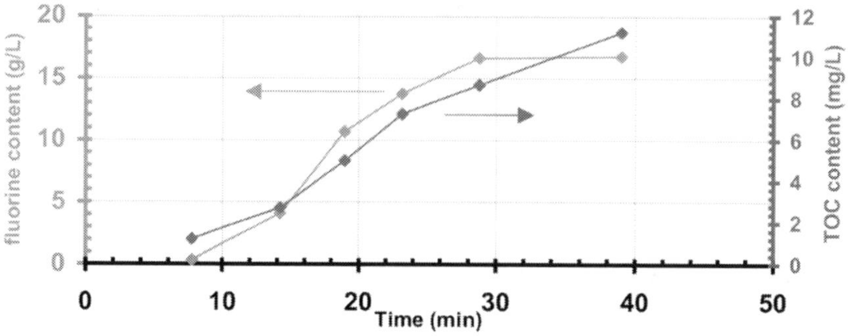

Figure 3. Fluorine and TOC contents

4.3. TCE

In the case of a highly chlorinated waste, following the Deacon equilibrium

$$2\ HCl + ½\ O2\ \square\ Cl2 + H2O \tag{4}$$

in order to trap the chlorine as HCl condensed in the solution rather than in the form of Cl_2 gas, we must shift the equilibrium in the endothermic direction and increase the partial pressure of water vapor, that we do by re-injecting a fraction of the solution (see figure 5).

Mass rate of the charge is ~ 2.9 kg/h (2 L/h) so 0.5 L are processed during the test run.

The combustion of TCE produces only 2.9 kW besides 31 kW effectively delivered by the plasma torch; so the total power is ~ 34 kW. 7 kW are transferred in the cooling stage, 7.2 kW in the reactor walls and condenser, the rest is found in the cooling circuit of the loop.

CO_2 content in the off-gas is calculated to be around 6.4 %. Measurements give 6.5 % for CO_2 (see figure 4 which highlights the stability of the process under steady injection conditions) and 0.04 % for CO: The combustion efficiency is ~ 99.38%.

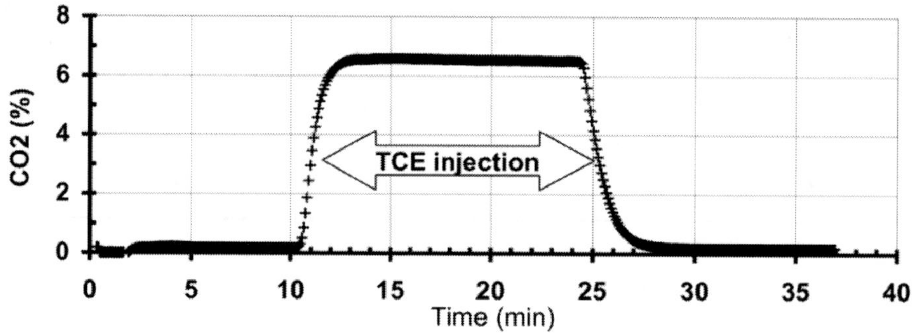

Figure 4. CO_2 evolution during TCE trial

At the end of the trial, the amount of TOC content in the whole solution is estimated at about 0.09 g; while 133 g of carbon were introduced: So we can inferred a destruction rate better than 99.93 %, within measurement accuracy.

The amount of chlorine content in the solution is ~ 586 g while the mass of chlorine introduced was 592 g. Thus, the trapping rate of chlorine is evaluated at 98.98 %, within measurement accuracy.

4.4. PFPE/TCE

In the practice, liquid wastes are rarely pure products, so it is important for evaluating the process efficiency to test it on a mixture of model products. The trial described here is focused on the incineration of a mixture of PFPE oil and TCE with neutralization of the solution using potassium carbonate K_2CO_3.

For this, we have included a tank filled with potassium carbonate in the recirculation loop (see Figure 5), a volume of K_2CO_3 have been periodically injected to neutralize the solution.

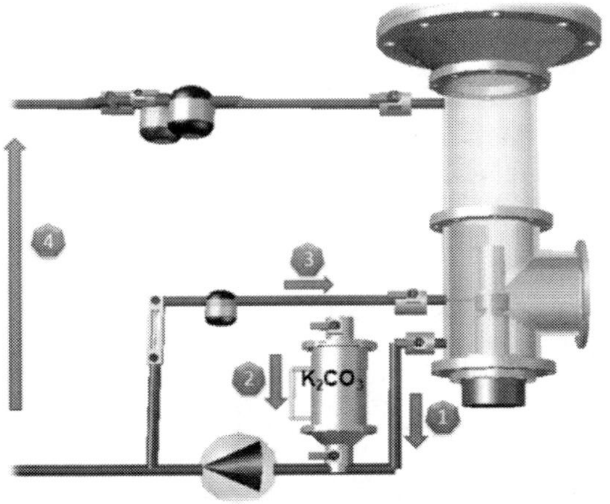

Figure 5. Neutralisation device.
1. *Solution racking*
2. *K_2CO_3 injection*
3. *Reinjection of a part of solution volume*
4. *Cooling loop*

So, if the injection flow rate of the mixture is 1 L/h and if we inject 1 L of K_2CO_3 every 10 min, number of moles of CO_2 in the off-gas is calculated to be around 43.5 moles. Measurements give 41.6 moles of CO_2. CO content is about 0.05 % but it increases (0.15%) each time we inject the solution of K_2CO_3.

We have injected in excess the volume of K_2CO_3 required for the neutralization of HCl and HF to finally get a basic solution: pH ~ 9.

The destruction rate is better than 99.88 % while the trapping rate of fluorine and chlorine are respectively better than 93.2 % and 76 % (see figure 6).

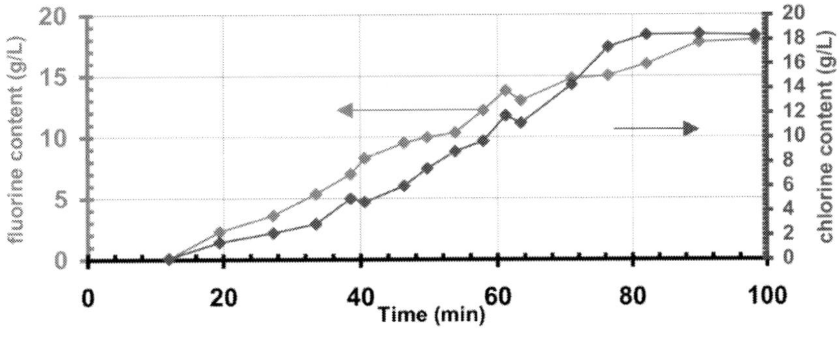

Figure 6. Fluorine and chlorine contents

5. Conclusion

Incineration of used TBP/dodecane with zirconium as plutonium surrogate was demonstrated without zirconium trace in the condensed residues. Process efficiency is also shown by the incineration of other wastes such as TCE and PFE, separately or mixed.

We have proved that:

- Mineralization rate better than 99 % for feed rate up to 4 L/h.
- High trapping rate for phosphorous as for fluorine and chlorine.
- Possibility of neutralization without great difficulty.

The next step consists to strengthen analysis of the solution and specify the measurement uncertainties and to compare thermodynamics and chemical kinetics.

6. Acknowledgments

The authors wish to express their thanks to P. Moisy and to J. Dauby of CEA/DEN/DRCP for their contribution to this work : to enrich TBP/dodecane with zirconium and the corresponding analysis.

References

[1] Leite O C 2002 Thermal oxidizers to abate chloro-Fluorocarbon compounds. In: *IT3'02 Conference*, (New orleans, Louisiana)

[2] Murphy A B 1999 *Heat and Mass Transfer under Plasma Conditions*, ed P Fauchais *et al.* (New York: New York Acad Sciences) pp 106-23

[3] Boudesocque N, Lemont F, Lafon C, Girold C, Vandensteendam C and Baronnet J M 2007 Organic liquid waste gasification by underwater thermal plasma. In: *18th International Symposium of Plasma Chemistry*, (Kyoto, Japan)

[4] Lemont F, Lafon C, Baronnet J M and Boudesocque N 2009 Procédé et dispositif pour le traitement des dechets par injection dans un plasma immergé. FR2953279, 30.11.09

[5] Alekseev N V, Samokhin A V, Belivtsev A N and Zhavoronkova V I 2000 Thermal-Plasma Jet Oxidation of Phenol in Aqueous Solutions *High Energy Chemistry* **34** 389-93

[6] Fortin L, Soucy G and al. 1999 The use of thermal plasma for waste water treatment. In: *ISPC 14*, (Prague) pp 2417-22

[7] Boudesocque N, Lemont F, Lafon C, Girold C and Baronnet J M 2007 Decontamination and gasification of aqueous organic waste by submerged thermal plasma. In: *IT3 Conference*, (Phoenix, AZ, USA)

[8] Moussa D and Brisset J-L 2003 Disposal of spent tributylphosphate by gliding arc plasma *Journal of Hazardous Materials* **102** 189-200

[9] Toshiki S, Tomoyuki S, Hayato K and Hiromi Y 2009 Steam-Assisted Pyrolysis System for Decontamination and Volume Reduction of Radioactive Organic Waste *J. Nucl. Sci. Technol.* **46** 232-8

[10] Gunzo U, Mitsuru M, Sachio F, Masayuki A, Katsuhide U and Mitsuhisa C 1994 Development of Spent Solvent Treatment Process by a Submerged Combustion Technique *J. Nucl. Sci. Technol.* **31** 228-39

[11] Luycx P and Deckers J 1994 Pebble bed pyrolysis for the processing of alpha contaminated organic effluent. In: *Waste Management Conference*, (Tucson, Arizona)

[12] Moisy P private communication. CEA/DEN/DRCP

[13] Suarez J A and al. 2000 Actinide and Fission Product Partitioning and Transmutation. In: *6th Information Exchange Meeting*, (Madrid, Spain)

[14] Doherty R E 2000 A history of the production and use of carbon tetrachloride, tetrachloroethylene, trichloroethylene and 1,1,1-trichloroethane in the United States: Part 2 - Trichloroethylene and 1,1,1-trichloroethane *Environ. Forensics* **1** 83-93

[15] Zhukov M F and al. 2007 *Thermal plasma torches*: Cambridge International Science Publishing

[16] Werner J H and Cool T A 2000 The kinetics of the combustion of trichloroethylene for low Cl/H ratios *Combustion and Flame* **120** 125-42

Decomposition of water-insoluble organic waste by water plasma at atmospheric pressure

Sooseok Choi and Takayuki Watanabe
Dept. Environmental Chemistry and Engineering, Tokyo Institute of Technology
4259-G1-22, Nagatsuta-cho, Midori-ku, Yokohama 226-8502, Japan

E-mail: choi@chemenv.titech.ac.jp

Abstract. The water plasma was generated in atmospheric pressure with the emulsion state of 1-decanol which is a source of soil and ground water pollution. In order to investigate effects of operating conditions on the decomposition of 1-decanol, generated gas and liquid from the water plasma treatment were analysed in different arc current and 1-decanol concentration. The 1-decanol was completely decomposed generating hydrogen, carbon monoxide, carbon dioxide, methane, treated liquid and solid carbon in all experimental conditions. The feeding rate of 1-decanol emulsion was increased with increasing the arc current in virtue of enhanced input power. The generation rate of gas and the ratio of carbon dioxide to carbon monoxide were increased in the high arc current, while the generation rate of solid carbon was decreased due to enhanced oxygen radicals in the high input power. Generation rates of gas and solid carbon were increased at the same time with increasing the concentration of 1-decanol, because carbon radicals were increased without enhancement of oxygen radicals in a constant power level. In addition, the ratio of carbon dioxide to carbon monoxide was increased along with the concentration of 1-decanol due to enhanced carbon radicals in the water plasma flame.

1. Introduction

Thermal plasma technology has been receiving attention as an advanced waste treatment method [1, 2]. The thermal plasma provides high temperature over thousands degrees and a large amount of chemically active species to decompose any kinds of non-degradable materials easily. Especially, gases used as an energy source such as hydrogen and carbon monoxide are synthesized in the thermal plasma treatment of organic wastes when oxygen radicals are existing in the thermal plasma flame by supplying oxygen or steam as the plasma forming gas [3-5]. In order to take advantage of abundant oxygen, hydrogen and hydroxyl radicals in the organic waste treatment process, a unique direct current (DC) arc water plasma torch has been introduced recently [4-7]. Besides the radicals that promote chemical reactions of the waste decomposition and the syngas production, the water plasma torch has benefits of high thermal efficiency and simple system because evaporated water is directly used as plasma supporting gas and coolant. Therefore, the water plasma is believed as a promising method to treat organic waste.

Although thermal plasma flame has high temperature, it also has a steep temperature gradient. In order to utilize the high temperature of arc discharge region, water-soluble organic waste can be supplied as a plasma forming liquid for the water plasma in a solution state [4]. On the contrary, water-insoluble waste is unfeasible to be used as plasma forming material directly. The emulsion including water-insoluble organic compound was used as plasma forming liquid instead of pure water in this work. As a model organic waste, 1-decanol ($C_{10}H_{22}O$) which is widely used in agricultural chemicals was mixed with water uniformly with the help of Tween60 ($C_{64}H_{126}O_{26}$) surfactant. With the 1-decanol emulsion, the demonstration of water plasma treatment of water-insoluble organic waste was conducted in the present work. Effects of the arc current and 1-decanol concentration on the gas and liquid products were investigated to understand the characteristics of 1-decanol emulsion treatment with the water plasma.

2. Experimental Setup

2.1. Water plasma system

The schematic diagram of the water plasma system used in this work is depicted in figure 1. The 1-decanol emulsion filled in a liquid reservoir is transferred to the arc discharge region through a liquid absorber. High heat flux from the DC arc plasma evaporates transferred emulsion, and then the emulsion is supplied to plasma forming gas. Therefore, 1-decanol is directly decomposed in the high temperature region of the arc plasma. Decomposed substance reacts further with abundant radicals in the plasma flame generating treated gases. The generated gas was analysed by a quadrupole mass spectrometry (QMS, Dycor-Proline, AMENTEK) and a gas chromatography (GC, GC-8ATP, Shimadzu). Condensed solid carbon and liquid were generated on the wall of a cooling reactor. The treated liquid was collected through a hole on the reactor bottom and analysed by a total organic carbon (TOC) analyser (TOC-V CSN, Shimadzu) and a high-performance gas chromatograpy (HPLC, pump: U-986, Jasco, column thermostat: CO-966, Jasco) with an ultraviolet- visible detector (UV-975, Jasco).

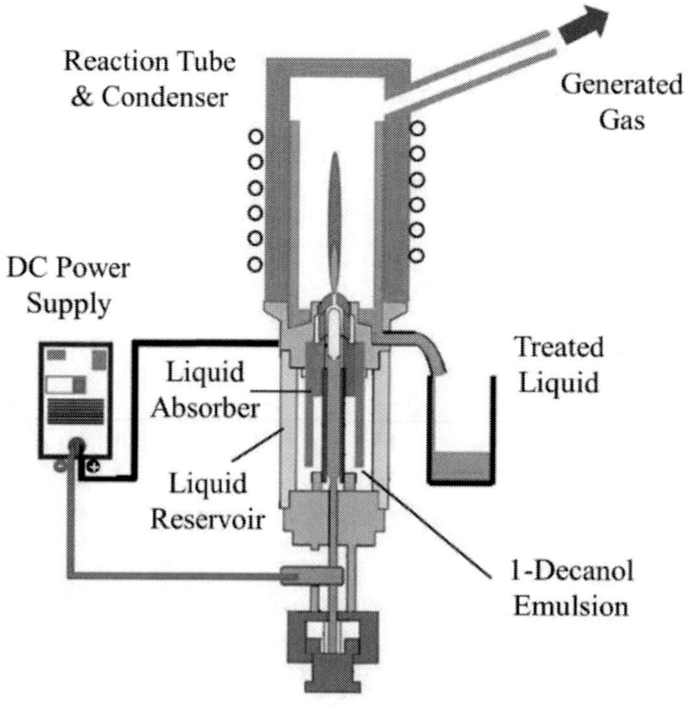

Figure 1. Schematic diagram of the water plasma system for the decomposition of 1-decanol emulsion.

The water plasma torch was operated at atmospheric pressure and the electrode gap distance was fixed at 2 mm because the arc length has a strong effect on the arc voltage and input power level. The arc current and the concentration of 1-decanol were controlled from 5 to 7 A and from 0.10 to 0.50mol%, respectively, in order to investigate their effects on the water plasma decomposition of 1-decanol emulsion. The average arc voltage was in the constant level of 100 V even in the change of the arc current and 1-decanol concentration. Therefore, the input power was increased with increasing the arc current, while it was not changed in different 1-decanol concentration. Table 1 shows operating conditions of the water plasma torch and 1-decanol emulsion.

Table 1. Operating conditions for the water plasma torch and 1-decanol emulsion

Water Plasma Torch	
Arc current	5, 6, 7 A
Average arc voltage	100 V
Electrode gap distance	2 mm
Operating pressure	101.3 kPa
1-Decanol Emulsion	
Concentration	0.10, 0.25, 0.50mol%
Tween60 : 1-decanol	1.0 : 26.5

2.2. Preparation of 1-decanol emulsion

A surfactant micelle has a hydrophilic group on one side and a hydrophobic group on the other side. Therefore, the surfactant can mix water-insoluble oil with water. In this work, Tween60 surfactant, 1-decanol and distilled water were mixed together to prepare 1-decanol emulsion. The composition ratio of 1-decanol to Tewwn60 was fixed at 26.5 to 1.0. The 1-decanol emulsion was stirred for 1 hour for the uniform dispersion of 1-decanol in water. Figures 2 (a) and (b) show the 1-decanol emulsion before the water plasma treatment and the generated liquid after the water plasma treatment. The decomposition rate and organic carbon removal rate were evaluated by comparing those two liquids.

Figure 2. (a) Raw 1-decanol emulsion with 0.5mol% and (b) generated liquid from the water plasma treatment.

2.3. Calculation of generation rates

In order to evaluate generation rate of solid, liquid and gas products, the balances of total substances and carbon are used as follows:

$$F_r = \frac{M_0 - (M_r + M_a)}{Time} = G_s + G_l + G_g \quad (1)$$

$$F_r x_r = G_s x_s + G_l x_l + G_g x_g \quad (2)$$

In equation (1), F_r is the feeding rate of the 1-decanol emulsion and G_s, G_l and G_g means generation rates of solid, liquid, and gas products, respectively. Masses of raw liquid before treatment (M_0), remained liquid in the reservoir after treatment (M_r), absorbed liquid after treatment (M_a) and operating time (*Time*) were used to evaluate the feeding rate of 1-decanol emulsion. x_r in the left hand side of equation (2) is carbon mass fraction in the feeding liquid. Likewise, x_s, x_l and x_g denote carbon mass fractions in the generating solid, liquid and gas products, respectively. Mass fractions of carbon in liquid and gas were measured by TOC analyser and GC, respectively. The generation rate of gas (G_g) was measured by a soap film flowmeter, and then the generation rate of liquid (G_l) was derived from equations (1) and (2) with the assumption of pure solid carbon product ($x_s = 1$) as follows:

$$G_l = \frac{F_r(1-x_r) - G_g(1-x_g)}{1-x_l} \quad (3)$$

Finally, the generation rate of gas was calculated based on equations (2) and (3) as follows:

$$G_s = F_r x_r - G_g - \frac{F_r(1-x_r) - G_g(1-x_g)}{1-x_l} \quad (4)$$

3. Results and discussion

Generation rates of gas, liquid and solid products by changing the arc current are presented in figure 3. The concentration of 1-decanol was fixed at 0.5mol% in the experiment for the arc current. The feeding rate shows a tendency to increase with increasing the arc current, because the input power level which is proportional to the arc current determines the evaporation rate of 1-decanol emulsion. The gas generation rate depicted as the gap between circle and diamond symbols in figure 1 is increased with increasing the arc current, while the generation of solid carbon is decreased at the same time. These results are explained by enhanced oxygen radicals in the high arc current condition. The high input power of the water plasma dissociates large number of water molecules resulting in the production of radicals. Among produced radicals, oxygen radicals actively react with carbon radicals that are dissociated from 1-decanol producing carbon oxides. Therefore, the gas generation rate is increased reducing the solid carbon in the high arc current condition.

Figure 4 shows the effects of 1-decanol concentration on the generation rates at the fixed arc current of 7 A. Different to the change of the arc current, both of gas and solid generation rates are increased with increasing the concentration of 1-decanol. Since the input power level was constant in the change of 1-decanol concentration from 0 to 0.5mol%, the number of oxygen radicals is also expected to be constant. Carbon radicals in the plasma flame, however, are increased with increasing the content of 1-decanol in the feeding liquid. Therefore, larger number of oxygen radicals reacts with abundant carbon radicals in the high 1-decanol concentration, and the rest carbon is converted to solid carbon on the inner wall of the cooling reactor.

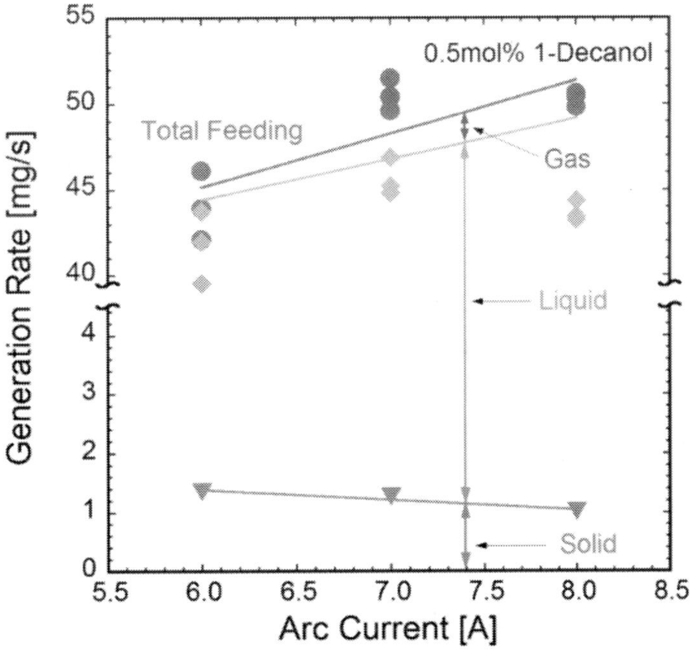

Figure 3. Generation rates of gas, liquid and solid products according to the arc current at the fixed 1-decanol concentration of 0.5mol%.

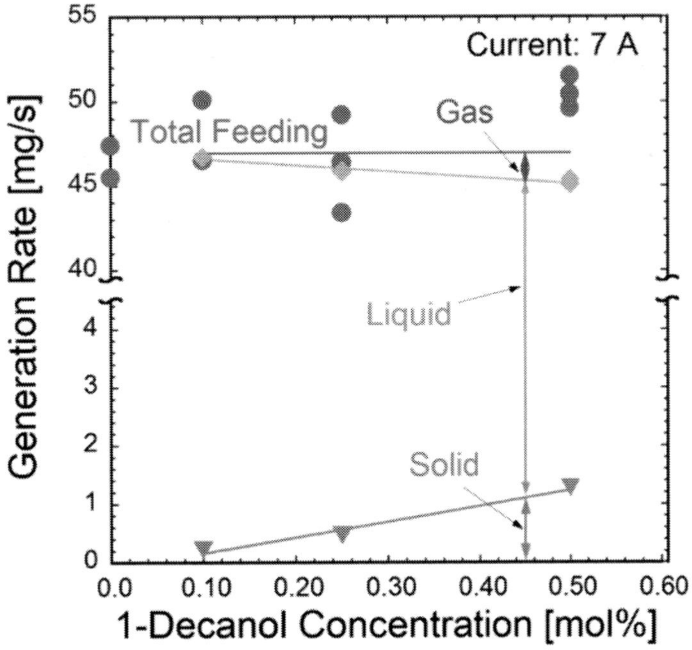

Figure 4. Generation rates of gas, liquid and solid products according to the concentration of 1-decanol at the fixed arc current of 7 A.

Generated gas species were monitored by QMS during the decomposition of 1-decanol by the water plasma. The main species are hydrogen, carbon monoxide and carbon dioxide as shown in figure 5. The QMS patterns were hard to distinguish even though the arc current and 1-decanol concentration were changed. Therefore, GC was used for the quantitative analysis on the generated gas product. Small amount of methane less than 0.1mol% was measured by GC in addition to hydrogen and carbon oxide gases. In the whole experimental conditions, the mole fractions of hydrogen and carbon monoxide were about 60mol% and 30mol%, respectively.

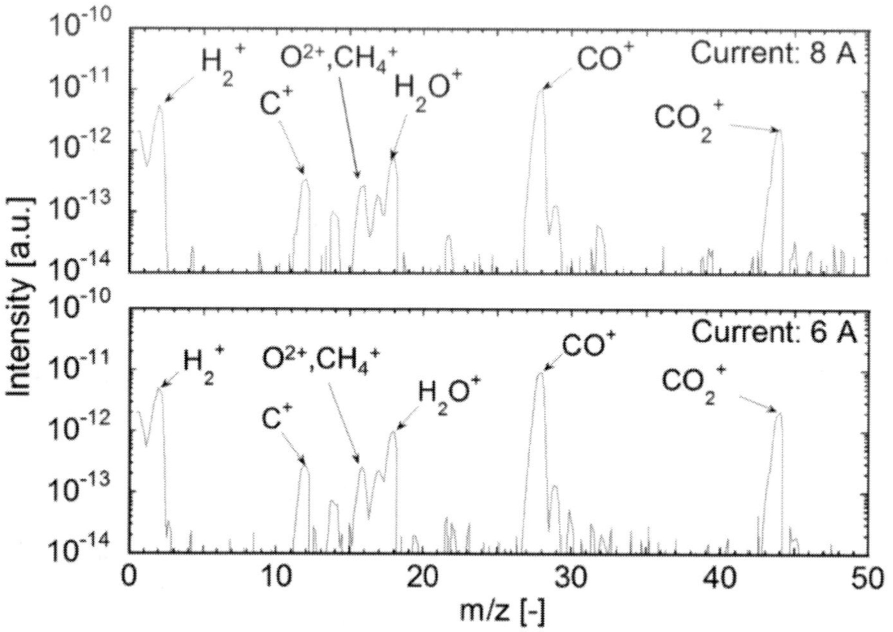

Figure 5. Analysis of the generated gas species by QMS in different arc current.

In order to investigate oxidation characteristics in the decomposition of 1-decanol with the water plasma, the mole fractions of each carbon oxides in the gas product were compared. The ratio of carbon dioxide to carbon monoxide is plotted in figures 6 and 7 based on GC measurement data. The value of carbon dioxide to carbon monoxide is increased with increasing the arc current, while that is decreased with increasing the concentration of 1-decanol. In the arc current change, enhanced oxygen radicals in the high current promote the generation of carbon dioxide. On the other hand, increased carbon radicals in the high 1-decanol concentration prefer to produce carbon monoxide rather than carbon dioxide. These oxidation characteristics for carbon oxide according to the arc current and 1-decanol concentration are in good agreement with results of the gas and solid generation rates in figures 3 and 4.

In HPLC analysis on the treated liquid, 1-decanol was not detected in the whole experimental conditions. Since a strong 1-decanol peak was found at 9.1 min in HPLC measurement for the raw feeding emulsion, the complete decomposition of 1-decanol was achieved by the water plasma treatment. The TOC that is expected to come from water-soluble byproducts, however, was measured by TOC analyzer. The TOC removal rates were evaluated with the measured TOC values for the raw 1-decanol emulsion ($n_{C,0}$) and for the treated liquid ($n_{C,l}$) as follows:

$$\eta_{TOC} = \frac{n_{C,0} - n_{C,l}}{n_{C,0}} \times 100 \tag{5}$$

As shown in figures 8 and 9, most carbons in the 1-decanol emulsion are removed by the water plasma treatment. This result means that the water plasma system can be effectively applied for the purification of contaminated water with organic wastes. The removal rate of TOC is increased with increasing the arc current as shown in figure 8, because enhanced oxygen radical promotes the generation of carbon oxide gases than water-soluble organic byproducts. Meanwhile in the change of 1-decanol concentration, the TOC removal rate does not have a distinct tendency. Therefore, oxygen radicals in the water plasma are believed as a major factor to control TOC level in the treated liquid.

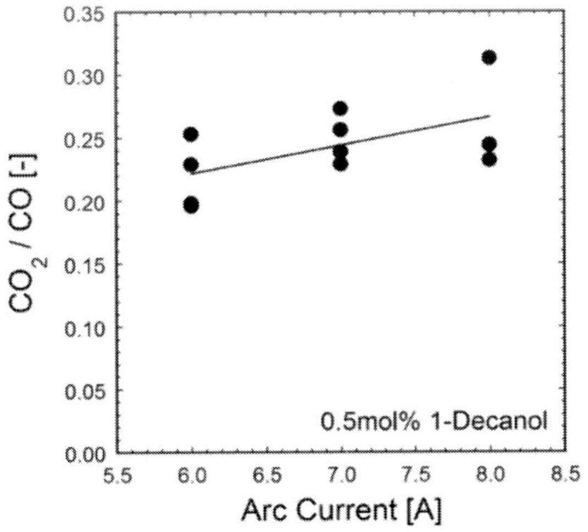

Figure 6. The ratio of CO_2 to CO according to the arc current at the fixed 1-decanol concentration of 0.5mol%.

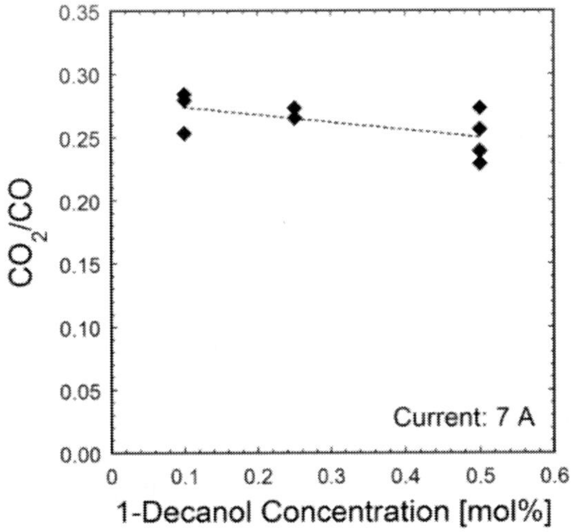

Figure 7. The ratio of CO_2 to CO according to the concentration of 1-decanol at the fixed arc current of 7 A.

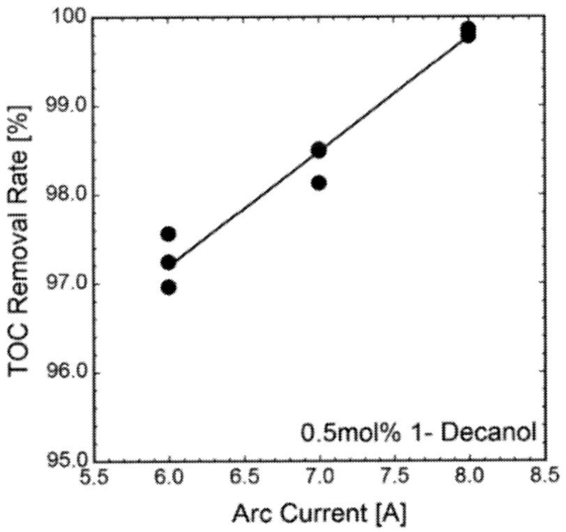

Figure 8. The removal rate of TOC in the generated liquid according to the arc current at the fixed 1-decanol concentration of 0.5mol%.

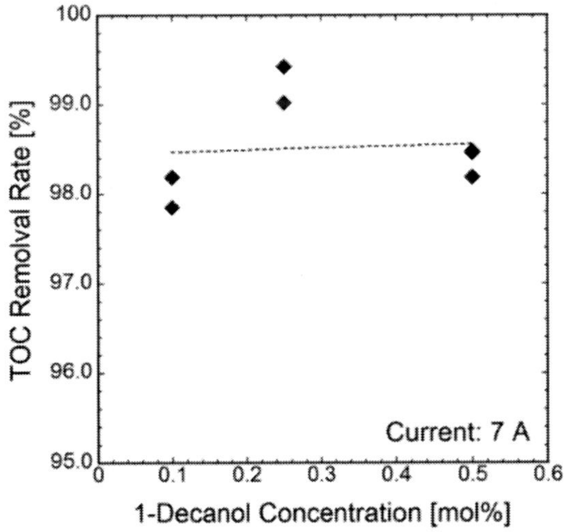

Figure 9. The removal rate of TOC in the generated liquid according to the concentration of 1-decanol at the fixed arc current of 7 A.

4. Conclusion

Water-insoluble 1-decanol in emulsion state was completely decomposed in the water plasma. Oxygen radicals in the water plasma are increased with increasing the arc current and input power. Therefore, the gas generation rate, the ratio of carbon dioxide to carbon monoxide and TOC removal rate are increased in the high arc current conditions, while the solid carbon generation is decreased at the same time. Different to the increase of the arc current, carbon radicals are enhanced with increasing the concentration of 1-decanol in the raw liquid. For this reason, both of gas and solid carbon generation rates are increased with increasing the concentration of 1-decanol. At the same time, the ratio of carbon dioxide to carbon monoxide is decreased due to relatively large amount of carbon radicals in the water plasma. Because the removal rate of TOC is mainly influenced by the population of oxygen radicals that are controlled by the input power level, the change of 1-decanol concentration does not have a significant effect on the TOC removal rate. Available organic wastes in the water plasma treatment are extended to water-insoluble materials through the present work. In addition, the role of oxygen radicals and carbon radicals were clarified in the water plasma treatment of organic wastes.

References

[1] Heberlein J and Murphy A B 2008 *J. Phys. D: Appl. Phys.* **41** 053001
[2] Gomez E, Rani D A, Cheeseman C R, Deegan D, Wise M and Boccaccini A R 2009 *J. Hazardous Materials* **161** 614
[3] Grigaitiene V, Snapkauskiene V, Valatkevicius P, Tamosiunas A, and Valincius V 2011 *Catalysis Today* **167** 135
[4] Yuan M -H, Narengerile, Watanabe T and Chang C -Y 2010 *Environ. Sci. Technol.* **44** 4710
[5] Narengerile, Saito H and Watanabe T 2010 *Plasma Chem. Plasma Process.* **30** 813
[6] Watanabe T 2005 *ASEAN J. Chem. Eng.* **5** 30
[7] Watanabe T, Narengerile and Nishioka H 2012 *Plasma Chem. Plasma Process.* **32** 123

Plasma Meets Chemistry: Combined Methods for Tailored Interface Design in Metal-Polymer Composites by Selective Chemical Reactions on Plasma Modified Surfaces

R Mix, S Hielscher, U Beck and J F Friedrich
Federal Institute for Materials Research and Testing, Berlin, Germany

E-mail: renate.mix@bam.de

Abstract. The applicability of polymer-metal composites is mainly determined by the durability of the adhesive strength between both components. Aluminium (Al) deposited on polypropylene (PP) exemplifies different options of interface design. By deposition of plasma polymers on PP the effect of the type of the functionality was investigated. Spacer insertion was accomplished to position the functional group away from the topmost surface. A further kind of interface design involved a partial condensation of functional groups. Hydroxyl and carboxyl groups were most effective to improve adhesion in Al-PP systems. Approximately 7-10 carboxyl or 25-27 hydroxyl groups per 100 C atoms were necessary to increase the peel strength up to ~700 N/m. In this range, the failure of the composite propagated along the interface Al-tape (no peeling of the metal). Spacer molecules between surface and functional groups provoked the effect that the number of needed functional groups for maximum adhesion was strongly reduced. Linking of the functional groups resulted in non-peelable Al-PP laminates. Two adhesion tests were applied - the peel test and the centrifuge technology. For PP foils modified with chemically bonded and additionally linked silanol groups (no peeling) an adhesive strength of $(2.5 \pm 0.2)\ N/mm^2$ was determined by centrifuge technology. XPS inspection of both fracture surfaces indicated a sub-surface failure in the polymer.

1. Introduction

The adhesion of metal onto polyolefin surfaces is one of the most sensitive features influencing the performance and durability of such composite materials found in various consumer goods. Aluminium (Al) thermally deposited on unmodified polypropylene (PP) shows no significant adhesion. Reason is the absence of any groups at the polymer surface which are able to interact with metallic Al. The poor adhesion properties of polyolefin surfaces have to be improved which was the motivation of numerous research efforts for modification of the polymer surface. [1-4] Plasma processes proved to be a useful tool to increase the adhesion between polymers and metals. Often, a simple oxygen plasma treatment of the polymer was used to produce several interactive and reactive O-containing functional groups, which are able to generate Al-polymer interactions as precondition for well-adherent Al-polyolefin systems. Nevertheless, it was observed that the introduced O-Al interactions were of different nature and weak. Therefore, increased efforts were made to overcome simple physical interactions at the metal-polymer interface by introduction of strong covalent bonds may be by inserted flexible spacer molecules. Following questions have to be answered:

1. Which type and which concentration of functional group are necessary for maximal adhesion?
2. Is it possible to replace numerous physical interactions by some covalent bonds with a profit in adhesion?
3. Which advantage produces the insertion of covalently bonded spacer molecules between polyolefin and metal?

Different options of interface design starting from plasma-treated and functionalized polymer surfaces were tested (Fig. 1). Pulsed plasma polymerization of functional group-carrying monomers was used to coat the PP surface with an adhesion-promoting layer, which is well-adherent to the polyolefin. Such films are pinhole-free with different functional groups such as hydroxyl, amino and carboxyl groups for anchoring the metal. The concentration of functional groups in the plasma polymer was varied applying the plasma copolymerization, thus linking comonomers with and without functional groups to layers with the desired concentrations of functional groups on its surface (type I in Fig. 1). Subsequent chemical reactions with bi- or multifunctional substances onto these functional groups were used for introducing spacer molecules. By this way the interface between metal and polymer was flexibilized (type II in Fig. 1). A third kind of interface design involved a partial condensation of functional groups (type III in Fig. 1). If the spacer molecule contains SiO_x elements the interface became hydrophobic and water molecules were repelled. Moreover, hydrolysis of SiOR groups formed adhesion-promoting SiOH groups and additionally SiO_x barrier layers.

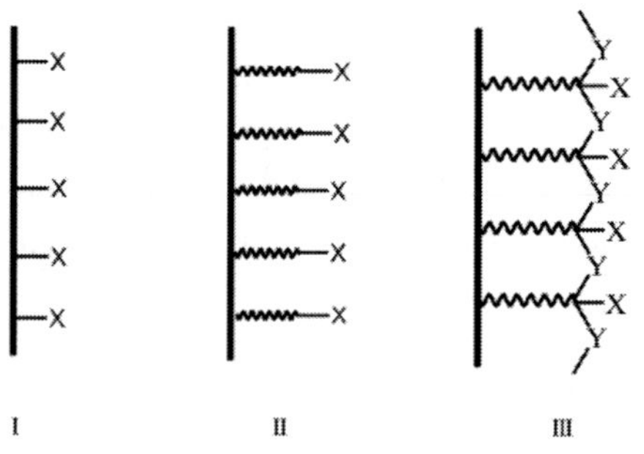

Fig. 1 Molecular design of the adhesion promotion interface (X – functional group (OH, NH_2, COOH, etc.; Y – brigde elements or groups (O, NH, …)

2. Experimental

2.1 Surface modification

Polypropylene foils (Goodfellow, UK, 100 μm thick), were ultrasonically cleaned in ether for 15 minutes. Plasma polymer layers were deposited in a cylindrical plasma reactor (Ilmvac, Germany) of 50 dm³ volume. The reactor was equipped with a radio-frequency (r.f. 13.56 MHz) generator (Advanced Energy Industries, Germany) with an automatic matching unit and a linear r.f. electrode (5 cm x 35 cm). The r.f. power was introduced to produce continuous-wave (cw) plasma or pulsed plasma with a variable duty cycle (0.1-1) and a repetition frequency of 1 kHz. Mass flow controllers were used for precise metering gases and vapours, liquid mass flow meters (Liqui-Flow®, Bronkhorst High-Tech B. V., The Netherlands) for liquid injection. A heated system of perforated metallic tubes distributed gases and vapours in the reactor. An adjustment of pressure and flow was performed by varying the speed of the turbomolecular pump. The standard pressure was adjusted to 10-25 Pa for depositing allyl alcohol, acrylic acid and allylamine as plasma polymers. These monomers were also used for copolymerization either with ethylene or butadiene in different molar ratios to vary the concentration of functional groups in the deposited plasma polymer (interface type I in Fig. 1).
A quartz microbalance was used for controlling the deposition rate.
PP films, used as substrates, were coated with 150 nm thick plasma polymer layers by applying the cw or the pulsed r.f. plasma modes (power input P = 30-300 W). The duty cycle (defined as $dc = t_{pulse\text{-}on} / (t_{pulse\text{-}on} + t_{pulse\text{-}off})$ characterizes the (plasma-on)/(plasma-off) period.
Spacer insertion was accomplished by the following chemical reactions: The samples with plasma polymerized allylamine layers deposited on PP were shaken in glutaraldehyde solution (1 % in water) for 2 hours at room temperature immediately after its removing from the reactor (to avoid auto-oxidation processes). Thus, aldehyde groups were introduced by reaction of the amino groups with glutaraldehyde, moreover, undesired oxidation at contact with ambient air was eliminated. Subsequently, chain-extension was performed using bi- or polyfunctional substances with at least one amino group. Variation of its chain length and structure produced different spacer length, end groups and flexibility as shown in Fig 2:

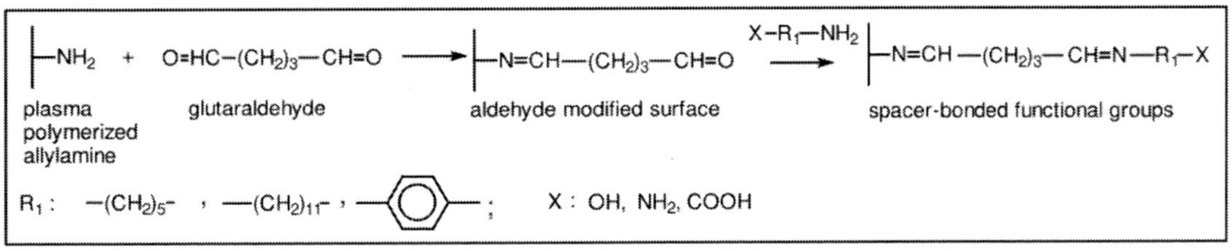

Fig. 2 Scheme of spacer introduction to the polyolefin surface (interface type II in Fig.1)

The concentration of the functional groups was determined by its chemical derivatization and measuring these derivates by photoelectron spectroscopy (XPS). The reactions of hydroxyl groups with trifluoroacetic anhydride, amino groups with pentafluorobenzaldehyde and carboxylic groups with trifluoroethanol in presence of carbodiimide are described elsewere [5]. The concentration of the hydroxyl and carboxyl groups, respectively were calculated by the following equation:

$$C_{OH,\,COOH} = \frac{100*[F]}{3*[C]-2*[F]}$$

derived from the derivatization reaction with trifluoroacetic anhydride (TFAA) (1) and trifluoroethanol (TFE) (2), respectively:

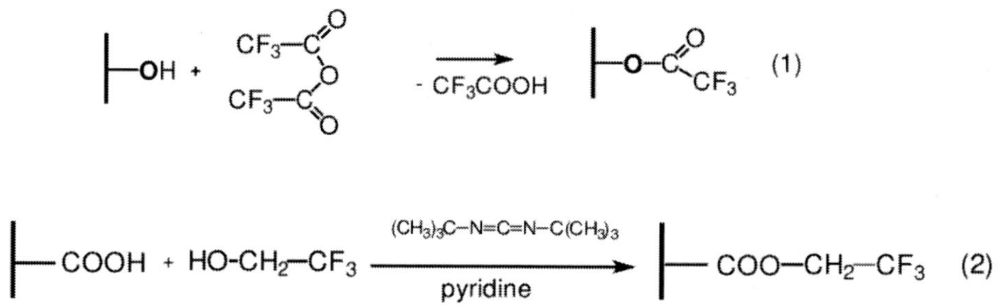

[C] and [F] indicate the concentrations of carbon and fluorine measured by XPS at the TFAA and TFE, respectively, derivatized surfaces.

Generation of functional groups at the topmost surface, additionally linked, was achieved in two ways (Figs. 3 and 4). A short spacer was introduced by reaction of 3-isocyanatopropyltrimethoxysilane with the primary amino groups of the plasma deposited allylamine layer. Subsequent stepwise hydrolysis produced -Si-OH groups on the surface. Simultaneously to hydrolysis a condensation of -Si-OH groups occurs to -O-Si-O- units (Fig. 3).

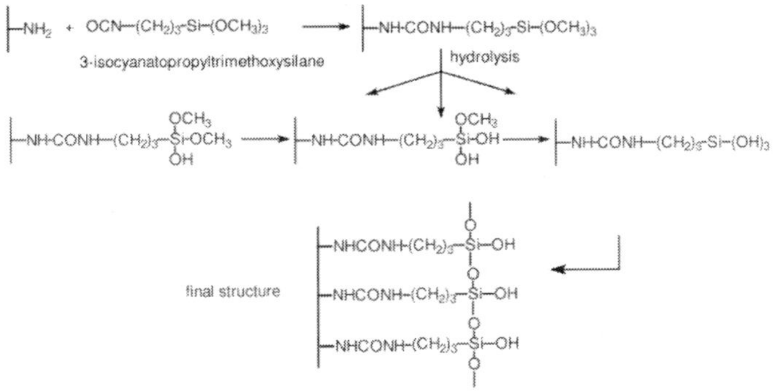

Fig. 3 Scheme of synthesis of surface-bonded –SiOH groups (short spacer), additionally linked (interface type III in Fig.1).

The other method started with bromination of PP or deposition of a Br-containing plasma polymer using the bromoform plasma. The brominated surface was reacted with a long-chain spacer (1.12-diaminododecan) followed by consumption of the residual amino group with 3-isocyanatopropyltrimethoxysilane and subsequent hydrolysis/condensation (Fig. 4):

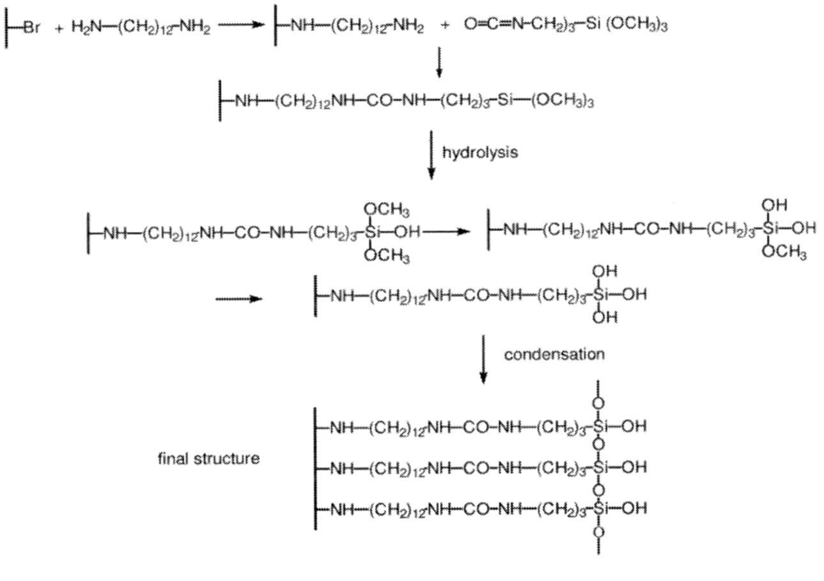

Fig. 4 Grafting of 1.12-diaminododecan and subsequent reaction with 3-isocyanatopropyltrimethoxy-silane followed by hydrolysis/condensation (interface type III in Fig.1)

Journal of Physics: Conference Series **406** (2012) 012004 doi:10.1088/1742-6596/406/1/012004

2.2. Test specimen assembly and composite characterization

The spacer-equipped polyolefin surfaces were Al coated using a thermal metallizer (Auto 306, Edwards, UK). The Al thickness was adjusted to 100 nm monitoring it by using a quartz microbalance. The metal-polymer adhesion was measured using the 90° peel test at a peeling speed of 25 mm/min. The peeled surfaces of the composites were inspected by XPS to obtain information on the origin of failure. These peel results were compared with those carried out by centrifuge technology [6]. The major problem of all tape-based peel tests was the limited adhesion between the adhesive tape and the aluminium layer. Often, the adhesive tape was peeled-off but the Al-polymer composite remained intact and thus metal adhesion could not be measured. Now, it is expected that the centrifuge technology will overcome this problem. The schematic diagram of this multiple sample test is shown in Fig. 5:

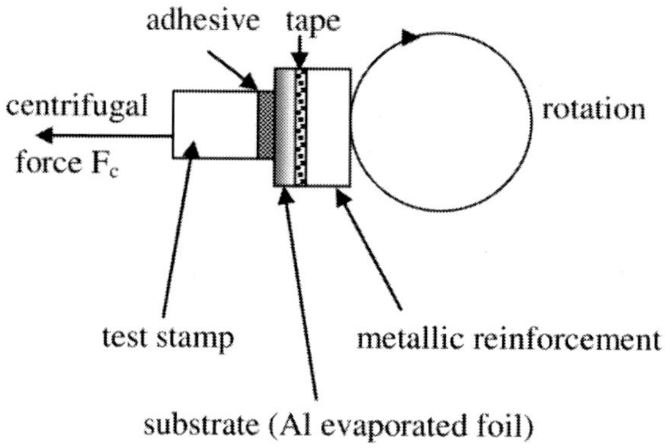

Fig. 5 Schematic diagram of the centrifuge technology

The test stamps were bonded on the Al-deposited side of the metallized PP foil using cyanoacrylate adhesive (Loctite 415). The non-metallized side of the foil was fixed by a double-sided adhesive tape (TESA 4964) on the metallic reinforcement. At rotation, individual rupture events occur at increasing rotational speed. All rupture events were detected on-line outside the centrifuge using a position-coded and rpm-correlated infrared data transmission.

The centrifuge technology uses the following equation for the calculation of adhesive strength $\sigma_{A/B}$:

$$\sigma_{A/B} = \frac{F_C}{A} = \frac{m \cdot \omega^2 \cdot r}{A} = \frac{m \cdot \left(2 \cdot \pi \cdot \frac{n}{60}\right)^2 \cdot r}{A}$$

with F_c - centrifugal force, A - fracture area, m - mass of the test stamp, n - critical revolutions per minute, r – radius of the rotational motion and ω – angular velocity ($F_c = m^* \omega^{2*} r$).

3. Results and Discussion

3.1 Functional groups at the top-most polymer surface

Functional groups like COOH, OH and SiOH at the polymer surface provoked a significant adhesion-promotion in the PP-Al composites. Amino groups and ethylene units (without any functional groups) did not show any effect. In general, the measured peel strength of aluminium deposits on PP increased linearly in dependence on the concentration of functional groups. Near the maximum concentration of OH groups (25-27 OH per 100 C atoms) and at concentration of 8-10 COOH per 100 C atoms, a constant level of peel strength was achieved characterized by a cohesive failure in these laminates. In the case of allyl alcohol homopolymer layer a dramatic decrease of the peel strength at a maximum concentration of OH groups (~30 OH per 100 C atoms) was observed (see Fig. 6). In this case the XPS inspection of the peeled surfaces showed on both fracture areas the typical shape of the C1s peak of the allyl alcohol plasma polymer. This result is characteristic for a cohesive failure within the poly (allyl alcohol) layer (see Fig. 6).

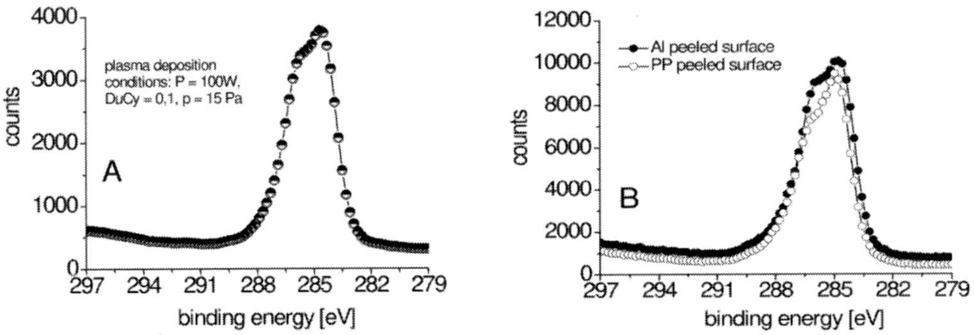

Fig. 6 C1s peak of plasma polymerized allyl alcohol (A) and C1s peaks of both peeled sides of Al-PP composite (B) with an allyl alcohol layer interface

3.2. Functional groups bonded at flexible spacer molecules far from the original surface

The grafting of spacers with adhesion-promoting endgroups onto the deposited allylamine plasma polymer layers improved the peel strength significantly. The spacer bonded hydroxyl endgroups $-N{=}CH{-}(CH_2)_3{-}CH{=}N{-}(CH_2)_{2(5)}{-}OH$ (number 3 and 4 in Fig. 7) and $-N{=}CH{-}(CH_2)_3{-}CH{=}N{-}(p{-}C_6H_4){-}OH$ (number 5 in Fig. 7) showed much higher peel strength in comparison to that OH groups, which were bonded at the surface without spacer. Here, the formation of Al-O-C bonds is assumed.

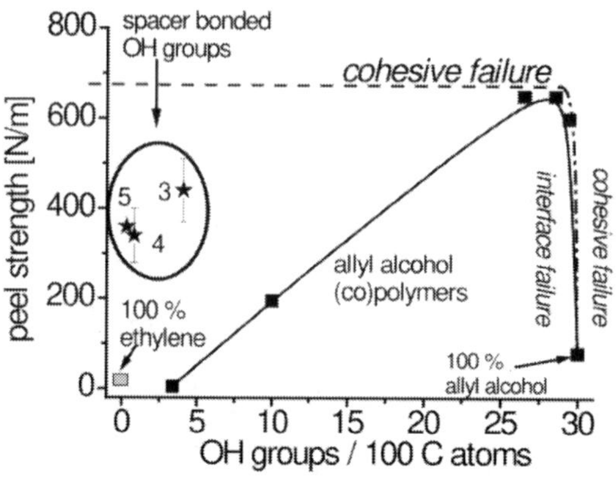

Fig. 7 Peel strength of aluminium-polypropylene laminattes arranged by plasma (co)polymers of allyl alcohol layers (directly bonded to the surface) and spacer-bonded hydroxyl groups

COOH groups are most efficient in adhesion promoting for the Al-PP system (cf. Fig. 8). Only 8-10 COOH groups per 100 C atoms were sufficient to produce maximal peel strength. The strong adhesion-promoting action of COOH groups was assumed to be caused by the formation of Al bidentate bondings as proposed by Alexander et al. [7]. Spacer-bonded COOH groups of the type $-N=CH-(CH_2)_3-CH=N-(CH_2)_{5(11)}-COOH$ (number 1 and 2 in Fig. 8)improved the peel strength stronger in relation to COOH bonded without spacer referenced to the same concentration.

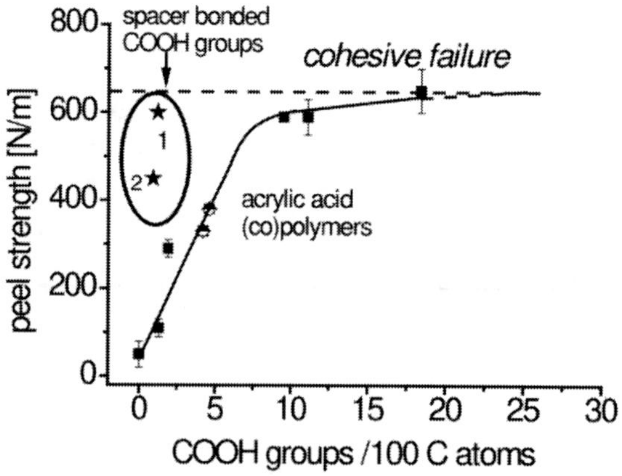

Fig. 8 Peel strength of aluminium-polypropylene laminates arranged by plasma (co)polymers of acrylic acid layers and spacer-bonded carboxyl groups

3.3. SiOH groups positioned in a short or long distance from the topmost surface, additionally linked

All peel tests of SiOR / SiOH-modified PP-Al composites did not give any result because the adhesive tape was peeled-off from the Al coating. Therefore, it was not possible to peel any aluminum from the PP surface. Thus, the real adhesion was higher than the measured peel strength. To overcome this hindrance the new developed centrifuge technology was applied. The results are represented in Figs. 9 and 10 for short- and long-chain spacer bonded SiOH groups, partially condensed, by using different hydrolysis times.

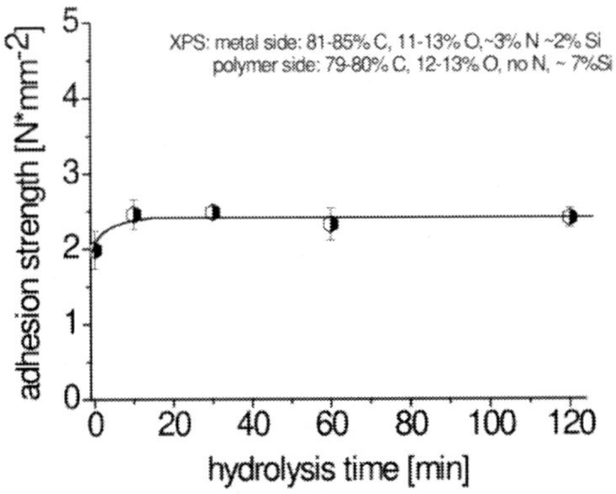

Fig. 9 Adhesion strength of Al-PP composites, PP equipped with short spacer bonded SiOH groups, additionally linked

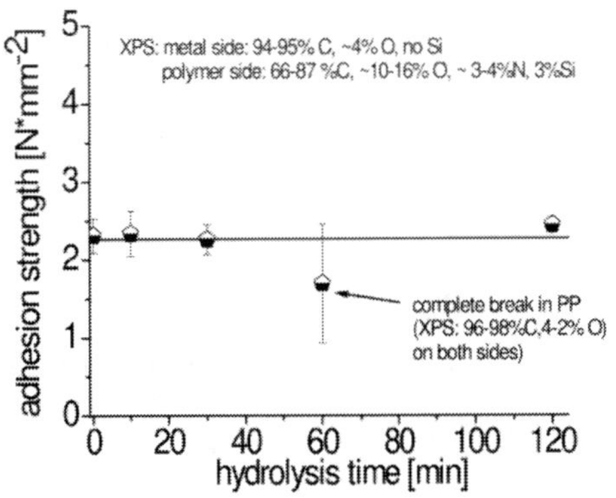

Fig 10 Adhesion strength of Al-PP composites, PP equipped with long spacer bonded SiOH groups, additionally linked

In principle constant adhesion strength (2.2-2.5 N/mm^2) was found in both cases independent on the hydrolysis time. On both fracture areas (named as metal and polymer side) their composition was estimated by XPS. In the case of short spacer-bonded SiOH groups a break in the polymer itself was detected because no nitrogen (N) on the polymer side was found. The reason for the relatively high Si content is not completely understood may be caused by a break within PP near the interface to the silane spacer. Long-chain spacer bonded SiOH groups exhibited a break in the surface layer of the PP near the interface to the silane spacer modification because it contains up to 4 % O. In one case (60 min hydrolysis times) the break occurred in deeper PP layers resulting in identical compositions of both fracture surfaces.

4. Summary

The nature of interactions between aluminium and different functional groups was reflected in the measured peel strengths. The succession of the adhesion-promoting effect was derived from the slope of the linear increase of the peel strengths vs. concentration of functional groups resulting in the following row: CH_2-CH_2 < NH_2 « OH < COOH

The mode of failure, either interface or cohesive, was inspected by microscopy and XPS. Most efficient groups for promoting the Al peel strength to PP were OH and any more COOH as seen in Figs. 7 and 8 probably due to formation of chemical bonds. This behaviour was confirmed when spacer bonded functional groups were applied. Spacers with stiffer chains and other end groups than OH and COOH had only slight influence on adhesion. The reason why should be the formation of covalent bidentate or ionic (salt-like) bonds between COOH groups and aluminium [8, 9] and Al-O-C (monodentate or alcoholate) bonds between OH groups and Al [10]. In comparison to high concentrations of directly to the surface bonded functional groups (8-10 COOH/100 C and 25-27 OH/100 C) a considerably lower concentration of spacer coupled functional groups (1-2 OH/100 C or 1-2 COOH/100 C) was necessary to achieve the same high level of peel strength of Al-PP composites characterized by cohesive failure in the polymer substrate.

The application of silanol groups bonded by short or long spacers to the PP surface was accomplished by reaction of 3-isocyanatopropyltrimethoxysilane with allylamine plasma polymer layers deposited on PP or by reaction of the brominated PP surface with 1.12 diaminododecane and subsequent reaction with the mentioned silane. The subsequent stepwise hydrolysis is associated with the formation of SiOH groups and their condensation to Si-O-Si units. For these modified surfaces, adhesion of Al was beyond the maximum peel strength of adhesive tapes, i.e. about 700 N/m. Using the centrifuge technology a constant adhesive strength of 2.2-2.5 N/mm^2 was measured. XPS inspection of both fracture surfaces indicated a sub-surface failure within the polymer may be near the interface.

5. References

[1] Moosheimer U and Bichler C 1999 *Surf. Coat. Technol.* **116–119** 812

[2] Arefi-Khonsari F, Kurdi J, Tatoulian M, and Amouroux J 2001 *Surf. Coat. Technol.* **142–144** 437

[3] André V, Arefi F, Amouroux J, De Puydt Y, Bertrand P, Lorang G and Delamar M 1989 *Thin Solid Films* **181** 451

[4] O'Kell S, Henshaw T, Farrow G, Aindow M and Jones C 1995 *Surf. Interf. Anal.* **23** 319

[5] Sabbatini L and Zambonin P G 1993 *Surface Characterization of Advanced Polymers*, (Weinheim: VCH Publishers) 221-56

[6] Beck U, Reiners G, Lerche D, Rietz U, Niederwald H 2011 *Surf. Coat. Technol.* **205** 182

[7] Alexander M R, Beamson G, Blomfield C J, Leggett G, and Duc T M 2001 *J. Electr. Spectr. Rel. Phenom.* **121** 19

[8] Schultz J and Lavielle L 1989 *Makromol. Chem. Macromol. Symp.* **23** 343

[9] Rancourt J D, Hollenhead J B and Taylor L T 1993 *J. Adhesion* **40** 267

[10] Akhter S, Zhou X-L and White J M 1989 *Appl. Surface Sci.* **37** 201

Plasma Spray-PVD: Plasma Characteristics and Impact on Coating Properties

G Mauer and R Vaßen

Forschungszentrum Jülich, Institute of Energy and Climate Research (IEK-1)
52425 Jülich, Germany

E-mail: g.mauer@fz-juelich.de

Abstract. Typical plasma characteristics of the plasma spray-physical vapour deposition (PS-PVD) process were investigated by optical emission spectroscopy. Electron temperatures were determined by Boltzmann plots while temperatures of the heavy species as well as electron densities were obtained by broadening analysis of spectral lines. The results show how the plasma properties and thermodynamic equilibrium conditions are affected by the admixture of hydrogen and the ambient chamber pressure. Some experimental examples of PS-PVD coatings demonstrate the impact on feedstock treatment and deposited microstructures.

1. Introduction

The very low pressure plasma spray (VLPPS) process has been developed with the aim of depositing uniform and thin coatings with large area coverage based on plasma spraying. At typical pressures below 500 Pa, the characteristics of the plasma jet change compared to conventional low pressure plasma spraying processes (LPPS, often called vacuum plasma spraying, VPS) operating at 5-20 kPa. By VLPPS, quite thin and dense ceramic coatings can be obtained for special applications like solid oxide fuel cells [1], gas separation membranes [2], and wear protection [3].

The combination of VLPPS with enhanced electrical input power has led to the development of the plasma spray-physical vapor deposition process (PS-PVD [4], initially called LPPS-TF process, TF = thin film). At appropriate parameters it is even possible to evaporate the powder feedstock material providing advanced microstructures and non-line of sight deposition e.g. of thermal barrier coatings for turbine components [5]. Since relative large deposition rates and areas can be achieved, PS-PVD has the potential to fill the gap between PVD on the one hand and conventional plasma spraying on the other hand.

To exploit the potential of such novel gas phase deposition, the plasma characteristics and their dependency on process conditions must be better understood. Besides power, spray distance and plasma flow conditions, the characteristics of the plasma gas are important parameters which are dependent particularly on its composition and chamber pressure. Thus, plasma conditions were investigated by optical emission spectroscopy. Based on the results, the distributions of plasma temperatures and electron densities were determined. Some experimental examples are given to demonstrate the impact on coating microstructures.

2. Experimental

Experiments were carried out on a Sulzer Metco Multicoat PS-PVD System (Sulzer Metco, Wohlen, Switzerland) which resulted from a comprehensive reconstruction of an existing conventional LPPS system. In particular, the system is equipped with an additional vacuum pumping unit, a large vacuum blower to provide sufficient pumping capacity at low pressures, enlarged cooling capacity, additional power sources, a new torch transfer system and new control units. In addition to a modified single cathode O3CP gun, which was used in this work, also the F4-VB torch as well as the three-cathode TriplexPro torch can be operated. Figure 1 shows the facility.

Journal of Physics: Conference Series **406** (2012) 012005 doi:10.1088/1742-6596/406/1/012005

Figure 1. PS-PVD facility at Forschungszentrum Jülich, IEK-1.

For the experiments an agglomerated 7-8 wt.% Yttria partially stabilized zirconia powder (YSZ, Sulzer Metco M6700) was used. The grain sizes measured by laser diffraction were $d_{10} = 2$ µm, $d_{50} = 8$ µm, and $d_{90} = 18$ µm. For PS-PVD operations, the powder grain sizes should be preferably small. The substrates used were polished stainless steel and were preheated by the plasma jet immediately before coating. Table 1 shows the appropriate plasma spray parameters.

Table 1. Plasma spray parameters.

	Parameter A	**Parameter B**
plasma gas	Ar 35 / He 60 slpm[a]	Ar 35 / He 60 / H_2 10 slpm[a]
current	2600 A	2200 A
net power	60 kW	60 kW
pressure	200-1000 Pa (varied)	200-1000 Pa (varied)
powder feed rate	1-20 g min^{-1} (varied)	1-20 g min^{-1} (varied)
spray distance	300-1400 mm (varied)	300-1400 mm (varied)

[a] slpm = standard liters per minute

The spectrometer applied for plasma characterization was an ARYELLE 200 (Laser Technik Berlin (LTB), Berlin, Germany) scanning a wavelength range of 381-786 nm. Plasma radiation was collected through a borosilicate glass window and an achromatic lens, transferred by an optical fiber to the 50 µm entrance slit and detected by a 1024x1024 CCD array. The system is equipped with an echelle grating and the spectral resolution obtained is 15.9-31.8 pm. Calibration was carried out with a spectral Hg lamp.

3. Plasma Characterization

Plasma characteristics were investigated by optical emission spectroscopy. Electron and excitation temperatures were determined by the Boltzmann plot method, respectively. Electron densities and heavy particle temperatures were calculated by peak broadening analysis. Some initial results based on the first two hydrogen spectral lines of the Balmer series H_α and H_β obtained at VLPPS conditions are given in [6, 7].

As the plasma jets are optically thin, the measurement is spatially integrated along the line of sight of the optical sampling setup [8]. To obtain the central characteristics of the plasma jet, generally a deconvolution of the measured two-dimensional projections of the integrated intensities must be performed to reconstruct the spatial plasma temperature field. This can be done by Abel inversion based on the assumption of axial symmetry [9]. However this method is very laborious and potentially increases the measurement errors due to high error transfer coefficients. Thus, alternative methods have also been developed to circumvent Abel inversion. In this work, a method proposed in [10] was applied to the evaluation of Boltzmann plots. However, the corrections of the temperatures obtained along the plasma jet axis were found to be small compared to the temperature bandwidths due to the standard errors of the linear fits. Hence, the uncorrected excitation temperatures were considered to be representative of the jet center zone. In contrast, a method to correct the spectral line broadening [11] yielded substantial corrections of the measured values und was thus applied for broadening analysis.

3.1. Boltzmann plot method

Based on spectroscopically measured emissions, plasma excitation temperatures can be determined by the atomic Boltzmann plot method [12, 13]. Applying the Boltzmann distribution, the absolute intensity I_{jk} of a spectral line emitted by the plasma due to the transition from an excited state j to a lower energy state k is

$$I_{jk} = \frac{Lhc}{4\pi\lambda_{jk}} A_{jk}\, n \frac{g_j}{Z}\, e^{\left(-E_j / k_B T_{exc}\right)} \tag{1}$$

where L is the emission source depth, h is the Planck constant, c is the velocity of light, A_{jk} is the transition probability, n is the density of emitting atoms/ions, g_j is the statistical weight (degeneracy) of the excited level j, λ_{jk} the wavelength of the emission, Z is the partition function, E_j the energy of the excited level, k_B the Boltzmann constant, and T_{exc} the excitation temperature. If the intensities of a set of emission lines of the same final state are measured a linear plot

$$\ln\left(\frac{I_{jk}\lambda_{jk}}{g_j A_{jk}}\right) = \frac{-E_j}{k_B T_{exc}} + C, \quad C = \ln\left(\frac{Lhcn}{4\pi Z}\right) \tag{2}$$

will yield a slope that is inversely proportional to the desired excitation temperature T_{exc}. The left side of this equation is called atomic-state distribution function (ASDF). The values of the energy of the excited level E_j, of the transition probability A_{jk} and the statistical weight g_j were taken from [14]. The intensity values were obtained from the areas under Voigt fits of the measured spectral emission line profiles. Error bars related to the temperature results were calculated based on the standard errors of the linear fits. They were found to be approximately ±5%. More details on the application of the Boltzmann plot method can be found elsewhere [15].

Thermodynamic equilibrium (TE) conditions affect the applicability of the Boltzmann plot method considerably. There are four types of balance mechanisms resulting in four corresponding distribution functions for particles and photons [16],

- Kinetic energy exchange and conservation (Maxwell distribution characterized by the kinetic temperature T)
- De-excitation and excitation (Boltzmann distribution characterized by the excitation temperature T_{exc})
- Recombination and ionization (Saha distribution characterized by the Saha temperature T_{Saha})
- Absorption and emission (Planck distribution characterized by the radiation temperature T_{rad})

The TE requires that all these temperatures are unique which is often not fulfilled, Figure 2. The weakest form of TE departure is when part of the emitted radiation is not reabsorbed in the plasma and escapes, but Maxwell, Boltzmann and Saha balances still can be maintained. As spatial and time-related gradients are the reason, this is called local thermodynamic equilibrium (LTE) [16]. A larger deviation from TE occurs if the kinetic energy transfer between electrons and heavy particles (atoms, ions) is no longer sufficient to balance energy evenly among them. This leads to the two-temperature (2T) model [17] with the kinetic electron temperature T_e, the kinetic temperature of the heavy particles T_h, and the non-equilibrium parameter $\theta = T_e/T_h$.

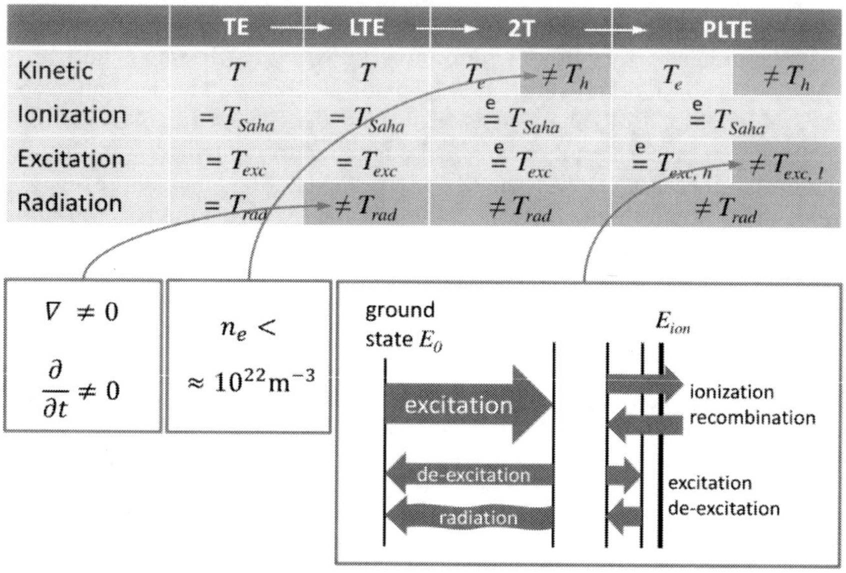

Figure 2. Different stages of departure from thermodynamic equilibrium in thermal plasmas; for abbreviations refer to the text.

The density distribution of excited atoms is assumed to be dominated by electron-induced transitions leading to a Boltzmann distribution where T_{exc} is equal to T_e. Thus, the results of Boltzmann plots can be regarded to represent the electron temperatures T_e [16]. Free electrons can also control the equilibrium between ionization and recombination so that T_{Saha} can be identified also with T_e. However, atoms in the energy levels close to the ground state do not de-excite predominantly by such electron-induced transitions, but rather radiatively [17], Figure 2. In case of recombining plasmas as given in expanding plasma jets, this leads to underpopulation of the lower energy levels and thus to lower values of the ASDF. This situation is referred to as partial local thermodynamic equilibrium (PLTE) [18]. As a consequence, the data points of the Boltzmann plots do not develop linearly [19] and only those at higher energy levels represent T_e.

Figure 3 gives an example of Boltzmann plot for recombining plasmas. Fifteen Ar I lines between 549.59 nm and 751.47 nm were evaluated. Four of them correspond to lower energy levels (4p – 4s). It is evident that they are not aligned linearly with respect to the eleven other data points of the higher energy levels (4d, 6s, 5d, and 6d – 4p). Thus, they were not included into the linear regression. The excitation temperature derived from its slope $T_{exc,\,h}$ is 5065 +466/-394 K which represents T_{Saha} and T_e as shown in Figure 2. The error bandwidth results from the standard error of the slope of the linear fit.

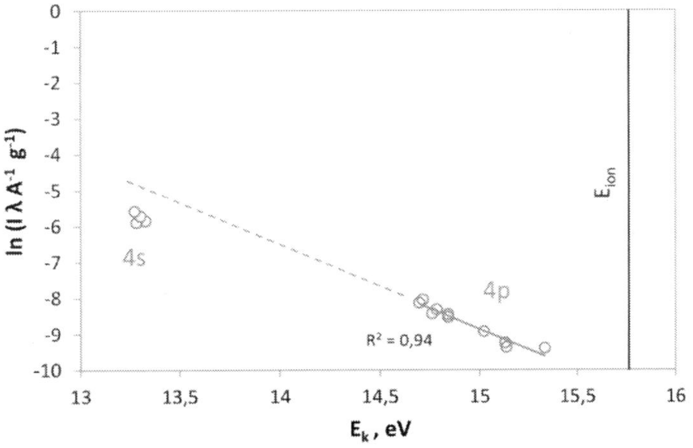

Figure 3. Example of Boltzmann plot for recombining plasmas with 15 evaluated Ar I spectral lines; parameter B, chamber pressure 200 Pa, and stand-off distance 400 mm.

Figure 4 gives the Boltzmann plot for the same plasma condition, but instead of Ar I lines eight He I multiplets were evaluated between 447.15 nm and 728.13 nm. For the same reasons as stated above, the data point for the transition being lowest to the ground state (3p – 2s) was not included into the linear fit since it slightly deviates. The excitation temperature obtained from the remaining seven data points (3s, 3d, 4s, and 4d - 2p) is 4344 +225/-204 K. It is not clear whether the difference to the excitation temperature obtained from the Ar I lines is substantial. However, considering He lines has practical advantage in this work since He is the largest plasma gas fraction in both investigated parameters. In particular at increasing stand-off distance the argon peaks become weak and the evaluation more uncertain.

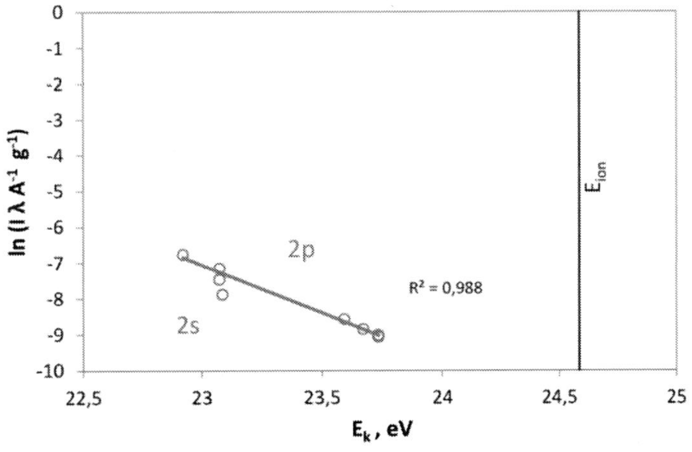

Figure 4. Example of Boltzmann plot for recombining plasmas with 8 evaluated He I spectral lines; parameter B, chamber pressure 200 Pa, and stand-off distance 400 mm.

3.2. Line broadening analysis

Spectral line broadening analysis was used to determine the electron density n_e as well as the kinetic temperature of the heavy species (atoms and ions) T_h. In principle, several broadening mechanisms exist [20],

- Stark broadening $\Delta\lambda_S$ due to collisions between charged particles with exited emitting atoms is significant in plasmas with $n_e > 10^{21}$ m^{-3}. At lower electron densities, lines of non-hydrogenic species are too narrow. Hydrogen lines are subjected to linear Stark effect whereas non-hydrogenic lines show quadratic Stark effect. For neutral atom lines, electron-impact broadening produces a symmetrical, shifted profile of the Lorentzian type, while ion-contribution introduces asymmetry as well as additional contribution to the width and the shift of the profile.
- Van der Waals broadening is produced by the collisions of neutral ground state atoms and exited emitting atoms. It is usually small and can be neglected for the conditions investigated here.
- Natural broadening arises from uncertainty in energy of the states involved in the transitions. It is rarely significant in atomic spectroscopy.
- Resonance broadening is confined to lines involving an energy level being coupled to the ground state which is not relevant for the lines investigated in this work.
- Doppler broadening $\Delta\lambda_D$ occurs due to the thermal motion of excited atoms. The line intensity distribution is Gaussian.
- The line shapes obtained by a spectrometer are usually approximately Gaussian profiles. Such instrument broadening $\Delta\lambda_I$ has to be determined by means of a laser source or spectral lamp.

Those effects which give rise to Gaussian line shapes tend to be independent from those producing Lorentzian broadening profiles. As such, the convolution of these two types of functions results in the Voigt function where the Gauss contribution determines the profile core while the Lorentzian part governs the far wings. In order to separate the Gauss and Lorentz contributions, a form of the Voigt function was used where the Gaussian and the Laurentzian half widths at full maximum (HWHM) $\Delta\lambda_G/2$ and $\Delta\lambda_L/2$ are direct parameters ($\Delta\lambda_G$ and $\Delta\lambda_L$ are the full widths at half maximum, FWHM);

$$I = \frac{a_0 a_3}{2\pi\sqrt{\pi a_2^2}} \int_{-\infty}^{\infty} \frac{e^{\left(-t^2\right)}}{\frac{a_3^2}{2a_2^2} + \left(\frac{x-a_1}{\sqrt{2}a_2} - t\right)^2} dt \tag{3}$$

where a_0 is the area, a_1 the centre wavelength, a_2 the Gaussian half width, and a_3 the Lorentzian half width. Thus, the fit of the measured intensity distribution yields immediately the Lorentzian as well as the Gaussian broadening FWHMs and analogically their standard errors by

$$\Delta\lambda_G = 2a_2\sqrt{2\ln(2)},\ \Delta\lambda_L = 2a_3 \tag{4}$$

To obtain the Doppler broadening the instrument broadening has to be subtracted by

$$\Delta\lambda_D = \sqrt{\Delta\lambda_G^2 - \Delta\lambda_I^2} \tag{5}$$

Since only Stark broadening causes a significant Lorentzian contribution they can be equated

$$\Delta\lambda_S = \Delta\lambda_L \tag{6}$$

The selection of the spectral lines for broadening analysis has to be done carefully. The most widely used for diagnosis purposes because of its great Stark width is the H_β line of the Balmer series of hydrogen at 486.13 nm [21]. It is relatively isolated, does not show significant self-absorption, and is almost independent on T_e and ion dynamics effects. The latter cannot be neglected if using other species like He I [22]. There are also some well investigated Ar I lines, e. g. at 696.54 nm [23, 24, 25]. However in this work, broadening analysis was only successful using the H_β line as it is supported by sufficient measured data points so that the Gaussian and Lorentzian broadening contributions could be separated reliably. Several non-hydrogenic lines of Ar I and He I were tried but found to be too narrow with respect to the spectrometer resolution. Thus, the determination of T_h and n_e could be performed only for the hydrogen containing parameter B.

Figure 5 shows three examples of Voigt fits and residuals of the H_β line profile at Parameter B and 400 mm stand-off distance. The quality of the fits is very similar for the three investigated chamber pressures 200, 500, and 1000 Pa. The coefficients of determination are better than 0.999 each. The developments of the residuals do not show any significant indication of asymmetry. The central dip is hardly discernible from the observed data.

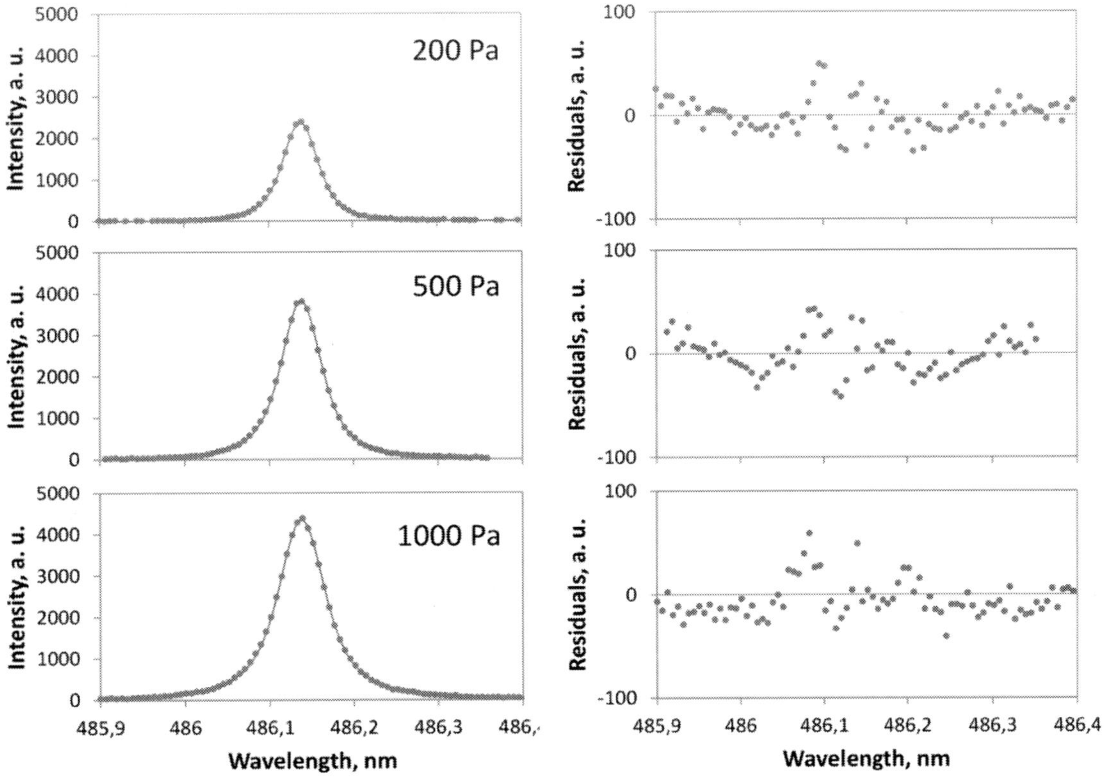

Figure 5. Voigt fits and residuals of H_β line at Parameter B and 400 mm stand-off distance, chamber pressures 200, 500, and 1000 Pa.

Electron density n_e (in m^{-3}) is calculated according to the Stark profile calculations in [26]. The method is independent of the assumption of PLTE. The FWHM (in nm) is calculated by

$$\Delta\lambda_S = 2.507 \cdot 10^{-4} \alpha \, n_e^{2/3} \tag{7}$$

The Stark broadening parameter α is tabulated depending on n_e and T_e; the latter dependence is only weak. Based on the Lorentzian broadening contribution separated from the measured broadening as described above, equation 6 was solved iteratively for n_e. Cross checks with other simplified approaches [27, 28] yielded similar results. Error bandwidths related to the electron density results were checked based on the standard errors of the Voigt fits. They were found to be small by approximately ±5% und hence not included in the graphs.

The FWHM due to Doppler broadening is calculated according to [29] by

$$\Delta\lambda_D = 2\lambda_0 \left(\frac{2k_B T_h \ln(2)}{M\, u\, c^2}\right)^{1/2} \tag{8}$$

where λ_0 is the central wavelength of the emission line, k_B is the Boltzmann constant, M is the relative atomic mass of the absorbing atom, u is the atomic mass unit, and c is the speed of light. Based on the Gaussian broadening contribution separated from the measured broadening and corrected for the instrumental broadening as described above, equation 7 was solved for T_h. Error bars related to the temperature results were calculated based on the standard errors of the linear fits. They were found to be approximately ±10 to 15%. Thus, the temperature determination based on Doppler broadening turned out to be prone to uncertainties.

The investigated plasma jets are supersonic when they exit the nozzle. The first transition to subsonic conditions is indicated by the so-called Mach disk. Further downstream, expansion and compression zones due to reflections at the jet fringes can be observed in particular at higher chamber pressures. In addition, the turbulent zones at the jet ends move closer to the torch exit with increasing chamber pressures. Thus, local pressure differences compared to the chamber pressure occur. However, these local static pressures p_{stat} were not determined. Hence, only for the relatively homogeneous jet at the lowest chamber pressure of 200 Pa but not for 500 Pa and 1000 Pa the neutral densities n_0 and the ionization degrees x_{ion} were determined assuming that the pressure at the measurement positions in the jet was equal to the chamber pressure by

$$n_0 = \left(\frac{p_{stat}}{k_B} - n_e(T_h + T_e)\right) T_h^{-1} \tag{9}$$

(only single ionization is considered). x_{ion} is then obtained by

$$x_{ion} = n_e/(n_o + n_e) \tag{10}$$

4. Results and Discussion

4.1. Electron temperatures

The results of the determination of the electron temperatures are given in Figure 6 for parameters A and B in dependence on the stand-off distance and the chamber pressure. For the hydrogen containing parameter B the values are on a lower level than for parameter A with He and Ar only. For both parameters the net power was the same. But the addition of almost 10% hydrogen for parameter B consumes energy for heating and dissociation. Furthermore the jet is broadened and the radial temperature profile is probably flattened. This is supposed to be the explanation for the lower temperature level.

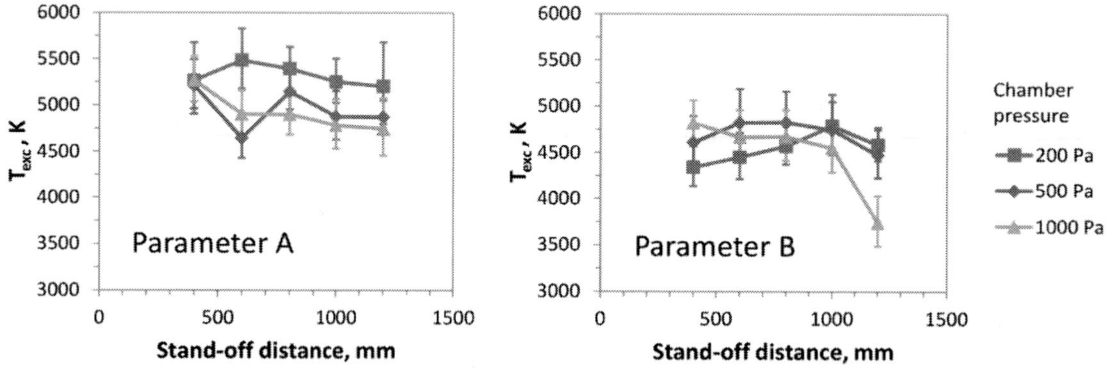

Figure 6. Electron temperatures determined for parameters A and B in dependence on the stand-off distance and the chamber pressure; the error bars result from the standard errors of the linear fit slopes of the Boltzmann plots.

For parameter B, the electron temperatures generally decrease with larger stand-off distances; only at 200 Pa and 500 Pa the densities increase initially. It is assumed that dissociation energy of recombining hydrogen atoms is released and transferred [30]. At 1000 Pa, this happens closer to the nozzle so that it was not covered by the measurements. At higher chamber pressures, the electron densities are slightly affected probably by the formation of expansion and compression zones due to the equilibration of the static pressure in the jet and the ambient pressure.

4.2. Electron densities and ionization rates

Electron densities and ionization rates determined for parameter B are given in Figure 7 in dependence on the stand-off distance and the chamber pressure. In principle, the electron densities decrease along the jet axis due to recombination and thus the ionization rate drops. The decrease is faster if the jet becomes shorter at higher chamber pressures of 500 Pa and 1000 Pa. Here, the electron densities are larger as the density in the jet is higher in general. The initial increase of the electron densities is assumed to be associated to static pressure and temperature alterations along the jet axis due to equilibration with the ambient pressure [30]. As recombination is already advanced at the applied measurement positions, the ionization rates are on low level. This is confirmed by the fact that no significant ionic lines of Ar II or He II are observed in the measured spectra.

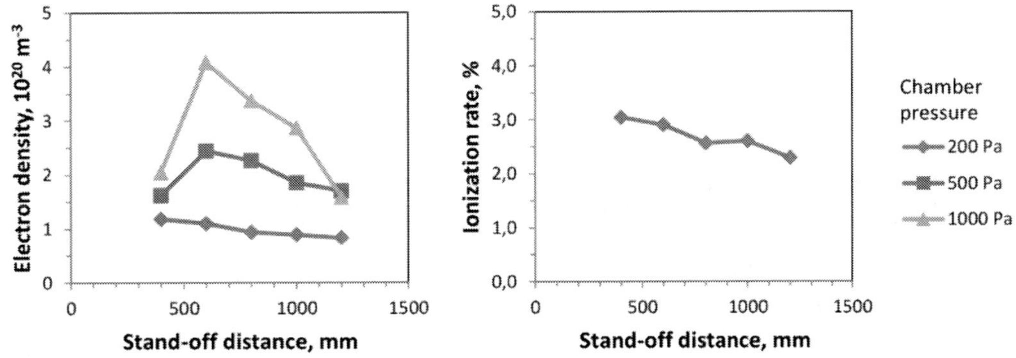

Figure 7. Electron densities and ionization rates determined for parameter B in dependence on the stand-off distance and the chamber pressure.

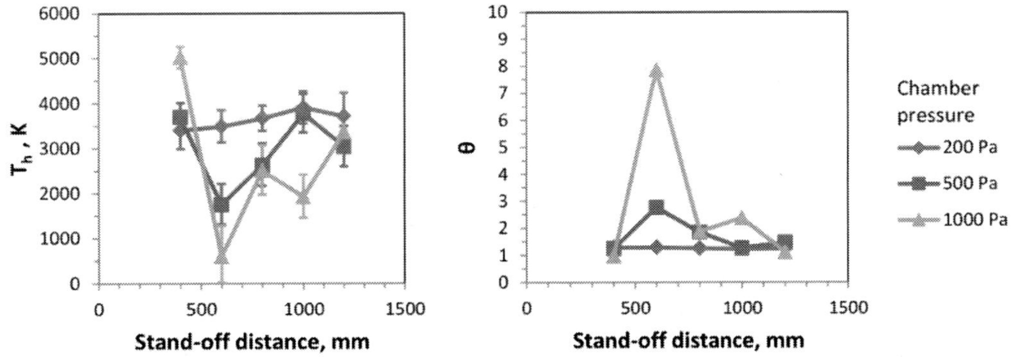

Figure 8. Temperatures of the heavy species and non-equilibrium parameters determined for parameter B in dependence on the stand-off distance and the chamber pressure; the error bars result from the standard errors of the Voigt fit parameters of the spectral line profile.

4.3. Temperatures of the heavy species

Temperatures of the heavy species (atoms and ions) and non-equilibrium parameters are given in Figure 8 determined for parameter B in dependence on the stand-off distance and the chamber pressure. For the lowest pressure of 200 Pa, T_h increases at larger stand-off distances in a similar manner as T_e (cp. Figure 6). As for the electrons, this can be explained by the release and transfer of dissociation energy of recombining hydrogen atoms. Here, θ is not far from unity (between 1.23 and 1.28) so that the departure from LTE is not large. However, at higher chamber pressures, large alterations of T_h occur due to static pressure and temperature changes along the jet axis and the formation of off-axial maxima in the expansion zones [30]. T_e is distinctly less affected (cp. Figure 6) as electrons do not experience a shock and their velocity is never supersonic due to their low mass [31]. This results in considerable kinetic non-equilibria between electrons and heavy particles.

5. PS-PVD coating microstructures

Some experimental examples are given to demonstrate the impact of plasma characteristics on PS-PVD coating microstructures [32]. The torch was equipped with a two-fold internal powder injection. The feedstock agglomerates are defragmented into sub-micron primary particles and heated already inside the nozzle where the plasma gas density is still high before exiting. This results in the formation of molten droplets and vapor with atomic species. The supersonic conditions effect that the plasma gas can exit the nozzle at a pressure which is different from the chamber pressure. As the flow is faster than the pressure waves traveling in the fluid at the local speed of sound, no information on the chamber pressure is carried inside the nozzle [33]. In the experiments, the development of clusters was also observed showing neither atomic nor liquid-like behavior at deposition. Such clusters may be the product of an incomplete evaporation or nucleation and growth. They play an important role for the formation either of columnar or dense coating microstructures. Vapor, cluster, and droplets are then transported to the substrate interacting with the plasma flux.

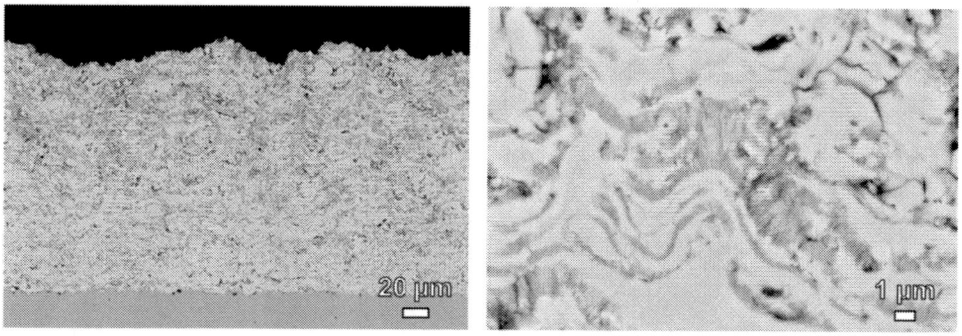

Figure 9. Mixed mode deposition from liquid splats, nano-sized clusters, and vapor phase; parameter B, chamber pressure 200 Pa.

Figure 10. Columnar (left) and dense structured (right) coatings from nano-sized clusters and vapor phase; parameter A, chamber pressure 200 Pa.

Figure 9 gives an example of a mixed mode deposition from liquid splats, nano-sized clusters, and vapor phase obtained from parameter B [34]. Here, the addition of hydrogen broadens the plasma jet and leads to energy consumption for dissociation. Thus, evaporation of the feedstock is incomplete. Figure 10 shows columnar and dense structured coatings only from nano-sized clusters and vapor phase obtained by parameter A [35]. The difference of these two structures is due to the plasma flow conditions along the substrate surface and shadowing effects [32].

6. Conclusion

The results of the optical emission spectroscopy show that the plasma jet at PS-PVD conditions consists of recombining plasma with a low degree of ionization. The admixture of hydrogen to the He/Ar plasma gas affects the plasma enthalpy due to initial consumption and later release of dissociation energy. Obviously, there is no significant effect on the transport properties. However, the hydrogen broadens the plasma jet and probably flattens the radial temperature profiles. Experimental examples show that this has impacts on the heating of the feedstock and the microstructure of the deposited coatings.

At the lowest investigated chamber pressure of 200 Pa, a moderate departure from LTE was found. Here, the distribution of temperatures and electron densities along the jet axis was smooth and homogeneous. In contrast, at higher chamber pressures large fluctuations of the plasma characteristics and thus distinct non-equilibrium conditions occur due to static pressure and temperature changes along the jet axis. Such conditions are assumed to be detrimental for treatment and transport of the feedstock. In future work this could be investigated more in detail by calculation of transport properties considering non-equilibria.

Acknowledgements

The authors would like to acknowledge gratefully the spectral emission spectroscopic measurements by Dr. Andreas Hospach and Dr. Nikolay Zotov, Forschungszentrum Jülich, IEK-1.

References

[1] Verdy C, Zhang C, Sokolov D, Liao H, Klein D, and Coddet C 2008 Gas-tight Coatings Produced by Very Low Pressure Plasma Spraying *Thermal Spray 2008: Thermal Spray Crossing Borders* on CD-ROM Lugscheider E Ed June 02-04, 2008 (Maastricht, The Netherlands) Verlag für Schweißen und verwandte Verfahren 2008 398-402

[2] Mauer G, Vaßen R, and Stöver D 2010 Thin and Dense Ceramic Coatings by Plasma Spraying at Very Low Pressure *J. Therm. Spray Technol.* **19** 495-501

[3] Zhu L, Zhang N, Zhang B, Sun F, Bolot R, Planche M-P, Liao H, and Coddet C 2011 Very low pressure plasma sprayed alumina and yttria-stabilized zirconia thin dense coatings using a modified transferred arc plasma torch *Appl. Surf. Sci.* **258** 1422-8

[4] von Niessen K and Gindrat M 2011 Plasma Spray-PVD: A New Thermal Spray Process to Deposit Out of the Vapor Phase *J. Therm. Spray Technol.* **20** 736-43

[5] von Niessen K, Gindrat M, and Refke A 2010 Vapor Phase Deposition Using Plasma Spray-PVD *J. Therm. Spray Technol.* **19** 502-9

[6] Zhang N, Sun F, Zhu L, Verdy C, Planche M P, Liao H, Dong C, and Coddet C 2011 Characteristics of Cu Film Deposited Using VLPPS, *J. Therm. Spray Technol.* **20** 351-7

[7] Zhang N, Sun F, Zhu L, Planche M P, Liao H, Dong C, and Coddet C 2011 Electron Temperature and Density of the Plasma Measured by Optical Emission Spectroscopy in VLPPS conditions, *J. Therm. Spray Technol.* **20** 1321-7

[8] Yugeswaran S and Selvarajan V 2009 Excitation temperature and electron number density behavior of atmospheric pressure D.C. argon plasma jet during spheroidization of nickel, *Vacuum* **83** 841-7

[9] Bockasten K 1961 Transformation of Observed Radiances into a Radial Distribution of the Emission of a Plasma, *J. Opt. Soc. Am.* **51** 943-7

[10] Kurskov A A, Ershov-Pavlov E A, and Chvyaleva L V 1986 Measurement of the local temperature in a plasma without the use of Abel inversion, *J. Appl. Spec.* **45** 1155-9

[11] Kurskov A A, Ershov-Pavlov E A, and Chvyaleva L V 1989 Determination of half-width of spectral lines of radiation of an inhomogeneous plasma, *J. Appl. Spec.* **51** 1135-8

[12] Semenov S and Cetegen B 2001 Spectroscopic Temperature Measurements in Direct Current Arc Plasma Jets Used in Thermal Spray Processing of Materials, *J. Therm. Spray Technol.* **20** 326-36

[13] Tu X, Chéron B G, Yan J H, and Cen K F 2007 Electrical and spectroscopic diagnostic of an atmospheric double arc argon plasma jet, *Plasma Sources Sci. Technol.* **16** 803-12
[14] Ralchenko Y, Kramida A, Reader J, and NIST ASD Team (2011). *NIST Atomic Spectra Database* (version 4.1), [Online]. Available: http://physics.nist.gov/asd [Friday, 15-Jun-2012 08:25:13 EDT]. National Institute of Standards and Technology, Gaithersburg, MD
[15] Mauer G, Vaßen R, and Stöver D 2011 Plasma and Particle Temperature Measurements in Thermal Spray: Approaches and Applications, *J. Therm. Spray Technol.* **20** 391-406
[16] van der Mullen J A M 1990 Excitation equilibria in plasmas; a classification, *Phys. Rep.* **191** 109-220
[17] Calzada M D, Moisan M, Gamero A, and Sola A 1996 Experimental investigation and characterization of the departure from local thermodynamic equilibrium along a surface-wave-sustained discharge at atmospheric pressure, *J. Appl. Phys.* **80** 46-55
[18] Rat V, Murphy A B, Aubreton J, Elchinger M F, and Fauchais P 2008 Treatment of non-equilibrium phenomena in thermal plasma flows, *J. Phys. D: Appl. Phys.* **41** 183001
[19] Quintero M C, Rodero A, García M C, and Sola A 1997 Determination of the Excitation Temperature in a Nonthermodynamic-Equilibrium High-Pressure Helium Microwave Plasma Torch, *Appl. Spectrosc.* **51** 778-84
[20] Konjević N 1999 Plasma Broadening and Shifting of Non-Hydrogenic Spectral Lines: Present Status and Applications, *Phys. Rep.* **316** p 339-401
[21] Gigosos M A and Cardeñoso V 1996 New plasma diagnosis tables of hydrogen Stark broadening including ion dynamics, *J. Phys. B: At. Mol. Opt. Phys.* **29** 4795-838
[22] Kelleher D E 1981 Stark broadening of visible neutral helium lines in a plasma, *J. Quant. Spectrosc. Radiat. Transfer* **25** 191-220
[23] Dong L, Ran J, and Mao Z 2005 Direct measurement of electron density in microdischarge at atmospheric pressure by Stark broadening, *Appl. Phys. Lett.* **86** 161501
[24] Pellerin S, Musiol K, Pokrzywka B, and Chapelle J 1996 Stark width of $4p'[^1/_2]$-$4s[^3/_2]^0$ Ar I transition (696.543 nm) *J. Phys. B: At. Mol. Opt. Phys.* **29** 3911-24
[25] Dimitrijević M S, Christova M, and Sahal-Bréchot S 2007 Stark broadening of visible Ar I spectral lines, *Phys. Scr.* **75** 809-19
[26] Kepple P and Griem H R 1968 Improved Stark Profile Calculations for the Hydrogen Lines H_α, H_β, H_γ, and H_δ, *Phys. Rev.* **173** 317-325
[27] Wiese W L, Kelleher D E, and Paquette D R 1972 Detailed Study of the Stark Broadening of Balmer Lines in a High-Density Plasma, *Phys. Rev. A* **6** 1132-53
[28] Balcon N, Aanesland A, and Boswell R 2007 Pulsed RF discharges, glow and filamentary mode at atmospheric pressure in argon, *Plasma Sources Sci. Technol.* **16** 217-25
[29] Griem H R *Plasma Spectroscopy*, McGraw-Hill, New York 1964
[30] Selezneva S E, Sember V, Gravelle D V, and Boulos M I 2002 Spectroscopic validation of the supersonic plasma jet model, *J. Phys. D: Appl. Phys.* **35** 1338-49
[31] Sember V, Gravelle D V, and Boulos M I 2002 Spectroscopic study of a supersonic jet generated by an ICP torch with a convergent-divergent nozzle, *J. Phys. D: Appl. Phys.* **35** 1350-61
[32] Mauer G, Zotov N, Hospach A, and Vaßen R 2012 Process Conditions and Microstructures of Ceramic Coatings by Gas Phase Deposition Based on Plasma Spraying, *Thermal Spray 2012: Proceedings from the International Thermal Spray Conference and Exposition*, May 21–24, 2012, Houston, Texas, USA, (Ed.) R.S. Lima, A. Agarwal, M.M. Hyland, Y.-C. Lau, C.-J. Li, A. McDonald, F.-L. Toma, ASM International, Materials Park, Ohio, USA, 2012, 828-33
[33] Jodoin B, Gindrat M, Dorier J-L, Hollenstein C, Loch M, and Barbezat G 2002 Modeling and Diagnostics of a Supersonic DC Plasma Jet Expanding at Low Pressure, *International Thermal Spray Conference*, E. Lugscheider, C.C. Berndt, Ed., March 04-06, 2002 (Essen, Germany), Verlag für Schweißen und verwandte Verfahren DVS-Verlag, 2002, 716-20
[34] Hospach A, Mauer G, Vaßen R, and Stöver D 2011 Columnar Structured Thermal Barrier Coatings (TBCs) by Thin Film Low Pressure Plasma Spraying (LPPS-TF), *J. Therm. Spray Technol.* **20** 116-20
[35] Hospach A, Mauer G, Vaßen R, and Stöver D 2012 Characteristics of Ceramic Coatings Made by Thin Film Low Pressure Plasma Spraying (LPPS-TF), *J. Therm. Spray Technol.* **21** 435-40

Influence of ultrasonic irradiation on ozone generation in a dielectric barrier discharge

Y Kusano[1], J Drews[1], F Leipold[2], A Fateev[3], A Bardenshtein[4] and N Krebs[5]
[1] Department of Wind Energy, Section of Composites and Materials Mechanics, Technical University of Denmark, Risø Campus, Frederiksborgvej 399, DK-4000 Roskilde, Denmark
[2] Department of Physics, Technical University of Denmark, Risø Campus, Frederiksborgvej 399, DK-4000 Roskilde, Denmark
[3] Department of Chemical Engineering, Technical University of Denmark, Risø Campus, Frederiksborgvej 399, DK-4000 Roskilde, Denmark
[4] Danish Technological Institute, Gregersensvej 3, DK-2630 Taastrup, Denmark
[5] FORCE Technology, DK-2605 Brøndby, Denmark

E-mail: yuki@dtu.dk

Abstract. An atmospheric pressure dielectric barrier discharge (DBD) was generated in an N_2/O_2 gas mixture at room temperature with and without ultrasonic irradiation to investigate ozone production. Powerful ultrasonic irradiation with the sound pressure level of approximately 150 dB into the DBD can enhance ozone production especially when the DBD was driven at a frequency of 15 kHz.

1. Introduction

Ozone is a strong oxidizing agent, decays without harmful by-products, and is useful for a variety of applications, including oxidation, ashing, sterilization, decontamination and cleaning [1]. Ozone is mainly produced by using an electrical gas discharge such as a dielectric barrier discharge (DBD) and corona discharge at around atmospheric pressure or above. These electrical gas discharge methods are practically useful for the production of high concentration ozone at high gas flow rates.

There are competing processes of formation and decomposition of ozone with the gas discharge methods. It is therefore important to adjust the plasma parameters so that the ozone formation is maximized while its decomposition is minimized. Here, the first important reaction for ozone generation is the generation of atomic oxygen by electron impact dissociation.

$$O_2 + e \rightarrow 2O + e. \quad (1)$$

In a second step, atomic oxygen reacts with molecular oxygen to form ozone.

$$O_2 + O + M \rightarrow O_3 + M \quad (2)$$

where M is a third collision partner. Detailed description of ozone production in discharges may be found in refs. [2,3]. It is noted that the ozone formation is very sensitive to temperature and a moisture content of the gas to be introduced into the discharge [3].

Improving ozone production efficiency is a key issue for developing new types of ozone generators. It is reported that the ozone production efficiency can be improved by introducing ultrasound into a gas discharge. Introduction of ultrasonic energy using a piezoelectric transducer enhances the ozone production when a negative direct current (DC) corona discharge is generated in a pure O_2 gas [4,5]. However the highest ozone concentration reported in [4,5] is less than 200 ppm only. In addition the use of the piezoelectric transducer is not energy-effective for introducing ultrasound into a certain gas-discharge volume due to the significant acoustic impedance mismatch between the transducer and the surrounding gas. On the other hand, high-power gas-jet ultrasonic transmitters generate acoustic waves directly in a gas without such a big mismatch. Furthermore they can potentially be integrated into large-scale setups easily. Introduction of ultrasonic waves using the high-power gas jet ultrasonic transmitters into atmospheric pressure plasmas has been studied for adhesion improvement of polymer

surfaces [6-11] and ozone production [12]. Pure O_2 gas was fed to a DBD in order to generate ozone [12]. It is reported that ozone production can be efficiently enhanced by ultrasonic irradiation into the DBD particularly when the DBD is generated at a frequency of 15 kHz. It is economically interesting to investigate if similar effect can be obtained in N_2/O_2 gas mixtures.
In the present work, an atmospheric pressure DBD is generated and a gas mixture of N_2/O_2 is fed to the DBD. The influence of ultrasonic irradiation into the DBD on ozone production is studied.

2. Experimental methods

Ozone was produced in the DBD both with and without ultrasonic irradiation. The DBD setup used in the present work is shown in figure 1. It is a modification of the DBDs introduced previously [13-20]. The DBD is generated between two parallel-plate water-cooled electrodes driven by either of alternating current (AC) power supplies; Power Amplifier 1140L (ENI) driven by a Sweep Generator (Model 164 Wavetek) or a high voltage generator (Model 6030 SOFTAL Electronic GmbH) driven by a Pulse/Delay Generator (BNC 555 Berkeley Nucleonics Corp.). The ENI power supply is used for applying 15 kHz sinusoidal AC voltage between the electrodes, while the SOFTAL power supply is used for ~ 40 kHz sinusoidal AC voltage. Voltage and current are measured with a high voltage probe (PPE20kV, LeCroy) and a 47 Ω resistor, respectively. The average electrical power applied to the DBD is calculated by time integrating the product of voltage and current over several periods and then dividing the result by the time corresponding to that number of periods. A change of the frequency from 40 to 36.8 kHz using the SOFTAL power supply corresponds to a discharge power from 30 W to 150 W. Figure 1 (b) shows a schematic diagram of the setup.

(a).

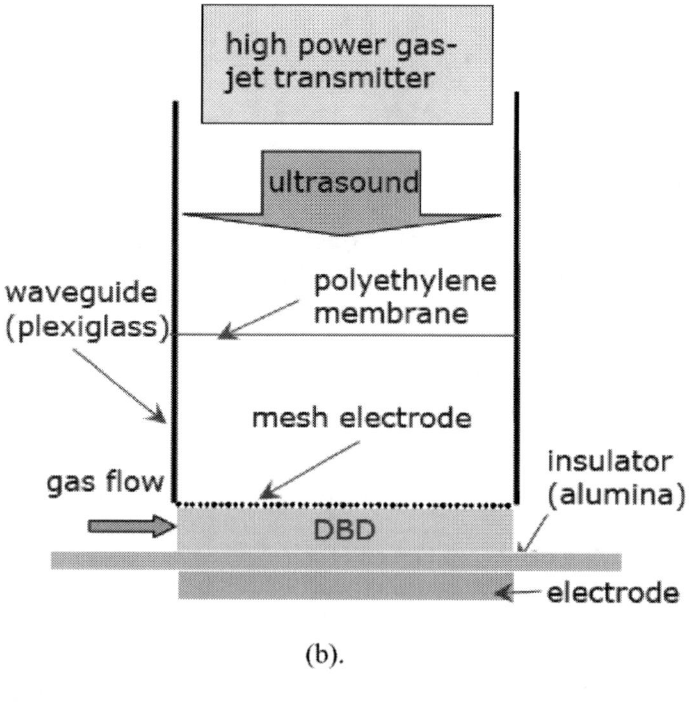

(b).

Figure 1. The experimental setup. The design (without the high power gas-jet transmitter) (a), and a schematic diagram (b).

The bottom aluminium electrode (50 mm × 50 mm) is covered with an alumina (Al_2O_3) plate (100 mm × 100 mm × 3 mm), while the upper aluminium electrode has a 42 mm diameter perforated hole covered with a stainless steel mesh (Wire diameter: 0.2 mm. Open area: ~ 70 %). The gap between the alumina plate and the mesh is adjusted to be ~ 0.5 mm. A 40-mm inner-diameter cylindrical acoustic waveguide (poly(methyl)methacrylate. PMMA) is attached above the upper electrode for the transmission of ultrasound. A high-power gas-jet ultrasonic generator (SonoSteam®, FORCE Technology) is placed near the top of the waveguide. The radiated acoustic field is probed using the B&K free-field ¼" microphone Type 4939 with the preamplifier Type 2670, and the signal is processed by means of the Portable Pulse Analyzer (Brüel&Kjær Sound and Vibrations, Denmark). The fundamental frequency of the ultrasound is ~ 30 kHz, and the sound pressure level (SPL) within the frequency range from 20 to 40 kHz is 165 dB. The DBD and the ambient air are separated using a thin polyethylene membrane clamped between the outer-wall of the waveguide and the upper part of the powered electrode. The ultrasound SPL measured at the discharge is 154 dB, i.e. the abatement of the SPL after passing the waveguide, the membrane and the mesh is 11 dB. A typical sound pressure spectrum is presented elsewhere [7].

A gas mixture of N_2 and O_2 (both purity of 99.999 %) is introduced from a side of the DBD. The gas flow rates are controlled by mass flow controllers. The gas passing the DBD is collected from the other side of the gas inlet, and the ozone concentration is measured by means of an ultraviolet (UV) absorption method [21,22]. In this method the sample gas is fed into a 5-mm long absorption cell through which UV light at approximately 254 nm is guided. The UV intensity after passing the absorption cell is detected by a UV sensitive photodiode. The ozone concentration is calculated by the ratio of the UV intensities measured with the sample gas and with a gas that does not contain ozone.

After measuring ozone concentration, the sample gas is further fed into a Fourier Transform Infrared spectroscopy (FTIR. MB-100 BOMEM) for detection of N_xO_y gas.

3. Results and discussion

Ozone concentrations were measured at various O_2 contents in the feeding gas, while fixing the O_2 flow rate at 10 SLM. The DBD was generated at a frequency of 15 kHz and the plasma power was fixed at ~ 120 W. The ozone concentration increases almost linearly as the O_2 content increases as shown in figure 2.

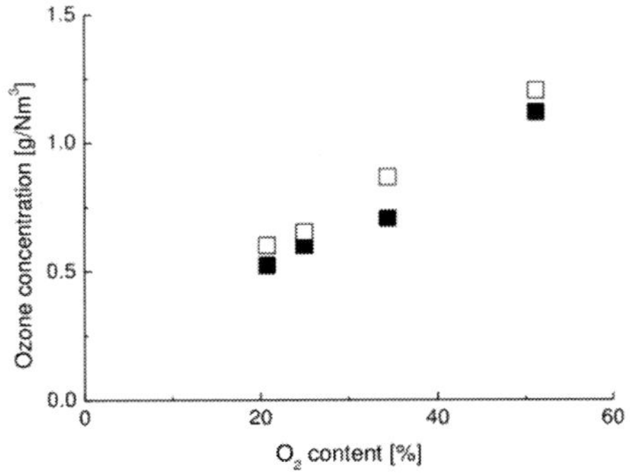

Figure 2. Ozone concentration at various O_2 gas contents in the feeding N_2/O_2 gas mixture. The O_2 gas flow rate: 10 SLM. The plasma power: ~ 120 W. Open and solid squares represent the ozone concentration with and without ultrasonic irradiation, respectively.

Ultrasonic irradiation consistently improves ozone production. It is indicated that ultrasonic irradiation can be useful for improving ozone production even in N_2/O_2 gas mixtures. However, it is noted that the ozone concentration measured in the present study was approximately one order lower than that using the O_2 DBD at the same O_2 flow rate and the plasma power [12]. It is partly because the energy introduced to the DBD was used not only for the ozone production but also production of N_xO_y which was detected by means of the FTIR.

Figure 3 shows measured ozone concentrations when the DBD was generated at 15 kHz (a) and ~ 40 kHz (b). The O_2 and N_2 flow rates were fixed at 10 SLM. In each case the ozone concentration increased as the input power to the plasma increased. Ultrasonic irradiation tends to improve ozone production when the DBD was driven at 15 kHz (Figure 3 (a)). On the other hand, when the DBD was generated at ~ 40 kHz, ultrasonic irradiation to the DBD resulted in no significant change of the ozone concentration (figure 3 (b)). A similar result is reported when a DBD is generated in a pure O_2 gas [12], but the measured ozone concentrations in the present work are again approximately one order lower than those in Ref. [12].

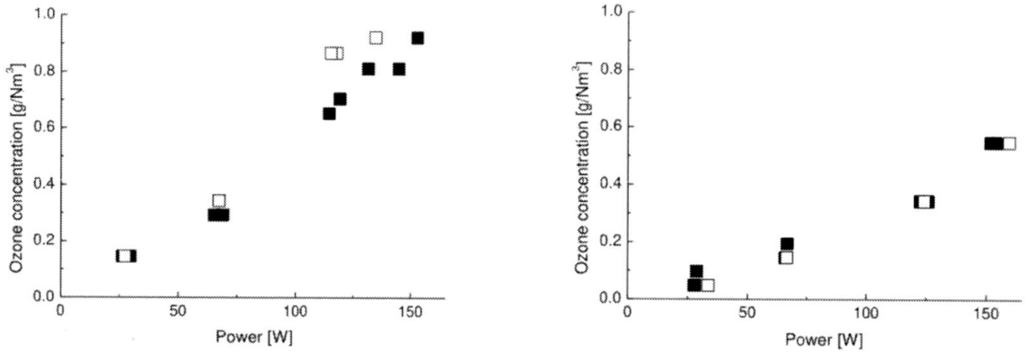

(a). (b).

Figure 3. Ozone concentration at different power to the plasma. The DBD was generated at 15 kHz (a) and ~ 40 kHz (b). The O_2 and N_2 flow rates are both 10 SLM.

Figures 4 and 5 show the voltage and current waveforms of the DBD driven at the frequencies of 15 kHz and ~ 40 kHz, respectively. Here, the voltage is the potential difference of the upper mesh electrode relative to the bottom electrode. The left panels (a) show the voltage and current waveforms without ultrasonic irradiation, while the right panels (b) show those with the ultrasonic irradiation. In both cases of the DBDs the ultrasonic irradiation does not influence the current waveforms significantly, i.e. when ultrasound is irradiated into the DBD, no significant change in the electrical properties of the discharge is observed.

The current waveforms of the DBD at 15 kHz are nearly symmetric between the positive and negative half-cycles as shown in figure 4. On the other hand, the intensity of the micro-discharge current pulses of the DBD driven at ~ 40 kHz was higher during the positive half-cycle of the current than during the negative one, although the voltage waveform is sinusoidal and symmetric (figure 5). The approximate numbers of the spiky peaks at positive and negative half-cycles are between 50 and 60 and between 60 and 70, when the driving frequencies are 15 and ~ 40 kHz, respectively. There are no clear differences at different polarizations of the voltages with/without ultrasonic irradiation. It is therefore indicated that the asymmetric current waveforms at ~ 40 kHz are attributed to the different intensities of the current peaks. This asymmetry is partly due to the asymmetric structure of the DBD setup in the present work. The same kind of asymmetric current waveforms is reported for a surface discharge where one of the electrodes is directly exposed in the discharge [23]. In order to apply same power to a DBD, application of higher voltage is generally necessary at a lower frequency, because the electrical impedance of the DBD is higher at the lower frequency. In fact, the peak-to-peak voltage at 15 kHz (figure 4) is approximately 20 % higher than that at ~ 40 kHz (figure 5). One possible explanation of the less asymmetric current waveforms for the DBD driven at 15 kHz is that when the upper electrode is negatively biased, the discharge ignition at 15 kHz might be easier than that driven at ~ 40 kHz because of the higher voltage applied at 15 kHz.

It is reported that in a corona discharge the electron density is higher near the positively biased electrode where the atomic oxygen is favourably produced [24-26]. In the present work, the upper mesh electrode is directly exposed in the DBD while the bottom electrode is covered with the alumina plate. This configuration is similar to a corona discharge in a sense of the asymmetric electrode system. When the positive and negative currents flow through the upper mesh electrode, the discharge is like positive and negative coronas, respectively.

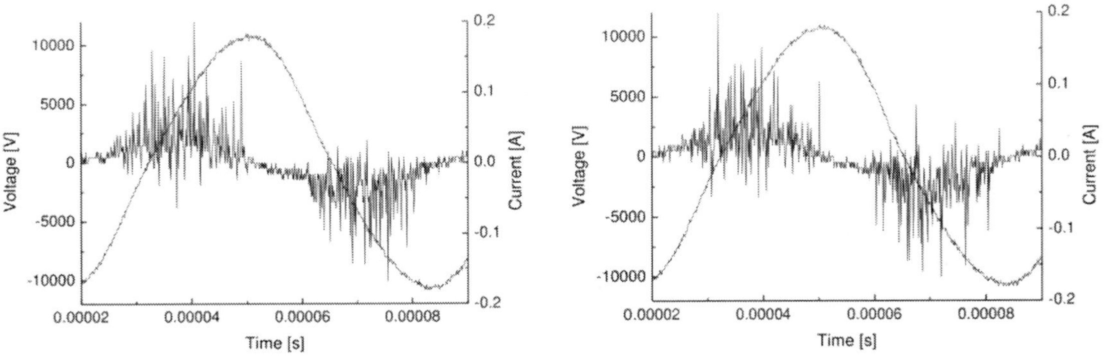

(a). (b).

Figure 4. Voltage (smooth) and current (spiky) waveforms of the DBD generated at 65 W at 15 kHz. (a) Without ultrasound, (b) with ultrasound. The O_2 and N_2 flow rates are both 10 SLM.

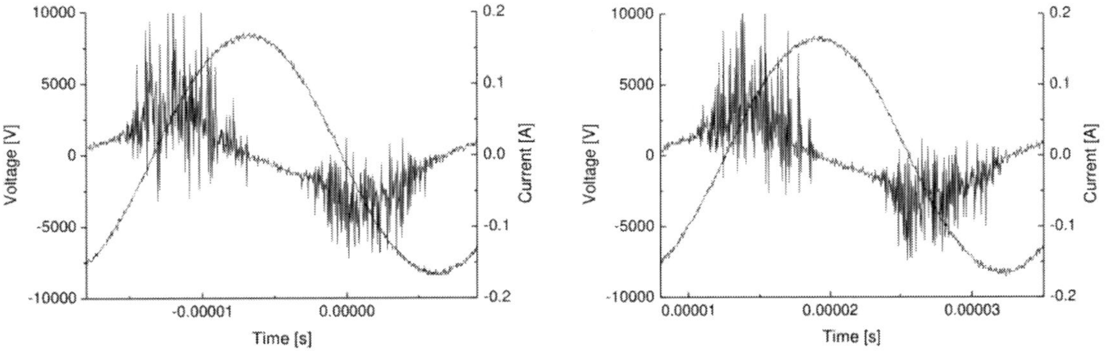

(a). (b).

Figure 5. Voltage (smooth) and current (spiky) waveforms of the DBD generated at 65 W at ~ 40 kHz. (a) Without ultrasound, (b) with ultrasound. The O_2 and N_2 flow rates are both 10 SLM.

A laminar boundary gas layer usually sticks at a material surface at atmospheric pressure. When powerful ultrasound is irradiated to the DBD, the thickness of the boundary gas layer at the alumina surface can be reduced [6-12,27]. This improves gas mixing and diffusion processes near the alumina surface. It is therefore suggested that at the frequency of 15 kHz the density of atomic oxygen can be higher near the alumina surface and thus the ozone production can be efficiently enhanced by ultrasonic irradiation. It is noted here that atomic oxygen, which is neutral and thus does not influence the electrical properties of DBD, can survive for more than 10^{-5} s after a micro-discharge [2,3]. This could be the reason why no significant change of the electrical properties was observed by the ultrasonic irradiation, while ozone production could be increased. It is reported that the ultrasound enhances ozone production only for the negative corona [4,5], showing a good agreement with the present result. Furthermore, the results in the present work show in good agreement with those obtained using the pure O_2 DBD [12].

4. Conclusions

Ozone production in a DBD in an N_2/O_2 gas mixture was investigated with and without ultrasonic irradiation. The ultrasonic irradiation did not significantly influence the ozone production for the DBD driven at ~ 40 kHz, while the ozone production was increased for the DBD driven at 15 kHz with ultrasonic irradiation. The difference can be attributed to the distribution of atomic oxygen, suggested by the difference in the current waveforms. It is indicated that at the lower frequency the concentration of the atomic oxygen would be higher near the alumina surface and that ultrasonic irradiation can efficiently enhance mixing and diffusion of the gas, resulting in the enhancement of ozone production. However, the typical measured ozone concentration was approximately one order lower than that produced in a pure O_2 DBD with similar conditions.

Acknowledgements

This work has been supported by the Ministry of Science, Technology and Innovation through the National Danish Proof of Concept (PoC) funding scheme (grant number 07-017766). Søren Nimb and Martin Jessen are gratefully acknowledged for the design and construction of the DBD setup.

References

[1] Kogelschatz U 2003 *Plasma Chem. Plasma Proc.* **23(1)** 1
[2] Eliasson B, Hirth M and Kogelschatz U 1987 *J. Phys. D Appl. Phys.* **20** 1421
[3] Teodoru S, Kusano Y and Bogaerts A 2012 *Plasm. Proc. Polym.* **9** 652
[4] Bálek R, Pekárek S and Bartáková Z 2007 *Ultrasonics* **46** 227
[5] Pekárek S and Bálek R 2004 *J. Phys. D Appl. Phys.* **37** 1214
[6] Kusano Y, Singh SV, Norrman K, Drews J , Leipold F, Rozlosnik N, Bardenshtein A and Krebs N 2012 *Surf. Eng.* **28(6)** 453
[7] Kusano Y, Singh SV, Norrman K, Leipold F, Drews J, Morgen P, Bardenshtein A and Krebs N 2011 *J. Adhesion* **87** 720
[8] Kusano Y, Norrman K, Drews J, Morgen P, Leipold F, Singh SV, Bardenshtein A and Krebs N 2011 *Surf. Coat. Technol.* **205** S490
[9] Kusano Y, Singh SV, Bardenshtein A, Krebs N and Rozlosnik N 2010 *J. Adhesion Sci. Technol.* **24** 1831
[10] Kusano Y 2009 *Surf. Eng.* **25(6)** 415
[11] Kusano Y, Andersen TL, Sørensen BF, Rozlosnik N, Mortensen H, Teodoru S, Kingshott P, Leipold F, Singh SV, Bilde-Sørensen J, Brøndsted P, Michelsen PK, Bindslev H, Bardenshtein A and Krebs N 2012 Adhesion improvement of glass fibre reinforced polyester composite by atmospheric pressure plasma treatment *Fiber-reinforced composites* ed G Cheng (New York: Nova Science Publishers, Inc.) pp 297-318
[12] Drews J, Kusano Y, Leipold F, Bardenshtein A and Krebs N 2011 *Ozone: Sci. Eng.* **33** 483
[13] Teodoru S, Kusano Y, Rozlosnik N and Michelsen PK 2009 *Plasm. Proc. Polym.* **6** S375
[14] Kusano Y, Teodoru S and Hansen CM 2011 *Surf. Coat. Technol.* **205** 2793
[15] Kusano Y, Andersen TL and Michelsen PK 2008 *J. Phys. Conf. Series.* **100** 012002
[16] Kusano Y, Mortensen H, Stenum B, Kingshott P, Andersen TL, Brøndsted P, Bilde-Sørensen JB, Sørensen BF and Bindslev H 2007 *Plasm. Proc. Polym.* **4(S1)** S455
[17] Kusano Y, Mortensen H, Stenum B, Goutianos S, Mitra B, Ghanbari-Siahkali A, Kingshott P, Sørensen BF and Bindslev H 2007 *Int. J. Adhesion and Adhesives*, **27(5)** 402

[18] Leipold F, Fateev A, Kusano Y, Stenum B and Bindslev H 2006 *Fuel* **85** 1383
[19] Kusano Y, Leipold F, Fateev A, Stenum B and Bindslev H 2005 *Surf. Coat. Technol.* **200** 846
[20] Fateev A, Leipold F, Kusano Y, Stenum B, Tsakadze E and Bindslev H 2005 *Plasm. Proc. Polym.*, **2** 193
[21] Leipold F, Kusano Y, Hansen F and Jacobsen T 2010 *Food Control* **21** 1194
[22] Leipold F, Schultz-Jensen N, Kusano Y, Bindslev H and Jacobsen T 2011 *Food Control* **22(8)** 1296
[23] Williamson JM, Trump DD, Bletzinger P and Ganguly BN 2006 *J. Phys. D Appl. Phys.* **39** 4400
[24] Cheng J and Davidson H 2002 *Plasma Chem. Plasma Proc.* **22(4)** 495
[25] Cheng J and Davidson H 2003 *Plasma Chem. Plasma Proc.* **23(3)** 501
[26] Wang P and Chen J 2009 *J. Phys. D Appl. Phys.* **42** 035202
[27] Borisov YuYa and Gynkina NM 1973 Acoustic drying *(part IX Physical principles of ultrasonic technology* vol 2) ed LD Rozenberg (New York: Plenum Press) pp 401-403

The investigation of movement dynamics of an AC electric arc attachment along the working surface of a hollow cylindrical electrode under the action of gas-dynamic and electromagnetic forces

A V Surov[1, 2], S D Popov[1, 2], EO Serba[1], G V Nakonechny[1], V A Spodobin[1], R V Ovchinnikov[1], I I Kumkova[1], S A Shabalin[1]

[1] *Institute for Electrophysics and Electric Power of Russian Academy of Sciences, 18 Dvortsovaya emb., 191186, St.Petersburg, Russia*
[2] *St.Petersburg State Polytechnical University, 29 Polytechnicheskaya, 195251, St.Petersburg, Russia*

E-mail: alex_surov@mail.ru

Abstract. Stationary electric arc alternating current plasma torches are used today for realization of plasma chemical technologies requiring relatively high energy input. Waste treatment is one these directions. The paper reports on experiment results directed towards the increase in the lifetime characteristics of electrode units of the powerful high-voltage electric-arc AC plasma torches. The solution to the problem of obtainment the uniform wear of a copper hollow cylindrical electrode achieved by the controlled movement of the arc attachment along the working surface was offered. Organization of gas supply in the near electrode area and application of alternating magnetic field ensured movement of arc attachment along the surface with average speed from 2 to 14 m/s. Arc current was about 47 A and 84 A, gas flow rate in near electrode area was about 5 and 4.5 g/s. Due to researches on the experimental prototype of a hollow cylindrical electrode, the erosion of its material reached only 3 μg/C, that enables production of the electrode assembly with life time above 1000 hours at currents in the arc up to 100-200 A.

Introduction

Currently, plasmachemical technologies are actively developed for efficient use of energy resources and implementation of ecologically safe methods for destruction and treatment of various wastes, including for the combustible syngas production [1-5]. The generator of thermal plasma is a key element of the majority of plasma chemical installations. Nowadays plasma torches are also actively used in power engineering for ignition and lightning of coal boilers [6]. The devices intended for operation in large-scale industrial installations should be powerful, efficiently transfer energy to the working gas, operate continuously for a long period of time and generally can operate in oxidizing media. Stationary electric arc plasma torches are very promising for these purposes. Westinghouse [7] and Europlasma [8] DC plasma torches are well-developed for industrial applications. The particular attention for a number of applications deserves the models using water or steam as a plasma forming gas [5, 9, 10]. A number of promising models of stationary AC plasma torches operating on various gases has been developed in IEE RAS. Among them are plasma torches using oxidizing media (including air): powerful single-chamber three-phase plasma torches with rod electrodes [11], single-phase and three-phase high-voltage plasma torches with power from 5 up to 100 kW with rod electrodes mounted in a separate cylindrical channels [12, 13] and single-chamber three-phase plasma torches with rail-type electrodes, with operating power up to 500 kW [14], and plasma torches working on a mixture of steam and air or another gas with power up to 100 kW were developed [15]. At present high-voltage AC plasma torches with hollow cylindrical electrodes are under development. Such plasma torches have the high arc voltage drop that enables essential current decreasing versus low-voltage models of the same power level and accordingly increase of the electrode lifetime. As a result these devices can provide continuous operation with high power and lifetime of electrodes more than 1000 hours. Prototypes of commercial plasma torches with power 200 and 400 kW are created. The devices with power of an order 1 MW are under development. This paper is devoted to investigation of movement dynamics of an AC electric arc attachment along the working surface of a hollow cylindrical electrode of such plasma torches. The purpose of this paper is presentation of results of the experimental researches obtained during development of new AC plasma torches.

Experimental setup

The investigations were carried out at the test bench of AC plasma torches equipped with required supply systems. The test bench comprises: power supply system, double-loop cooling system, system of preparation and supply of plasma forming gas and system of treatment and discharge of exhaust gases. Prototypes of AC plasma torches with cylindrical (hollow) electrodes developed by IEE RAS were used. Figure 1 shows the schematic diagram of the installation. The high-voltage three-phase plasma torch is connected to the AC power supply source of industrial frequency (50 Hz). While the plasma torch operates the arcs burn between three electrodes and have the general point behind the nozzle exit section.

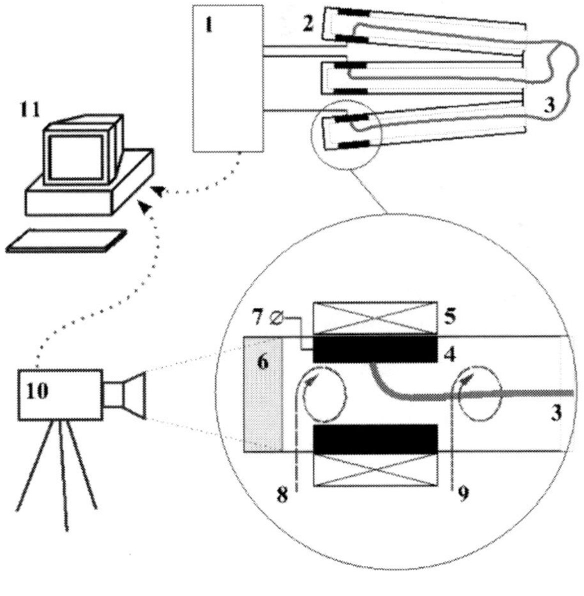

Figure 1. Schematic diagram of the experimental installation. 1 - Power supply, 2 - high-voltage three-phase plasma torch with cylindrical (hollow) electrodes, 3 - arc column, 4 - hollow cylindrical electrode, 5 - solenoid, 6 - pipe hole for photo and video shooting, 7 - terminal for connection of the power supply, 8 and 9 - upstream and downstream loops of tangential supply of plasma forming gas, 10 - high-speed video camera, 11 - recording industrial computer.

In the bottom of Figure 1 (zoomed in) is shown the schematic diagram of one of three plasma torch channels (one of three phases) with electrode, solenoid, and peep hole. From both sides of a hollow electrode (upstream and downstream) are designed loops of tangential plasma forming gas supply. Rotation of the arc attachment in each of the electrodes is provided due to the tangential gas supply and the magnetic field of the solenoid installed coaxially around the electrode. The current and voltage analogue sensing transducers are incorporated for measurement of arc currents and voltages in the power supply. The industrial computer with an acquisition boards (with bandwidth of 10 kHz per channel) is incorporated in the measuring system for data recording. Power of the plasma torch and effective values of electric parameters are calculated on their instantaneous values. There are automated pressure regulators - the flow meters which are also connected to the industrial computer for supply of the required plasma forming gas flow rate into the plasma torch gas inlet loops. The peep holes are provided for measurement of speed of the arc attachments on the electrode surface by means of a high-speed video shooting. Three monochrome high-speed video cameras which are a part of the measuring system, allow simultaneous synchronous recording of arc attachment movement along the three electrodes with speed of shooting 4000 fps.

Description of experiments and main results

Experiments were carried out on prototypes of the plasma torch working models. Considering the practical importance of the carried-out investigations, the principal criterion for the flow rate range selection for a gas supply mode variation and solenoid parameters for the magnetic field change were expediency and feasibility of providing these parameters during the plasma torch work in the industrial plants. Investigations were carried out on modes of the stable arc burning in the plasma torch electrode channels for the arc currents about 40A and 80A. Longitudinal and rotary moving of the arc attachment – in the axial and tangential directions was required for providing the long-lasting lifetime of electrodes. Figure 2 shows the photo of the arc attachment. The photo made through the peep hole under some angle to the axis of the plasma torch channel in order to see the electrode working surface.

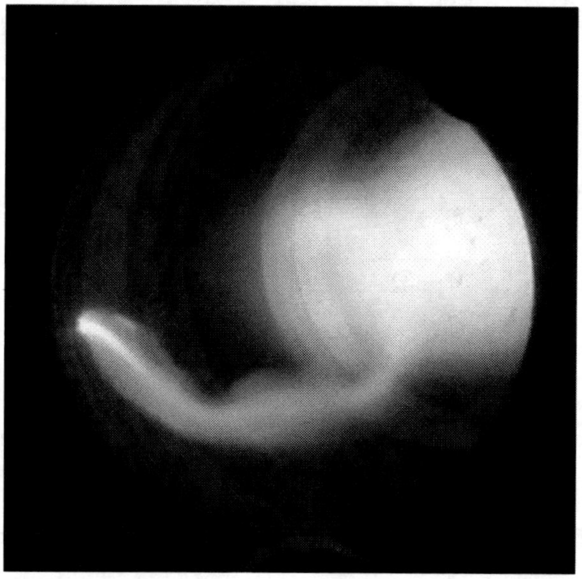

Figure 2. Photo of an AC electric arc attachment on the working surface of the cylindrical electrode. shutter speed 1/4000 s.

The internal part of the cylindrical electrode surface, where during the plasma torch work the arc attachment leaves the trace, we will call a working surface. The arc attachment position on the electrode working surface in the longitudinal direction was possible to set by regulation of the flow rates in near-electrode loops of the plasma forming gas supply. The rotation speed of the arc attachment depends both on the gas supply mode, and on characteristics of the impressed magnetic field.

The AC arc attachment rotates under the action of the gas-dynamic forces, sliding over the electrode working surface during the plasma torch work. Plasma of the arc column does not have enough time to relax at the current zero crossing, and the arc is ignited in the same channel in the each following half-period. The solenoid, which is forming a longitudinal magnetic field in the hollow electrode volume, is coaxially put on the electrode for the intensification of the arc attachment rotation. The solenoid coil is powered from the AC power supply, so that at the arc current polarity changing, the current direction in the coil changes. Thus, despite of the arc current polarity changing, the force direction, acting on the arc attachment, and promoting the intensification of its rotational movement remains the same. The magnitude of force acting on the charged particles of the arc column in the arc attachment area changes due to the sinusoidal change of the current in the coil. This force reaches the maximum value at the peak arc current and decreases at the moment of the current transition through zero. The connection of the AC coil in the opposite direction can cause the braking or even arc attachment moving towards to the gas flow, which is undesirable, as it leads to increase of the electrode material erosion and the reduction of its lifetime.

The magnetic field produced by a solenoid in the electrode channel was obtained using computer simulation. Figure 4 shows the schematic diagram of the electrode with the solenoid (a) and characteristic graphs of distribution of value of magnetic inductance on a working surface of an electrode in the longitudinal direction (b) and its distributions along the diameter of the electrode chamber (c). Values of magnetic inductance during the moments of time matching to maxima of currents in the solenoid windings, for the points of space laying on lines L and D are built in the graphs. Line L represents the generating line of the working surface, and line D coincides with the diameter.

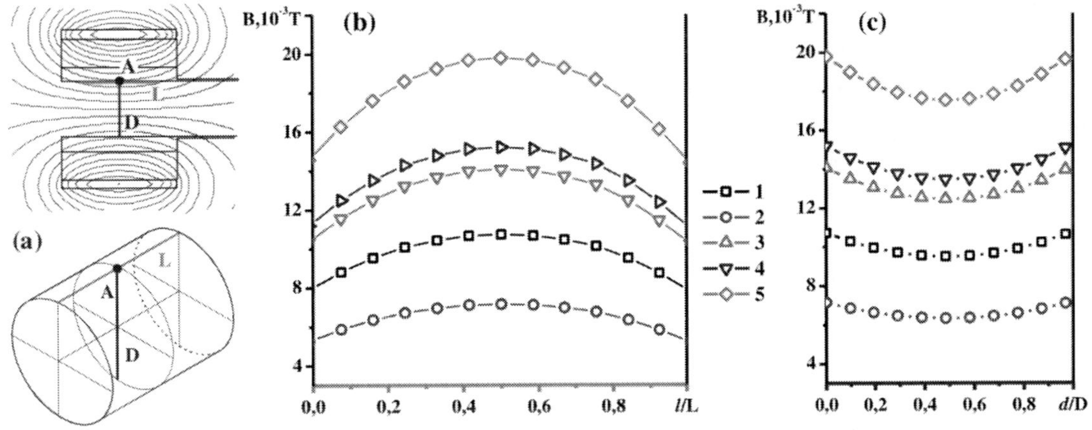

Figure 3. Schematic diagram of the hollow cylindrical electrode with solenoid (a) and characteristic graphs of distribution of the value of the magnetic inductance in the electrode channel at the moment of current maximum in the solenoid windings. Lines D and L – lines, along which are calculated the values of magnetic inductance, presented in graphs: (b) - longitudinal distribution of the magnetic inductance along the electrode working surface; (c) - distribution of the magnetic inductance on the diameter of the electrode chamber. Point A belongs to ring area with the maximum magnetic field density on the electrode working surface. Curve 1 shows magnetic inductance for plasma torch working with arc current of 47 A, curves 2-5 correspond the plasma torch working modes with arc current of 85 A.

Analysis of materials of a high-speed video shooting of the arc attachment movement along the electrode surface shows dependences of the arc attachment versus plasma torch operating parameters and characteristics of applied magnetic field. Shooting was carried out through the peep holes located in a back part of the electrode channel, with use of High Speed Video Cameras “Citius Imaging C100 Centurio” with a speed 4000 fps and exposure 2 μs. Speed of the arc attachment movement round a circle is non-uniform, it increases under the influence of electromagnetic forces at current maximum of a coil and decreases with its transition through zero. Mean values of tangential speed of the arc attachment were calculated on 50 complete revolutions for different regimes, showed in figure 3. For the arc current about 47 A without applied magnetic field the attachment velocity is about 2.5 m/s and it is rising twice when applied magnetic field density achieves 10^{-2}T. For the arc current about 85 A without applied magnetic field the attachment velocity is about 3.2 m/s and it is rising from 4 to 14 m/s when applied magnetic field density accordingly changes from 7×10^{-3}T to 20×10^{-3}T.

Numerical calculations of gas-dynamic parameters in the near electrode area for comparison the arc attachment rotation speed and the gas flow speed for the examined modes of the plasma forming gas feeding were carried out. Considering the fact, that the main flow in the back part of the channel is formed by the cold gas, submitted tangentially, and that the part of the arc column, situated there, occupies rather small share of the channel volume, so that parameters of a cold gas flow rate (without arc presence) were calculated in a first approximation. The task was solved in the stationary statement for the viscous compressed air. The approach, developed for high-speed compressible flows, was chosen. Besides the Navier-Stokes equations and the continuity equations, the system was supplemented by the energy equation and a set of the empirical constants, accompanying the connection of the turbulence model (k-ε), at calculations. The Figure 5 shows the calculation data.

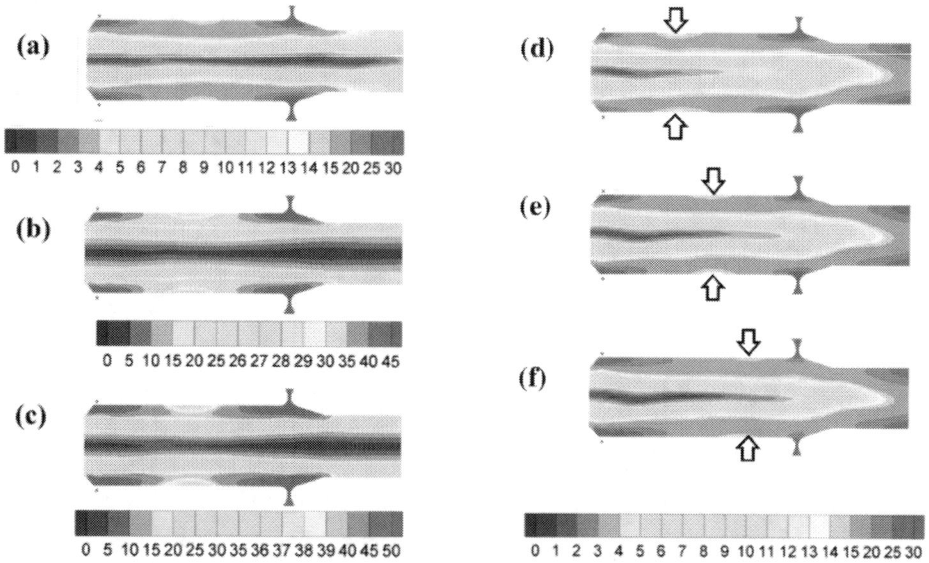

Figure 5. Plasma forming gas magnitude speed fields (m/s) in the electrode channel of the plasma torch in dependence of a total flow rate in near electrode loops of the plasma forming gas (air): (a) - 5 g/s, (b) - 9.4 g/s, (c) - 12.2 g/s; in the upstream and the downstream feeding circuits of the plasma forming gas (air). (d)-(f) - Representation of velocity fields for different distribution of flow rates between upstream and downstream gas feeding circuits. Arrows show the arc attachment areas: (d) - arc attachment is situated in back end of electrode, (e) - in the middle area and (f) - in the downstream end. Total gas flow rate - 4.5 g/s.

Figure 5 shows calculated gas velocity fields for plasma torch operation modes with different air flow rates supplied to the near electrode area. On each diagram can be seen location of slower near-wall flow, bounded on both sides (upstream and downstream) of a faster flow. Such areas are marked by arrows in diagrams (d)-(f). The electric arc attachment of working plasma torch stabilized in these locations, where the tangential velocity component makes the main contribution to the velocity magnitude (>99%). Diagrams (a)-(c) show velocity fields calculated for plasma torch operation modes with different air flow rates supplied to the near electrode area. Flow rates were from 5 to 12.2 g/s. Average speeds of the arc attachment moving under the influence of a magnetic field with an

Journal of Physics: Conference Series **406** (2012) 012007 doi:10.1088/1742-6596/406/1/012007

inductance of about 10^{-2} T were experimentally obtained for the same flow rate modes. The arc current was 47 A. The average speed of the arc attachment under these conditions is approximately 3 times less than the gas velocity in the near-wall electrode channel area.

Figure 5 (d)-(f) shows gas magnitude velocity fields for three modes of gas feeding with different ratios of its flow rates through upstream and downstream near electrode loops. The arc rotates in regions bounded on both sides of a faster flow marked by arrows. In these regions the tangential velocity is more than 98% of the velocity magnitude. The arrows in the diagrams (d), (e) and (f) comply with the trace of the arc at the beginning middle and end of the working electrode surface (in the longitudinal direction). Thus, the position of the arc attachment can be adjusted on the working electrode surface in the longitudinal direction. This allows organizing a permanent rotational and reciprocating movement of the arc in the long cylindrical electrode over its entire surface by changing the ratio of flow rates in upstream and downstream near-electrode feeding loops. For represented on these diagrams modes were experimentally determined speed of the arc attachment moving under the influence of a magnetic field with an inductance of about $14x10^{-3}$ T. The average speed of the arc attachment with a current about 84A slightly lags behind the velocity of the gas flow in the near electrode area and is about 80-90% of it. In reviewing should take into account that the values of the arc attachment speed are obtained experimentally, and the calculated values of flow rates are approximate and were obtained for the cold flow (excluding the arc in the working volume of the electrode) and that the real value of the flow velocity in the examined area may be different from calculated.

The speed of the arc attachment longitudinal motion depends on the rate of change of the flow rate proportion. Based on the results of a number of experiments it was offered for our device to organize continuous reciprocating motion of the arc attachment in the longitudinal direction (scanning) with longitudinal speed about 1 cm/s, it is successfully realized by gas supplying system, equipped with industrial automatic pressure controllers. Long-term 300-hour experiments showed that longitudinal scanning of the arc attachment with its gas-dynamic and magnetic rotation allow providing the uniform wear of the electrode surface. The specific erosion value is about 1-3 μg/C. It is 3 times less than that of the stationary AC plasma torches with the rail-type electrodes, working at the same power level with arc currents about 900 A. That enables production of the 1 MW industrial plasma torch with the hollow cylindrical electrode assembly, with lifetime above 1000 hours at currents in the arc up to 100-200 A.

Conclusion

Results of experiment directed towards the increase in the lifetime characteristics of electrode units of the powerful high-voltage electric-arc AC plasma torches are presented. The solution to the problem of obtainment the uniform wear of a copper hollow cylindrical electrode achieved by the controlled movement of the arc attachment along the working surface was offered. Organization of gas supply in the near electrode area and application of alternating magnetic field ensured movement of arc attachment along the surface with average speed from 2 to 14 m/s. Arc current was about 47A and 84A. Gas flow rate in near electrode area was about 4.5 g/s. The findings of influence methods on the electric arc attachment by electromagnetic and gas-dynamic forces are presented. The erosion of a hollow cylindrical electrode reached only 3 μg/C that enables production of the electrode assembly with life time above 1000 hours at currents in the arc up to 100-200 A.

Acnowledgements

The work is supported by Russian academy of sciences and also RFBR grants 11-08-01022-a and 12-08-00916-a. Authors also express gratitude to Anton Surov for the provided data of gas-dynamic calculations.

References

[1] Rutberg Ph G 2003 *Plasma Phys. Control. Fusion* 45 957-69
[2] Rutberg Ph G, Bratsev A N, Safronov A A, Surov A V, Shegolev V V 2002 *IEEE Transactions on Plasma Science*, v.30, 4, ISSN 0093-3813, pp. 1445-48

[3] Popov V E, Bratsev A N, Kuznetsov V A, Shtengel S V and Ufimtsev A A 2011 *Journal of Physics: Conference Series* 275 012015 doi:10.1088/1742-6596/275/1/012015
[4] Heberlin J, Murphy A B 2008 *J. Phys. D: Appl. Phys.* 41 053001
[5] Jeni˘sta J, Takana H, Nishiyama H, Bartlov´a M, Aubrecht V, K˘renek P, Hrabovsk´y M, Kavka T, Sember V and Ma˘sl´ani A 2011 *J. Phys. D: Appl. Phys.* 44 435204
[6] Messerle V E, Ustimenko A B, Askarova A S and Nagibin A O 2010 *Thermophysics and Aeromechanics* Vol. 17, N.3, 435-444, DOI: 10.1134/S0869864310030145
[7] Internet link: http://www.westinghouse-plasma.com/technology/plasma-torches (access date: 22.06.2012)1-14
[8] Hacala A, Michon U 2009, *ISPC19, Bochum, Germany, proceedings* 39 P2.14.04
[9] Ni Guo-Hua, Meng Yue-Dong, Cheng Cheng, Lan Yan 2010 *CHIN. PHYS. LETT.* Vol. 27, No. 5 055203
[10] Guohua Ni, Peng Zhao, Cheng Cheng, Ye Song, Hirotaka Toyoda and Yuedong Meng 2012 *Plasma Sources Sci. Technol.* 21 015009
[11] Rutberg Ph 2009 *Physics and Technology of High-Current Discharges in Dense Gas Media and Flows*, Nova Science Publishers, Inc., New York, ISBN 978-1-60692-232-3, 214 p
[12] Rutberg Ph G, Kumkova I I, Kuznetsov V E, Popov S D, Rutberg A Ph, Safronov A A, Shiryaev V N and Surov A V 2007 *IEEE Pulsed Power Conference (PPPS-2007) Digest of Technical Papers* 1976-2007, June 17-20, 2007, Albuquerque, New Mexico, USA, IEEE Catalog Number 07CH37864C, ISBN 1-4244-0914-4, Library of Congress 81-644315, Omnipress, Madison p. 1556-59
[13] Rutberg Ph G, Popov S D, Safronov A A, Serba E O, Nakonechny Gh V 2007 *XXVIII Internatuional Conference on Phenomena in Ionized Gases (ICPIG 2007) Proceedings,* Prague, Czech Republic, ISBN 978-80-87026-01-4, Published by Institute of Plasma Physics AS CR, v.v.i. (4P11-07) p.1762-65
[14] Rutberg Ph G, Safronov A A, Popov S D, Surov A V and Nakonechny Gh V 2005 *Plasma Phys. Control Fusion* 47,1681-96
[15] Rutberg Ph G, Lukyanov S A, Kiselev A A, Kuschev S A, Nakonechny Gh V, Nikonov A V, Popov S D, Serba E O, Spodobin V A and Surov A V 2011 *Journal of Physics: Conference Series* 275 012006 doi:10.1088/1742-6596/275/1/012006

Electrode temperature measurements of multi-phase AC arc by high-speed video camera

M Tanaka, T Ikeba, Y Liu, T Matsuura and T Watanabe
Dept. Environmental Chemistry and Engineering, Tokyo Institute of Technology
Yokohama, Kanagawa 226-8502, Japan

E-mail: mtanaka@chemenv.titech.ac.jp

Abstract. The multi-phase AC arc plasma has been applied in the glass melting technology as a promising heat source. The electrode erosion of the multi-phase arc is one of the most important issues to be solved. In the present work, the theory of the two-colour pyrometry by using the high-speed video camera with band-pass filters was applied to the electrode temperature measurements. First, the spectroscopic measurements of the electrode region in the multi-phase arc were conducted to select the appropriate wavelengths of band-pass filters. Then, the electrode temperatures of 2-phase, 6-phase, and 12-phase arcs were measured. The electrode tip temperature and the molten area were evaluated from the obtained 2-dimensional temperature distributions. Results indicated that the increase of the number of the phases leads to the lower tip temperature and the larger molten area. Observation of the multiple arcs further revealed the particular characteristics of the multi-phase arc, such as the periodical arc motion has important role on the electrode molten state.

1. Introduction

The glass industry is a large global industry that annually produces more than 100 million tons of glass products such as sheet glass, container glass, fiber glass, optical glass, and so on. Most glass has been produced by typical Siemens-type melter fired in air with heavy oil or natural gas as the fuel. This type of melter has been used for more than 140 years because of its good large-scale performance and continuous melting system. In the air-fuel fired furnace, the heat transfer from above burner flame to glass melt is so low that the conventional melting technology is energy intensive and time consuming, especially in the melting and the refining processes. With the rapid growth of glass usage and the increased energy and environment issues, it is crucial to develop a new glass melting technology to solve these problems.

An innovative in-flight glass melting technology with thermal plasmas was developed to solve above problems [1-3]. The granulated raw material with small diameter is dispersed in thermal plasmas and the powders contact fully with the plasma and/or burner flame. The high heat-transfer and temperatures of the plasma will melt the raw material quickly. In addition, the decomposed gas of carbonates is removed during the in-flight treatment to reduce the fining time considerably. Compared with the traditional glass production, the total vitrification time is evaluated only 2-3 h at the same productivity as the fuel-fired melter.

Arc plasmas as an energy source with high energy efficiency have been applied to the in-flight glass production. In particular, the multi-phase AC arc is one of the most suitable heat sources for the in-flight glass melting because it possesses many advantages such as high energy efficiency, large plasma volume (about 100 mm in diameter), low velocity (5-20 m/s) [4, 5]. In spite of the recent experimental works, which revealed the importance of the number of the phases on the discharge characteristics and the arc stability of the multi-phase arc, it remains to be explored. Electrode erosion is one of the considerable issues for the practical use of the multi-phase arc system in glass melting process.

Electrode temperature is important to understand erosion phenomena of the electrode material and the interactions between the arc and the electrodes. Two-colour pyrometry is a useful tool to measure the surface temperature of solid and/or liquid. Therefore, some experimental works have been reported to evaluate the electrode surface temperature of the DC [8] or AC arcs [9]. In the present work, the combination between the two-colour pyrometry and the high-speed camera observation was applied to investigate the dynamic behaviour of the molten electrode on the millisecond time scale. The purpose of the present study is to measure the electrode temperature by the high-speed camera and to investigate the influence of the number of the phases on the electrode temperature and the molten state in the multi-phase arc.

2. Experimental Procedure

2.1. Experimental setup

The schematic diagram of the experimental setup is shown in figure 1 (a). It consisted of 12 electrodes, arc chamber, and AC power supply. The electrodes were symmetrically arranged by the angle of 30 degree and were divided two layers, upper six electrodes and lower six electrodes. The electrodes were made of tungsten (98wt%) and thoria (2wt%) with diameter of 6 mm. City water was used to cool the electrodes at 3 slpm of water flow rate for each electrode and 99.99% argon was injected around electrodes to prevent them from oxidation at 5 slpm of gas flow rate. The arc chamber was made of stainless steel and cooled by water. 24 Sets of arc welding transformers with single-phase AC (DAIHEN B-300, Japan) were used to generate the multi-phase AC arc discharge. The more detail about the power supply for the generation of the multi-phase AC arc discharge was explained [4]. In the present work, arc current was adjusted at AC 100 A for each electrode.

Figure 1 (b) shows the cross-sectional view of the electrode tips for the multi-phase AC arc. The applied voltage between each electrode and the neutral point of the coil of the transformer can be calculated by the following equation:

$$V_i^N = V_m^N \sin\left[\omega t - 2\pi(i-1)/12\right], \text{ (i = 1, 2...12)} \quad (1)$$

where V_i^N indicates the applied non-load voltage for each electrode number i and V_m indicates the amplitude of the non-load voltage (about 220 V). The voltage was applied to only electrode No. 1 and No. 7 for the 2-phase arc discharge. In the case of the 12-phase arc discharge, the voltage was applied to electrode No. 1, 2...12.

2.2. Temperature measurements by high-speed video camera

According to the Wien's approximation of Planck's law, the spectrum of thermal radiation from a blackbody is used to determine the electrode temperature at the applied wavelengths. A real body that emits less thermal radiation than the blackbody has surface emissivity ε less than 1. Since the emissivity is unknown for many applications, the grey body assumption $\varepsilon(\lambda_1) = \varepsilon(\lambda_2)$ is used where it is hypothesized that the surface emissivity is independent of wavelength.

$$TR_{\lambda 1} = K\varepsilon(\lambda_1)d^2\lambda_1^{-5}\left(e^{k2/\lambda_1 T} - 1\right)^{-1} \quad (2)$$

Based on these approximations, the electrode temperature T is obtained from the ratio of the radiation intensity,

$$T = \frac{K_2(\lambda_1 - \lambda_2)}{\lambda_1 \lambda_2}\left[\ln\left(\frac{TR_{\lambda 1}}{TR_{\lambda 2}}\right) + 5\ln\left(\frac{\lambda_1}{\lambda_2}\right)\right]^{-1} \tag{3}$$

where K_2 is the second radiation constant in Plank's law which value is 1.439×10^{-2} m·K, $TR_{\lambda 1}$ and $TR_{\lambda 2}$ are the thermal radiations emitted by the electrode at λ_1 and λ_2, respectively.

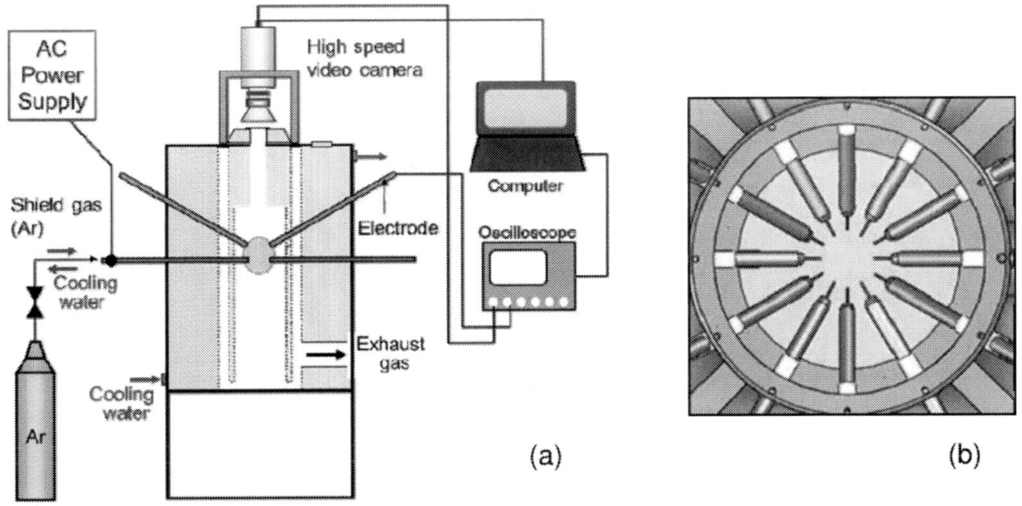

Figure 1. Schematic of the generating system of the multi-phase arc (a) and the top view of the electrodes (b).

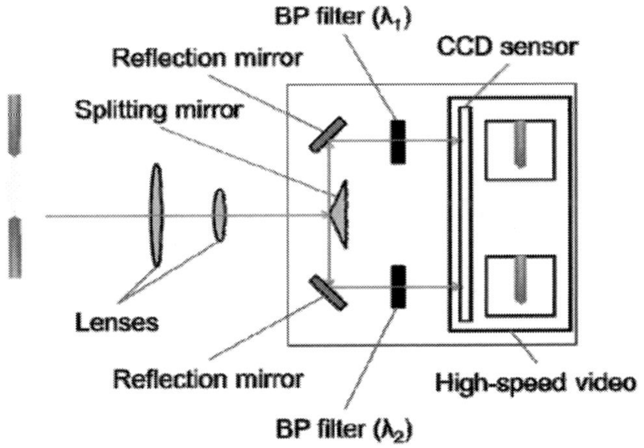

Figure 2. Schematic of the two colour pyrometry by using the high-speed video camera with the band-pass filters.

Spectroscopic measurements (iHR550, Horiba Jobin Yvon) were conducted to determine appropriate wavelengths for two-colour radiations. Electrode temperature of the multi-phase arc was then measured from the thermal radiation at the two wavelengths which determined by the spectroscopic results. High-speed video camera (FASTCAM SA-WTI, Photron) with the band-pass filter system was used to measure the radiation intensities as shown in figure 2. Frame rate and exposure time of the measurements were 10,000 fps and 20 μs, respectively. The voltage of each electrode was recorded at 1 MHz by an oscilloscope (Scope Corder DL 850, Yokogawa).

3. Results and Discussion

Figure 3 shows the emission spectrum of the multi-phase arc at electrode tip. Line emissions from argon and oxygen atoms were observed while continuum emission from the electrode surface was detected. From this spectrum, two wavelengths, 785 nm and 880 nm, were selected for the temperature measurements in order to avoid the direct plasma emission.

The surface temperature of the electrode for the multi-phase arc was measured with above selected wavelengths. Figure 4 shows the snapshot of the electrode obtained by the high-speed camera and the calculated temperature distribution of the electrode for the 12-phase arc. The temperature around the electrode tip was higher than the melting point of tungsten (3,695 K). The electrode tip temperature and the molten area were estimated from the obtained temperature distributions.

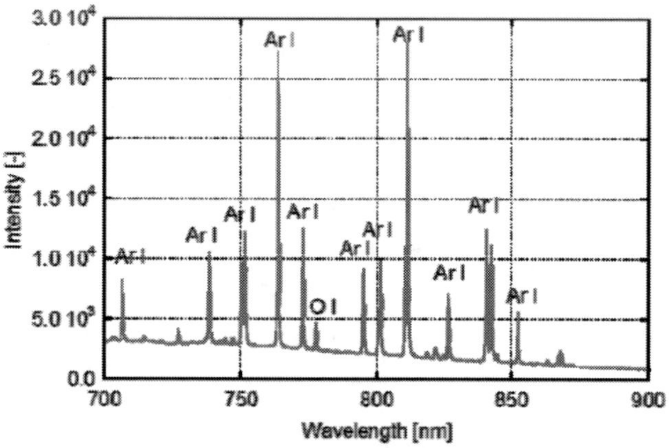

Figure 3. Emission spectrum of the multi-phase ac arc in electrode region.

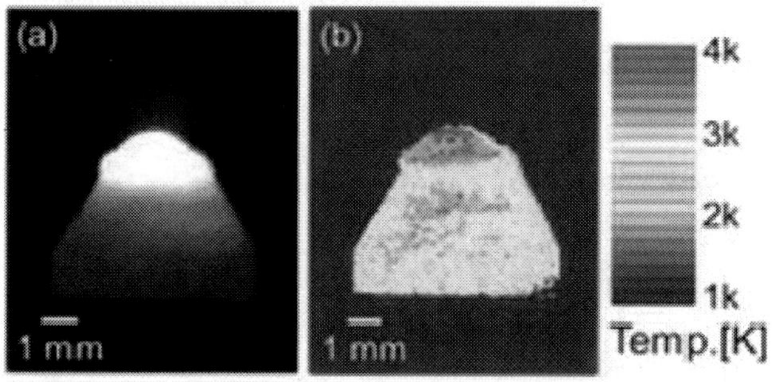

Figure 4. Snapshot of the electrode taken by the high-speed camera (a) and electrode temperature distribution in 12-phase ac arc (b).

Figure 5 shows the time dependence of the electrode tip temperature for the 2-phase and the 12-phase AC arcs in a cycle (20 ms). Each plot corresponds to the averaged value more than three data. Two peaks of the tip temperature were found in both cases for the 2-phase and the 12-phase arcs. These peaks originated in the sinusoidal current waveform in AC cycle, resulting in the peak values of the heat transfer from the arc to the electrode. The temperature variation range of the 12-phase arc was smaller than that of the 2-phase arc. This could originate in the existence of the multiple arcs at the same time in the case of the 12-phase arc, while only single arc exists in the case of the 2-phase arc.

Molten area was evaluated from the obtained temperature distribution to characterize the molten state of the electrode surface quantitatively. Number of the pixels, in where the temperature was more than melting point, was counted and then converted into area. Figure 6 shows the time variation of the molten area for 2-phase and 12-phase AC arcs. The variation range of the molten area of the 12-phase arc was larger than that of the 2-phase arc. This result will be discussed in the following paragraph.

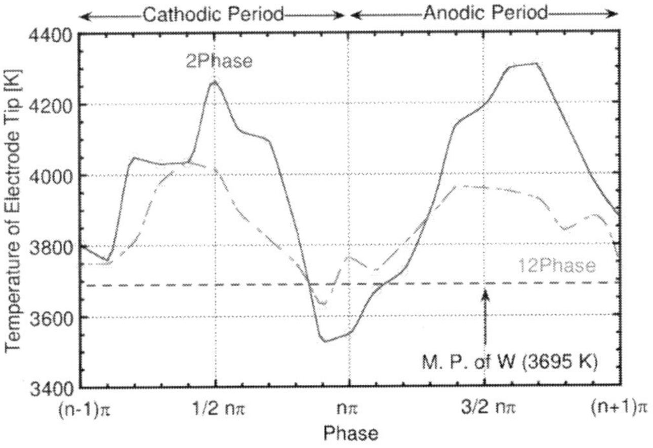

Figure 5. Time variation of the electrode tip temperature for the 2-phase and the 12-phase arcs in a cycle (20 ms).

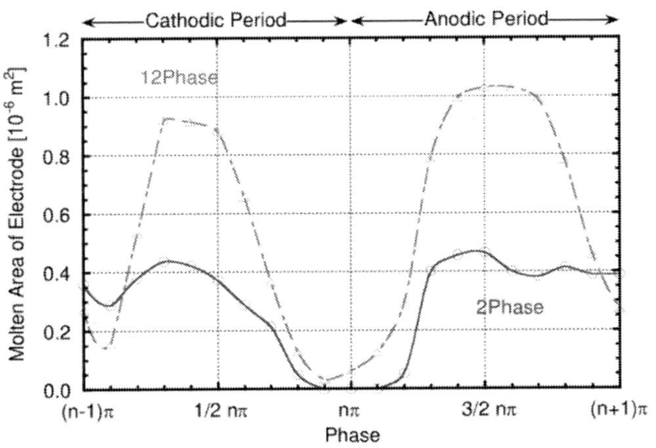

Figure 6. Time variation of the molten surface area for the 2-phase and the 12-phase arcs in a cycle (20 ms).

Figure 7 shows the relationship between the number of the phases and the electrode tip temperature. An increase of the number of the phases leads to the decrease of the electrode tip temperature. Figure 8 shows the relationship between the number of the phases and the molten area. In contrast to the tip temperature, an increase of the number of the phases leads to the increase of the molten area. The molten area at anodic period was larger than that at cathodic period. This is because the total heat transfer from the arc to the electrode at anodic period was larger than that at cathodic period, resulting from the electron condensation.

In order to investigate the effects of the number of the phases on the tip temperature and the electrode molten state, the movement of the arc spot was considered. Figures 9 and 10 represent the snapshots of the high-speed video images without band-pass filters for the 2-phase arc and the 12-phase arc, respectively. The 2-phase arc spot only moves slightly, whereas the arc spot for the 12-phase arc moves from the right to the left and then returns to the right side. This periodic swinging motion would be attributed to the rotational electro-magnetic field of the multi-phase ac arc.

Because of the periodical motion of the arc spot, the molten state of the electrodes can be predicted as indicated in the figure 11, which shows the model images of the electrode cross sections. These model images imply the following characteristics of the molten state; (a) the molten surface area for the 12-phase arc is larger than that for the 2-phase arc and (b) the molten depth for the 2-phase arc is deeper than that for the 12-phase arc. Note that the molten volume of the electrodes for the 2-phase and the 12-phase arcs were assumed to be almost the same because the total input powers for each electrode were adjusted to the same level.

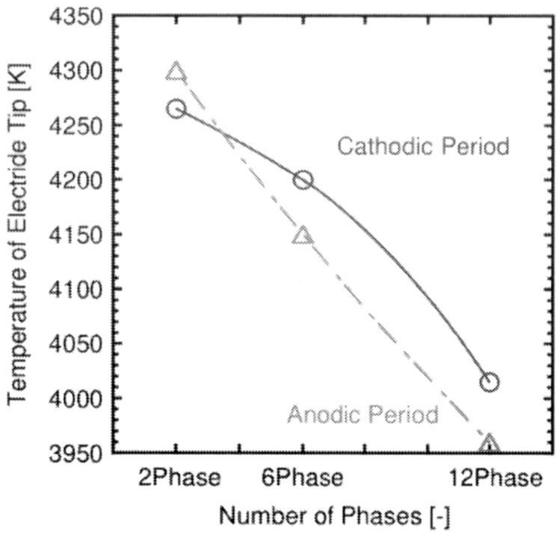

Figure 7. The relationship between the electrode tip temperature and the number of the phases.

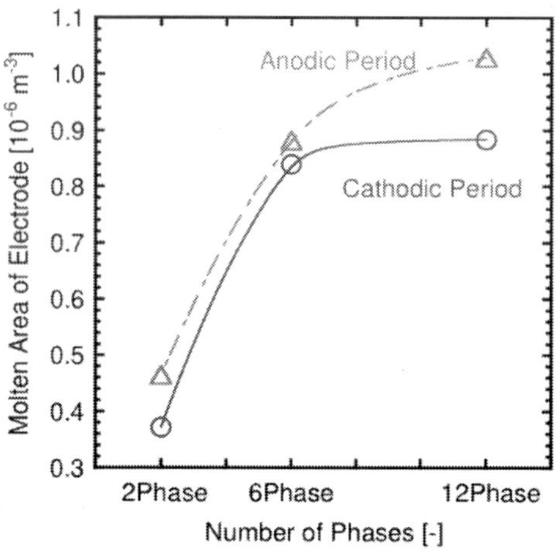

Figure 8. The relationship between the electrode molten surface area and the number of the phases.

The obtained experimental results can be explained by the above model. Wider area of the arc spot motion in the 12-phase arc leads to the larger molten surface area, resulting in the lower tip temperature. Wide and shallow molten region for the 12-phase arc compared with the 2-phase arc leads to the larger boundary area between the solid phase and the liquid phase inside the electrode, resulting in the rapid solidification of the molten electrode for the 12-phase arc. These characteristics of the electrode molten state would have important role on the electrode erosion of the multi-phase arc. Further experimental studies are needed to understand the electrode erosion mechanism of the multi-phase AC arc.

4. Conclusions

The temperature measurements system by using the high-speed video camera with the band-pass filters was established. From the surface temperature distributions of the electrodes, the time variation of the tip temperature and the molten area were obtained to investigate the effects of the number of the phases on the electrode molten state. Results indicated that the increase of the number of the phases leads to the decrease of the tip temperature and the increase of the molten surface area. These results were also explained by the periodical arc spot motion, resulting in the different molten state between the 2-phase arc and the multi-phase arc. The obtained results suggest that the particular characteristics of the multi-phase arc, such as arc motion, have the strong effects on the electrode erosion mechanism.

Acknowledgments

The financial support provided by the Strategic Development of Energy Conservation Technology Project of NEDO (New Energy and Industry Technology Development Organization, Japan) is gratefully acknowledged.

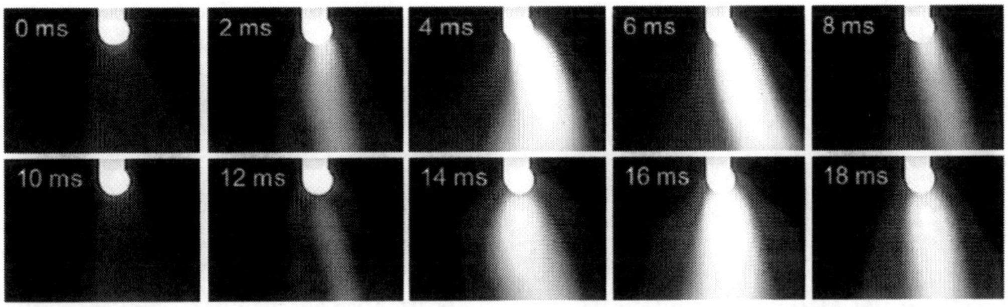

Figure 9. The snapshots of the high-speed video images for the 2-phase arc at electrode region.

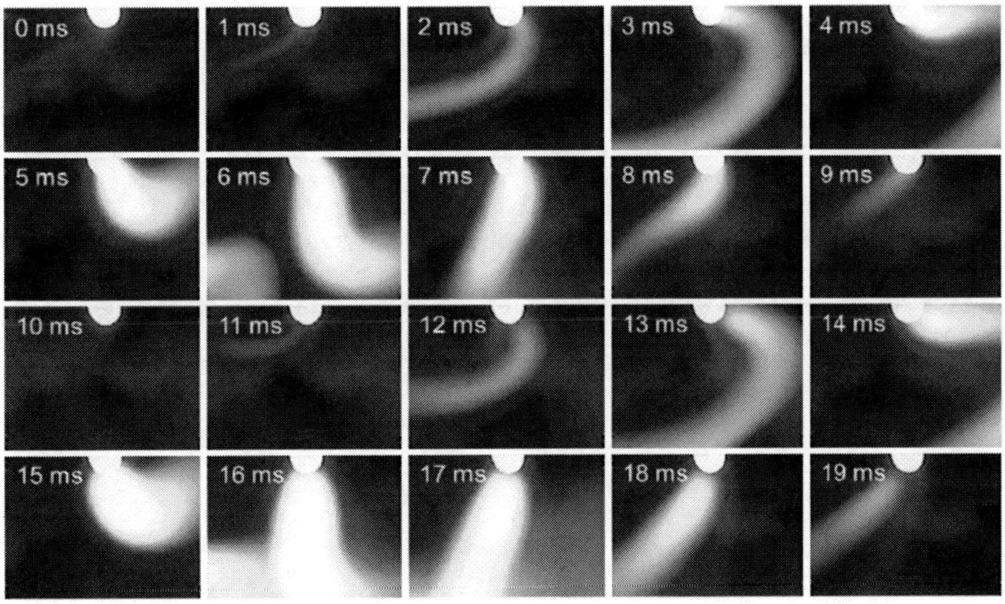

Figure 10. The snapshots of the high-speed video images for the 12-phase arc at electrode region.

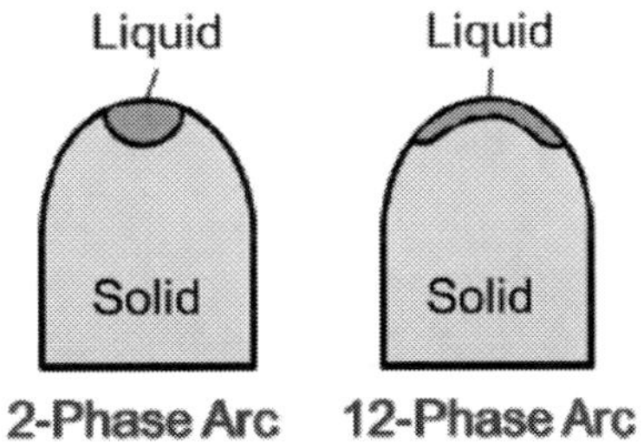

Figure 11. Cross sectional images of the electrodes for the 2-phase and the 12-phase arcs.

Journal of Physics: Conference Series **406** (2012) 012008 doi:10.1088/1742-6596/406/1/012008

References

[1] Yao Y, Watanabe T, Yano T, Iseda T, Sakamoto O, Iwamoto M and Inoue S 2008 An innovative energy-saving in-flight melting technology and its application to glass production *Sci. Technol. Adv. Mater.* **9** 025013

[2] Yao Y, Yatsuda K, Watanabe T, Funabiki F and Yano T 2008 Investigation on in-flight melting behavior of granulated alkali-free glass raw material under different conditions with 12-phase AC arc *Chem. Eng. J.* **144** 317-23

[3] Watanabe T, Yatsuda K, Yao Y, Yano T and Matsuura T 2010 Innovative in-flight glass-melting technology using thermal plasmas *Pure Appl. Chem.* **82** 1337-51

[4] Liu Y, Tsuruoka Y, Tanaka M, Ichihashi T, Yano T and Watanabe T 2011 In-flight melting behavior of different glass raw materials by hybrid heating of twelve-phase AC arc with oxygen burner *Thin Solid Films* **519** 7005-8

[5] Tanaka M, Tsuruoka Y, Liu Y and Watanabe T 2011 Investigation on in-flight melting behaviour of granulated glass raw material by multi-phase AC arc plasma and hybrid plasma *IOP Conf. Ser.: Mater. Sci. Eng.* **18** 112010

[6] Tanaka M, Tsuruoka Y, Liu Y, Matsuura T and Watanabe T 2011 Investigation of multiphase AC arc behaviour by high-speed video observation *IEEE Trans. Plasma Sci.* **39** 2904-5

[7] Tanaka M, Tsuruoka Y, Liu Y, Matsuura T and Watanabe T 2011 Stability analysis of multi-phase AC arc discharge for in-flight glass melting *Current Appl. Phys.* **11** S35-9

[8] Zhou X, Ding B J and Herberlein J 1996 Temperature measurement and metallurgical study of cathodes in DC arcs *IEEE Trans. Comp. Pack. Technol.* **19** 320-8

[9] Reinelt J, Westermeier M, Ruhrmann C, Bergner A, Awakowicz P and Mentel J 2011 Investigating the dependence of the temperature of high-intensity discharge (HID) lamp electrodes on the operating frequency by pyrometric measurements *J. Phys. D: Appl. Phys.* **44** 095204

Experimental observations of arc-anode attachment in steam-argon-air environment

O. M. Chumak, A. Mašláni, M. Hrabovský

Institute of Plasma Physics AS CR, v.v.i., Za Slovankou 1782/3, 182 00 Prague 8, Czech Republic

chumak@ipp.cas.cz

Abstract. DC arc plasma torches performance depends strongly on processes near anode. Arc-anode interaction not only defines lifetime of the electrode but also influences power distribution in the plasma and flow structure of the jet. This paper presents an experimental investigation of effects of gas properties and flow conditions on an arc-anode attachment. Several optical and electric diagnostic means were used to provide complex image of behaviour of the attachment and resulting state of anode surface. It is shown that even in one particular regime of the arc-anode attachment, which is called a restrike mode, dynamic behaviour and resulting erosion of anode can be significantly different. For example with argon in arc-anode gap the anode surface melt more intensively than with air environment. Despite increased melting weaker radiation of copper vapour was found in plasma. It was also shown that control of arc-anode attachment position in air environment had substantial effect on anode erosion pattern and plasma flow behaviour.

1. Introduction

An arc-anode attachment is one of the least known and predictable phenomena in non-transferred dc arc plasma torches. In such torches an anode surface is parallel to an arc column and dynamics of the attachment is controlled by a balance of a drag by boundary gas flow and magnetic force acting on the anode attachment with heating and electrochemical processes influencing conductivity in arc-anode gap [1-3]. The anode attachment strongly influences flow of generated plasma and its position controls the arc power. In this way it affects all main parameters of generated plasma and influences a process of plasma treatment, which was observed in experimental studies [4, 5] and in simulations [6].

In classic non-transferred torches anode surface is hidden inside the torch and is not accessible easily. Authors of work [7] simulated conditions in the torch by special geometrical and flow configurations. They observed three typical regimes of arc-anode attachment behavior: steady, takeover and restrike modes. Different types of anode erosion for different modes were reported. The restrike mode is the most frequent in commercial torches as this mode is typical for the highest power and enthalpy of plasma. Restrike mode features repeated process of movement of the attachment stream wise due to drag by boundary gas flow and magnetic force and followed appearance (restrike) of new current path in upstream position leading to decay of the old attachment.

In this work a special plasma torch was used. It features an external anode and allows access of diagnostics and observations in anode region. Here the authors examine restrike modes at different flow conditions and with different gasses in anode region. Distinct behaviors as well as different types of anode erosion were observed for the same mode in dependence on environment conditions.

Journal of Physics: Conference Series **406** (2012) 012009 doi:10.1088/1742-6596/406/1/012009

2. Experimental set-up and diagnostics

2.1. Plasma torch

The plasma jet generated by dc arc plasma torch WSP®H-2000 with a combination of gas and liquid arc stabilization [8] was used for the investigation. Water is utilized as the primary plasma forming medium and it mainly defines heat transfer processes in the torch. Argon is secondary medium for plasma generation. An anode of the torch is created by the rotating water-cooled copper disc, which touches the arc from the bottom. The anode surface moves perpendicular to the flow with the velocity 47 m/s, which is several times lower than the arc-anode attachment velocity and two orders lower than plasma velocity.

The arc consists of three parts: an argon stabilized cathode region, water stabilized arc column in the main arc chamber, and free arc in the anode region. The length of the stabilized parts is approximately 50 mm while the anode part of the arc can vary in the range of 16 mm, restricted by the width of the anode. The arc-anode attachment usually does not move in the whole possible range and the arc power fluctuations reach values of approximately 20 % for the total arc power of 110 kW. Plasma jet and anode attachment is shown in figure 2. The length of the jet is approximately 60 mm. Two types of arc-anode connection can be seen in the picture.

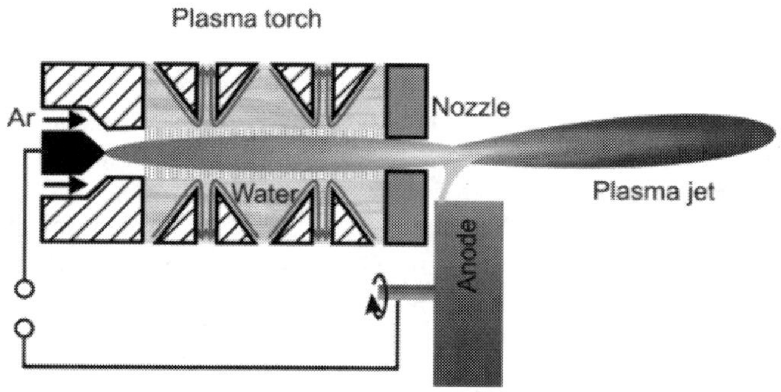

Figure 1. Sketch of dc arc plasma torch WSP®H-2000.

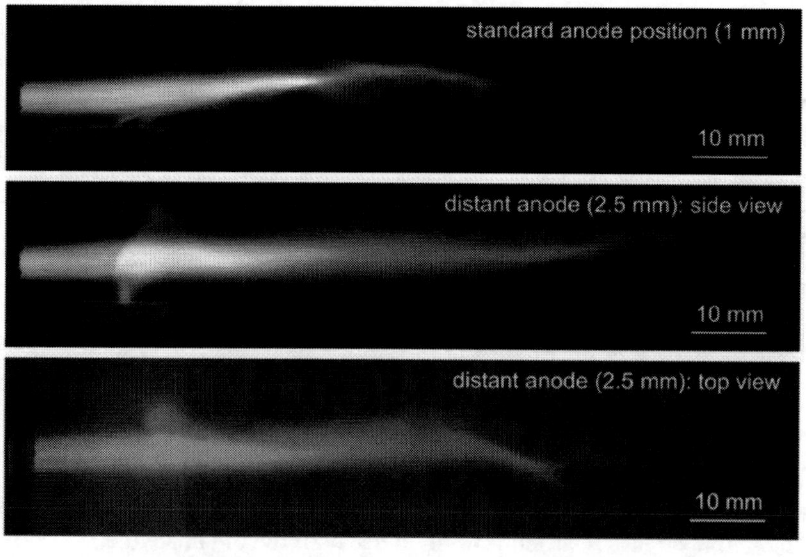

Figure 2. Plasma jet in two operation regimes dependent on anode position.

2.2. Conditions of the experiments

In this investigation arc current and argon flow rate, which are the main operation parameters, were kept at values 400 A and 20 slm respectively. Thus, parameters of the plasma jet at the exit of the arc chamber were constant. Anode attachment behavior was defined by a distance of the anode from the jet edge and by control of ambient gas flow in anode region. Four tests with the most prominent results were chosen to show three types of influences: an effect of the anode position, an effect of properties of gas in the arc-anode gap and an effect of air flow pattern in the arc-anode gap.

Anode position was changed by 1 mm from standard position to further one. It was expected that increase of distance would reduce temperature in the arc-anode gap and heat transfer to anode surface while amount of air dragged by anode surface to arc-anode gap will increase. The only exception in conditions is in figure 2 where gap was increased by 1.5 mm for better distinction.

Injection of argon to anode region was done as it is shown in figure 3a. Weak flow of argon of 8 slm through blower tube with outlet cross section 1x16 mm impinged upon anode surface in vicinity of plasma flow core. Anode surface dragged argon instead of air into the arc-anode gap. By comparison of experiments with and without argon blowing effect of properties of gas in the arc-anode gap was analyzed.

Injecting of air was done against anode surface movement (Figure 3b). The aim of this configuration was to compensate air drag by the anode surface and prevent anode attachment from movement across the plasma jet. Optimal flow was found to be 15 slm and it was checked that flow in such position of blower distributes in anode vicinity rather and does not influence plasma jet.

Thus four experiments are presented: 1) standard configuration; 2) wider arc-anode gap by 1 mm; 3) blowing of air into arc-anode gap; 4) blowing of argon into arc-anode gap. And three effects (three cases) will be discussed: effect of anode distance, effect of gas in arc electrode gap, effect of flow pattern in the arc-anode gap.

2.3. Diagnostic methods

Several diagnostics were used for arc characterization. Anode region was observed with a fast shutter camera from the top and the side (Figure 4a). Top view was applied as the attachment in some regimes traveled to the side of the jet as it is seen in the photograph. Statistical processing was used for visualization of fluctuating regions. Brightness distribution in every point was analyzed and effective width of the distribution was derived via Shannon entropy parameter [9, 10]. Resulting map (Figure 4b) is in relative units, shown in a color scale so that red relates to the most intensive brightness fluctuation and blue show most stable brightness areas.

a **b**

Figure 3. Position of argon blower (a) and air blower (b), anode surface moves clockwise.

Journal of Physics: Conference Series **406** (2012) 012009 doi:10.1088/1742-6596/406/1/012009

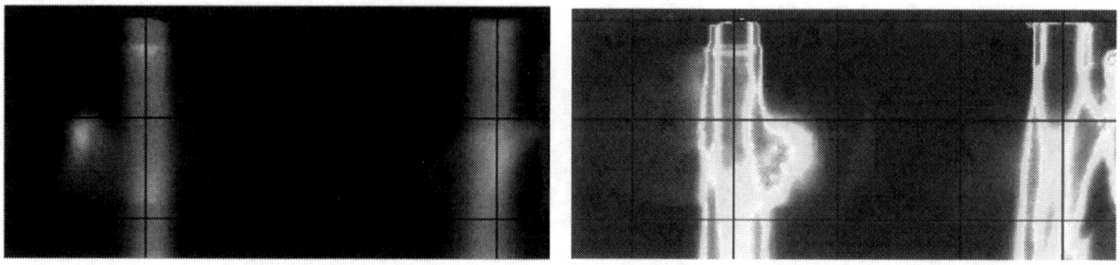

Figure 4. Photograph of plasma jet in top and side views (a), fluctuation intensity map (b).

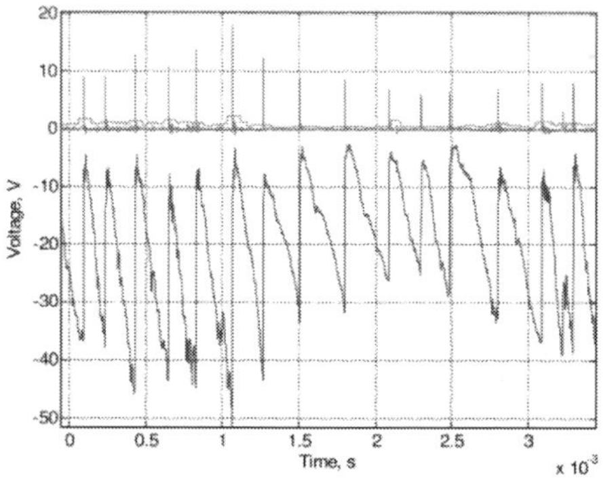

Figure 5. Curve of voltage on free part of arc and its differential for restrike occurrence identification.

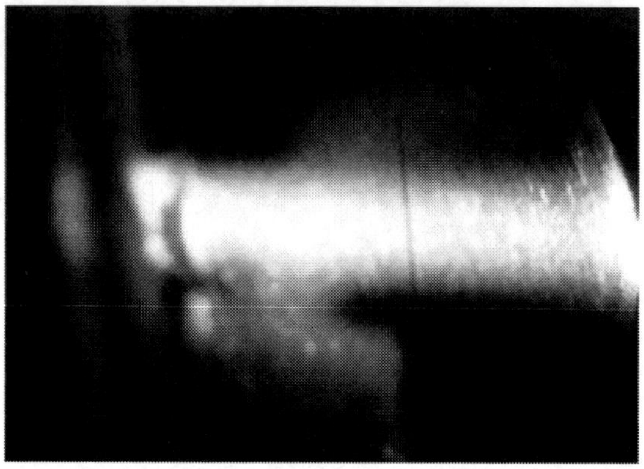

Figure 6. Projection of plasma jet on head of emission spectrometer.

The arc-anode attachment was in the restrike mode during the measurement and voltage on the open part of arc had typical saw tooth pattern (Figure 5). A special processing was applied to characterize saw tooth fluctuations. Sudden voltage drops were identified by signal derivative using an adaptive threshold. A distribution of voltage drops during the restrike and periods between the drops were analyzed. As the distribution had close to Gaussian shapes only mean values are discussed.

Emission spectroscopy was done. Optical fibre bunch head was positioned so that plasma jet section on position just after the anode was projected on a slit of the head. Spectra across the plasma jet were recorded in the band from 499 to 525 nm. This range includes argon, nitrogen and copper lines and allows considerations about elements distribution in plasma jet and its vicinity.

3. Results and discussion

In the next figures (7-10) a comparison of results of different diagnostics are shown for the studied experimental conditions as it was numbered in section 2.2. They are grouped by diagnostic used. The experiment 2 is referential as in all cases only one parameters was changed with respect to the case 2: 2-1 – arc-anode distance, 2-3 arc-anode gap gas (air to argon), 2-4 arc-anode gap flow pattern (air blowing compensated boundary layer drag by anode surface). In the following text the results will be discussed according to the cases rather than grouped by diagnostics as in the figures.

3.1. Effect of arc-anode gap width (conditions 2 and 1)

Maps of fluctuations (Figure 7) of the top view show intensive local fluctuations on the side of the jet for the wider arc-electrode gap. It means travelling of the anode attachment to the side of the jet following the anode surface motion. The side view shows more intensive and localised fluctuations for the wider gap. Together it leads to a conclusion that wider gap lead to travelling of the attachment across the jet with anode surface while for close arc position travel direction is along the jet forced by plasma boundary flows. Voltage parameters (Figure 8) changed substantially also. Wider gap led to increased period (a lifetime of the attachment) and increased amplitude of the saw tooth fluctuations while mean voltage did not change critically. In addition, figure 2 shows that wider gap change the shape of anode attachment from constricted to a wide current channel. As a result surface erosion pattern (Figure 9) changes from 0.2 mm droplets to 0.8 mm stains including small scale structures. Anode surface is also covered with a crack network which testifies lower melting intensity. However spectra of emission (Figure 10) show intensive copper lines in both cases. The difference is in propagation of copper vapour. For wider gap copper radiates also above the jet carried by anodic jet on the side of the main plasma flow.

3.2. Effect of air flow pattern in arc-anode gap width (conditions 2 and 3)

Maps of fluctuations (Figure 7) of the top view show that blowing with air in against anode surface movement forced attachment to stay under the plasma jet. Fluctuations in the view from the top became more symmetrical and in the view from the side very well distributed along the jet and anode surface. Analysis of voltage fluctuations (Figure 8) shows increased amplitude of saw tooth pattern claiming an increased range of attachment movement too. The anode surface (Figure 9) shows exceptionally low erosion. Inclined paths are visible on the surface and only at the beginning there some small structured droplets indicating the melting. Spectra of plasma radiation (Figure 10) prove drastically reduced evaporation copper as copper lines fully disappeared.

3.3. Effect of argon in arc-anode gap (conditions 2 and 4)

Maps of fluctuations (Figure 7) of the top view show that substitution of air with argon led to restoration of longitudinal travelling of the attachment. It is not dragged to the side of the jet anymore and side view also indicates very uniform distribution of fluctuation in anode vicinity. Amplitude of voltage fluctuations decreased drastically with argon but the period remained the same. Anode surface pattern also changed radically. Surface roughness increased claiming intensive melting. Nevertheless copper lines in spectra almost disappeared.

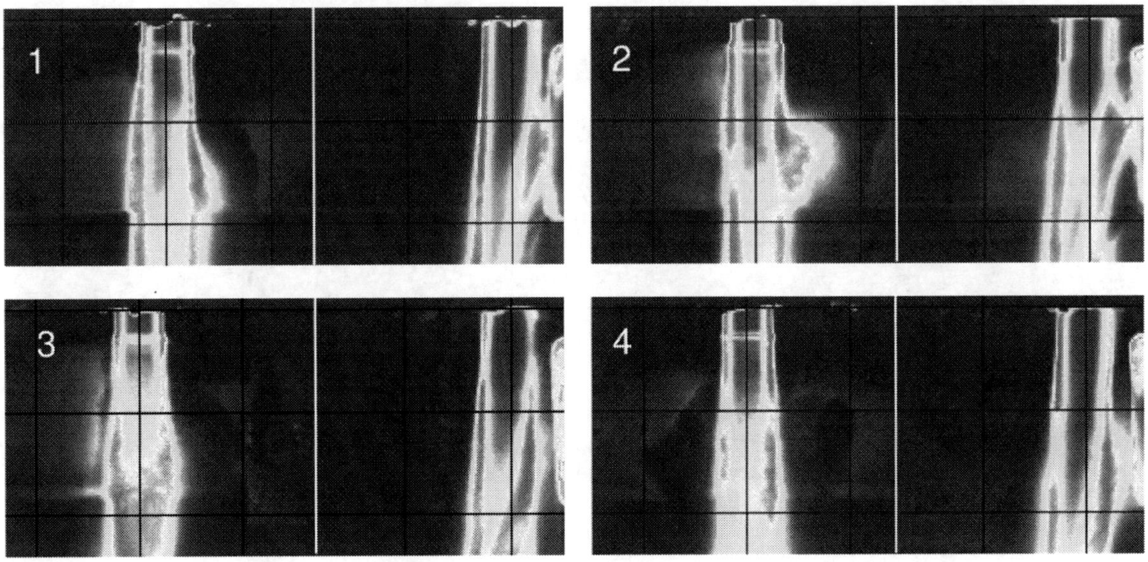

Figure 7. Fluctuation intensity maps for the studied experimental conditions. Views from the top and the side of the jet. Grid step is 10 mm.

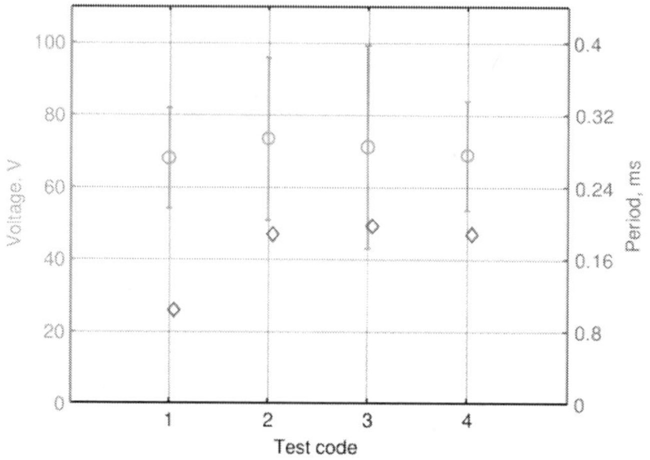

Figure 8. Mean voltage on the open arc (circles) mean amplitude of saw tooth fluctuation (ranges) and mean period of saw tooth fluctuation of voltage (diamonds) for the studied experimental conditions.

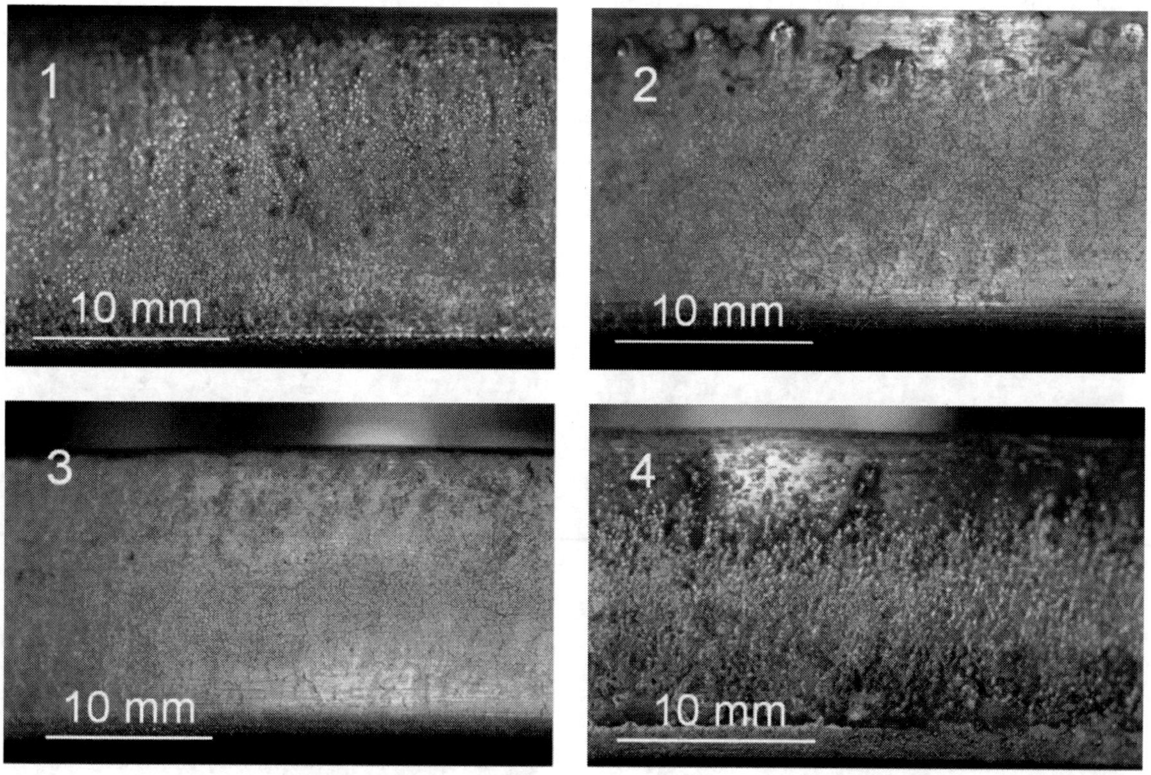

Figure 9. Anode surface pattern for the studied experimental conditions.

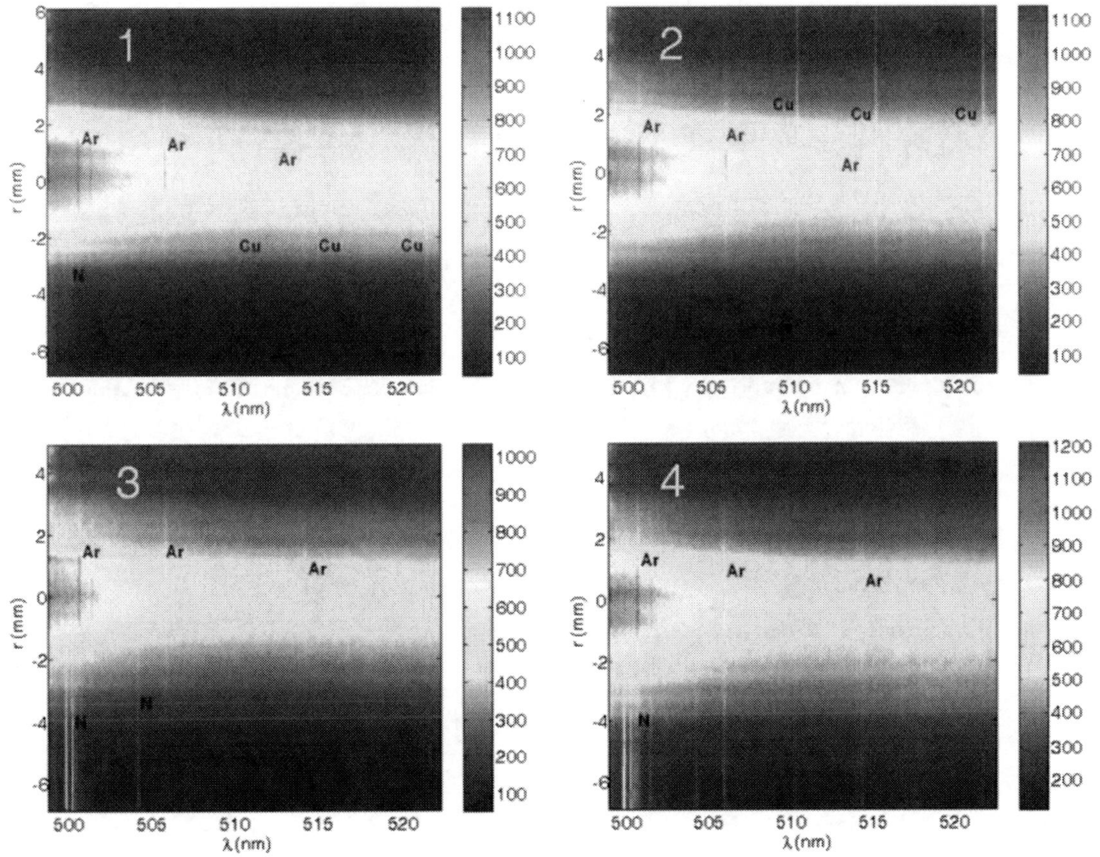

Figure 10. Spectra of plasma radiation across the jet for the studied experimental conditions.

4. Conclusion

This work demonstrates behavior of the arc-anode attachment in restrike mode and resulting erosion of the anode surface. The restrike mode was studied at different gas and flow conditions in the arc-anode gap. Gap width between the arc and anode was first factor which influences its behavior. For the narrow gap plasma jet boundary layer forced fast stream-wise travel of the attachment and provoked concentration of the current path at the anode to a small dot. For wider gap broader radiation channel reveal better distribution of current density and boundary flows created by anode movement controlled attachment behavior. If these boundary flows were compensated by air blowing the flows of plasma boundary took advantage again. Attachment moves stream wise but preserve its wider channel shape. It reduced melting and evaporation of anode material drastically.

Blowing of argon into arc-electrode gap reduced the effect of anode surface movement on behavior of anode attachment. It was noted that despite intensified melting of anode surface vapor is radiation significantly weakened.

The work showed that in the frame of restrike mode the attachment can behave in substantially different way, structure and resulting erosion intensity and type differs drastically. It shows importance and potential of control of flow pattern and boundary layer properties on processes in the anode region of arc and on the lifetime of anode.

Acknowledgments

The research was supported by Grant Agency of the Czech Republic in the frame of grant GAP205/11/2070.

References

[1] Wutzke S A Pfender E Eckert E R G 1968 *AIAA J.* **6** 1474
[2] Chumak O Hrabovsky M Kopecky V 2003 *Proc. of XVI Int. Sym. on Plasma Chem. Taormina Italy* p 619
[3] Brilhac J F et al. 1995 *Plasma Chem. and Plasma Process* **15** 231
[4] Fauchais P 2004 *J. Phys. D: Appl. Phys.* **37** R86
[5] Dussoubs B Mariaux G Vardelle A Vardelle M Fauchais P 1999 *J. of High Temp. Material Processes* 235
[6] Baudry C Vardelle M Mariaux G 2005 *J. High Temp. Material Processes* **9** 1
[7] Yang G Cronin P Bendix P Heberlein J V 2005 *Plasma Science, IEEE Transactions on* **33** 1123
[8] Brezina V Hrabovsky M Konrad M Kopecky V and Sember V 2001 *Proc. of 15th Int. Symp. on Plasma Chemistry (ed. A. Bouchoule et al.) Orleans* **3** p 1021
[9] Chumak O Hrabovsky M 2006 *Czechoslovak J. of Physics* **56** p B767
[10] Chumak O Kavka T Hrabovsky M 2008 *Plasma Science, IEEE Transactions on* **36** p 1062

Journal of Physics: Conference Series **406** (2012) 012010 doi:10.1088/1742-6596/406/1/012010

Calculation of the net emission coefficient of an air thermal plasma at very high pressure

T Billoux[1], Y Cressault[1], Ph Teulet[1] and A Gleizes[1]

[1] *Université de Toulouse; UPS, INPT; LAPLACE (Laboratoire Plasma et Conversion d'Energie), 118 route de Narbonne, F-31062 Toulouse cedex 9, France*

E-mail : yann.cressault@laplace.univ-tlse.fr

Abstract

The aim of this paper is to present an accurate evaluation of the phenomena appearing for high pressure air plasmas supposed to be in local thermodynamic equilibrium (LTE). In the past, we already calculated the net emission coefficient for air mixtures at atmospheric pressure and for temperatures up to 30kK (molecular contribution being restricted to 10kK). Unfortunately, the existence of high pressures does not allow us to use this database due to the non-ideality of the plasma (Viriel and Debye corrections, energy cut-off…), and due to the significant shifts of molecular reactions towards upper temperatures. Consequently, this paper proposes an improvement of our previous works with a consideration of high pressure corrections in the composition algorithm in order to take into account the pressure effects, and with a new calculation of all the contributions of the plasma radiation (atomic lines and continuum, molecular continuum, and molecular bands) using an updated database. A particular attention is paid to calculate the contribution of all the major molecular band systems to the radiation: O_2 (Schumann-Runge), N_2 (VUV, 1st and 2nd positive), NO (IR, β, γ, δ, ε) and N_2^+ (1st negative and Meinel). The discrete atomic lines and molecular bands radiation including the overlapping are calculated by a line-by-line method up to 30kK and 100 bar. This updated database is validated in the case of optically thin plasmas and pressure of 1bar by the comparison of our integrated emission strength with the published results. Finally, this work shows the necessity to extend the molecular radiation database up to 15kK at high pressure (bands and continuum) since their corresponding contributions could not be neglected at high temperature.

1. Introduction

In the earlier works of our team, the calculation of the net emission coefficient of air thermal plasmas was performed at atmospheric pressure for the temperature range 300-15000K [1]. The molecular contribution was limited to temperatures up to 10kK. In fact, several applications of air arcs may involve very high pressures such as special plasma torches (electric arc heaters) used in aerospace for the simulation of atmospheric re-entry behaviour of thermal protective materials [2]. Furthermore, another application in aerospace concerns the high enthalpy wind tunnel which provides flows at high velocity and pressures of several hundreds atmospheres [3]. Therefore, our previous results are not available in these conditions for two main reasons: the plasma is not ideal at pressures up to 100 bar and the plasma compositions did not take into account the pressure effects such as the shift of the dissociation of the diatomic molecules to higher temperatures. The role of the most important molecular species must thus be included for temperatures above 10kK. Consequently, we present here an improvement of the radiation data for air thermal plasmas at high pressures (up to a few hundreds bars) and rather high temperatures (up to 30000K) paying a particular attention to the role of molecular species in the total plasma radiation. The first step of this work is an upgrade of the plasma composition algorithm used until now in order to consider the high pressure effects on the plasma behaviour. The second part concerns the calculation of the total atomic radiation which is the sum of the atomic continuum (photo-ionization, radiative recombination and atomic bremsstrahlung) and of the atomic lines due to transitions between two discrete states of the atomic species. The third part is devoted to the calculation of the molecular contribution to the total radiation. The molecular continuum is achieved by the compilation of the photoabsorption cross-sections of all the main diatomic molecules constituting air plasmas (O_2, N_2, NO, N_2^+). A particular attention is also devoted to the calculation of each electronic band system which mainly contributes to the radiation of air plasmas. A systematic line-by-line calculation is achieved for each system, taking in account the overlapping of molecular discrete radiation lines. Finally, the total emission and absorption spectra of air plasmas is obtained by the summation of all the contributions (atomic continuum, atomic lines, molecular continuum and molecular bands) and using a line-by-line method with 4,470,000 wavelengths distributed in the range [30 – 4500 nm], the overlapping being accurately treated. It was then possible to evaluate the influence of each process on the total net emission for different temperatures and pressures. The total emission or absorption spectra can then be used to solve the radiative transfer equation either by an exact calculation or using some approached methods as the NEC (Net Emission Coefficient) [4] or the Mean Absorption Coefficients method (P_n models).

2. Plasma composition

In order to characterize the radiative properties of thermal plasmas, the preliminary step consisted in an accurate determination of the mixture composition for each couple temperature / pressure. The computation of equilibrium compositions of air plasmas in the temperature range 300K – 30kK and pressure between 1 and 100 bar was performed according to the method described by Godin and Trépanier [5]. This calculation was based on the mass action law and the chemical base concept. At very high pressure, deviations from the ideal gas model had to be implemented, especially for the atomic species. Based on the works of Drellishak [6] and D'Angola [7] we added the Fermi criterion (high pressure density effect limiting the maximum atomic radius) and the Griem criterion (interactions between charged particles) in order to determine the total energy cut-off limiting the number of atomic levels in the calculation of the internal partition functions. More details are available in [8].

In this work, 31 gaseous species were considered in the calculation of the mixture composition: e^-, O, O^-, O^+, O^{2+}, O^{3+}, O_2, O_2^+, O_2^-, N, N^+, N^{2+}, N^{3+}, N_2, N_2^+, NO, NO^+, NO_2, NO_2^-, N_2O, N_2O^+, O_3, N_3, NO_3, N_2O_3, N_2O_4, N_2O_5, Ar, Ar^+, Ar^{2+}, Ar^{3+}. *Figure 1* shows the composition of air plasmas at 100 bar.

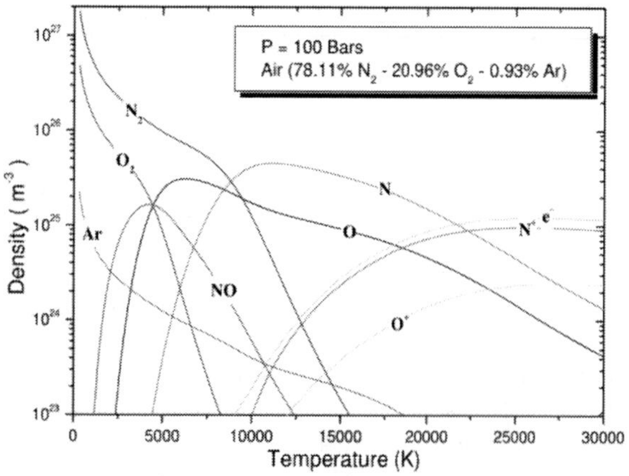

Figure 1. Composition of air plasmas (P=100 bar).

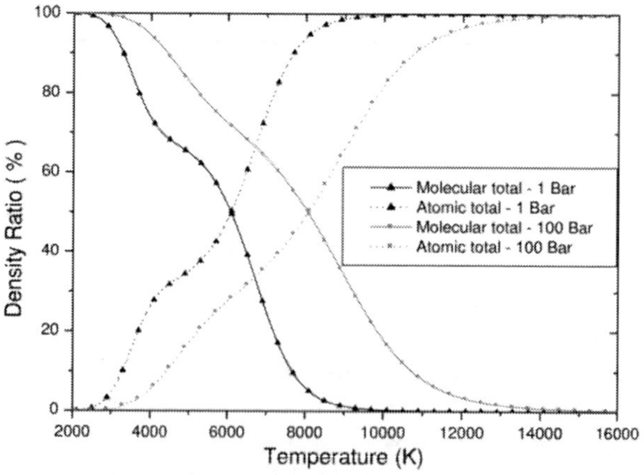

Figure 2. Total number densities of molecular and atomic species in air (P=1 and 100 bar).

Journal of Physics: Conference Series **406** (2012) 012010 doi:10.1088/1742-6596/406/1/012010

It is well known that the increase of the pressure leads to shift the temperature dissociation of the molecules (around 2000 K in the case of N_2), and consequently gives a more important role to these species in a larger temperature range as illustrated in *Figure 2*. As consequence of that, we have to extend and improve our previous molecular database following the new pressure conditions.

3. Total absorption coefficient

The determination of the NECs for air plasmas and pressures from 1 bar to 100 bar required an accurate description of the total emission and absorption spectra which were obtained by adding all the radiative phenomena occurring in the mixture: the atomic continuum, the molecular continuum, the atomic lines and the molecular bands. These last radiative mechanisms are briefly described in the following sections.

3.1. Atomic continuum

The radiation of the atomic continuum results from the combination of three main phenomena: the radiative attachment, the radiative recombination and the Bremsstrahlung. The atomic continuum of air plasmas was thoroughly studied in previous papers [8,9].

3.2. Molecular continuum

This continuum contribution plays an important role at low and intermediate temperatures, where the molecular densities are predominant. We thus selected the major molecular species in air plasmas (O_2, N_2, NO, N_2O, ...) and we elaborated an exhaustive review of the most accurate photoabsorption cross-sections published in the literature. Over a hundred partial photoabsorption cross-sections were found in the literature. The most exhaustive data were those of Fennely and Torr [10], Lee *et al.* [11] and the critical review of Hudson [12]. For the calculation, we assumed that the photoabsorption cross-sections did not depend on temperature. Consequently, the molecular continuum was approximated by the multiplication of each cross-section with molecular number density of the corresponding species. *Figure 3* shows the molecular continuum of the main molecular species in air plasmas (O_2, N_2 and NO) for atmospheric pressure and temperature of 300K.

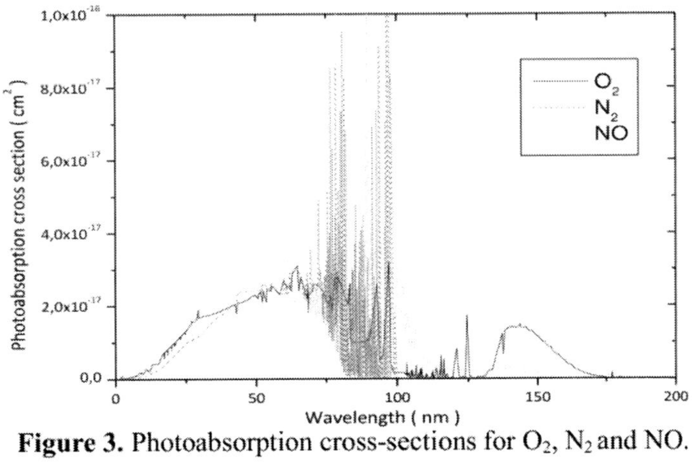

Figure 3. Photoabsorption cross-sections for O_2, N_2 and NO.

3.3. Atomic lines

The calculation of atomic lines emissions required a complete data bank including the properties of 15530 lines (6217 lines for oxygen species and 9313 lines for nitrogen species) whose characteristics (energy levels, quantum numbers, transition probabilities, oscillator strengths, ...) were issued from [13-15]. In a previous work [1], we assumed that the lines' overlapping had a negligible influence on the radiation allowing us to treat each line by means of an escape factor. In this work, we decided to use the exact description of the atomic line spectrum and to compare the results with those obtained using the escape factor. We then calculated the atomic line radiation for isothermal plasmas of thickness R_p as follows:

$$\varepsilon_\lambda^{line}(T,R_p)=\left(\frac{e^2}{4\pi\varepsilon_o}\right)\cdot\left(\frac{\pi}{m_e c}\right)\cdot L_\lambda^o(T)\cdot f_{\ell u}\cdot\bar{n}_\ell(T)\cdot\exp(-K'_\lambda R_p)\cdot\left(1-\exp\frac{-hc}{\lambda k_b T}\right) \quad (1)$$

where $L_\lambda^o(T)$ is the Planck function at the centre of the line, e is the electron charge, λ is the wavelength corresponding to the transition between the upper energy level E_u and the lower energy level E_l, m_e is the mass of the electron, $\bar{n}_\ell$ is the number density of the lower energy level, f_{lu} is the oscillator strength of this transition and K_λ is the total absorption coefficient in m^{-1}. As the spectral description of the atomic lines' radiation required the knowledge of the line profiles, each line was described by a Voigt profile:

$$P_\lambda^{Voigt}=\frac{\gamma_L \ln 2}{\gamma_D^2\pi^{3/2}}\int_0^\infty\frac{\exp(-y^2)}{\left(\frac{\gamma_L\sqrt{\ln 2}}{\gamma_D}\right)^2+\left(\frac{(\lambda-\lambda_0)\sqrt{\ln 2}}{\gamma_D}-y\right)^2}dy \quad (2)$$

with γ_D and γ_L the Doppler and Lorentz (natural, resonance, Van Der Waal and Stark) broadenings respectively. The HWHM (Half Widths at Half Maximum) were given by Laux [16] for the Doppler effect, from Ali and Griem [17] for the resonance broadening, from Walkup [18] for the Van der Waals broadening and finally from Griem [19] for the Stark broadening including the corrections of Griem [19,20,21] to take into account the ionic influence. Due to the overlapping and a line by line treatment, we used two algorithms to spare time: a quicksort-insertion algorithm to sort the data before summation, and the Drayson algorithm [22] to integrate the relation (2).

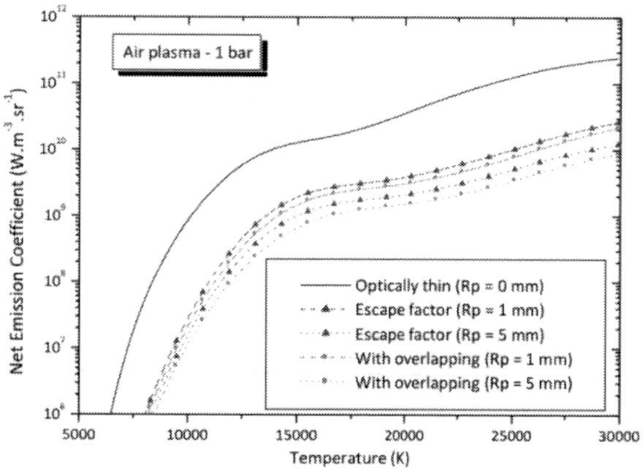

Figure 4. Influence of the atomic line's overlapping for different R_p and 1 bar.

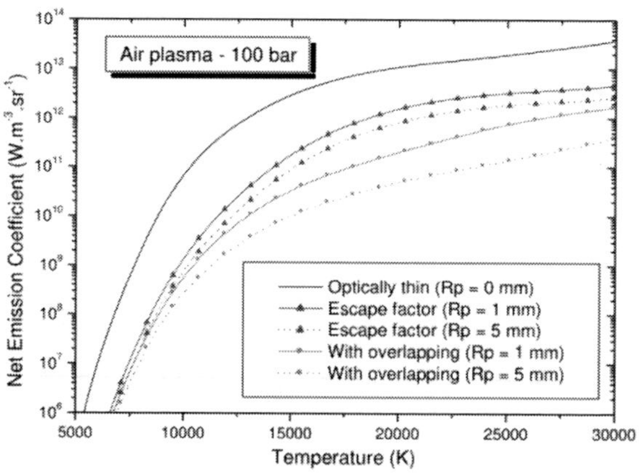

Figure 5. Influence of the atomic line's overlapping for optically thin air plasmas at 100 bar.

The last step to describe the total emission spectrum was a direct summation in the wavelengths range (30-4500 nm) with a 10^{-3} nm spectral step. The latter was validated by a comparison between the integration of the emission spectrum and the sum of the integrated atomic lines. As shown in *Figure 4*, the NECs obtained with and without overlapping are similar in the case of optically thin plasmas. We can also remark that the escape factor does not lead to an important error at low temperatures and low plasma's thicknesses R_p. At contrary, the use of this factor can lead to an overestimation of factor 2 for higher thicknesses at 1 bar. Moreover, the *Figure 5* highlights that discrepancies could overtake a factor 10 with the rise of pressure and/or thickness. This difference is mainly due to the total HWHM of the atomic lines which increases with temperature and/or pressure, causing a more important overlapping phenomenon.

Table 1. Molecular electronic systems considered in this study

	Electronic States	Calculated bands (v',v")	Number of rotational lines		Electronic States	Calculated bands (v',v")	Number of rotational lines
O_2	Schumann-Runge $B^3\Sigma_u^- - X^3\Sigma_g^-$	(19,21)	≈ 200.000	N_2	1st positive $B^3\Pi_g - A^3\Sigma_u^+$	(21,21)	≈ 300.000
NO	Beta $B^2\Pi_r - X^2\Pi_r$	(15,22)	≈ 550.000		2nd positive $C^3\Pi_u - B^3\Pi_g$	(4,19)	≈50.000
	Gamma $A^2\Sigma^+ - X^2\Pi_r$	(8,20)	≈ 300.000		Birge-Hopfield 1 $b^1\Pi_u - X^1\Sigma_g^+$	(19,0)	≈ 1.000
	Delta $C^2\Pi_r - X^2\Pi_r$	(4,21)	≈ 160.000		Birge-Hopfield 2 $b'^1\Sigma_u^+ - X^1\Sigma_g^+$	(28,0)	≈ 1.000
	Epsilon $C^2\Sigma^+ - X^2\Pi_r$	(5,22)	≈ 90.000		Caroll-Yoshino $c_4'\Sigma_u^+ - X^1\Sigma_g^+$	(8,0)	≈ 400
	Infrared $X^2\Pi_r - X^2\Pi_r$	(26,25)	≈ 120.000		Worley-Jenkins $c_3^1\Pi_u - X^1\Sigma_g^+$	(4,0)	≈ 200
N_2^+	1st negative $B^2\Sigma_u^+ - X^2\Sigma_g^+$	(12,20)	≈ 200.000		Worley $o_3^1\Pi_u - X^1\Sigma_g^+$	(4,0)	≈ 200
	Meinel $A^2\Pi_u - X^2\Sigma_g^+$	(9,10)	≈ 110.000				

3.4. Molecular bands

Finally, a particular attention was devoted to the radiation of the molecular bands. In one of our previous works [1], 9 molecular electronic systems (1788 vibration bands, 10^6 rotational lines) were considered in the visible and UV parts of the spectrum, taking into account 9 bound-bound molecular systems (Schumann-Runge system for O_2, first and second positive systems for N_2, β, γ, ε and δ systems for NO, first negative and Meinel systems for N_2^+). This work highlighted the fundamental role of molecules in the total emission of air plasmas. However, it was restricted to temperatures below 10kK, to optically thin plasmas, to Doppler broadening and to atmospheric pressure. Therefore, it was indispensable to extend this work so as to consider the radiation of molecular bands for higher pressures, higher temperatures and assuming or not the self-absorption. In this work we updated our spectroscopic database by adding the contribution of N_2 VUV (Birge-Hopfield 1 and 2, Carroll-Yoshino, Worley-Jenkins and Worley systems) and the NO Infrared system. All molecular systems were treated in the coupling Hund's case b. We have to precise here that this case is valid for Σ states, and tends to be valid when the rotational level increases for Π states. The molecular systems were then described by the quantum numbers Λ, S, N and J [23]. The total discrete molecular bands radiation was calculated from the main diatomic molecular systems referenced in Table 1 [24]. Therefore, we considered 15 bound-bound diatomic molecular systems in this study.

In order to describe the radiation spectrum of molecular bands by a line-by-line method, it was necessary to calculate the corresponding spectral emission coefficient for each wavelength as follows:

$$\varepsilon_\lambda = \frac{hc(E_{up}^{e'v'N'J'} - E_{down}^{e''v''N''J''})}{4\pi}\Phi\left(\frac{N(T)\exp(-\dfrac{hcE_{up}^{e'v'N'J'}}{kT})}{Q_{\text{int}}^Z(T)}\right)A_{e''v''}^{e'v'}S_{N''J''}^{N'J'}P_\lambda^{Voigt} \quad (3)$$

where $E_{up}^{e'v'N'J'}$ arre the total energies expressed in cm^{-1}, $N(T)$ is the total population number density of the molecule in m^{-3}, $Q_{\text{int}}^Z(T)$ is the total internal partition, Φ is the nuclear spin degeneracy factor of the emitting state [1], $A_{e"v"}^{e'v'}$ is the vibrational transition probability in s^{-1}, $S_{N"J"}^{N'J'}$ is the Hönl-London factor corresponding to the rotational transition involved and P_{λ}^{Voigt} is the Voigt profile in m^{-1}. In the formalism of Hund's case b, the numbers e, v, N and J correspond to electronic, vibrational, rotational and sub-rotational levels (spin splitting) respectively.

The line-by-line method used to describe the emission spectrum consists in determining for each rotational transition the emitting population number density, the line position, its integrated intensity and the line shape.

The population number density of the rotational states was assumed to be a Boltzmann distribution since the plasma was supposed to be in LTE (Local Thermodynamic Equilibrium).

The line position is related (in cm^{-1}) to the energy difference between two rotational levels and was calculated in the Klein-Dunham formalism (equilibrium spectroscopic constants for each electronic state). In the Born-Oppenheimer approximation, the expression of the energy levels is given by the expression:

$$E_{e,v,N,J} = \sum_{i,j} Y_{ij}(v+1/2)^i [F(N,J)]^j \tag{4}$$

The rotational term dependency F(N,J) = N(N+1) only depends on N in the case of a singlet transition (S=0). For multiplet states (spin-splitting) it not only depends on the parameter N but also on the sublevel J. Expressions for each electronic state were given in the literature [1,16,24]. Lambda-doubling was neglected for the Π states. For each electronic state (Table 1), we chose the most accurate and most recent spectroscopic constants given by Naghizadeh-Kashani [1], Chauveau [24] and Laux [16].

The integrated line intensity is proportional to the product of the vibrational transition probability (Einstein coefficient) and the Hönl-London factor which defines the relative rotational intensity between transitions allowed by the selection rules. The Hönl-London factors used in this work were those of Tatum [25], Schadee [26] and Hertzberg [23]. All factors were normalized according to the Whiting convention:

$$\sum_{J"} S_{N"J"}^{N'J'} = (2-\delta_{0,\Lambda'+\Lambda"})(2S'+1)(2N'+1) \tag{5}$$

We used the Einstein coefficients published by Laux [16] for most of the molecular systems. For the N_2 VUV systems (Birge-Hopfield 1 and 2, Worley-Jenkins, Worley and Caroll-Yoshino), we used the absorption oscillator strengths compiled by Chauveau [24]. The table published by Langhoff [27] was valuable for the infrared system of NO. Finally, we had to develop the classical method for the calculation of the vibrational transition probabilities for the electronic systems of the O_2 Schumann-Runge and NO-gamma systems since we obtained significant differences with the results published for these cases. Accordingly, the first step was the reconstruction by the RKR method [28] of the internuclear potential curves for each state. We extrapolated the RKR potential with a repulsive potential $V_{rep}(r)=a.r^{-b}$ for the shorter internuclear distances, and with a Hulburt-Hirschfelder potential (HH) for the longer internuclear distances. In order to calculate the vibrational wave functions $\psi_v(r)$, this potential was injected in the radial Schrödinger equation and the differential equation was then resolved by a method proposed by Cooley [29] using a Numerov iteration algorithm (two-step). This calculation was achieved for all couple (v',v"). Knowing the upper and lower rotationless vibrational wave functions, the vibrational transition probabilities were obtained as follows:

$$A_{e"v"}^{e'v'} = 2.026.10^{-6}(\sigma_{v'v"})^3\left(\frac{2-\delta_{0,\Lambda'+\Lambda"}}{2-\delta_{0,\Lambda'}}\right)\left[\int_0^{\infty}\psi_{v'}(r)R_e(r)\psi_{v"}(r)dr\right]^2 \tag{6}$$

The electronic transition dipole moments $R_e(r)$ used in the calculation were published by Friedman for the O_2 Schumann-Runge [30] and Langhoff for NO-gamma systems [31]. Our Einstein coefficients

were compared with published values [16], and the NECs obtained with our tables of coefficients were in good agreement with the literature for those two molecular systems [24].
Finally, **the line shape** of each line was calculated by a Voigt profile taking into account the Doppler and the collisional broadenings. The Doppler broadening for molecules obeys to the same formula as atomic case [16]. For collisional broadenings, we did not find enough information to cover our large pressure and temperature ranges. Consequently, for all diatomic molecular bands, we systematically used the semi-empirical approximation law to determine the molecular collisional broadenings HWHM (Half widths at Half Maximum) as a function of pressure:

$$\gamma_C(T,P)=\gamma(T_{ref}).P.(T_{ref}/T)^n \tag{7}$$

with $\gamma_{T_{ref}}=0.1$ and n=0.7 for all molecules as recommended by Breene [32].

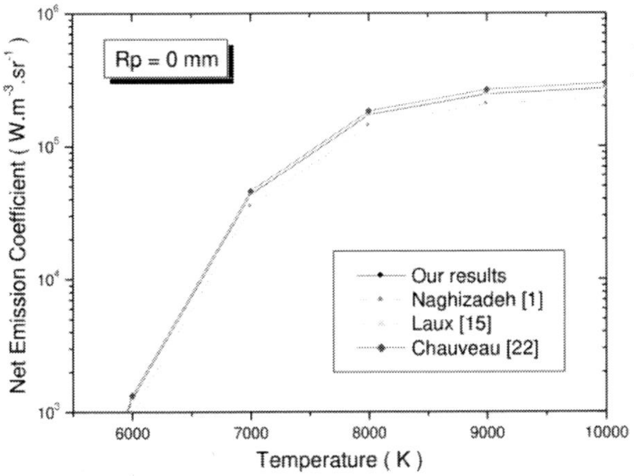

Figure 6. Comparison with the literature for the particular system N_2^+ 1st Negative system

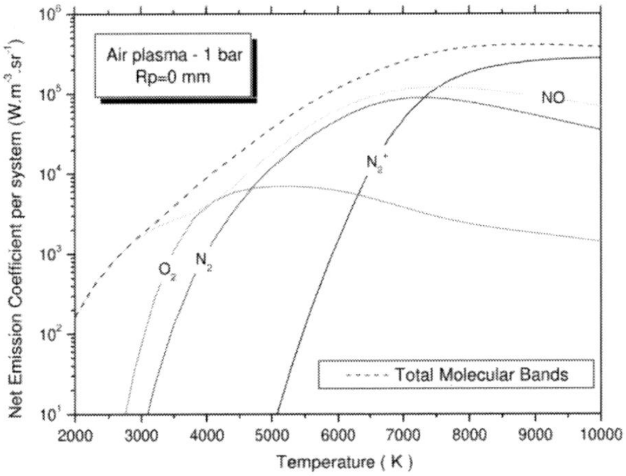

Figure 7. Contribution of the molecular diatomic species to the NEC.

The NECs of all molecular bands systems were validated by comparing them to previous works for a pressure of 1 bar (*Figure 6*). At most 80% of discrepancy was observed with Naghizadeh-Kashani [1], 30% with Chauveau [24] and 70% with Laux [16]. Our results almost systematically overestimated those of Naghizadeh and Laux and underestimated those of Chauveau. These discrepancies are mainly caused by the number of vibrational levels considered in the Tables of the Einstein coefficients. Indeed, this number differs according to the used methods for the limitation of the rotational levels for each vibrational level, and/or according to the spectroscopic constants chosen to describe the energy levels. In this study, we defined this maximum number by comparing the vibrational energy with the dissociation limit. The limitation of the rotational levels for each vibrational state was performed by comparison of the total rotational states energies $E_{e,v,N,J}$ with the local maximum of the rotational-vibrational potential curve [33]. *Figure 7* highlights that N_2^+ 1st negative system is highly responsible for the molecular radiation at 10kK whereas O_2, NO and N_2 systems are above 7kK.

4. Results obtained with the Net Emission Coefficient (NEC)

The plasma's radiation was obtained according to the Net Emission Coefficient (NEC) which consists in calculating the net emission at the center of an isothermal and homogenous sphere of radius R_p.

$$\varepsilon_N(T,R_p)=\int_0^\infty L_\lambda^o(T)\cdot K'_\lambda(T)\cdot\exp(-K'_\lambda(T)\cdot R_P)\cdot d\lambda \tag{8}$$

where λ (m) is the wavelength, T and P are the local temperature and pressure respectively, $L^o_\lambda(T)$ (W/m^2/sr/m) is the Planck function, R_p is the size of the sphere assimilated to the plasma's size and $K'_\lambda(T)$ (m^{-1}) is the monochromatic absorption coefficient corrected by the induced emission and correlated with the local emission coefficient by the Kirchhoff law. Furthermore, due to the exponential term, radiation's self-absorption in the emitting regions is rather well estimated.

In numerical modelling, the NEC is a good approximation to characterize the radiation losses in hot regions while taking into consideration the absorption. Our previous studies also showed that this coefficient not only depended on temperature and pressure but also on the plasma size.

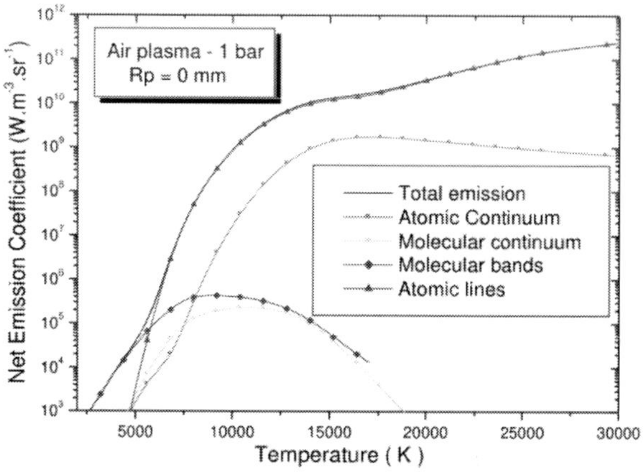

Figure 8. NEC for optically thin air plasmas at 1 bar.

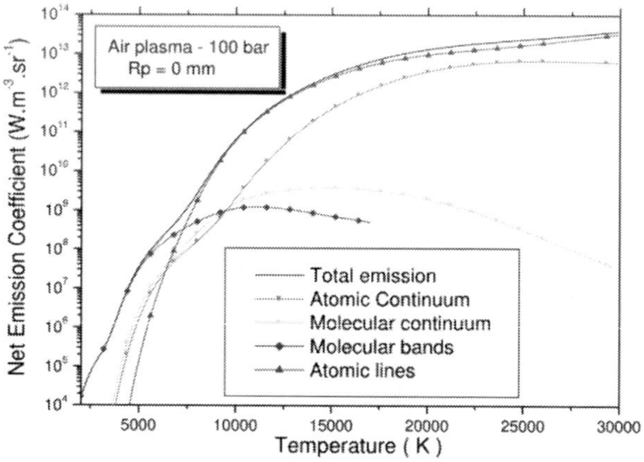

Figure 9. NEC for optically thin air plasmas at 100 bar.

Figures 8 and 9 represent the NEC and the contribution of each radiative mechanism for optically thin air plasmas at 1 bar and 100 bar. We observe that the pressure increase does not significantly modify the relative contribution of each process. We also remark that molecular bands are predominant at low temperatures, up to 5kK for 1 bar and up to 7.5kK for 100 bar. In this case, atomic lines are not self-absorbed and consequently contribute to the radiation at high temperature.

Figure 10 shows the contribution of each process to the NEC for various plasma sizes at 100 bar. We note that the self-absorption of the atomic lines hardly impacts on the NEC due to the absorption coefficient in the exponential term (8) which increases with pressure. The contribution of the atomic lines is hardly predominant for optically thin plasmas at high temperatures and low pressures. When the pressure increases, the absorption coefficient and the broadening parameters of each line increases too. The self-absorption becomes so important that atomic continuum and molecular bands significantly extend their ascendancy domains when the thickness of plasma increases.

Figure 11 illustrates this phenomenon showing the contribution of all molecular bands to the NEC. We observe a shift of the molecular contribution towards higher temperatures depending on the pressure and the plasma's thickness. The molecular bound-bound contribution could be neglected above 10kK only for optically thin plasmas. For those reasons, a particular attention has to be paid to realise an accurate calculation of each diatomic electronic system involved in the radiation of air plasmas.

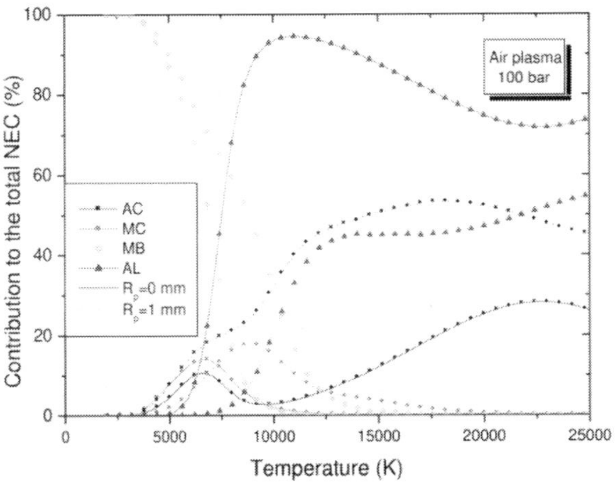

Figure 10. Contribution of the radiative processes to the NEC for two R_p and P=100bar. (*AC : Atomic continuum, MC : Molecular continuum, MB : Molecular Bands, AL : Atomic lines*)

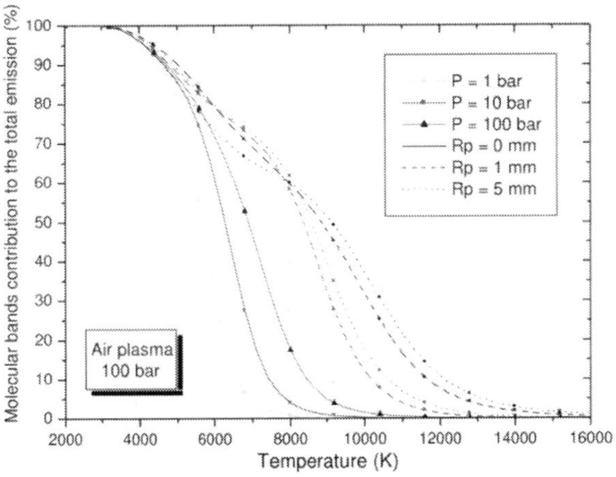

Figure 11. Total contribution of the molecular bands to the NEC for different pressures and plasma thicknesses (R_p).

5. Conclusions

The aim of this paper was to calculate the Net Emission Coefficient of air plasmas for pressures up to 100 bar. First, we upgraded our composition algorithm to better estimate the effects of high pressures. Then, we compared our NECs obtained by a line-by-line calculation or using the escape factor. We showed that the differences between these two approaches rise with the pressure due to the overlapping of the atomic lines. Finally, we showed that the molecular contribution, often neglected in this kind of works, has to be accurately calculated until approximately 15kK. A particular attention has to be paid to model these bound-bound molecular bands in the case of air plasmas.

6. References

[1] Naghisadeh-Kashani Y, Cressault Y and Gleizes A 2002 *J. Phys. D: Appl. Phys. 35* 2925
[2] Felderman E J, MacDermott W N and Fisher C J 1996 *Journal of Propulsion and Power* 12 (6) 1084
[3] Sagnier P and Vérant J L 1998 *Aerospace Science and Technology* 7 425
[4] Lowke J J 1974 *J.Q.S.R.T* **14** (2) 111
[5] Godin D and Trépanier J Y 2004 *Plasma Chem. Plasma Proc.* **24** 447
[6] Drellishak K S, Aeschliman D P and Bulent Cambel A 1965 *Physics of Fluids* **8** 1590
[7] D'Angola A, Colonna G, Gorse C and Capitelli M 2008 *The European Physical Journal D* **46** 129
[8] Cressault Y, Hannachi R, Teulet Ph, Gleizes A, Gonnet J P and Battandier J Y 2008 *Plasma Sources Sci. Technol.* **17 035016**
[9] Gleizes A, Gonzalez J J, Liani B and Raynal G 1993 *J. Phys. D: Appl. Phys.* **26** 1921
[10] Fennelly J A and Torr D G 1992 *Atomic Data and Nuclear Data Tables* **51** (2) 321
[11] Lee L C, Carlson R W, Judge D L and Ogawa M 1973 *J.Q.S.R.T.* **13** 1023
[12] Hudson R D 1971 *Reviews of Geophysics* **9** (2) 305
[13] Moore C E 1949 *Atomic energy levels, Vol I.* (Washington, Circular of the National Bureau of Standards)
[14] Ralchenko Y, Kramida A E, Reader J and NIST ASD team (2008) 2009 *NIST Atomic spectra database* (version 3.1.5), available online at http://physics.nist.gov/asd3, National Institute of Standards and Technology, Gaithersburg, MD

Journal of Physics: Conference Series **406** (2012) 012010 doi:10.1088/1742-6596/406/1/012010

[15] Kurucz R L and Peytremann E 1995 *Atomic line data* (Kurucz R L and Bell B) Kurucz CD-ROM N0.23, Cambridge, Mass: Smithsonian Astrophysical Observatory, available online at http://cfa-www.harvard.edu/amp/ampdata/kurucz23/sekur.html

[16] Laux C O 1993 *Optical diagnostics and radiative emission of air plasmas* (Ph.D Thesis, HTGL report T-288, Stanford University, California)

[17] Ali A W and Griem H R 1965 *Physical Review* **140** 4A 1044

[18] Walkup R, Stewart B and Pritchard D E 1984 *Physical Review A* **29** 1 169

[19] Griem H R 1974 *Spectral line broadening by Plasmas* (New York, London, Academic press)

[20] Traving G 1968 *Interpretation of line broadening and line shift in Plasma diagnostics* (North-Holland, W. Lochte-Holtgreven, 66-134)

[21] Griem H R 1964 *Plasma spectroscopy* (New York, McGraw-Hill)

[22] Drayson S R 1976 *J.Q.S.R.T.* **16** 611

[23] Hertzberg G 1950 *Molecular spectra and molecular structure I. Spectra of diatomic molecules* (New York, Van-Nostrand Company)

[24] Chauveau S, Perrin M Y, Rivière Ph and Soufiani A 2002 *J.Q.S.R.T.* **72** (4) 503

[25] Tatum J B 1966 *Canadian Journal of Physics* **44** (11) 2944

[26] Schadee A 1964 *Communication from the Observatory at Utrecht* B.A.N. **17** (5) 331

[27] Langhoff S R, Bauschlicher C W Jr and Patridge H 1994 *Chemical Physics Letters* **223** 416

[28] Rydberg R 1931 *Ann. Physik* **73** (376), Klein O *Z. Physik* **76** (226), Rees A L G 1947 *Proc. Phys. Soc.* **A59** (998)

[29] Cooley J W 1961 *Math. Computation* **15** (76) 363

[30] Friedman R S 1989 *JQSRT* **43** 3

[31] Langhoff S R, Bauschlicher C W Jr and Patridge H 1988 *Chemical Physics Letters* **89** 4909

[32] Breene R G 1967 Applied Optics **6** 141

[33] Babou Y, Rivière Ph, Perrin M Y and Soufiani A 2009 *International Journal of Thermophysics* **30** (2) 416

Detailed Numerical Simulation of Single-Walled Carbon Nanotube Synthesis in a Radio-Frequency Induction Thermal Plasma System

S Arabzadeh Esfarjani[1], S B Dworkin[2], J Mostaghimi[1], K S Kim[3], C T Kingston[3], B Simard[3], G Soucy[4]

[1] Department of Mechanical and Industrial Engineering, University of Toronto, M5S 3G8, Canada
[2] Department of Mechanical and Industrial Engineering, Ryerson University, M5B 1G8, Canada
[3]Emerging Technologies Division, Security and Disruptive Technologies Portfolio, National Research Council of Canada, Ottawa, ON K1A 0R6, Canada
[4]Department of Chemical Engineering, Université de Sherbrooke, J1K 2R1, Canada

Email: mostag@mie.utoronto.ca

Abstract. 2D axisymmetric numerical calculations are conducted to model the thermo-fluid fields and chemical reactions leading to the formation of SWCNTs in an RF plasma system. A modified version of the SWCNT "reduced" chemical model is used to estimate the formation of SWCNT in an RF plasma system for the first time. The "reduced" model incorporates 14 species and 36 chemical reactions to predict the formation of metal and carbon clusters and SWCNTs. By combing the chemistry model into the RF plasma CFD code, the formation and development of carbon and metal catalyst clusters and their reactions which produce SWCNTs are shown. The chemistry model is shown to under-predict the yield rate of SWCNT. In order to better predict the yield rate, a sensitivity analysis is performed to modify the dominant reaction rates. The modified model predicts the yield of SWCNTs correctly within the range reported experimentally. However, more studies should be conducted to validate the accuracy of the model for different operating conditions.

1. Introduction

Carbon nanotubes have been at the forefront of nanotechnology since their discovery in 1991. These nanomaterials are expected to make a major impact particularly in the composite material fields such as aerospace, automobile, and personal protective industries, where strength and weight of materials are important factors. A new method for the production of SWCNTs at large-scale that is receiving much attention is the induction thermal plasma process, developed initially for fullerene synthesis [1-3].

Figure 1 shows a schematic of an induction thermal plasma system for the continuous production of SWCNTs. The synthesis system mainly consists of a plasma torch, a reaction chamber and a filtration system. Three different gas streams of powder, central, and sheath gases are introduced into the induction plasma torch. The central gas has a swirl component to stabilize the plasma. The mixture of carbon black and metallic catalyst powder (nickel) is carried into the plasma region by the powder gas. The reaction chamber includes a graphite liner, heating elements, and a thermal insulator, which are employed for an active control of maximum achievable temperature, background temperature, and cooling rate. The final products are then collected in a filtration chamber, which consists of three porous metallic filters.

A detailed kinetic mechanism for SWCNT production in numerical simulations of an RF plasma system is required for understanding the process and identifying the parameters that mainly influence nanotube production. In the present study, a "reduced" form of a fullerene/soot/metal cluster model developed for the arc discharge and laser ablation processes [4] has been employed and modified to describe the SWCNT formation kinetics in this RF plasma process, and applied to a CFD code which models the 2-D thermo-fluid field of an RF plasma system with chemical reactions [5, 6].

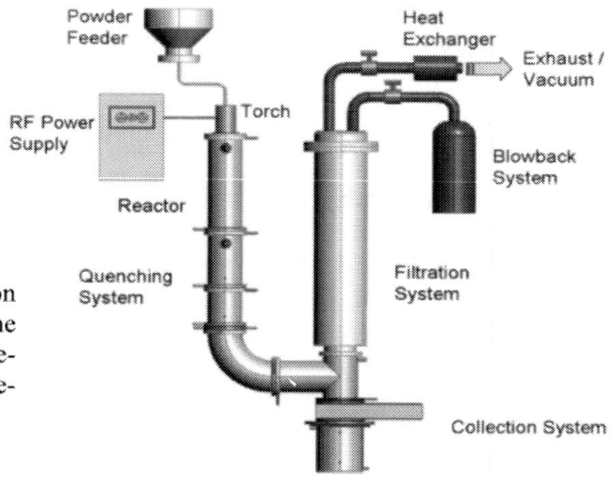

Figure 1. Schematic layout of an induction thermal plasma processing system for the continuous production of high quality single-walled carbon nanotubes (SWNT) at large-scale

2. Model Description

In the present work, the models of the RF plasma generation, and the interaction between the plasma and injected feedstock material. i.e., carbon and nickel particles are considered in a 2D axisymmetric numerical framework. The kinetic mechanism of the formation of SWCNTs and several intermediate species was approximated by using the carbon/metal cluster model (see section 2.6) developed by Scott [4]. In order to integrate the kinetic model into the reaction chamber's thermo-flow field model, a separate in-house gas phase parallel computational fluid dynamics code [7-8] was used. The parallel computing code was primarily developed for computing the detailed chemical kinetic mechanisms of soot formation in flames [7-8], and can perform numerous fast computations on multiple computer processing units (CPUs).

2.1. Radio Frequency Plasma Model

A 2-dimensional axi-symmetric model for simulations of plasma generation by an RF induction plasma torch has been developed based on the procedure described by Mostaghimi et al. [9] and Proulx et al. [10]. The following assumptions were considered in modeling the induction plasma: (i) two-dimensional, axisymmetric and steady-state flow; (ii) quasi-neutral and optically thin plasma in a local thermodynamic equilibrium (LTE) state; (iii) negligible displacement current and viscous dissipation; (iv) planar coils; (v) turbulent flow; (vi) consideration of the momentum and energy source terms due to the particle loading. Based on these assumptions, the plasma equations are described by steady-state conservation equations for the transport of mass, momentum, energy, concentration of species and the k-$\odot$ turbulence model coupled with electromagnetic field equations.

The axisymmetric cylindrical coordinate system (r, z) was adopted in this model. The governing equations are similar to those in previous works [11, 12].

2.2. Plasma-Injected Particle Interaction Model

In the induction thermal plasma process, the carbon precursor material is directly evaporated in the plasma plume, and therefore, the yield rate of SWCNT in the reactor can be affected by the evaporation efficiency of the injected particles. The Lagrangian approach was used to describe the injected particles inside the plasma [13, 14]. The following simplifications were assumed in this work: (i) spherical particles were assumed; (ii) a uniform temperature distribution inside a particle was assumed; (iii) particle sizes were assumed to follow the Rosin-Rammler distribution function; (iv) the rarefaction effect (Knudsen effect) was not taken into account [15]. The detailed formulations can be found in previous works [11, 12].

2.3. Computational domain and operating condition

The computational domain in the present work, figure 2, consists of two zones; an RF plasma torch, and a reaction chamber. A commercial RF plasma torch considered in this work (PS-50, TEKNA Plasma Systems, Inc.), is driven by a plate power of 45 kW with the coupling efficiency of about 70% which is operated at an oscillator frequency of 3 MHz.

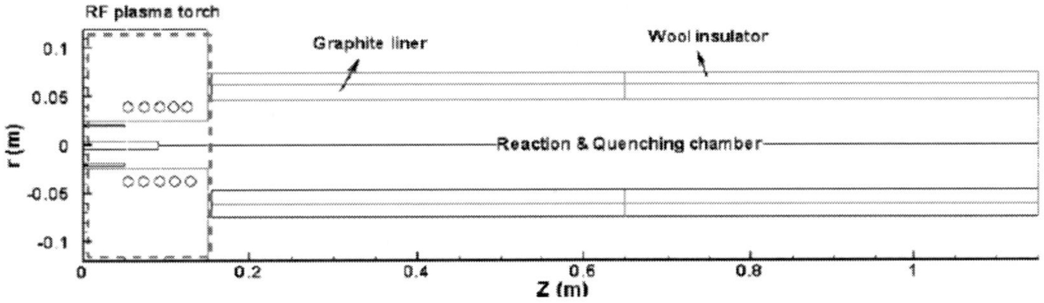

Figure 2. The computational domain including RF plasma torch and the reaction chamber

The detailed dimensions of the reactor system can be found in the previous studies [11,16]. In this work, the length of the graphite liner and insulator (wool) in the reaction and quenching chamber was extended from 50 cm to 1 m in accordance with the new experimental setup being used at NRC. A typical operating condition used in this numerical study is summarized in table 1.

Table 1 A typical operating condition employed in this numerical study

Net plasma power (kW)	Pressure (kPa)	Carrier gas (slpm)	Central gas (slpm)	Sheath gas (slpm)	Feed rate (g/min)
31.5	66	3 (Ar)	23 (Ar)	135 (He)	1.5 (CB/Ni: 91/9wt.%)

A mixture of carbon black (CB), and nickel was considered as a feedstock material, with a ratio of CB/Ni = 91.0 wt%/9.0 wt%, mean diameter of 2 μm, and minimum and maximum diameters of 1 and 3 μm, respectively.

2.4. Chemistry model

The fullerene/soot/metal clusters model was developed by Scott [4] in predicting the production of nanotubes in arc vaporization and laser ablation systems [18-20]. Scott's proposed model is based on the fullerene model of Krestinin et al. [20] with a modification on the soot formation component. Their model is based on the formation of clusters of carbon that are built up from atoms to carbon chains, rings and polycycles, fullerene shells, C_{60} and C_{70}, and then soot. They give detailed gas-phase reactions for producing carbon clusters and fullerenes as well as for fragmentation and evaporation of clusters. The final production of carbon cluster growth is soot that grows via heterogeneous reactions according to its surface area. Scott [4] added combined carbon-nickel clusters to the fullerene model of Krestinin et al. [20].

In Scott's carbon-nickel clusters model, the soot-nickel clusters (ZNiC) are nuclei for SWCNT formation and growth. In the model, the formation of SWCNTs was assumed to occur from Ni_{1024} clusters (NiC) reacting with soot (Z) and fullerene (C_{60F}) as feedstocks. The size of the nickel cluster was approximately chosen based on the diameter of SWCNT. After the formation, the clusters are assumed to combine with large carbon clusters, namely soot or fullerene, to form SWCNTs. The model cannot represent the real number of nanotubes of different lengths, but the total amount of carbon in SWCNTs (the mass fraction) can be meaningful in the post-processing analysis.

The "reduced" model includes 14 species and 35 reactions. The reduced kinetics model has been previously used in CFD simulations of a multi-cycle laser ablation method in the production of nanotubes and the model was shown to be computationally inexpensive and robust compared to the full CNT model [4]. The reactions in this work are from "reduced" chemistry model by Scott [4], reproduced in table 2. The model, as described by Scott [4], is phenomenological and assumes constant reactions rates of formation of nickel-carbon clusters, approximating their kinetic rates, since there was no available data on temperature-dependency. All SWCNTs in the model are assumed to consist of 999 carbon atoms. This assumption means the model cannot represent the real number of nanotubes of different lengths, but the total amount of carbon in SWCNTs (the mass fraction) can be meaningful in the post-processing analysis.

Table 2. Reduced carbon vapor/nickel catalysts model [4]

$k = A\ T^b\ exp(-E/RT)$

E [K], A [mole-cm-sec-K]

	Reactions	A	b	E
1	$C + C = C_2$	2.00E+14	0	0
2	$C + C_2 = C_3$	2.00E+14	0	0
3	$C_2 + C_2 = C_3 + C$	2.00E+15	0	9040
4	$C_3 + C_3$ □ 0.15CC	2.00E+14	0	0
5	CC+C=>1.025CC	2.30E+14	0	0
6	CC + C_2 □ 1.05CC	2.30E+14	0	0
7	CC + C_3 □ 1.075CC	2.30E+14	0	0
8	CC □ 0.95CC + C_2	3.20E+13	0	61900
9	CC + C_3 □ 0.70C_{60F} + C	2.00E+13	0	0
10	CC + C_2 □ 0.70C_{60F}	4.00E+09	0	-30196
11	CC + C □ 0.6833333C_{60F}	2.00E+13	0	0
12	C_{60F} □ 1.45CC + C_2	8.00E+13	0	61900
13	$C_{60F} + C_3$ □ 0.7875Z	2.00E+13	0	10065
14	CC + CC □ Z	4.00E+13	0	0
15	Z + C_3 = 1.0375Z	4.00E+12	0	0
16	Z + C_2 = 1.0250Z	4.00E+12	0	0
17	Z + C = 1.0125Z	4.00E+12	0	0
18	Z □ DZ	1.00E+02	0	0
19	Ni + Ni □ Ni_2	3.70E+12	0.5	0
20	Ni_2 + Ni □ Ni_3	4.10E+12	0.5	0
21	$Ni_2 + Ni_2$ □ 0.0039062NiC	4.40E+12	0.5	0
22	Ni_3 + Ni □ 0.0039062NiC	4.51E+12	0.5	0
23	$Ni_3 + Ni_2$ □ 0.0048828NiC	4.37E+12	0.5	0
24	$Ni_3 + Ni_3$ □ 0.0058593NiC	4.45E+12	0.5	0
25	Ni + NiC □ 1.0009766NiC	8.04E+13	0.5	0

26	Ni_2 + NiC □ 1.0019531NiC	5.96E+13	0.5	0
27	Ni_3 + NiC □ 1.0029297NiC	5.03E+13	0.5	0
28	Ni_2 □ Ni + Ni	4.30E+17	-0.5	24506.6
29	Ni_3 □ Ni_2 + Ni	3.37E+17	-0.5	28145.3
30	2NiC □ NiC + 1024Ni	6.61E+18	-0.5	44780
31	ZNiC □ 0.9990234ZNiC + Ni+ 0.0009766Z	6.61E+18	-0.5	44780
32	Z + NiC □ ZNiC	1.00E+16	0.5	0
33	ZNiC □ DZNiC	1.00E+04	0	0
34	Z + ZNiC □ 0.08008008008CNT+ZNiC	1.00E+12	0.5	0
35	C_{60F} + ZNiC □ 0.06006006006CNT + ZNiC	1.00E+13	0.5	0
36	C_2 + ZNiC □ 0.002002002002CNT + ZNiC	1.00E+12	0.5	0

Table 2-1. Nomenclature and definitions of species in reaction mechanisms [4]

NiC	Nickel cluster (Ni_{1024})
Z	soot (C_{80})
DZ	dead soot (C_{80})
C_{60F}	fullerene
ZNiC	soot-nickel cluster ($C_{80}Ni_{1024}$)
DZNiC	dead soot-nickel cluster ($C_{80}Ni_{1024}$)
CNT	carbon nanotube (C_{999})

2.5. Computational Scheme and Boundary Conditions

In the computational domain shown in figure 3, the no-slip boundary conditions are applied to the torch and reaction chamber walls. The temperature on the outside of the reaction chamber is set to room temperature, i.e., 300K. At the torch and reactor axes, the radial velocity and the radial gradients of all other variables were set zero, while axial gradients of the variables were assumed zero at the reactor exit. Lastly, the boundary conditions needed for the electromagnetic field calculation inside the torch were based on Xue's approach [6].

First, the 2D axisymmetric thermo-fluid fields inside the plasma torch and reaction chamber were modeled by the commercial code FLUENT [21]. The thermo-flow field equations, solved inside the reactor, were continuity, conservation of momentum, conservation of energy, and conservation of species. The k-ε turbulent model was implemented for the flow inside the reactor. FLUENT solves the discretized forms of the governing equations by a SIMPLE-like algorithm proposed by Patankar [22]. The vector potential equation for the electromagnetic field was solved by using the User Defined Functions (UDFs) approach. The Discrete Phase Module (DPM) of FLUENT was used to consider the interaction between plasma and the injected particles. In total, 70 particles of each nickel and carbon black were released from the probe located at the center of the torch and the injection velocity of the particles was assumed to be equal to the axial velocity of the carrier gas. For the given operating conditions in table 1, all the injected particles were completely evaporated before leaving the reactor. In FLUENT the thermodynamic and transport data for the plasma gases (i.e., argon and helium) were obtained from the tabulated data [23], which were calculated with the assumption of LTE. The effects of the metallic and carbon vapors on the plasma properties - except on the radiation and electrical properties - were taken into account using the available data for carbon, and nickel vapors [25]. In the second part of the simulations, only the thermo-flow field results from FLUENT simulations for the reaction chamber is interpolated into the mesh generated by the parallel computing code [7] where the chemical kinetic mechanism, described in section 2.4, for the formation of SWCNT is included. In the parallel computing code, the finite volume method is used to discretize the equations of conservation of species mass and energy. A staggered mesh is used with a semi-implicit scheme to solve the discretized equations [22]. The diffusive terms are discretized using a second-order central difference scheme while the convective terms are discretized using a power law scheme [22]. At each time step, all the gaseous species including those generated from the reactions in the chemical kinetic model were solved in a coupled way at each control volume to handle the stiffness of the system. In order to speed up convergence, only the species equations were iterated and after every few thousands iterations, the temperature field was updated. In the parallel computing code, the thermal properties of the gaseous species and chemical reaction rates are obtained using CHEMKIN subroutines. Distributed-memory parallelization with strip-domain decomposition is employed. The computational domain is divided uniformly into 64 subdomains with 3 columns of control volumes per CPU as shown in figure 3. The algorithm uses the Message Passing Interface (MPI) library to parallelize the code. The computations are performed on the General Purpose Cluster (GPC) systems of SciNet [26], on eight 8-processor nodes on the infiniband network with 2.53 GHz chip speeds.

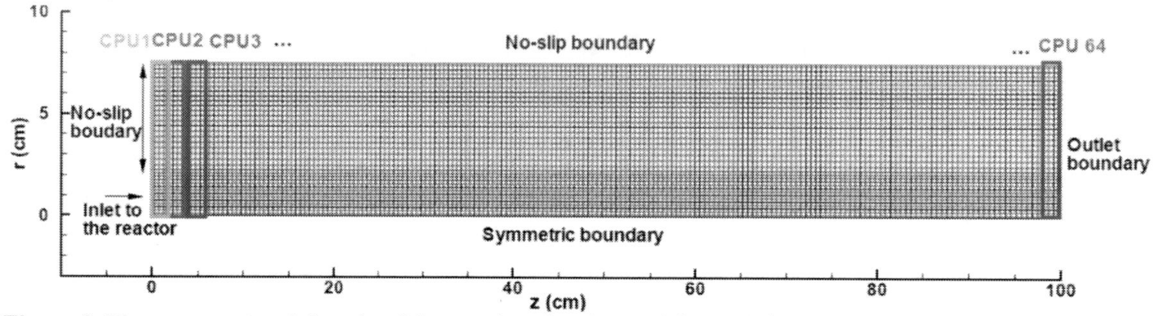

Figure 3. The computational domain of the reaction chamber and the mesh in the parallel computing code

The results are divided into two main sections; in the first section the simulations of the thermo-fluid filed inside the RF plasma torch and the reaction-quenching chamber are shown and the results from the implementation of the chemical kinetics model into the CFD code for predicting the formation of carbon-nickel clusters and carbon nanotube (specifically SWCNT) are presented. A sensitivity analysis on the chemistry model was conducted to adjust (within the uncertainty range) the rate reactions of the most dominant reactions in order to compare the SWCNT yield with the experimental data. In the second section (sec 3.3), the CFD simulations were repeated using the modified chemistry model and the results were compared with the experimental data.

3. Results and Discussion

3.1. CFD results with original chemistry model

The distributions of the temperature and species concentration fields are shown in figures 4. The off-axis profiles of the high temperature zones are clearly observed for both cases due to the skin effect. The central region is also cooled down by the presence of injected powders. Once the plasma flow leaves the torch it expands into the reaction chamber and enhances the mixing phenomena inside the reaction chamber, the off-axis temperature peak begins to disappear. The presence of the graphite liners throughout the entire length of the chamber helps to maintain the higher temperature region (>1400 K) inside the reactor longer and therefore lengthen the favorable temperature window for the growth of SWCNTs. Figure 4b shows the distribution of the carbon and nickel mass fractions inside the reactor system. The carbon and nickel vapours produced in the hot regions are transported to the cooler regions via diffusion and convection, and the carbon and nickel atoms start to form various carbon clusters, nickel clusters, nickel-carbon composite clusters, and SWCNTs.

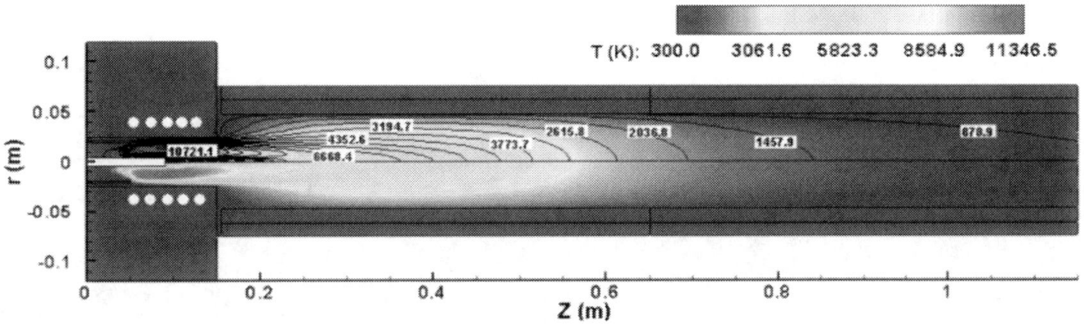

Figure 4a. Temperature field in the torch and the reaction chamber

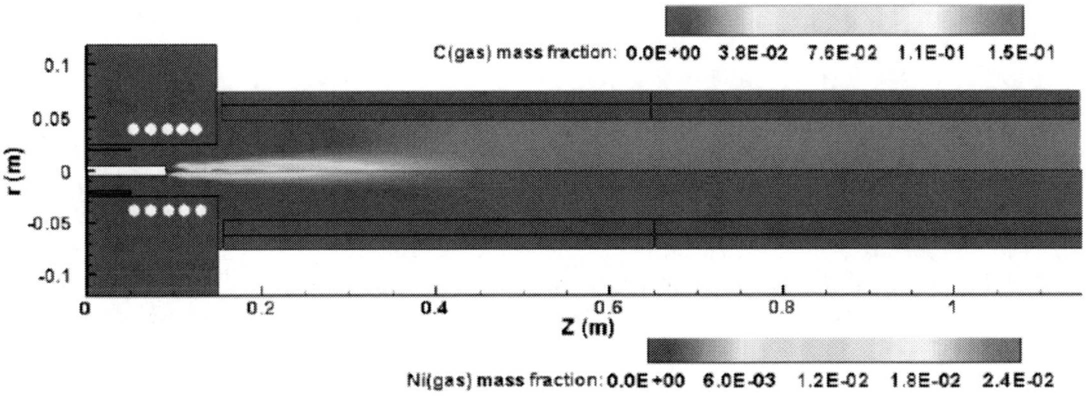

Figure 4b. Carbon and Nickel vapor mass concentrations in the reaction chamber

In the next step in the modeling, the calculated thermo-fluid fields inside the reaction chamber are used as initial conditions and interpolated over the grid generated by the parallel computing code (see figure 3). Similarly, the data extracted at the exit of the torch are used as the boundary condition at the reaction chamber's inlet in the parallel computing code and the chemical kinetic model for the formation of intermediate species and SWCNT, table 2, was added. Then, the thermo-fluid fields with the chemical reaction mechanism were iterated until a converged solution was obtained.

Figure 5 illustrates the mass fraction of carbon nanotubes (specifically SWCNTs) generated inside the reaction chamber. The main concentration is located away from the centreline and close to the reactor's wall where the temperature gradient is steep. The production of SWCNTs in the RF plasma processing system is estimated about 10~20 wt% of the injected carbon black material in the experiments. According to operating conditions of this work, with a feed rate of 81.9 g/h of carbon black, the SWCNT mass flow rate obtained from the experiments is between 8.6-17.2 g/h. To evaluate the predicted results from the chemistry model, the mass flow rate of SWCNTs at the exit of the reaction chamber was calculated as follows:

$$\dot{m}_{CNT}\left[\frac{g}{h}\right] = 3600\left[\frac{s}{h}\right] \times \sum_{j}\left(A_j \times U_j \times \rho_j \times Y_{CNT\,j}\right) \quad (16)$$

where A_j is the cross section area of cell j at the exit of the reactor; $Y_{CNT\,j}$ is the mass fraction of carbon nanotubes in the cell j at the exit of the reactor; U_j and ρ_j are the axial velocity and density of the mixture in cell j. The yield rate of SWCNT predicted by the model was 0.548 g/h. It is assumed that this under-prediction of yield relates to the chemistry model and the wide uncertainty ranges of reaction rates. In order to better predict the yield of SWCNTs, in the next section a sensitivity analysis was performed on the chemistry model to identify the dominant reactions and their role in production of carbon nanotubes in the simulations.

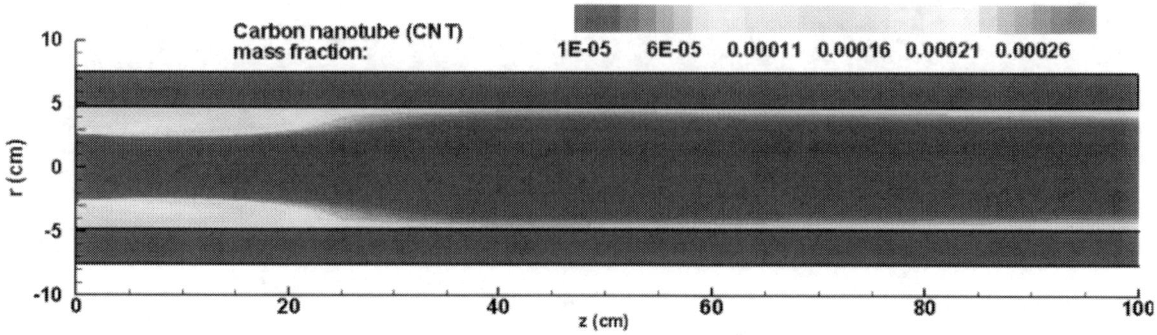

Figure 5. SWCNT mass concentration in the reaction chamber

3.2. Sensitivity Analysis of the chemistry model

In sensitivity analysis, the sensitivity of the model's results to a small perturbation of a single input is studied. Sensitivity analysis and absolute rate of production parameters provide useful tools to assess the importance of different reactions and pathways on the formation of desired final products in the system. A zero-dimensional test calculation and a sensitivity analysis on the original reduced chemistry model of Scott [4] were performed in CHEMKIN-PRO [27] to identify the most important reactions in the chemistry model that influence the yield rate of CNTs. The chemical rate equations given in table 2 were calculated using the Perfectly Stirred Reactor (PSR) model in CHEMKIN-PRO. The mass-weighted average of the temperature and species concentrations and the area-weighted average of the axial velocity profiles at the inlet of the reaction chamber in an RF plasma processing system were calculated and employed as the inlet boundary conditions in the zero-dimensional. A steady state condition was assumed to analyze the reaction rates in different temperature regimes, and the temperature inside the PSR model was reduced from 4000 K to 1200 K. The residence time in each case was set to 30 seconds. Table 3 shows the sensitivity of SWCNT formation to the most three dominant reactions in different temperature regimes. It can be seen that at temperatures higher than 2000 K, the most dominant mechanisms are the formation of Ni_2 from nickel atoms and the reverse reaction of evaporation of Ni_2 to nickel atoms. However, since the production rate of nanotubes at this temperature range is negligible, the sensitivity to the reaction Ni_2 □ 2Ni is less important. As the temperature drops under 2000 K, the dominant mechanisms are the formation of CNT from fullerene (C_{60F}) and nickel-carbon cluster (ZNiC), i.e., C_{60F} + ZNiC □ 0.06 CNT + ZNiC , and the nickel-carbon clusters inerting, i.e., ZNiC □ DZNiC. The latter reaction has an adverse effect on the formation rate of CNTs by deactivating nickel-carbon clusters. Based on the present analysis, one may conclude that the nanotube formation in the chemistry model is mainly dominated by the one-way reaction: C_{60F} + ZNiC □ 0.06 CNT + ZNiC and the limiting parameter in the formation rate of nanotubes is the concentration of nickel- carbon clusters (ZNiC).

Table 3. The dominant reactions and the sensitivity analysis results

Reactions / T (K)	1200	1250	1300	1400	1500	2000	2500	3000	3500	4000
Ni + Ni □ Ni_2						1.18	2.5	8.7	0.3	98
Ni_2 □ Ni + Ni						-1.18	-2.5	-8.7	-0.3	-98
ZNiC □ DZNiC	-0.945	-0.95	-0.956	-0.968	-0.979				0.998	
C_{60F} + ZNiC □ 0.06CNT + ZNiC	0.945	0.95	0.956	0.968	0.979					

The actual reaction rate coefficients for SWCNT formation and for the rate at which carbon and carbon clusters become over coated with carbon and become poisoned are not accurately known. Therefore in the following test cases the reaction rates of those mechanisms believed to be the most important, were adjusted within their range of uncertainty to gain more realistic production of SWCNT. The most noticeable change in SWCNT mass fraction at temperatures lower than 2000 K was by decreasing the Arrhenius coefficient by one order of magnitude from 10^4 to 10^3 in the rate reaction of ZNiC inerting. This change resulted in higher concentration of active carbon-nickel clusters (ZNiC) that can participate in the reaction mechanisms for producing SWCNTs.

3.3. CFD results with the modified chemistry model

To assess the effect of the change in the reaction rate of carbon-nickel clusters deactivation on SWCNT yield, the CFD calculations in the reaction chamber were repeated with the modified reaction rate. Figure 6 shows the mass concentration of SWCNTs in the reaction chamber with the modified rate. The maximum concentration of nanotubes was increased compared to the previous simulation as there are more active carbon-nickel clusters present in the system.

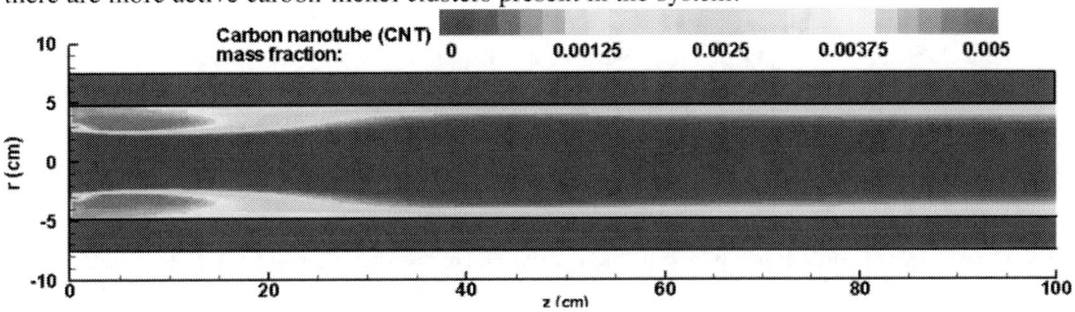

Figure 6. SWCNT mass concentration by the modified chemistry model in the reaction chamber

As can be seen from figure 6, SWCNTs start to form in a small area near the reactor's inlet and close to the graphite liners. The lower temperature region generated in this area, due to the expansion of the flow entering the reaction chamber, promotes the formation of carbon-nickel clusters (ZNiC) as shown in figure 7. The calculated value of the mass flow rate of SWCNTs at the exit of the reactor in the new simulation is about 12 (gr/min) which is in the range of 8.6-17.2 g/h of the experiments.

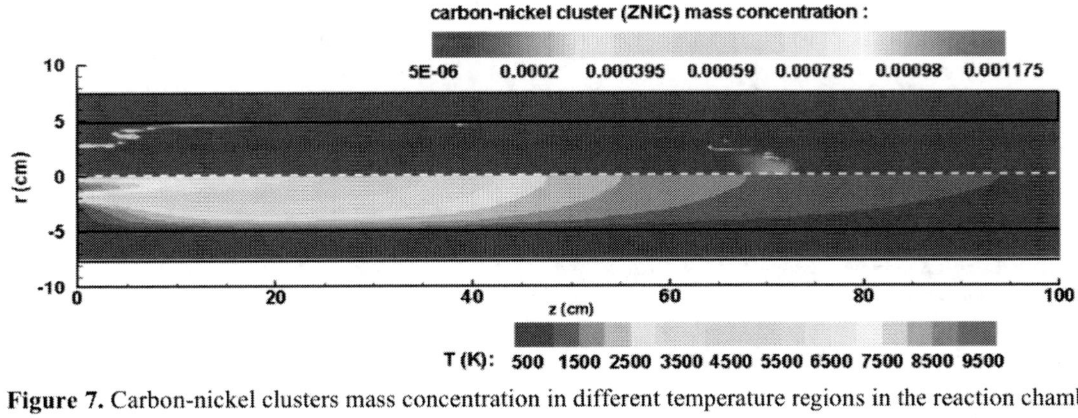

Figure 7. Carbon-nickel clusters mass concentration in different temperature regions in the reaction chamber

4. Conclusion

In the present study, a chemistry model of SWCNT formation from the intermediate species developed for the arc discharge and laser ablation processes [4, 17, 18] was used for the computation of SWCNTs yield in an RF plasma processing system. The principle contribution of the present work is in the adaptation of the chemistry model to a CFD code coupled with an electromagnetic model in order to numerically describe the entire process, in macroscopic scales, from the injection of the feedstock material into the plasma, to the final products containing the SWCNTs. A sensitivity analysis was performed which demonstrated that reactions forming ZNiC were the current bottleneck in SWCNT formation. When key reaction rates were adjusted within their known uncertainty, CNT yield could be well predicted for a single isolated case. Currently, the model, as explained in the previous sections, has deficiencies in accuracy and further work will be done to investigate accurate chemical the reaction rates in different operating conditions.

Acknowledgement

This work was supported by the Natural Science and Engineering Research Council (NSERC) of Canada and the Ontario Ministry of Research and Innovation. The authors acknowledge Dr. Carl D. Scott and Dr. Samir Farhat for providing the CHEMKIN chemical files and SciNet HPC Consortium for the supercomputer resources.

References

[1] Cota-Sanchez G. 2003 *Synthesis of Carbon Nanostructures Using a High Frequency Induction Plasma Reactor* (PhD thesis, Sherbrooke, QC, Canada: Université de Sherbrooke)
[2] Cota-Sanchez G, Soucy G, Huczko A, Beauvais J and Drouin D 2004 *J. Phys. Chem. B.* 108(50), 19210-19217
[3] Cota-Sanchez G, Soucy G, Huczko A, and Lange H 2005 Carbon 43(15), 3153-3166
[4] Scott C D 2004 *J. Nanosci. Nanotechnol.* 4, 368-376
[5] Mostaghimi, J and Boulos M I 1989 *Plasma Chem. Plasma Process.* 9, 25-44.
[6] Xue S W, Proulx P and Boulos M I 2001 *J. Phys. D: Appl. Phys.* 34, 1897-1906.
[7] Zhang Q, Guo H, Liu F, Smallwood G J and Thomson M J 2009 *Proc. Combust. Inst.* 32, 761–768.
[8] Dworkin S B,Cooke J A, Bennett B A V, Smooke M D, Hall R J and Colket M B 2009 *Combust. Theor. Model.* 13 (5), 795–822
[9] Mostaghimi J, Proulx P and Boulos M I 1984 *Plasma Chem. Plasma Process.* 4(3), 199-217
[10] Proulx P, Mostaghimi J and Boulos M I 1985 *Int. J. Heat Mass Tran.* 28(7), 1327-1336
[11] Kim K S, Cota-Sanchez G, Kingston C T, Imris M, Simard B and Soucy G 2007 *J. Phys. D: Appl. Phys.* 40, 2375-2387
[12] Arabzadeh Esfarjani S, Dworkin S B, Mostaghimi J, Kim K S, Simard B, Shahverdi A, and Soucy G 2011 *Proc.42nd AIAA Plasmadynamics and Lases Conf. (Hawaii)* vol 1, 601-612 ISBN: 978-1-61839-170-4
[13] Proulx P, Mostaghimi J and Boulos M I 1987 *Plasma Chem. Plasma Process.* 7(1), 29-52
[14] Shigeta M, Watanabe T 2008 *J. Appl. Phys.* 103(7), 074903-074918
[15] Chen X 1988 *Pure Appl. Chem.* 60(5), 651-662
[16] Kim K S, Moradian A, Mostaghimi J, Alinejad Y, Shahverdi A, Simard B and Soucy G 2009 *Nano Res.* 2, 800-817
[17] Greendyke R B, Creel J, Payne B, Scott C D and Arepalli S 2007 *Collection of Technical Papers – 45th AIAA Aerospace Sciences Meeting* 18, 12285-12313
[18] Greendyke R B, Stein T H, Ray L A, Creel J, Payne B and Scott C D 2005 *43rd AIAA Aerospace Sciences Meeting and Exhibit – Meeting Papers*, 8473-8486
[19] Cau M, Dorval N, Attal-Trétout B, Cochon J L, Loiseau A, Farhat S and Scott C D 2006 *J. Nanosci. Nanotechnol.* 6(5), 1298-1308
[20] Krestinin A V and Moravsky A P 1998 *Chem. Phys. Lett.* 286, 479-484
[21] Fluent user's manual, http://www.fluentusers.com, Version 12.1.4, 2009-12-22
[22] Patankar S V 1980 *Numerical Fluid Flow and Heat Transfer* (New York :Hemisphere)
[23] Boulos M I, Fauchais P and Pfender E 1994 *Thermal Plasmas, Fundamentals and Applications*, 1 (New York: Plenum)
[24] Pousse J, Chervy B, Bilodeau J F, Gleizes A 1996 *Plasma Chem. Plasma Process.* 16(4),605- 634
[25] Gale W F, Totemeier T C 2009 *Smithells metals reference book*, 8th edn. (Oxford: Elsevier Butterworth-Heinemann)
[26] http://www.scinet.utoronto.ca/
[27] Reaction Design: http://www.reactiondesign.com/products/open/chemkin-pro.html

Exploring Chemical and Thermal Non-equilibrium in Nitrogen Arcs

S Ghorui and A.K. Das
[1] Bhabha Atomic Research Centre, Trombay, Mumbai-400085, India

E-mail: srikumarghorui@yahoo.com

Abstract Plasma torches operating with nitrogen are of special importance as they can operate with usual tungsten based refractory electrodes and offer radical rich non-oxidizing high temperature environment for plasma chemistry. Strong gradients in temperature as well as species densities and huge convective fluxes lead to varying degrees of chemical non-equilibrium in associated regions. An axi-symmetric two-temperature chemical non-equilibrium model of a nitrogen plasma torch has been developed to understand the effects of thermal and chemical non-equilibrium in arcs. A 2-D finite volume CFD code in association with a non-equilibrium property routine enabled extraction of steady state self-consistent distributions of various plasma quantities inside the torch under various thermal and chemical non-equilibrium conditions. Chemical non-equilibrium has been incorporated through computation of diffusive and convective fluxes in each finite volume cell in every iteration and associating corresponding thermodynamic and transport properties through the scheme of 'chemical non-equilibrium parameter' introduced by Ghorui et.al. Recombination coefficient data from Nahar et. al. and radiation data from Krey and Morris have been used in the simulation. Results are presented for distributions of temperature, pressure, velocity, current density, electric potential, species densities and chemical non-equilibrium effects. Obtained results are compared with similar results under LTE.

Introduction

Industrial plasma torches operating with nitrogen have the advantages of operating with usual tungsten based refractory electrodes and radical rich non-oxidizing high temperature environment for plasma chemistry and processing. Strong gradients in temperature as well as species densities and huge convective fluxes are very common in such devices. While the regions away from the core are most affected by thermal non-equilibrium, convective and diffusive fluxes results in varying degrees of chemical non-equilibrium in the associated regions. Existence of steep gradients in plasma quantities, extremely high temperature at the core, high fluid dynamic flow fields, highly nonlinear variation of plasma properties and mutual interactions among thermal, chemical, electric, magnetic and fluid dynamic fields make the region difficult to be modelled numerically. The species densities computed under the assumption of chemical equilibrium deviate significantly from the actual densities due to presence of huge diffusive and convective fluxes. As a consequence, thermodynamic and transport properties of associated regions also get affected and final distribution of plasma quantities differ significantly from that under equilibrium. Several modeling works on nitrogen plasma under LTE considerations have been reported. Efforts have been made to model such phenomenon in oxygen cutting torches using two-temperature non-equilibrium model [1]. However, no effort has been made to explore the thermal and chemical non-equilibrium in dc nitrogen arcs to the best of the authors' knowledge.

The paper considers an axi-symmetric two-temperature chemical non-equilibrium model of a nitrogen plasma torch to understand the effects of thermal and chemical non-equilibrium. A 2-D finite volume CFD code in association with a non-equilibrium property routine enables extraction of steady state self-consistent distributions of various plasma quantities inside the torch for different thermal and chemical non-equilibrium conditions.

Chemical non-equilibrium has been incorporated by computing diffusive and convective fluxes in each finite volume cell in every iteration and associating corresponding thermodynamic and transport properties through the scheme of 'chemical non-quilibrium parameter' [2]. Recombination coefficient data from Nahar et. al. [3] and radiation data from Krey and Morris [4] have been used in the simulation. Results are presented for distributions of temperature, pressure, velocity, current density, electric potential, species densities and chemical non-equilibrium effects. Obtained results are compared with similar results under LTE.

Computational Model

The plasma torch, computational domain and the finite volume grid considered in the present study are shown in Fig.1. The cathode is made of copper with a hafnium insert and the anode is made of copper. Both the electrodes are appropriately water-cooled to maintain the system integrity under intense heat flux. Pure nitrogen is considered as the plasma gas. Under two-fluid model, the plasma is considered to be composed of two interpenetrating fluids: one formed of the electrons and the other formed of all other heavy particles including ions, atoms and molecules. Locally, the electrons are in thermal equilibrium among themselves at a temperature T_e while the rests are in similar equilibrium at different temperature T_h. In general, T_e and T_h are not equal for most of the regions inside the plasma.

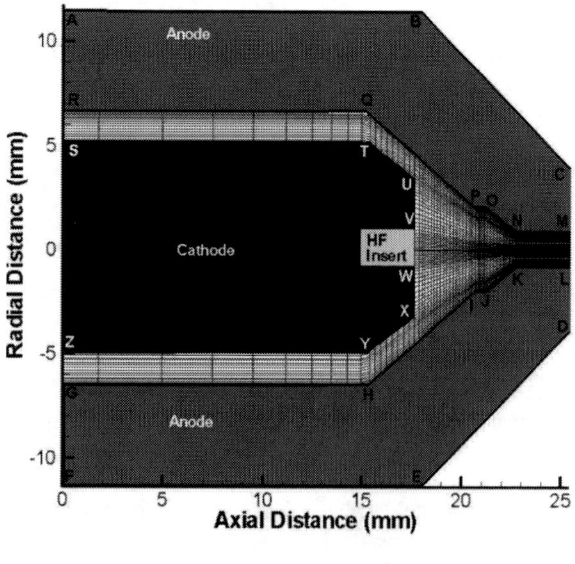

Fig.1. The plasma torch, the computational domain and the computational grid

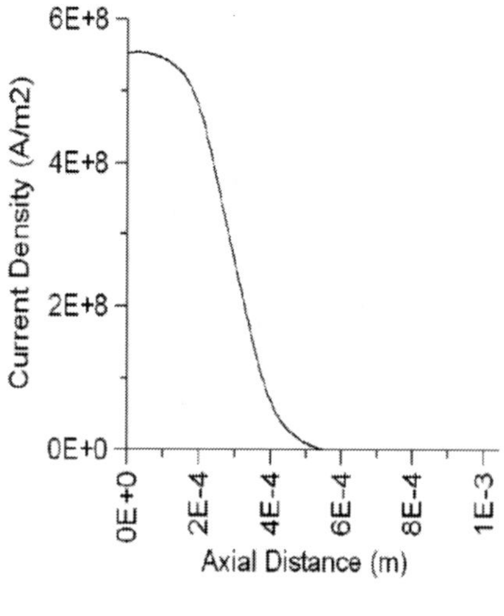

Fig.2. Assumed distribution of current density over the Cathode surface

To extract electro-fluid dynamic and thermal behaviour, the plasma is considered to be continuum with varying degree of electrical and thermal conductivity under the framework of two-fluid model.

Usual Navier-Stokes equation appropriately modified by additional source terms originating from electromagnetic interaction with conducting fluid describes the system with appreciable accuracy. Sets of equations and associated boundary conditions are similar to those used in Ref. [1]. The study does not include any turbulence. Plasma being a viscous medium, flow inside the torch is mostly laminar and inclusion of turbulence does not significantly change the scenario [1]. The set of differential equations representing conservation of mass, momentum and energy are solved under finite volume formulation with SIMPLE [5] like algorithm. Separate energy equations are considered for electrons and heavy particles. Ohmic heating and radiation are considered to be part of electron energy equation alone. Inter-fluid exchange of energy is implemented through the inclusion of appropriate exchange terms involving volumetric collision coefficient

For the considered torch the simulation is carried out for a current of 200A distributed over the cathode surface with a current density profile as shown in Fig2. Gas injection port, located on the top of the torch near the edge of the annular channel, makes an angle such that the gas enters with no radial velocity but with an axial velocity of 20.60 m/s and an azimuthal velocity of 6.77 m/s. The torch operates in transferred arc mode so that no current transfer happens with the nozzle. While the cathode stays at negative potential, the nozzle and the plasma at the location of exit from the nozzle are assumed to be at ground potential.

3. Thermodynamic and Transport Properties of Nitrogen Plasma for CFD Modelling Under Chemical and Thermal Non-equilibrium

Chemically, the generated plasma is modeled with four simultaneous reactions:
(I) $N_2 = N + N$, (II) $N^{+} = N-\ e$, (III) $N^{++} = N-\ 2e$ and (IV) $N^{+++}=N-3e$
The system involves six species: molecules (M), atoms (A), ions (I), double ions (D), triple ions (τ) and electrons (e). While, electron dynamics, electronic excitations and ionization of atoms and ions are governed by electron temperature (T_e), the heavy particle dynamics, molecular dissociation, rotational and vibrational level populations are governed by heavy particle temperature (T_h). Under two temperature formalism, number densities for a given T_e and T_h are determined by solving associated two temperature Saha equations [2,6] together with the equations for total pressure, charge and mass conservation. Once the number densities are computed, associated thermodynamic properties are determined from standard thermodynamic relations.

Transport properties are determined following the first order perturbation theory of Chapmann and Enskog. The properties appear as ratios of infinite determinants, elements of which are functions of collision integrals [6]. Required collision integrals for the present computation are taken from ref [7]. While the details of the computation of associated thermodynamic and transport properties are being published elsewhere, a comparison of the computed properties with those published in literature under thermal equilibrium is presented in Fig.3. The properties are computed for T_e ranging from 0K to 50,000K, T_e/T_h ranging from 1 to 20 and pressure ranging from 0.1 atm to 7 atm. It may be noted from Fig.3 that the computed properties matches quite well with those published in literature under thermal equilibrium.

4. Scheme for Handing Chemical Non-equilibrium

Regions inside the plasma torches are exposed to extremely high gradients in velocity, temperature and species densities. Depending on realized temperature, specific reactions take place at specific spatial zones and localized generation of newer species starts playing dominant role in determining the thermo-physical properties. The process in association with sharp gradients in temperature results in spatial spiking of species densities. Convection and diffusion start playing significant role in determining the number densities in the nearby regions. The phenomenon, termed as chemical non-equilibrium, quantifies itself through the following equation [8] for the kth species:

$$\frac{\partial n_k}{\partial t} + \nabla.(n_k \vec{u}) + \nabla.\vec{g}_k = \dot{n}_k \tag{1}$$

$n_k \vec{u}$ is the convective flux and $\vec{g}_k$ is the diffusive flux for species k.

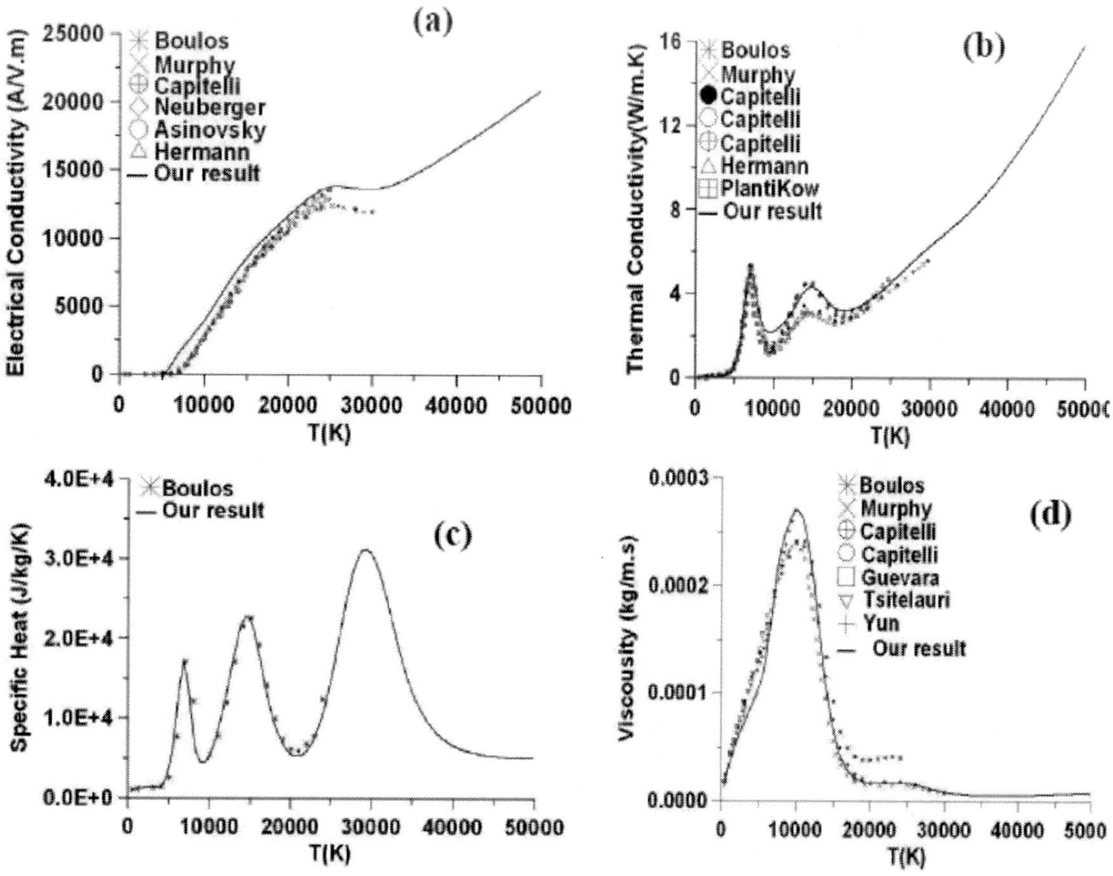

Fig.3. Comparison of computed properties of nitrogen plasma with those published in literature under thermal equilibrium (T_e=T_h): (a) electrical conductivity (b) thermal conductivity (c) specific heat (d) viscosity.

In a system containing ν different species, the expression for $\vec{g}_k$ in terms of overall pressure gradient (∇p), individual species gradient (∇x_j), species temperature gradient (∇T_k) and external electric field (E^X) can be written as:

$$g_k = \frac{n^2}{\rho}\sum_{j=1}^{\nu} m_j D_{kj}^{a}\left[\nabla x_j + (x_j - \frac{\rho_j}{\rho})\nabla \ln p\right] - \frac{n}{\rho}\sum_{j=1}^{\nu}\left(\frac{n_j Z_j e D_{kj}}{kT}\right)E^X - \frac{D_k^{aT}}{m_k}\nabla \ln T_k \tag{2}$$

Where, ρ is the total mass density, k_B is the Boltzmann constant, x_j and ρ_j are, respectively the mole fraction and the mass density of species j, $D_{kj,} D_{kj}^{a}, D_{k}^{aT}$ are respectively the general diffusion coefficient, the general ambipolar diffusion coefficient and the thermal ambipolar diffusion coefficients for the kth species [6].

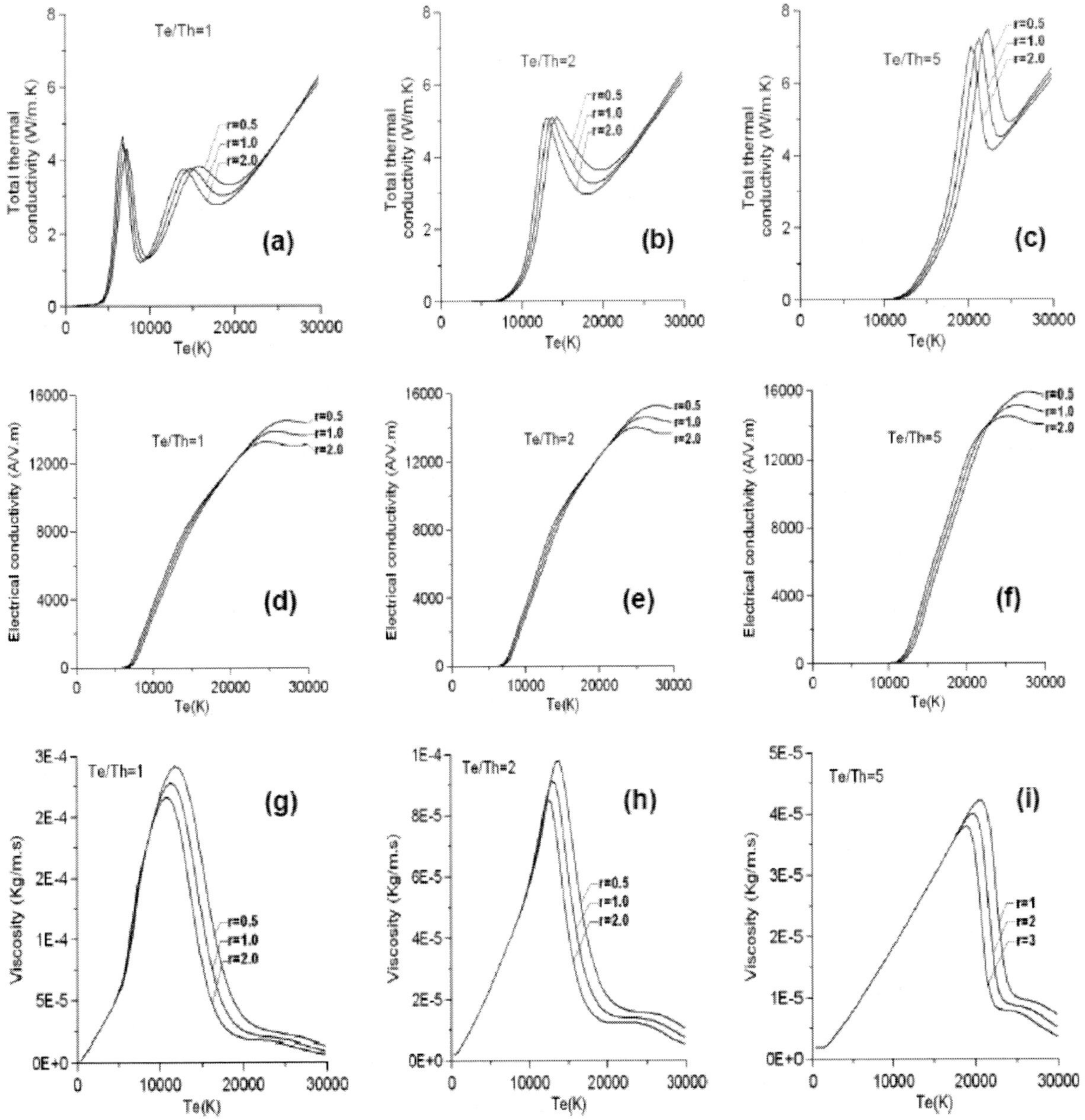

Fig.4. Effect of chemical non-equilibrium on transport properties under different thermal non-equilibrium. r=0.5 corresponds to $r_1=r_2=r_3=r_4=0.5$ and so on.(a)-(c): total thermal conductivity; (d)-(f): Electrical conductivity; (g)-(i): viscosity

Journal of Physics: Conference Series **406** (2012) 012012 doi:10.1088/1742-6596/406/1/012012

In terms of recombination coefficients and equilibrium rate constants, the right hand side of equation (1) can be expressed as following for the second reaction [8]:

$$\dot{n}_I = \alpha_2 n_a \left[S_2 - \frac{n_e n_I}{n_a} \right] \tag{3}$$

where, α_2 and S_2 are the recombination coefficient and equilibrium rate constant for reaction-2. n_e, n_I, and n_a are the actual steady state densities of electron, ion and atom including non-equilibrium. Combining equation (1) and (3) it is possible to write the revised rate equation as following for steady state under chemical non-equilibrium:

$$\frac{n_e n_I}{n_a} = R_2 \tag{4}$$

$$\text{with } R_2 = S_2 - \Delta S_2 \text{ , and } \Delta S_2 = \frac{1}{\alpha_I n_a}\left[\nabla.(n_I \vec{u}) + \nabla.\vec{g}_I\right] \tag{5}$$

In similar way, the revised rate constant R_i for any i^{th} reaction can be written down.
The effect of chemical non-equilibrium can thus be incorporated in a thermal plasma system just by modifying the equilibrium rate constant in an appropriate manner. From simulation point of view, in every iteration the associated convective and diffusive fluxes in a control volume are computed by associated CFD routine and utilized by corresponding thermodynamic and transport property routine (TTP) to determine the actual number density of species through equations (4) and (5). Once the updated number densities are obtained associated thermodynamic and transport properties are recomputed by TTP routine and fed back to the CFD routine for recalculating the fluid dynamic fields based on the updated property values. The cycle goes on and on. The CFD and the TTP routine communicate sequentially with each other until a final self-consistent converged solution is obtained.

The degree of chemical non-equilibrium for the ith reaction is quantified by chemical non-equilibrium parameter (r_i) as following [2]:

$$r_i = \frac{S_i - \Delta S_i}{S_i} \tag{6}$$

Figure 4 presents the effect of chemical non-equilibrium on various transport properties under different thermal non-equilibrium. For ease of presentations, the graphs are presented with identical r_i for all reactions. It is noted that the effect depends on temperature and may cause a deviation from equilibrium value by as high as 30%.

4. Results and Conclusions

Results of simulation under thermal and chemical non-equilibrium and the way it differs from the results under thermal equilibrium are presented in Fig.5. Inclusion of non-equilibrium results in lower potential drop, lower exit velocity but higher exit temperature compared to the LTE case. Under non-LTE the heavy particle temperature near the exit is lower compared to LTE temperature but the electron temperature is much higher than LTE temperature and heavy particle temperature. Exit velocity is more realistic when non-equilibrium is included. Distribution of chemical non-equilibrium parameter, electron, single ion and double ion densities are presented in Fig.6.Significant variation of chemical non-equilibrium parameter is observed near the core region [Fig.6a] where sharp gradients in particle densities occur and sudden increase in velocity takes place [Fig.6(b)-(d)].

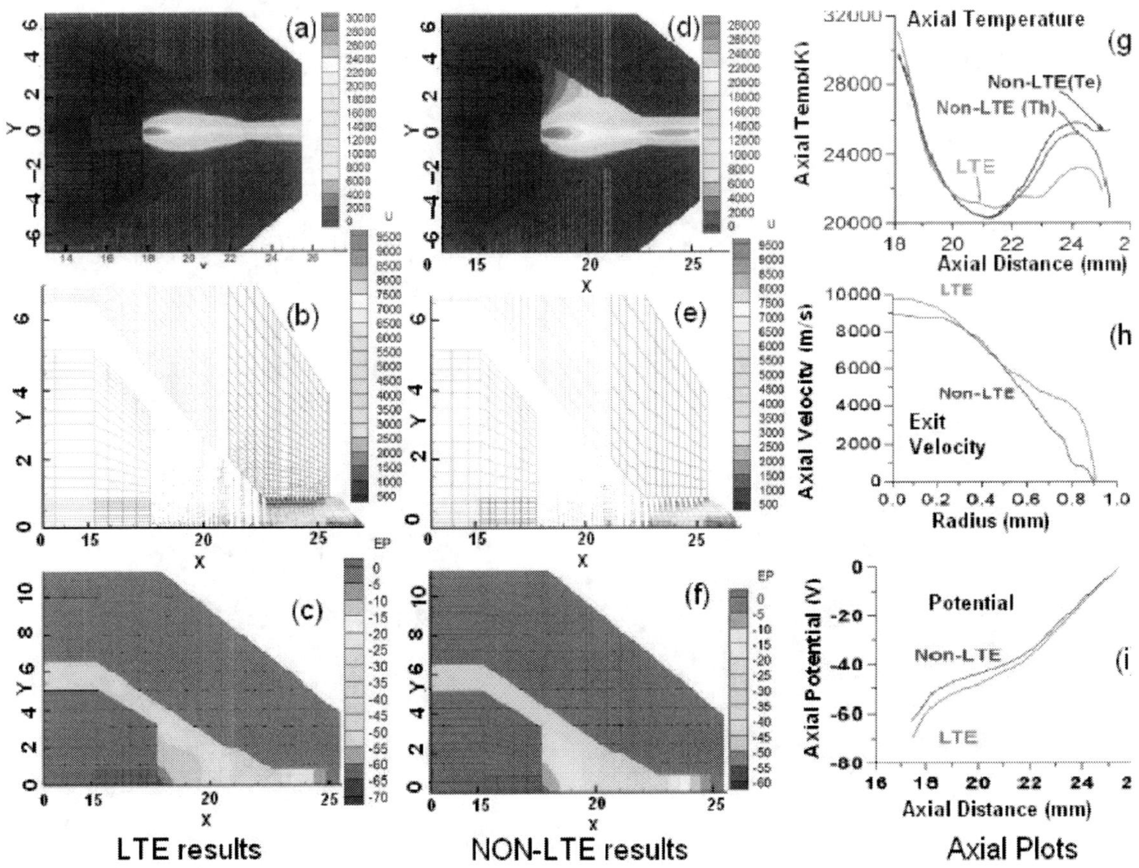

Fig.5 Temperature, velocity and potential distribution under LTE [(a) ,(b) and (c)], non-LTE [(d), (e) and (f)] and axial variation of them [(g), (h) and (i)]. [X and Y dimensions are in mm]

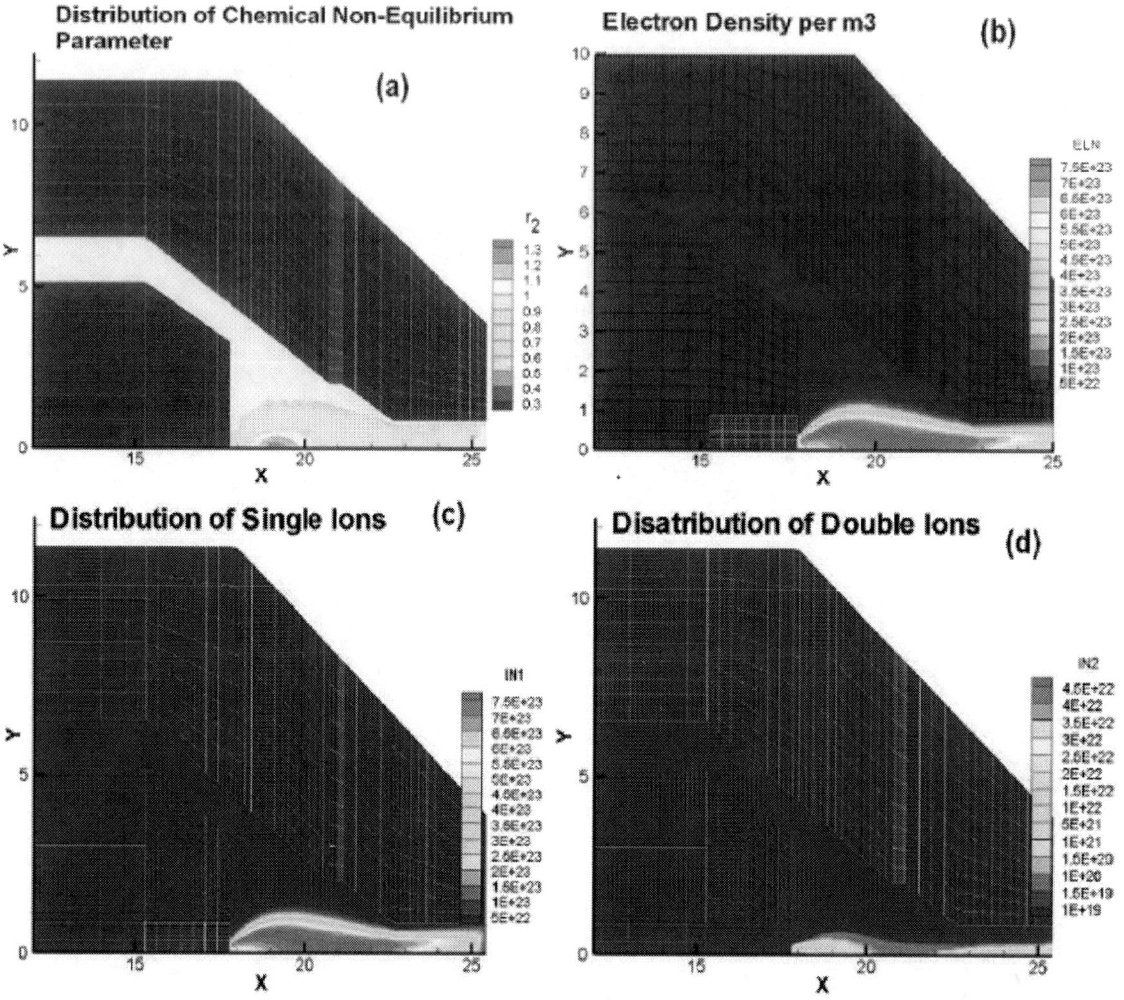

Fig.6. Distribution of (a) chemical non-equilibrium parameter for single ions (r_2), (b) electron density, (c) ion density and (d) double ion density. [X and Y dimensions are in mm]

Acknowledgements: The authors are thankful to Prof. J.V.R. Heberlein and Prof. E. Pfender of University of Minnesota for many useful discussions during initial stages of the study.

References

1. Ghorui S, Heberlein J V R and Pfender E 2007 *J. Phys. D: Appl. Phys.* **40** 1966
2. Ghorui S, Heberlein J V R and Pfender E 2007 *Plasma Chem Plasma Process* **27**267
3. Nahar S N 1999 *Astr. Phys. J. Suppl. Series,* **120** 131
4. Krey R U and Morris J C 1970 *Phys. Fluids.* **13** 1483
5. Patankar S V 1980 *Numerical Heat Transfer and Fluid Flow* (New York: McGraw-Hill)
6. Ghorui S, Heberlein J V R and Pfender E 2008 *Plasma Chem Plasma Process* **28** 553
7. Capitelli M, Gorse C, Longo S 2000 *J Thermo Phys Heat Transfer* **14** 259
8. Mitchner M, Kruger CH 1973 *Partially Ionized Gases* Wiley, New York

Time-resolved characterization of a pulsed discharge in a stationary bubble

Patrick Vanraes, Anton Nikiforov, Mario Lessiak, Christophe Leys

Ghent University, Jozef Plateaustraat 22, 9000 Ghent, Belgium

E-mail: patrick.vanraes@ugent.be

Abstract. In recent years, plasma generation in water has been proposed for the application of water treatment. The process efficiency is believed to be improved by the introduction of bubbles in the plasma active region. For further optimization, the initiating and developmental mechanisms of plasma inside bubbles need to be understood to a greater extent. In order to meet this necessity, we investigated pulsed electrical discharge inside a stationary bubble in water. This paper deals with the evolution of the discharge and of the bubble shape during discharge, investigated by electrical characterization and fast imaging. Only several microseconds after the application of the voltage pulse, plasma light is observed. Different phases are observed during plasma formation. The plasma is strongest at the bubble surface, causing the surrounding water to evaporate. This leads to both the formation of propagating streamers into the water and the expansion and collapse of the bubble. These observations show that plasma inside a bubble has the strongest activity at the bubble surface, making it attractive for water treatment.

1. Introduction

Plasma in or in contact with water generates high-energetic radicals, ions and molecular species as well as UV radiation and shock waves, which makes it suitable for a wide range of applications like water treatment [1–5]. Plasma has a strong oxidation effect, caused by generated OH radicals and O_3 directly in gas phase and H_2O_2 in water. Therefore, it provides a more versatile and more cost-efficient alternative for existing water treatment techniques like ozonation [6,7]. The investigation and optimization of the chemical effect of plasma with liquid electrodes for water treatment has led to a wide number of papers (see e.g. [8] and references therein) and to the recent industrial implementation of plasma treatment of tertiary-treated wastewater effluent and contaminated groundwater [9].

When plasma is generated inside bubbles in water, the plasma-liquid interface is enlarged and the needed input power is decreased in comparison to discharge in water without bubbles. The introduction of externally generated gas bubbles in the region of discharge is therefore expected to improve the efficiency of plasma chemical processes

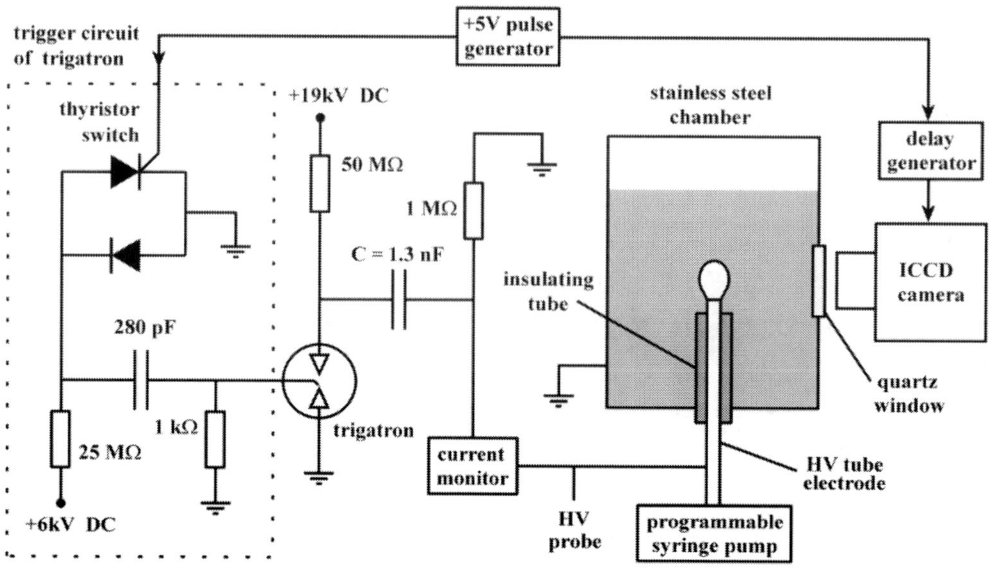

Figure 1. The setup consists of a negative HV pulse generator connected to a metal tube electrode submerged in distilled water inside a stainless steel chamber. A stationary bubble on the metal tube is produced by pumping a gas through the tube with a programmable syringe pump.

at the plasma-liquid interface, where the bubble size and gas flow rate play important roles [7, 10].

The present work deals with pulsed plasma discharge in a stationary bubble on a metal tube electrode in water inside a grounded chamber. Electric characterization and fast-imaging of discharge in helium, argon, nitrogen, oxygen and air bubbles with two different sizes are carried out and compared.

2. Experimental setup and procedure

The electrical bubble discharge is generated inside a grounded cylindrical chamber made of stainless steel with a quartz window of 30 mm diameter, described in [7]. It is filled with distilled water with a conductivity of 22 μS/cm up to 12 cm from the bottom. The stainless steel tube electrode with inner diameter of 0.8 mm and outer diameter of 1.2 mm is submerged vertically in the water with its flattened orifice at the same height as the center of the quartz window. To insulate the tube below its orifice from the water and the chamber, it is placed in a glass tube.

A gas bubble is generated by pumping a gas through the tube electrode with a programmable syringe pump (New Era, model NE-1000). From photographs, the volume of the introduced bubble of 45 μl or 100 μl is determined (corresponding to prolate spheroidal bubbles with a horizontal minor axes length of 2.2 mm or 2.7 mm and a vertical major axis length of 2.3 mm or 3.3 mm respectively). Unless mentioned otherwise, the larger bubble size is used, with helium, argon, nitrogen, oxygen or air as

Table 1. Spectral filters

filter name	λ_{line} (nm)	λ_L (nm)	λ_U (nm)
H_α filter	656	652.5	662.5
O filter	777	770.4	791.0
He filter	588	585.4	595.1
Ar filter	751	747.8	758.2

bubble gas.

A triggered HV power supply with an output capacitance of $C = 1.3$ nF was adjusted to generate negative high voltage (HV) pulses with a rise time of 72 ns, an amplitude of 19 kV and a repetition rate of 0.1 Hz. The schematic of the used power supply and chamber is presented on figure 1. The trigger circuit of the trigatron is controlled by a HTS thyristor switch (Behlke, model 300-100-SCR). The water is grounded in all experiments. Voltage and current waveforms are measured with a Tektronix high voltage probe (model P6015A) and a Pearson current monitor (model 2877) connected to a Tektronix oscilloscope (model DTS 1002).

The discharge is photographed through the circular quartz window with a Hamamatsu ICCD camera (model C8484), with camera exposure times of 50 ns or 5 μs. The camera is triggered with adjustable delay time, using the same trigger signal as the trigger circuit of the trigatron. In some measurements, spectral filters were placed in front of the camera. The lower and upper cutoff wavelengths λ_L and λ_U of the filters are mentioned in table 1, as well as the wavelength λ_{line} of the atomic line the filters were used for.

3. Results

3.1. Electrical characterization

Figure 2 shows the voltage and current waveforms of discharge in an air bubble on two different time scales. The voltage decays exponentially with a time constant of 312 μs. A delay time Δt after the voltage pulse application is determined as the time between the onset of two current pulses. No light emission is observed during this delay time, up to about 100 ns before the second current pulse oscillations start. Therefore, the oscillations indicate the moment of plasma discharge. During one series of experiments, Δt varied with a standard deviation of 0.5 μs. The average value of Δt has an order of magnitude of 10 μs. It is independent of the bubble gas, but decreases with decreasing bubble size. For unprocessed distilled water, for example, Δt has an average value of 12.3 μs and 15.4 μs for bubble volumes of respectively 45 μl and 100 μl.

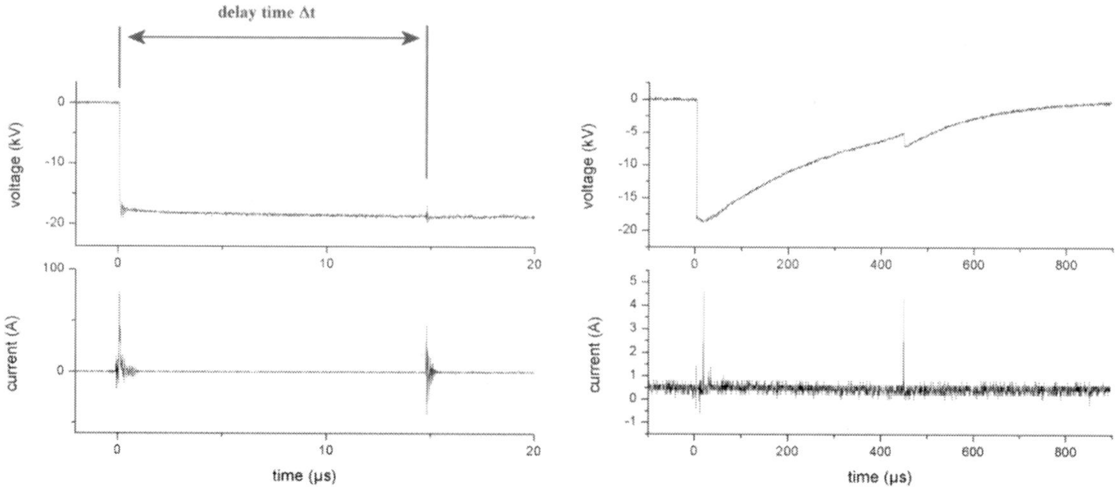

Figure 2. Voltage and current waveforms during discharge in a 100 μl air bubble in distilled water. Oscillations in the waveforms occur a delay time $\Delta t = 14.8$ μs after the application of the HV pulse. During the delay time, there was no plasma light emission.

3.2. Time-resolved characterization

An example of the evolution of plasma geometry in the first two microseconds after the delay time Δt is depicted in figure 3. The times indicated on the figure are relative to the point in time where the oscillations on the waveforms begin. Until 100 ns before these oscillations, no plasma light is emitted. Five phases in the development of the plasma geometry are seen. In a time interval of the order of 10 ns or less, 72 ± 18 ns before the voltage oscillations start, intense discharge occurs inside the whole bubble. In a second phase up to 72 ± 30 ns after the start of the oscillations, the plasma diminishes towards the electrode. The third phase is called the *dark phase*, since no emitted light from the bubble bulk is observed during this period of time. Still, plasma remains in proximity of the metal tube. During the fourth phase, the *bubble filling phase*, plasma extends towards the bubble top, either homogeneously for helium bubbles or in filamentary structures along the bubble surface for the other bubble gases. For convenience, these structures are called in short *filaments* henceforth.

The lenght of the dark phase and the bubble filling phase depends strongly on the bubble gas (table 2) and is independent of the bubble size within the accuracy of the measurements. The lenght of the other phases are independent on the bubble gas and size.

From 12.5 ± 2.5 μs after voltage pulse application, the bubble surface blurs due to evaporation of the surrounding water (figure 4). At 22.5 ± 2.5 μs after voltage application, streamer-like structures start to form at the bubble surface, some of which are relatively intense. In order not to confuse these structures with the filaments inside the bubble, they are called henceforth *streamers* in short. The longest streamers that are observed have a branched shape and an estimated velocity with an order of magnitude of

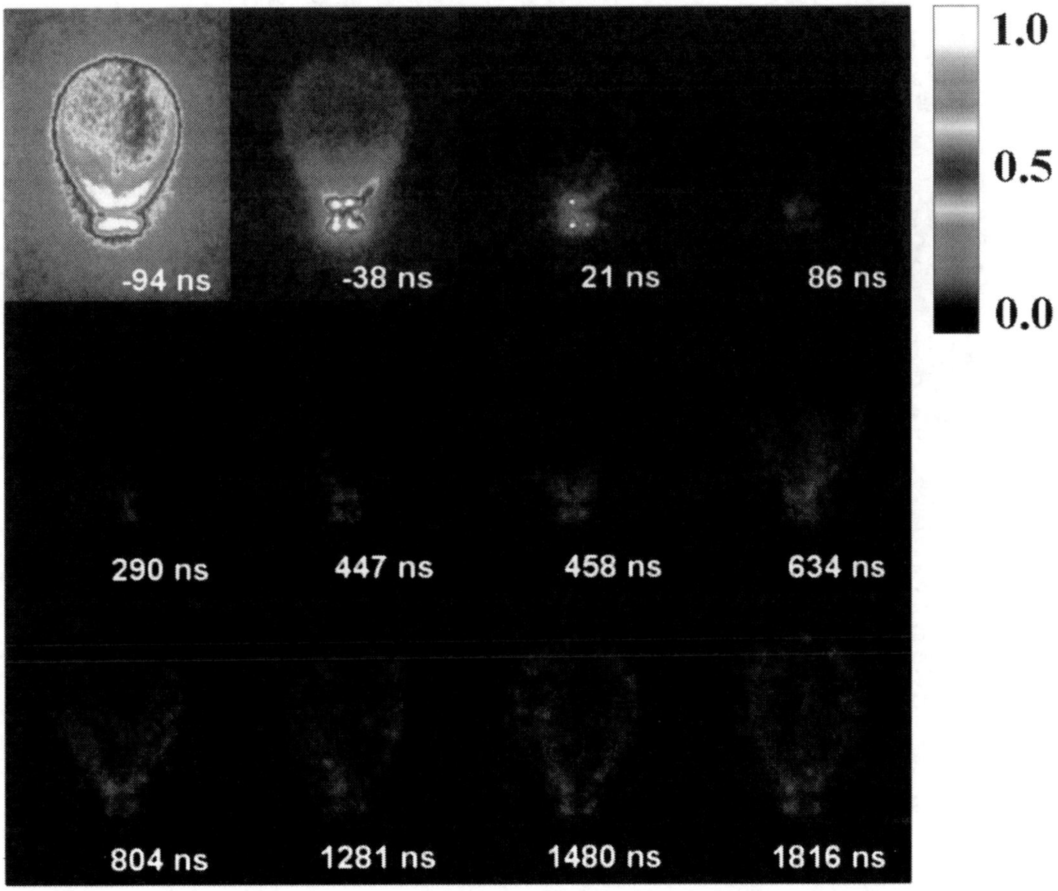

Figure 3. Images of the discharge in a 100 μl air bubble captured by the ICCD camera with an exposure time of 50 ns. Every indicated time is the start time of the camera subtracted from the start time of the voltage oscillations.

Table 2. Phase length

bubble gas	dark phase length (ns)	bubble filling phase length (ns)
helium	176 ± 5	81 ± 21
argon	116 ± 29	80 ± 5
nitrogen	25 ± 17	711 ± 30
oxygen	194 ± 8	196 ± 10
air	394 ± 40	1008 ± 78

10^2 m/s or more. Although the plasma intensity and homogeneity is strongly influenced by the bubble gas, the process of bubble deformation is independent of bubble gas. In bubbles with a volume of 45 μl, streamers are formed at 17.5 ± 2.5 μs after voltage application and they are more intense in comparison to 100 μl bubbles, as seen in figure 5.

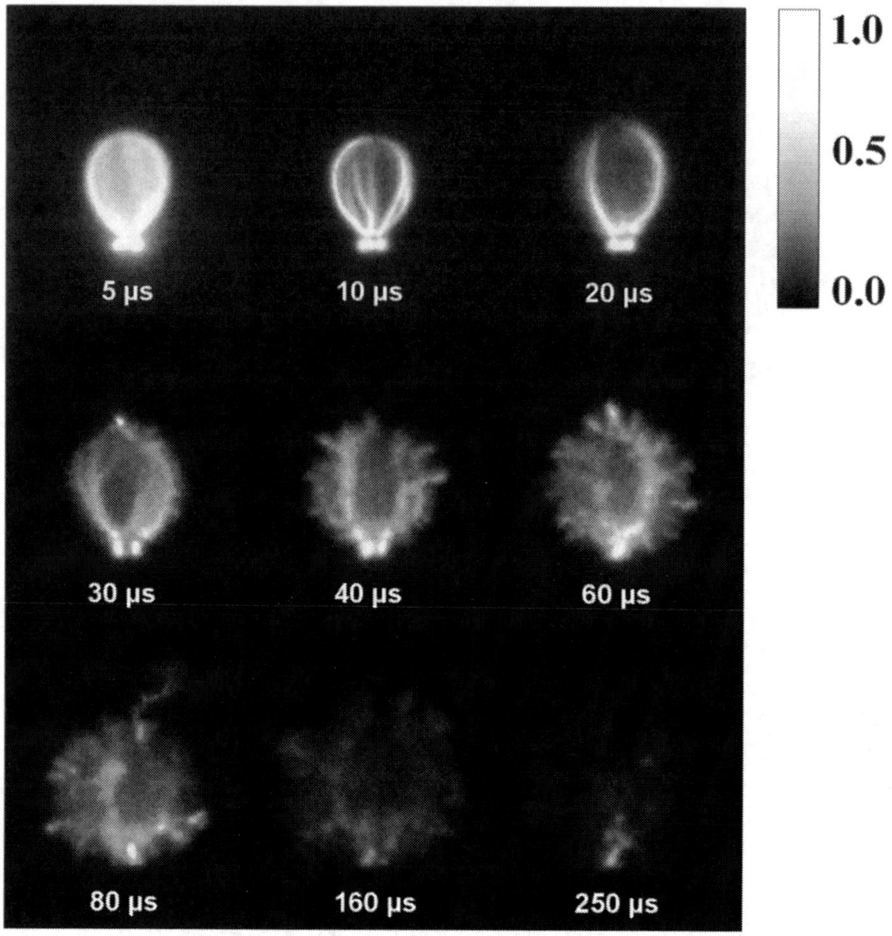

Figure 4. Images of the discharge in a 100 μl air bubble captured by the ICCD camera with an exposure time of 5 mus. The delay time of the camera relative to the time of voltage pulse application is indicated in every image.

The filaments in the bubble emit a spectrum that contains the H_α line, the oxygen line at 777 nm and, for helium and argon bubbles, the corresponding bubble gas element line at 588 nm or 751 nm respectively (figure 6). The filaments in an oxygen bubble are an exception, since H_α is absent or very dim in their spectrum. Relatively far from the bubble surface, H_α is dominant in the streamer spectrum, while the oxygen line and bubble gas element line are absent or very dim.

4. Discussion

The first current peak at the moment of voltage application is caused mostly by displacement current. Also, Joule heating by current from electrolysis contributes to this peak, but this contribution is expected to be very low due to the low water conductivity. Due to the closing of the HTS thyristor switch, a third current peak that corresponds to a negative bump in the voltage waveform is seen more than 400 μs after the voltage

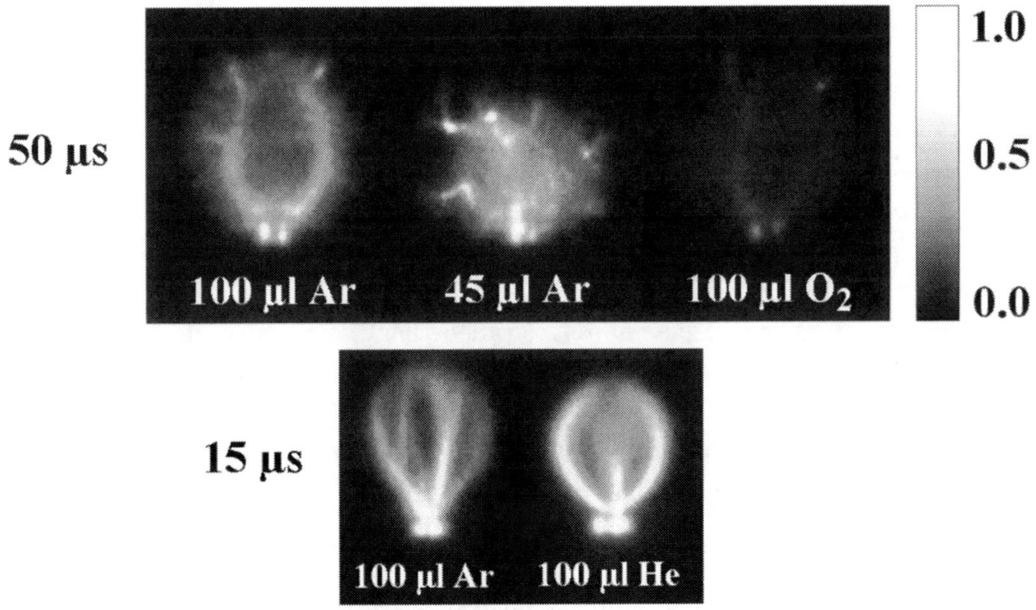

Figure 5. Images of the discharge in a bubble captured by the ICCD camera with an exposure time of 5 μs and a delay time of 50 μs (top) or 15 μs (bottom) relative to the time of voltage pulse application. The bubble gas and size are indicated in every image.

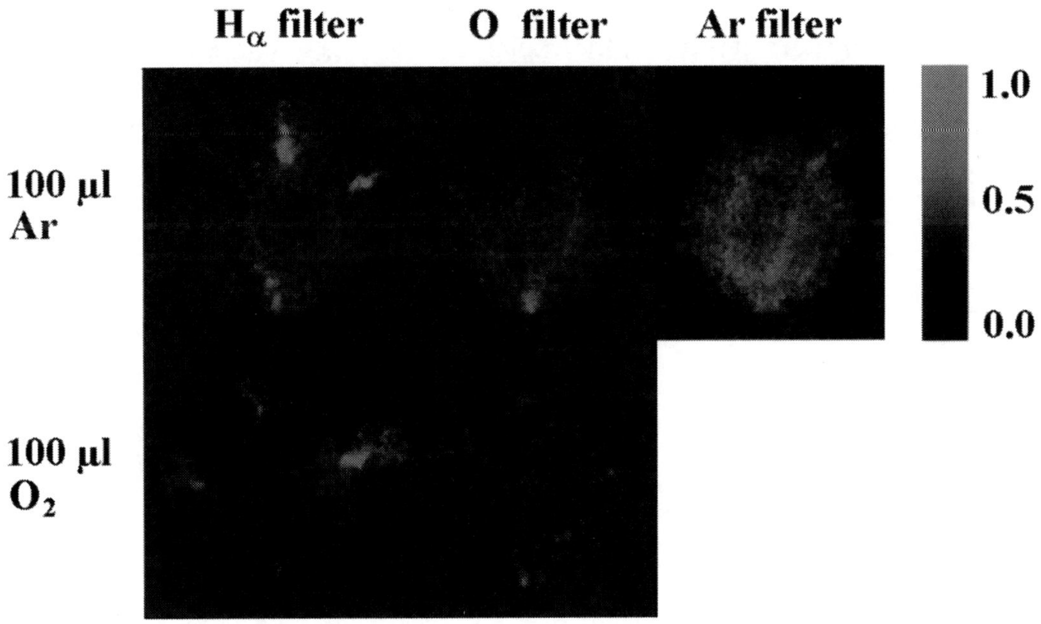

Figure 6. Images of the discharge in a 100 μl argon (top) or oxygen (bottom) bubble captured by the ICCD camera with an exposure time of 5 μs and a delay time of 60 μs relative to the time of voltage pulse application when a spectral filter was placed in front of the camera. The type of spectral filter is indicated above the image.

pulse application. Light emission is no longer observed when this peak occurs, indicating that it is not related to plasma discharge in water.

Plasma light is observed only after a delay time Δt of the order of 10 μs, which indicates that the electric field in the bubble is too small for discharge to occur during this period. The delay time is long enough for ions and dipolar molecules in the water in proximity of the bubble to reorganize. Under the influence of the electric field, a positive net charge accumulates on the bubble surface, enhancing the electric field in the bubble. Once the electric field reaches the critical value necessary for breakdown in the bubble, the first phase of plasma development begins. During the first phase, the bubble is filled with plasma in a time of the order of 10 ns or less, which is too short for atomic and molecular species to travel distances of the order of millimeters. Therefore, the mechanism behind plasma onset is explained by electron avalanches.

The charge accumulation on the bubble surface points out that a capacitive feature of the bubble has to be taken into account. This can also explain the next phases in the development of the plasma geometry. In the first phase, electrons from the HV electrode reach the bubble surface, which leads to the neutralization of the positive net charge and, subsequently, the accumulation of a negative net charge. The electric field in the bulk of the bubble drops rapidly, by which the plasma diminishes towards the electrode. During the dark phase, the negative ions at the bubble surface travel away from the bubble by the repulsive Lorentz force caused by the negative high voltage electrode. The electric field inside the bubble increases, leading to the bubble filling phase.

If we consider the bubble, the surrounding water and the external pulse circuit as a series RC circuit with constant resistance, the difference in average delay time Δt for bubbles of 45 μl and 100 μl can be explained. On the assumption that a prolate spherical bubble with horizontal axes length r and vertical axis length d can be modelled as a classic capacitor that consists of two parallel conductive plates with surface $A \propto r^2$ separated by a distance d by the bubble gas with permittivity $\epsilon \approx \epsilon_0$, its capacitance can be written as

$$C = \epsilon \frac{A}{d} \propto \frac{r^2}{d} \tag{1}$$

Therefore, the time constant $\tau = RC$ is directly proportional to

$$\tau \propto \frac{r^2}{d} \tag{2}$$

Breakdown in the bubble occurs when the electric field E in the bubble gas reaches a critical value E_c. For a charging plate capacitor in a RC circuit, the critical electric field is given by

$$E_c = \frac{V_c}{d} = \frac{V_0\left(1 - e^{-t/\tau}\right)}{d} \tag{3}$$

where V_c is the critical voltage over the capacitor, V_0 is the constant voltage over the RC circuit and t is the time needed to charge the capacitor up to the critical voltage.

Denoting the parameters of a 45 μl and 100 μl bubble respectively with an index 1 and 2 and equating the critical electric field for both bubble sizes leads to

$$-\frac{t_2}{\tau_2} = \ln\left(1 - \frac{d_2}{d_1}\left(1 - e^{-t_1/\tau_1}\right)\right) \tag{4}$$

On the assumption that $t_1 \ll \tau_1$ and applying the approximations $\ln(1-x) \approx -x$ and $1 - \exp(-x) \approx x$ for $x \ll 1$, equation 4 is reduced to

$$\frac{t_2}{t_1} = \frac{d_2\tau_2}{d_1\tau_1} = \frac{r_2^2}{r_1^2} \tag{5}$$

where equation 2 was used. Substituting the average experimental values of Δt in t_1 and t_2, we find that $t_2/t_1 = 1.25$, which deviates 17.2% from $r_2^2/r_1^2 = 1.51$. Considering the simplicity of the model, this is a good agreement.

The branched structure and the estimated velocity of the streamers in our measurements are in agreement with the lower values mentioned in literature for pulsed streamers in water without externally generated bubbles [11–14]. Also, typical spectra of pulsed streamer discharge in water contain a strong H_α line and a weak oxygen line at 777 nm, which indicates that the streamers in our measurements are similar in nature [12, 14].

5. Conclusion

Plasma light emission is recorded only a delay time Δt of the order of 10 μs after the HV pulse application. The plasma geometry evolves in five phases. First, plasma rapidly fills the whole bubble. Next, the plasma decays in 144 ± 35 ns and a dark phase follows. Subsequently, during the bubble filling phase, the plasma extends again towards the top of the bubble. The length of the dark phase and the bubble filling phase depends strongly on the bubble gas. After 22.5 ± 2.5 μs, branching streamer-like structures start to propagate from the bubble surface into the water. Their spectrum contains a strong H_α line and a weak or absent oxygen line at 777 nm.

The delay time Δt and the five phases in the plasma geometry evolution are explained by a capacitive feature of the bubble. A simple model that considers the bubble as a classic plate capacitor is in relatively good agreement with the experimental values in our experiments. The investigated features of the streamers in our measurements resemble the features of pulsed streamer discharge in water, indicating they are similar in nature.

In future research, the electron density and temperature and the plasma chemistry need to be investigated with time-resolved spectroscopy, in order to gain better insight in the time-resolved characterization of a pulsed discharge in a bubble.

References

[1] Kurahashi M, Katsura S and Mizuno A 1997 *Journal of Electrostatics* **42** 93–105

[2] Sunka P, Babicky V, Clupek M, Lukes P, Simek M, Schmidt J and Cernak M 1999 *Plasma Sources Science & Technology* **8** 258–265
[3] Miichi T, Hayashi N, Ihara S, Satoh S and Yamabe C 2002 *Ozone-Science & Engineering* **24** 471–477
[4] Yamatake A, Katayama H, Yasuoka K and Ishii S 2007 *International Journal of Plasma Environmental Science & Technology* **1**
[5] Sato K, Yasuoka K and Ishii S 2010 *Electrical Engineering in Japan* **170** 1–7
[6] Ihara S, Miichi T, Satoh S, Yamabe C and Sakai E 1999 *Japanese Journal of Applied Physics Part 1-Regular Papers Short Notes & Review Papers* **38** 4601–4604
[7] Vanraes P, Nikiforov A and Leys C 2012 *Journal of Physics D-Applied Physics* **45** 245206
[8] Malik M A 2010 *Plasma Chemistry and Plasma Processing* **30** 21–31
[9] Even-Ezra I, Mizrahi A, Gerrity D, Snyder S, Salveson A and Lahav O 2009 *Desalination and Water Treatment* **11** 236–244
[10] Aoki H, Kitano K and Hamaguchi S 2008 *Plasma Sources Science & Technology* **17** 025006
[11] Beroual A, Zahn M, Badent A, Kist K, Schwabe A J, Yamashita H, Yamazawa K, Danikas M, Chadband W G and Torshin Y 1998 *IEEE Electrical Insulation Magazine* **14** 6–17
[12] Akiyama H 2000 *Ieee Transactions on Dielectrics and Electrical Insulation* **7** 646–653
[13] van Veldhuizen E and Rutgers W 2001 *15th International Symposium on Plasma Chemistry* 3245–3250
[14] An W, Baumung K and Bluhm H 2007 *Journal of Applied Physics* **101**

Journal of Physics: Conference Series **406** (2012) 012014 doi:10.1088/1742-6596/406/1/012014

Vacuum laser-produced plasma for analytical application in fusion technologies

S Almaviva[1], L Caneve[1], F Colao[1], R Fantoni[1], G Maddaluno[2]

[1]ENEA, UTAPRAD-DIM, C.R. Frascati, Via Enrico Fermi, 45, 00044 Frascati (RM) Italy

[2] Associazione EURATOM-ENEA sulla Fusione, C. R. Frascati, C.P. 65, 00044 Frascati, Roma, Italy

E-mail: salvatore.almaviva@enea.it

Abstract. LIBS is a well established tool for qualitative, semi-quantitative and quantitative analysis of surfaces, with micro-destructive characteristics and some capabilities for stratigraphy. In this work, the depth profiling capabilities of LIBS has been checked by determining the composition of multilayered samples simulating the plasma facing components fusion device covered with co-deposited impurity layers. A new experimental setup has been designed and realized in order to optimize the characteristics of a LIBS system working at low pressure and remotely.

1. Introduction

High temperature plasmas of hydrogen isotopes are peculiar of thermonuclear fusion devices. The study of plasma-wall interaction [1-3] is of paramount importance for avoiding both damage of plasma facing components (PFCs) and pollution of the plasma. PFCs are eroded and the material is deposited by the edge plasma in different areas of the tokamak. So the composition of PFCs' surface changes strongly from the original. To ensure safety operation a regular monitoring of the tritium inventory and the deposited material has to be applied. To this respect Laser-Induced-Breakdown-Spectroscopy (LIBS) [4-5] may be suitable for *in-situ* characterization of the conditions of PFCs.

The characterization of first wall surfaces by LIBS, which means the local measurement of thickness and composition of deposited layers including tritium inventory, could be carried out by probing with the laser beam different wall areas between tokamak discharges. Depending on the power density and pulse duration of the laser radiation, wall particles are released by ablation and are spectroscopically analysed. The spectral emission of such LIBS plasma is governed by atomic physics processes involving collisional excitation and decay rates of several states of the neutrals and ions in the plume. For optically thin plasma the line intensity for transition between the energy states k and i can be given by [5]

$$I_\alpha = FC_\alpha \frac{g_k A_{ki} e^{-E_k/K_B T}}{U_\alpha(T)} \qquad (1)$$

where g_k is the state degeneracy and A_{ki} the Einstein coefficient for spontaneous emission, T is the plasma temperature, $U_α(T)$ is the partition function, K_B is the Boltzmann constant and F is a constant to account for particular experimental conditions (optical arrangement and photon counting efficiencies) and plasma density (depending on material and absorbed laser energy).

As an analytical technique, LIBS has demonstrated its unique versatility [5-7], allowing fast contact-less analysis of almost any type of material, from solids to liquids and gases also examined at low pressure [8], each time allowing to adapt the technique to the spectral requirements of diverse practical analytical problems. The spectroscopy of the radiation emitted by the Laser Induced Plasma (LIP) may be used to obtain the characteristic physical parameters, such as the temperature, the electron number density, the atom and ion number densities, and the relative concentration of different element present in the target.

In the framework of the European Fusion Development Agreement (EFDA) task WP11-ETS_DTM-01 "Dust and Tritium Management" the objective of this work is to advance and qualify the operation of LIBS on a fusion device (tokamak) by performing Deuterium inventory in targets specifically prepared to simulate PFCs covered with impurity layers codeposited with fuel

The optimization criteria here considered are the capability of a LIBS system in working at low pressure and remotely, at a distance of some meters from the target.
The optical system used is sketched in Figure 1.

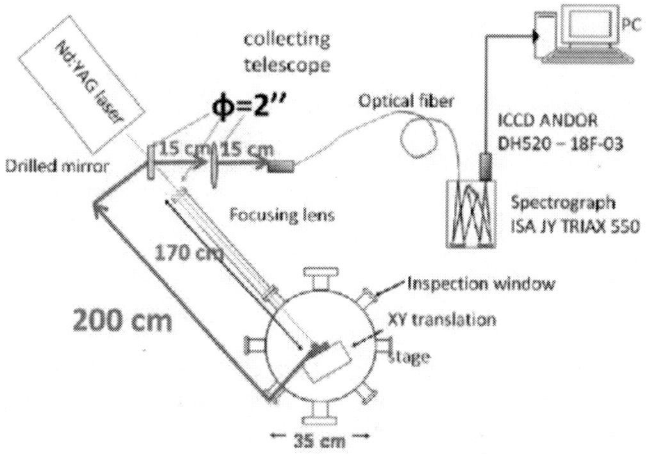

Figure 1. sketch of the LIBS apparatus.

The apparatus is composed of a vacuum chamber, 35 cm diameter in which the sample is placed vertically on a two axis motorized translation stage remotely controlled for changing the measurement point. LIPs were obtained by using a Q-switched Nd:YAG laser (model handy, Quanta System) emitting at the fundamental wavelength of 1064 nm. Each laser pulse has a duration of 8 ns, while repetition rate is 10 Hz. The laser beam was focused onto the sample by a 200 cm focal length quartz lens through a 150 cm length vacuum arm closed at the end by an optical window highly transparent to infrared radiation. With an average power of 140 mJ and a spot diameter of 1.3 mm on the sample surface the energy fluence released was about 11 J/cm^2 per single shot.

To realize remote plasma light detection and simulate the dimensional constraints found in a real tokamak a plane mirror with a 10 mm hole in the center was placed in front of the focusing lens and aligned with the laser beam (passing trough the hole). In this way the plasma light exiting from the optical window, collinear with the laser beam, is reflected by the mirror at an a angle of 45° respect to the laser beam and is collected at a distance of 200 cm from the sample surface.

All the measurements were carried out at a residual pressure of $1\text{-}3\cdot10^{-5}$ mbar The collection optics is composed by a 15-cm focal length fused quartz lens, 50 mm aperture. The lens focuses the plasma light into an optical fiber bundle, composed of 37 fibers, 100 µm core, f# 4, connected to a 100 µm entrance slit of a TRIAX 550 ISA JOBIN-IVON spectrograph, f# 6.4. For a diffractive grating of 2400 g/mm the wavelength resolution at 500 nm is around 0.1 Å. Spectra were recorded using a gated ICCD (DH532-18F, Andor), whose gate aperture was synchronized with the laser burst by mean of an electronic trigger.

1.1 Samples

A set of diamond-like carbon (DLC) and mixed W:Al:C (with nominal concentrations W(5-10%):Al(45-48%):C(45-48%)) layers were codeposited with D on W substrates by vacuum arc deposition method.

The mixed sample simulates ITER-like PFCs surface composition in the redeposition zones [9]. Al was used as proxy for Be (not used in these samples due to its toxicity) that, together with W and C will be one of the principal constituent of ITER PFCs [9]. The nominal thickness of the superficial layer was ≥ 3 µm. The samples, round shaped, with a diameter of 25 mm, were first characterized with Thermal Desorption Spectroscopy (TDS) [10] and Nuclear Reaction Analysis (NRA) [12] to determine the concentration of each element.

2. Results

2.1. Depth profiling

To highligth the depth profiling capabilities of LIBS we have repeatedly hit the same point of the sample with subsequent laser shots ablating the material coming from deeper and deeper layers of the sample and recording at the same time the evolution of the spectral signal as a function of the shot number.

To do this the spectral region between 425 and 445 nm has been selected because in this range it is possible to find intense spectral emissions of both C (in particular CII at 426.73 nm) and W. Figure 2 shows the sequence of spectra obtained for successive laser shots on the sample DLC; it is possible to clearly recognize the evolution of the spectral signal that shows the intense spectral emission of the CII up to about the 21° laser shot where subsequently emerges a very different spectral track, ascribable to the W substrate, to indicate that the surface layer of codeposited carbon was completely ablated.

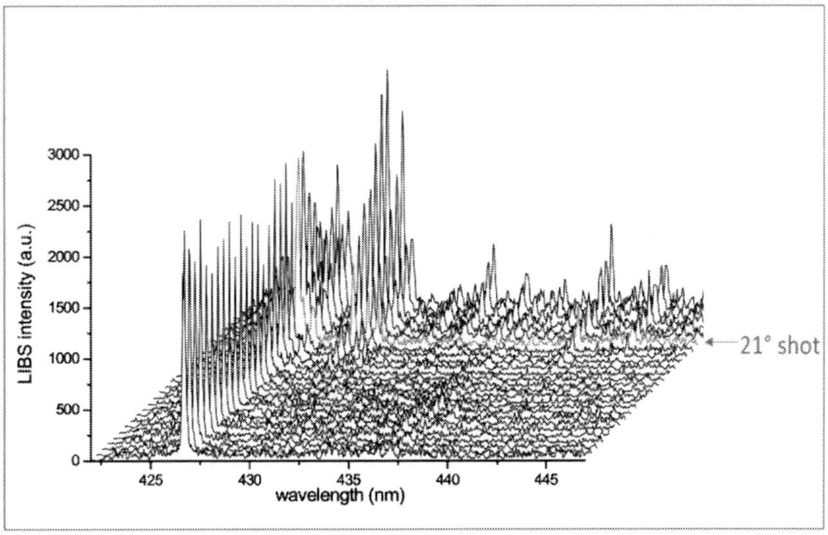

Figure 2. Sequence of spectra in the spectral region between 425 and 445 nm.

Similar results were also obtained on the mixed sample W: Al: C, not only in the spectral region above but also in the spectral region between 392 and 405 nm, where strong spectral emissions attributable to tungsten and aluminum are detectable.

The obtained results suggest that monitoring the intensity of the spectral lines of each species of interest (e.g. Al I at 394.4 and 396.2 nm and WI at 400.9 nm for the substrate) allows to control its concentration in the layer under investigation. Figure 3 shows a graph of some spectral emissions of C and W in the spectral range of figure 2 as a function of laser shot number. The transition from the mixed layer to the bulk material is clearly seen around the laser shots # 20-25.

Considering a depth of the surface layer of 3 μm (as shown by the NRA measurements) an ablation rate of 0.13 μm/laser pulse is obtained.

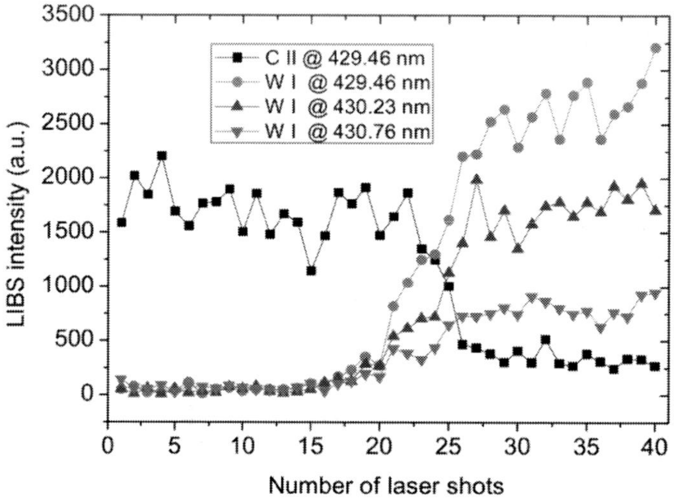

Figure 3. LIBS intensity as a function of laser shot for some of the C and W atomic and ionic lines in the spectral region 425-445 nm.

This value depends on many factors including the laser power, the profile of the beam, the focusing optics, the duration of the laser shot [13-14], and can not be regarded as an absolute reference.

However is worth remembering that a further increase in depth profiling resolution could be obtained simply decreasing the laser fluence (provided that the ablation treshold is still reached and the emission signals emerge from the background noise), while a greater laser power could provide a higher ablation rate and a higher spectral signal at least until effects of saturation of the spectral lines, mainly due to reabsorption of the laser energy by the same plasma are not observed [15], limiting the interaction with the target and the ablated mass per single shot.

2.2. Detection of trace elements.

The evaluation of the relative and absolute abundance of deuterium as a function of depth in the bulk of the tiles is of principal interest for the present study.

It has been possible to detect the D_α spectral line at 656.1 nm, well distinguished from H_α spectral line at 656.29 [16]. The graph of Figure 4a shows the LIBS signal in the spectral region of interest (between 653 and 659 nm) on sample DLC. H_α and D_α emissions are clearly visible and well resolved, together with the spectral emissions of two ionic C lines at 657.8 nm and 658. 29 nm respectively.

Through the preliminary NRA and TDS measurements we have mentioned before, the atomic concentration of D in the codeposited surface layer was estimated to be around 3%, a value in agreement with what obtained in the present LIBS measurements once the calibration free method was applied [17].In the graph of Figure 4b the evolution of the spectral emission as a function of laser shots is reported, it shows a reduction of the spectral signal around shot 20-25, consistently with the evolution observed between 425 nm and 445 nm.

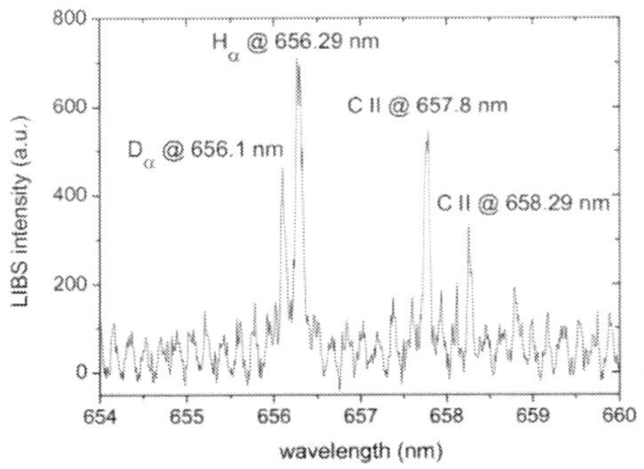

Figure 4a. LIBS spectrum of sample DLC in the region between 654 and 660 nm, the spectral emission of hydrogen (H_α) and deuterium (D_α) is shown, in addition with the emission of two C ionic lines

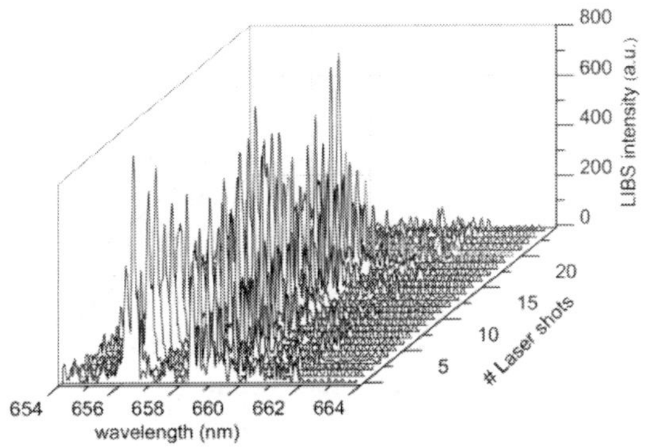

Figure 4b. Evolution of the LIBS signal as a function of laser shot in the same spectral region of Figure 4a. As in Figure 2 the red spectrum was acquired at the 21th laser shot

3. Conclusions

We set up a LIBS apparatus able to excite and detect plasma emission at long distances and in vacuum.

The targets used were prepared in order to reproduce the surface layers of PFCs of current and new generation fusion devices.

It was possible to reveal both the atomic composition of the targets surface coating and the presence of deuterium in trace amounts on the surface layer, as confirmed by indipendent measurements.

From this study emerges that the LIBS technique is suitable to operate as stand-off diagnostics, both for monitoring the PFCs erosion and the fuel deposition on the fusion device wall.

Further developments of this study aim to determine the relative concentrations of the detected elements and the influence of a toroidal magnetic fields (<3 T) on the plasma plume shape and performance, testing the apparatus in a real tokamak like FTU (Frascati Tokamak Upgrade).

Acknowledgments
This work was carried out in the frame of Europea Fusion Development Agreement (EFDA) task WP11-ETS_DTM-01 "Dust and Tritium Management".

References
[1] Rubel M, Wienhold P, Hildebrandt D 2001 *J. Nucl. Mater.* **290–293** 473–7
[2] Roth J *et al.*, *J. Nucl. Mater.* 390–391 (2009) 1–9
[3] Stangeby P C 2000 *The Plasma Boundary of Magnetic Fusion Devices* (Bristol and Philadelphia: Institute of Physics Publishing)
[4] Miziolek A W, Schechter I, Palleschi V 2006 *Laser-Induced Breakdown Spectroscopy (LIBS): Fundamental and Applications* (Cambridge University Press).
[5] Caneve L, Colao F, Fantoni R, Spizzichino V 2006 *Appl. Phys. A* **86** (2006) 151–7.
[6] Fujimoto K, McWhirter R W P 1991 *Phys Rev A* **42** (11) 6588–6601.

[7] Barbini R, Colao F, Fantoni R, Palucci A, Capitelli F 1999 *Appl. Phys. A* **69** S175–8.
[8] Vacillo J M, Fernandez Romero J M, Rodriguez C, Laserna J J 1999 *Surf. Inter. Anal.* **27** 1009–15.
[9] Loarte A *et al* 2007 *Nucl. Fusion* **47** S203
[10] Gallagher P K 1998 *Handbook on thermal analysis and calorimetry* vol.1 (Amsterdam: Elsevier)
[11] Feldman L C, Mayer J W 1986 *Fundamentals of Surface and Thin Film Analysis*, (North-Holland)
[12] Cremers D A, Radziemski L J 2006 *Handbook of Laser-Induced Breakdown Spectroscopy* (New York: Wiley-Interscience)
[13] Dimitrov G, Zheleva T 1998 *Spectrochim. Acta* B **39** 1209
[14] Allemand C D 1972 *Spectrochim. Acta* B **27** 185
[15] Bogaerts A, Chen Z 2004 *J. Anal. Atomic Spectrom.* **19** 1169
[16] NIST atomic spectra database, http://www.nist.gov/pml/data/asd.cfm (last access 16.06.12).
[17] Ciucci A, Corsi M, Palleschi V, Rastelli S, Solvetti A, Rognoni E 1999 *Appl. Spectr.* **53** 960–4.

Generation of nano roughness on fibrous materials by atmospheric plasma

I Kulyk[1], M Scapinello[2] and M Stefan[3]
[1]Manager Atmospheric Plasma Technology Unit, *Veneto Nanotech SCpA*, Venice, Italy
[2]On leave from *CIVEN Inter University Association*, Venice, Italy
[3]Researcher Atmospheric Plasma Technology Unit, *Veneto Nanotech SCpA*, Venice, Italy

E-mail: illya.kulyk@venetonanotech.it

Abstract. Atmospheric plasma technology finds novel applications in textile industry. It eliminates the usage of water and of hazard liquid chemicals, making production much more eco-friendly and economically convenient. Due to chemical effects of atmospheric plasma, it permits to optimize dyeing and laminating affinity of fabrics, as well as anti-microbial treatments. Other important applications such as increase of mechanical resistance of fiber sleeves and of yarns, anti-pilling properties of fabrics and anti-shrinking property of wool fabrics were studied in this work. These results could be attributed to the generation of nano roughness on fibers surface by atmospheric plasma. Nano roughness generation is extensively studied at different conditions. Alternative explanations for the important practical results on textile materials and discussed.

1. Introduction

Applications of atmospheric plasma for light industry are going to substitute, in many areas, conventional liquid chemical treatments as well as low pressure plasma technology [1]. In particular, atmospheric plasma finds many important applications in textile industry. There could be noted the possibilities for improvement of fibers cohesion, of fiber sleeves and yarns breaking force, the anti-pilling property of fabrics and the anti-shrinking property of wool fabrics. Probably the most interesting application of atmospheric plasma in textile field is producing of anti-shrinking property of wool. Wool shrinking during washings occurs due the presence of scales on the fibers (of ca. 20 μm in diameter) surface. The scales are positioned on fiber similarly to roof tiles, oriented all in the same direction along the fiber. This structure creates different friction coefficients between two fibers in the cases when they are parallel or anti-parallel to each other. When anti-parallel fibers slide one over another during washing, differential friction brings to the irreversible shrinking of wool fabric. Conventional chemical anti-shrinking treatments (Basolan, Hercosett) are based on smoothing or elimination of scales from fibers surface. In the result of such strong modification (levelling) of fibers at μm level, they could slide one over the other with much less (differential) friction, thus producing minor shrinking of fabric. Plasma treatment, instead, doesn't eliminate the scales on the fibers.

Actually, anti-shrinking effect generated by plasma on wool fabrics is not clearly attributed.

[1] To whom any correspondence should be addressed.

Application of plasma produces different effects on fibers [1, 2]:

- partial elimination of adhered and covalently bounded lipids and other adhered substances (cleaning effect);
- generation of active sites on fibers, such as carbonyl CO, carboxyl COOH, amino groups and sulphonated SO_3 groups in particular case of wool fibers (activation effect);
- generation of nano roughness on fibers (morphologic effect [1]).

Presented work is dedicated to the study of morphologic effect produced by atmospheric plasma on fibers. Obtained results prove that nano structuring of wool fibers is responsible for the modified mechanical properties of wool, such as anti-shrinking, anti-pilling and sleeves strength.

2. Instruments and methods

2.1. Apparatus for Dielectric Barrier Discharge treatments of fibrous materials

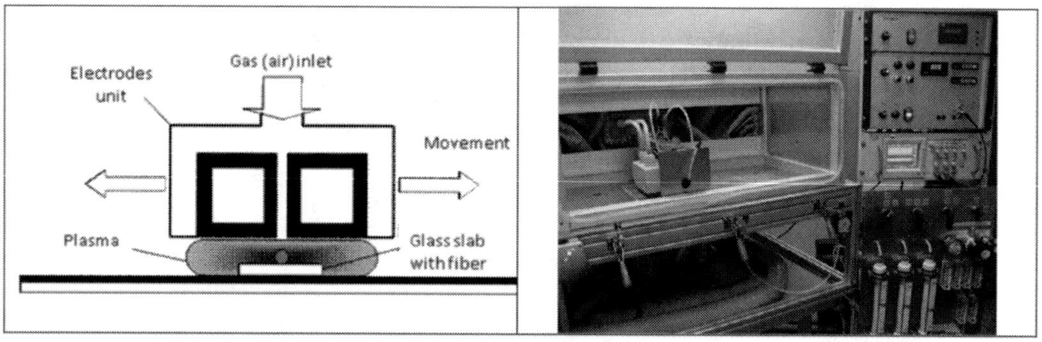

Figure 1. Scheme and layout of the apparatus for DBD treatment of textile materials.

Dielectric Barrier Discharge (DBD) treatments on wool and cotton fabrics were performed with original apparatus (figure 1) produced by Tigres GmbH (Rellingen, Germany). The discharge is applied in 2,0 mm gap between moving electrodes unit (2 electrodes covered by ceramic) and the grounded stainless steel plate covered with silicone sheet (ca. 1 mm). Conditioned air (RH 5 ÷ 30%) or special gas mixture flows at max 50 slpm between the electrodes and grounded plate. Current Wave and Pulsed mode (duty cycle 10 ÷ 80%) can be applied at max power 500 W and specific power 6,7 W/cm^2, at frequency ca. 40 kHz. The treatment can be performed in air, in pure gases or in mixtures (O_2; N_2; He; Ar; CO_2) with O_2 traces below 10 ppm.

Treatment of wool, cotton and polypropylene fiber sleeves were performed in-line by original pre-industrial scale machine “Nanofabia”. Treatments of other different fibers (TiO_2, basalt fibers) were performed with remote atmospheric plasma jet produced by Plasma Treat GmbH (Steinhagen, Germany).

2.2. Preparation and characterization of samples

Industrially scored knitted wool fabric was used. Before plasma treatments fabrics were additionally cleaned with dichloromethane for 24 hours (Soxhlet extraction) [3], washed twice with ethanol and twice with deionized water and finally dried.

Chemical extraction in alcoholic basic solution was applied in different experiments on wool for partial elimination of covalently bounded lipids by hydrolysis [4].

Washings of fabrics were performed in domestic washing machine according to internal test standard, analogous to Woolmark test method TM 31. Shrinking of wool fabrics (in area) were determined after test washing by standard ISO method.

Wicking experiment (rise of water in vertical sample of fabric) was used for the evaluation of the wettability of fabrics. The accuracy of wicking measurements is of ± 1 cm due to the variations of ambient humidity. All reported values of wicking were obtained with sampling time of 5 min.
AFM images of fibers were obtained with Scanning Probe Microscope Veeco D-3100 in tapping mode, using silicon tips NT-MDT NSG 10. The analyzed fibers were fixed on an adhesive tape before scanning. The images were elaborated with Gwyddion software and a polynomial functions were applied to exclude the fibers curvature.
SEM images were obtained with Tescan VEGA TS 5130 LM under 30 kV acceleration. Prior to scanning, the fibers were deposited with ca. 5 nm Au layer by means of sputtering in Cressington Sputter Coater 108 auto.
Raman spectra of fibers were obtained by scanning near field optical microscope Witec Alpha SNOM (Ar 514.5 nm).
FTIR-ATR spectra of fabrics samples were obtained by Perkin Elmer Spectrum One.

3. Experimental results of atmospheric plasma treatments

3.1. DBD treatments of single fibers and wool fabrics in air

The effect of nano roughness generation ("cauliflower" effect) by atmospheric plasma jet was reported in [5]. AFM study of DBD treatment in air applied on PVC films [6] has proved the chemical (not thermal) origin of the effect. It was observed the generation of nano roughness (RMS ca. 5 nm), superimposed on the original morphology of the polymer surface.

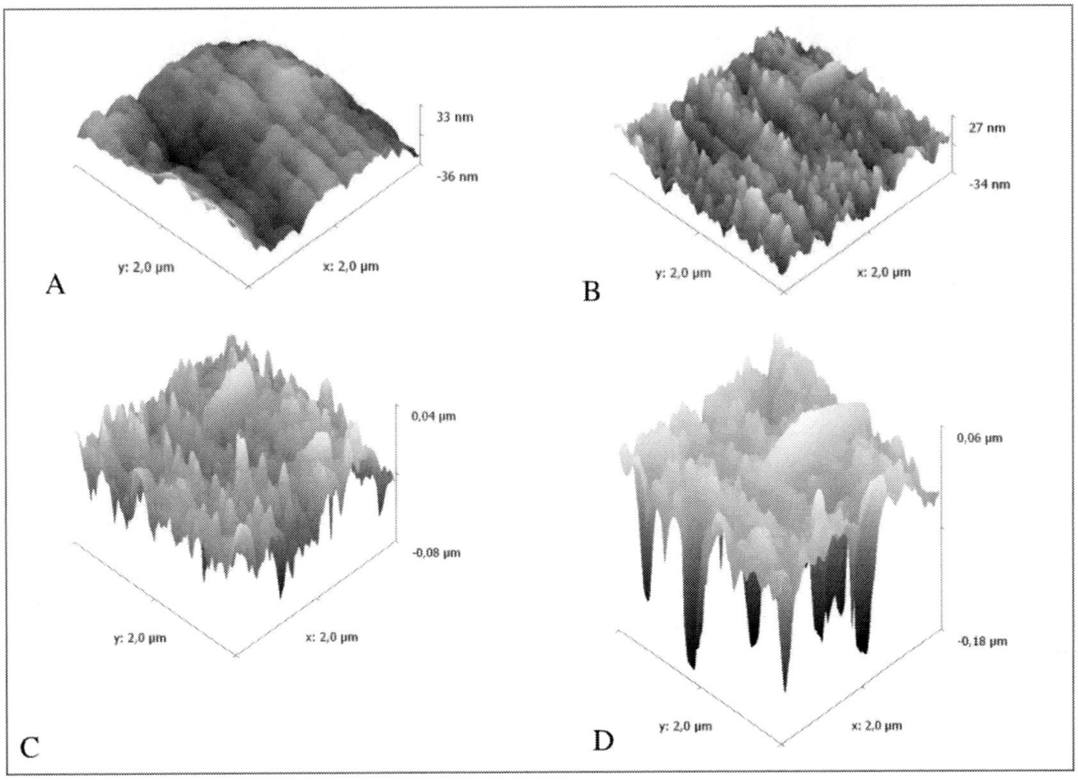

Figure 2. AFM images of the same zone of single wool fiber not treated (a) and treated by DBD in air with applied energy 60 J/cm^2 (b), 120 J/cm^2 (c) and 240 J/cm^2(d).

Subsequently, AFM studies of nano roughness effect was performed on single wool fibers (figure 2) extracted from fabric before DBD application. Analysis performed on precisely detected zone of

single fiber (2 μm x 2 μm) have demonstrated the increasing of produced nano roughness with applied treatment energy. Not treated fiber presents characteristic "striations" structure (figure 2a). Treatment by air DBD with the energy of 60 J/cm^2 produces nano roughness on fibers with RMS value of 10 nm. SEM images of fibers treated at the energy of 60 J/cm^2, that produce requested by Woolmark anti-shrinking property on wool fabric, present no difference respect to not treated fibers. Subsequent application of higher energies (120 J/cm^2 and 240 J/cm^2) on the same single fiber increase roughness to RMS 50 nm (figure 2c, d). SEM images of the fibers treated at 120 J/cm^2 shows the erosion of fibers at μm level, and at 240 J/cm^2 the pin holes on fibers could be observed.
Weight losses were measured on wool fabric samples treated at 60 J/cm^2 and 240 J/cm^2. After post-treatment conditioning (RH 65%, 24 hours), there were observed weight losses of 0,5% and 1,1%, correspondingly.
Post treatment monitoring (5 months) of treated fabric samples presents the decay of wicking (from 11 cm to 3 cm), while produced anti-shrinking property was not decayed.
After additional extraction of covalently bounded lipids from wool fabric, residual amount of covalently bound lipids was detected by gas chromatography measurements (GC-FID) [7]. Chemical extraction of 90% of covalently bounded lipids has not generated nano erosion of the fibers (figure 3) and has brought to fabric shrinking of - 12%. Instead, air DBD treatment, eliminating only 30% of covalently bounded lipids, has brought to fabric shrinking of - 8%.

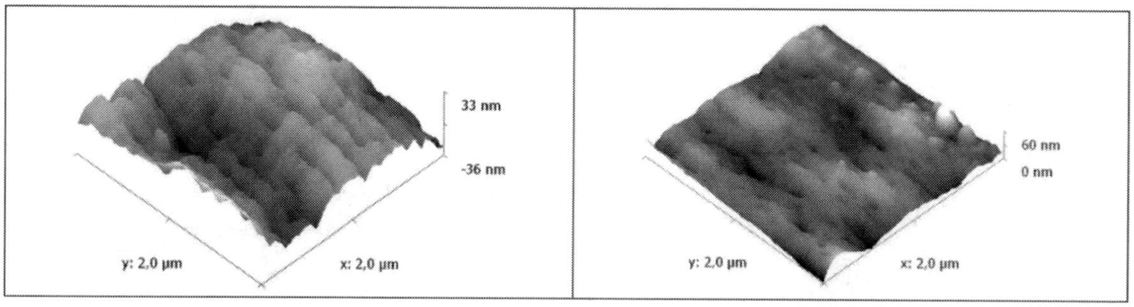

Figure 3. AFM images of wool fibers of industrially scoured fabric: after Soxhlet extraction (left) and after additional chemical extraction of covalently bounded lipids (right).

Anti-pilling effect of DBD treatment on different fabrics was studied. Air DBD treatment of knitted wool fabric improves anti-pilling property from 2÷3 to 3÷4 (at scale of 5). Application on woven fabric (wool 65%, polyamide 33%, elastan 2%) improves anti-pilling property from 2 to 4 (at scale of 5).
In line air DBD treatments of wool and cotton non-scoured fiber sleeves were performed by innovative pre-industrial machine "Nanofabia" [8]. Increasing of wettability and of breaking force of sleeves were observed. Subsequently produced yarns present increase of breaking force for ca. 20%.
Similar values for increasing of breaking force were obtained by in line air DBD treatment of cotton yarns.
Generation of nano erosion proportional to applied energy of air DBD was observed on single cotton fibers as well (figure 4).

Journal of Physics: Conference Series **406** (2012) 012015
doi:10.1088/1742-6596/406/1/012015

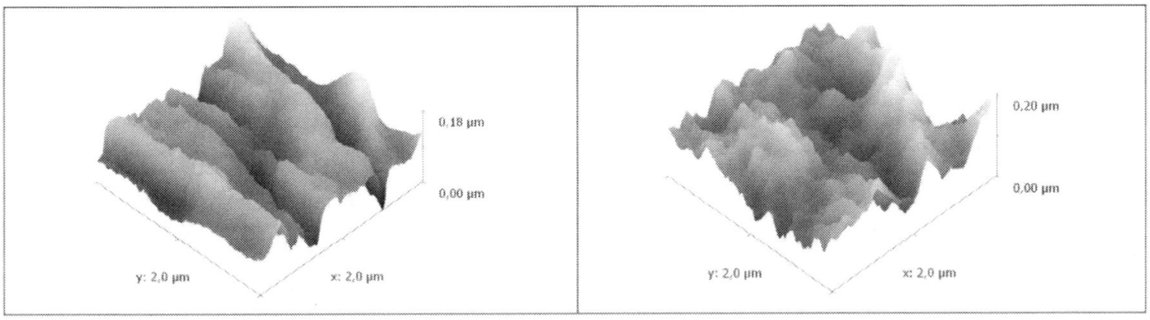

Figure 4. AFM images of single cotton fibers not treated (left) and treated by DBD in air with applied energy of 300 J/cm^2 (right).

3.2. DBD treatment of single wool fibers and fabrics in N_2

The effect of nano scale erosion of single wool fibers was studied for DBD treatment in N_2 at atmospheric pressure (figure 5), with O_2 in traces (5 ÷ 20 ppm).

Micro Raman analysis at characteristic bands of sulphonated groups reveals no oxidation effect of N_2 plasma on wool single fibers, at a difference with air DBD treatment.

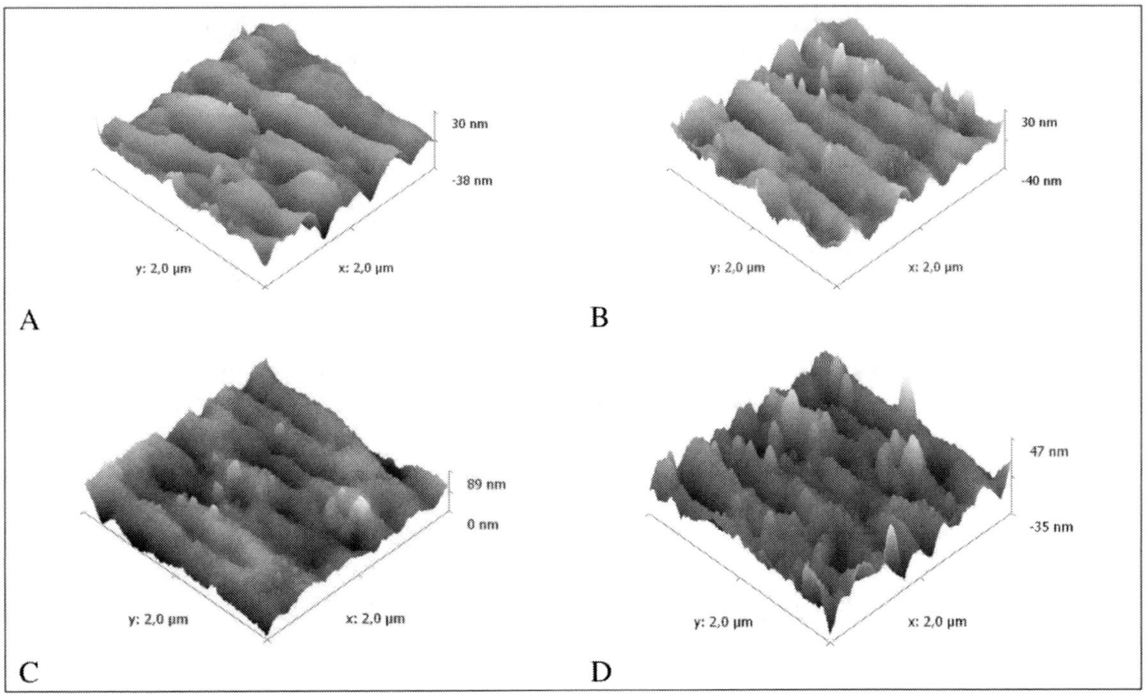

Figure 5. AFM images of single wool fiber not treated (a) and treated by DBD in N_2 atmosphere with applied energy 60 J/cm^2 (b), 120 J/cm^2 (c) and 240 J/cm^2 (d). Images (a, b) and (d) refer to the same zone of single fiber, image (c) refers to the adjacent zone of the same fiber.

The wicking effect and produced anti-shrinking properties of wool fabrics treated in N_2 DBD are at small extent lower respect treatments in air: wicking of 8 cm in N_2 DBD instead of 10 cm for air DBD, final shrinking of – 7% in the result of N_2 DBD application, instead of -4% in the result of air DBD treatment of the same energy.

4. Discussion

4.1. Modified properties of wool fabrics observed in wet conditions

Anti-shrinking property of wool fabrics measured by standard washings can be related to different effects:

- Elimination of lipids (cleaning of fibers).
  Chemical extraction of lipids doesn't produce nano roughness on fibers (figure 3) and is not sufficient for strong anti-shrinking effect on wool fabrics. It can generate active sites on fibers that bring to medium level of anti-shrinking properties.
- Generation of active sites (activation of fibers).
  Decay of wool activation was reported for low pressure plasma treatments [9]. This phenomenon was attributed to the conversion of covalently bounded lipids on the fibers surface. Observed by us permanent anti-shrinking effect on wool fabric indicates that anti-shrinking properties are not directly related to the activation of fibers by atmospheric plasma.
- Generation of nano roughness on fibers.
  DBD treatment etches the fibers, creating the nano roughness structure on their surface. Surface nano roughness increases the friction between fibers and limits their relative movement in wet conditions, thus reducing the probability for fibers scales to make fabric shrinking.
  In particular case of DBD treatment in N_2, there was not revealed the oxidation of wool. Observed anti-shrinking effect in this case can not be attributed neither to the elimination of lipids nor to the oxidative activation of fibers. Nano roughness effect on fibers and anti-shrinking properties on fabric were found less pronounced in this case respect to the DBD treatment in air.

4.2. Modified properties of textile materials observed in dry conditions

Increase of differential friction was observed in dry conditions between plasma treated fibers [10]. Optimized properties of wool fabrics, observed in dry conditions, can't be attributed to the generation of active groups and to fiber-fiber interactions. Instead, they are related to nano roughness produced on fibers by plasma treatment. Observed improvement of breaking force of fiber sleeves and of yarns, as well as improvement of anti-pilling property of fabrics are due to the nano roughness effect produced on fibers by atmospheric plasma treatment.

5. Conclusions

Atmospheric plasma technology is going to substitute environmentally harmful conventional liquid chemistry technology applied nowadays in textile industry. Besides of impressive practical results and of ecological and economical advantages of this emerging technology, applications of atmospheric plasma preserve fine structure of fibers, producing modifications at nano level.

Generation of nano roughness on fibrous materials by atmospheric plasma appears to be a basic phenomenon on this kind of materials, that makes possible such important industrial applications as increasing of breaking force of fiber sleeves and of yarns, and improving of anti-pilling property of fabrics.

With the all probability, the effect of nano roughness generation plays basic role also for the production of anti-shrinking properties of wool fabrics in the result of atmospheric plasma treatments.

6. References

[1] Thomas H 2007 Plasma modification of wool *Plasma Technologies for Fabrics* ed Shishoo R (Woodhead Publ. Ltd) ch. **9** 228-46

[2] Molina R, Jovancic P, Comelles F, Bertrán E and Erra P J 2002 *Adhesion Sci. Technol.* **16** 1469-85

[3] Brack N, Lamb RN, Pham D, Phillips T and Turner P 1999 *Surf. Interface Anal.* **27** 1050

[4] Meade S J, Dyer J M, Caldwell J P and Bryson W G 2008 *Textile Res. J.* **78** 943

[5] Noeske M, Degenhardt J, Strudthoff S and Lommatzsch U 2004 Plasma jet treatment of five polymers at atmospheric pressure: surface modifications and the relevance for adhesion. *Int. Journal of Adhesion & Adhesives* **24** 171

[6] Kulyk I, Scopece P and Schiavuta P 2008 Nano structuring of different polymer surfaces by atmospheric plasma treatment *Proc. Federchimica Workshop on Nanomaterials (Milan, Italy)*

[7] Negri A P, Cornell H J and Rivets D E 1991 *Aust. J. Agric. Res.* **42** 1285

[8] Kulyk I and Stefan M 2012 Novel Machine for Atmospheric Plasma Roll-to-Roll Treatment of Continuous Materials *Proc. Int. Conf. NanotechItaly 2012 (Venice, Italy, 21-23 November 2012)* in press

[9] Canal C, Molina R, Bertrán E and Erra P 2008 *Fibers and Polymers* **9** 444-9

[10] Mori M and Inagaki N 2006 Relationship between anti-felting properties and physicochemical properties of wool fibers treated with Ar-plasma *Textile Res. J.* **76** No. 9 687-94

Study of RF-excited Diethylene Glycol Dimethyl Ether Plasmas by Mass Spectrometry

M. A. Algatti, R. P. Mota, P. W. P. Moreira Júnior, R. Y. Honda, M. E. Kayama, K. G. Kostov

FEG-DFQ-UNESP, Av. Ariberto Pereira da Cunha 333, 12516-410, Guaratinguetá, SP, Brazil

E-mail: algatti@feg.unesp.br; mauricio.algatti@gmail.com

Abstract. This paper deals with the study of the fragmentation process of diethylene glycol dimethyl ether ($CH_3O(CH_2CH_2O)_2CH_3$) (diglyme here in) molecule in low pressure RF excited plasma discharges. The study was carried out using mass spectrometry. The results showed that for a fixed pressure, the increase of the RF power coupled to the plasma chamber from 1 to 35 W produced a plasma environment much more reactive which increases the population of the ionized species like CH_3^+ (15 amu), $C_2H_4^+$ (28 amu), CH_3O^+ (31 amu), $C_2H_4O^+$ (44 amu), $CH_3OCH_2CH_2^+$ (59 amu) and $CH_3OCH_2CH_2O^+$ (75 amu). This fact may be attributed to the increase of the electronic temperature that makes predominant the occurrence of inelastic processes that promotes molecular fragmentation. For a fixed value of RF power the increase of pressure from 50 mTorr to 100 mTorr produces the decreasing of the above mentioned chemical species due the lower electronic mean free path. These results suggest that if one wants to keep the monomer's functionality within the plasma deposited films resulting from such kind of discharges one must operate in low power conditions.

1. Introduction

Plasma materials processing by low pressure RF excited glow discharges is of key importance in many current technological issues encompassing microelectronics and biomaterials industry [1]-[4]. The main reason is that within such kind of plasmas electrons may attain an energy excess of some eV in comparison with the heavy particles present in the discharge. This unusual thermal non-equilibrium situation is extremely profitable in molecular fragmentation by electronic impact, giving rise to a very reactive chemistry in a relative cold environment whose kinetics is not easily controlled [2, 4]. Therefore is of paramount importance to set many plasma diagnostics in order to probe the trends of chemical species as well as the electrons for different plasma parameters such RF power coupled to the plasma chamber, gas pressure, gas flux and so on [1]-[4]. In the field of biomaterials science, plasma polymerized poly-ethylene glycol dimethyl ether is a material that has been keeping the attention of the scientific community due to its non fouling characteristics [5]-[11]. If the appropriate plasma parameters are set these films may be synthesized keeping a molecular structure similar to the polyethylene oxide-like (PEO-like) with the advantage that these films are not soluble in water. The aqueous solubility of PEO makes it less appropriate for many biomaterials applications. In order to retain the monomer structure within the plasma deposited films and consequently its functionality, many different issues have been addressed in recent literature as for instance the film deposition under low mean RF power level by controlling the power supply on/off ratio [12, 13], the decreasing of monomer residence time and consequently the reduction of its interaction with the plasma environment [14], the cooling of substratum with liquid nitrogen [15], the energy reduction of the ions reaching the substratum [16], and so on. In order to set the appropriate experimental parameters that would result customized film structures in plasma polymerization of PEO-like coatings this paper deals with the study of the RF power dependence of different ionized chemical species resulting from diglyme fragmentation for different values of the pressure inside the plasma reactor. The trends of different chemical species were followed by mass spectrometry [2, 4].

2. Experimental setup and plasma diagnostics

The glow discharges were generated by a RF power supply operating in the range from 5 to 45 W in diglyme atmospheres ranging from 50 mTorr to 100 mTorr within a stainless steel cylindrical, 210 mm of internal diameter and 225 mm long, parallel plate electrodes plasma reactor. The chamber is provided with eight lateral entrances, positioned at the mid plane between the electrodes, that may be used for setting optical, electrical and mass diagnostics and the low (mechanical pump) and high (turbo-molecular pump) vacuum systems. The vacuum inside the plasma chamber is monitored by pirani^{TM} (thermocouple) and penning^{TM} (inverse magnetrom) gauges. The turbo-molecular pump is coupled to the chamber through a gate valve and is used for cleanness purposes. The pressure is pumped down to 10^{-6} Torr, being the chamber purged with argon several times before each running of the experiment. The inner side of the plasma chamber was polished up to the optical quality (roughness of 0.5 μm or less) in order to minimize the retention of impurities and facilitate the cleaning process. The plasma chamber walls were heated with a temperature controlled belt in order to minimize the monomer's condensation as well as the humidity. Diglyme was placed inside a stainless steel bottle and was fed into the plasma chamber through a needle valve. The plasma was excited by a RF power supply operating in 13.56 MHz whose output intensity could be varied from 0 W to 300 W (Tokyo HY-Power model RF-300^{TM}). The RF power was coupled to the plasma reactor through an appropriate matching network (Tokyo HY-Power model MB-300^{TM}) that allows one to minimize the reflected RF power. The mass spectrometry was performed using a mass spectrometer and energy analyser (Hiden Analytical model EQP-300^{TM}), operating in the mass and energy range from 1 amu to 300 amu and from 0 eV to 100 eV respectively. The block diagram of the experimental setup is presented in Figure 1.

3. Results and discussion

It is presented in table 1 all the possible primary fragments of diglyme molecule (134 amu) resulting from the disruption of the chemical bonds between carbon-carbon and carbon-oxygen atoms. Since the molecule is central symmetric to the oxygen atom located in ethylene-glycol group one may identify only four different points of molecular disruption involving the above mentioned chemical bonds which results the fragments listed in Table 1. It must be pointed out that all the molecular fragments presented in table 1, with the exception of the number 8 (119 amu), may result from secondary or higher fragmentation process of the more heavy fragments of the diglyme molecule. The number of possibilities contributing for the population of a fragment increases inversely proportional to its mass. From the point of view of the non-fouling characteristic of the plasma deposited diglyme films the ethylene-glycol structure, e.g., CH_2CH_2O (44 amu), is one the most important structure to be retained within the film composition if one wants to keep its functionality [2]. Therefore is important to keep under control the plasma parameters that enhance the presence of heavy fragments within the discharge. Figure 2 shows the RF power dependence of diglyme fragments within the plasma, for a fixed pressure of 140 mTorr. It can be seen that as the RF power coupled to the plasma chamber increases the concentration of the ionized species increases indicating the predominance of electronic inelastic collisions resulting in molecular fragmentation. Figure 3 shows a typical

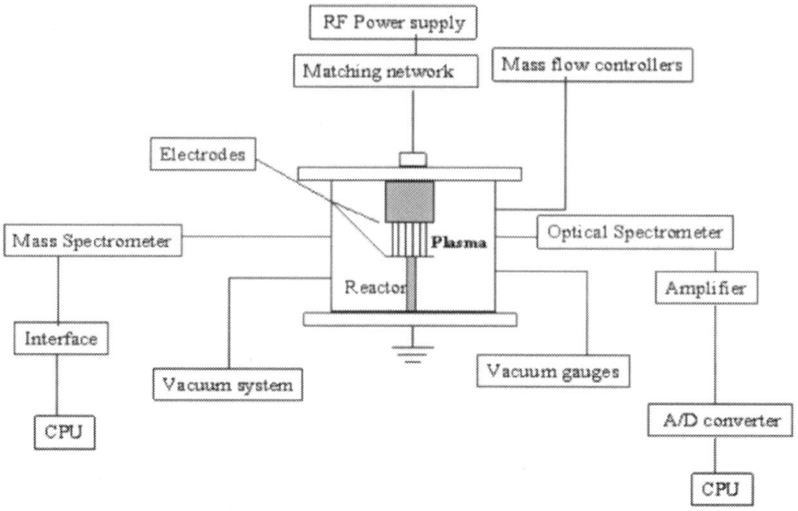

Figure 1. Block diagram of the experimental set-up

Table 1. Primary Diethylene-glycol-dimethyl-ether fragments and the monomer molecule

fragment	chemical structure	mass (amu)
1	CH_3	15
2	CH_3O	31
3	CH_3OCH_2	45
4	$CH_3OCH_2CH_2$	59
5	$CH_3OCH_2CH_2O$	75
6	$CH_3OCH_2CH_2OCH_2$	89
7	$CH_3OCH_2CH_2OCH_2CH_2$	103
8	$CH_3OCH_2CH_2OCH_2CH_2O$	119
monomer	$CH_3OCH_2CH_2OCH_2CH_2OCH_3$	134

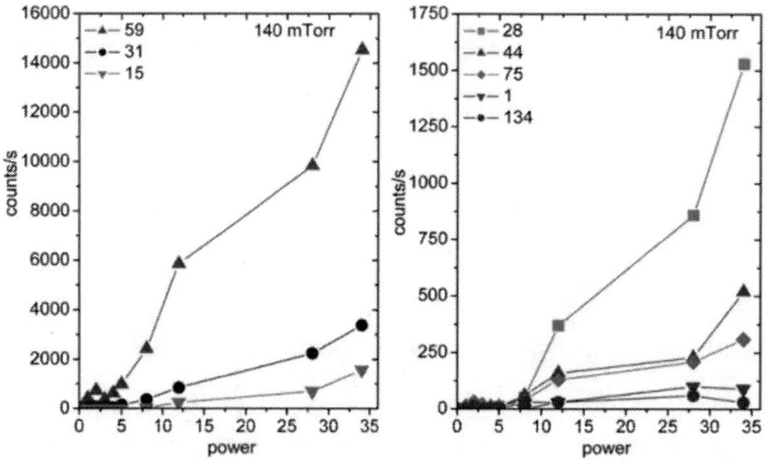

Figure 2. RF power dependence of diglyme fragments for a fixed pressure of 140 mTorr

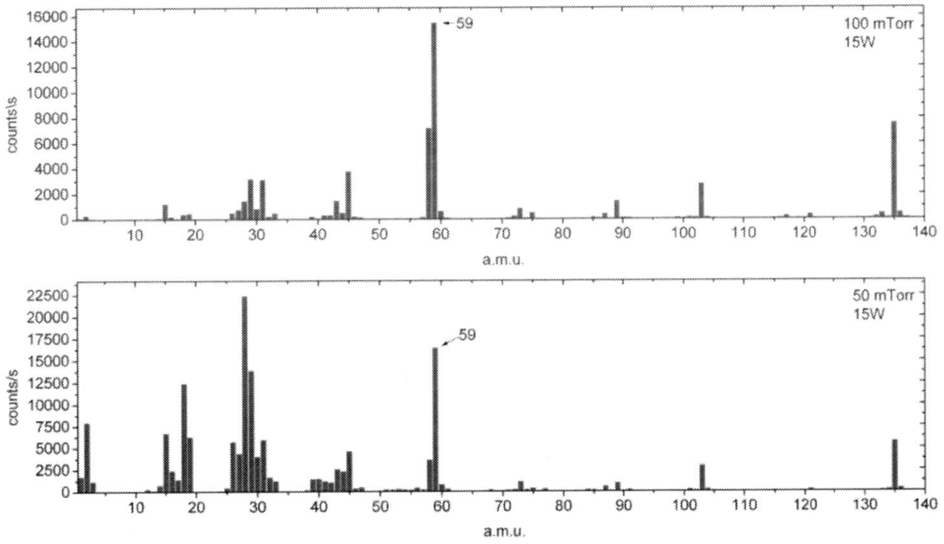

Figure 3. Typical diglyme mass spectra from 1 amu (atomic mass unit) to 134 amu, at 100 mTorr (top) and at 50 mTorr (bottom), for 15 W of RF

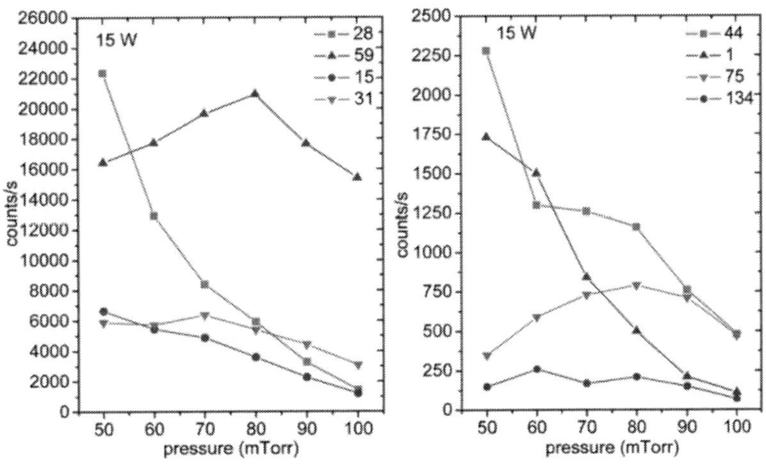

Figure 4. Pressure dependence of diglyme fragments for 15 W of RF power

diglyme mass spectra run at 100 mTorr and 50 mTorr for the RF power kept fixed on 15 W. It may be seen that the number of lighter chemical species increases considerably. This behaviour of discharge can be explained if one considers that for lower pressure the molecular fragmentation is enhanced due the higher values of electronic mean free path and mean temperature. Figure 4 shows the pressure dependence of diglyme ionized fragments for a fixed RF power coupled to the plasma chamber. It can be appreciated that the number of ions decreases with pressure mainly due to molecular recombination process that is enhanced in such conditions. This fact is corroborated by the neutral species mass spectra data, not shown in the present paper.

4. Conclusions

One can conclude from the presented results that the operation of the plasma reactor at low RF power levels is appropriate if one wants to preserve the monomer functionality into the plasma deposited thin films, since it enhances the predominance of heavy chemical species within the discharge reducing the fragmentation processes.

Acknowledgements

Authors would like to thank FAPESP, CAPES and CNPq for financial support.

References

[1] 2008 *Advanced plasma technology* ed R D'Agostino, P Favia, Y Kawai, H Ikagami, N. Sato and N. Arefi-Khonsari (Weinheim: Wiley-VCH)
[2] 2008 *Low temperature plasmas: fundamentals, technologies, and techniques* ed R Hippler, H Kersten, M Schmidt and K Shoenbach (Weinheim: Wiley-VCH)
[3] Ostrikov K 2008 *Plasma nanoscience: basic concepts and applications of deterministic nanofabrication* (Weinheim: Wiley-VCH)
[4] 2000 *Handbook of advanced plasma processing techniques* ed R J Shul and S J Pearton (Berlin: Springer)
[5] Hendricks S. K., Kwok K., Shen M., Horbett T. A., Ratner B. D., and Bryers J. D 2000 *J. Biomed. Mater. Res.* **50** 160
[6] Mar M. N., Ratner B. D. and S. S. Yee S. S 1999 *Sensors and Actuators B* **54** 125
[7] Kwok C. S., Horbett T. A. and B. D. Ratner B. D 1999 *J. Control. Release* **62** 301
[8] Lopez G. P., Ratner B. D., Tidwell C. D., Haycox C. L., Rapoza R. J. and Horbett T. A 1992 *J. Biomed. Mater. Res.* **26** 415
[9] Algatti M. A., Mota R. P., Honda R. Y., Kayama M. E., Kostov K. G., Fernandes F. S., Azevedo T. C. A. M. and Cruz N. C 2009 *Eur. Phys. J. D* **54** 325
[10] Cheng Q., Li S., Komvopoulos K 2009 *Biomaterials* **30** 4203
[11] Cheng Q. and Komvopoulos K 2009 *J. Phys. Chem. C* **113** 213
[12] Cicala G., Creatore M., Favia P., Lamendola R. and D'Agostino R 1999 *App. Phys. Lett.* **75** 37
[13] Oehr C., Müller M., Elkin B., Hegemann D. and Vohrer U 1999 *Surf. Coat. Technol.* **116-119** 25
[14] Favia P. and D'Agostino R 1998 *Surf. Coat. Technol.* **98** 1102
[15] Biederman H. and Slavínská D 2000 *Surf. Coat. Technol.* **125** 371
[16] Itatani R 1998 *Plasm. and Ions* **1** 37

Journal of Physics: Conference Series **406** (2012) 012017 doi:10.1088/1742-6596/406/1/012017

Surface Modification of Polyester Fabric by Non-Thermal Plasma Treatment

R C Lima da Silva[1,4], C Alves Jr[2], J H Nascimento[3], J R O Neves[4] and V Teixeira[5]

[1]Mechanical and Materials Department, Federal Institute of Rio Grande do Norte, Mossoró/Brazil.
[2] Mechanical Engineering Department, Federal University of Rio Grande do Norte, Natal/Brazil.
[3] Textile Engineering Department, Federal University of Rio Grande do Norte, Natal/Brazil.
[4] Textile Engineering Department, Minho University, Guimarães/Portugal
[5] Physics Department, Minho University, Braga/Portugal

E-mail: [1]ruthilima@yahoo.com.br

Abstract. In the present study it was carried out a surface modification of polyester fabric by plasma treatment with aim of providing hidrophilicity to fabric. In the process it were used three different gaseous atmosphere with mixtures of argon, nitrogen and/or oxygen [(1) Ar + N_2 (4:2), (2) Ar + N_2 +O_2 (4:2:2) and N_2 + O_2 (2:0.6)] and maintained others parameters such as pressure, current and time of treatment fixed at 1.35 mBar, 0.08 A and 30 minutes, respectively. The plasma treatment was monitored by optical emission spectroscopy (OES) to identify the species presents in the plasma reactor. Chemical changes in the fabric surface after plasma treatments were determined by Fourier Transform Infrared (FTIR) and X-ray photoelectron spectroscopy (XPS). Wicking properties were used to evaluate hydrophilicity of the fabric. Wicking measurement showed that the hydrophilicity of polyester fabric was remarkably improved after treatment and was explained by XPS analysis indicating higher presence of functional groups containing nitrogen and oxygen.

1. Introduction

Non-thermal plasma treatment is a clean, dry and environmental friendly method of textile materials modification since it improves the surface characteristic of polymers without affecting their bulk properties and does not require water or chemical reagents dispensing effluents treatment (Fang and Zhang 2009[1]; Wang *et al.* 2008[2]; Barni *et al.* 2005[3]). Plasma is produced by an electric discharge combined with gas, by applying a voltage between two electrodes inserted in a chamber at low pressures. Due to this voltage, electrons are accelerated and collide with plasma, forming neutral particles, ions, excited particles and electrons. Thus, it can possibly break molecule chains, generate new functional groups and/or morphological alterations in the polymers surface, such as microporosity (Yaman *et al.* 2009[4]). The introduction of water compatible functional groups (–COOH, –OH, –NH_2, etc.) increases the hydrophilicity of polyester (Morent *et al.* 2008[5]). Therefore, oxygen and nitrogen are the type of gas chosen to be used in the study of this textile material modification.

The influence of the process parameter in the Plasma treatment of textile materials has been extensively studied. For instance, Sun *et al.* (2011)[6] used an atmospheric pressure plasma jet (APPJ) to treat cotton yarns with helium/oxygen (He/O2) gas for the hydrophilicity and sizing property improvement. Kale and Desai (2009)[7] studied the use of non-polymerizing gas in atmospheric pressure plasma to treat different textile materials. Navaneetha Pandiyaraj and Selvarajan (2007)[8] investigated the effect of the exposure time, discharge potential and pressure level in the plasma treatment on cotton fabric, and they concluded that the dyeability of this fabric has been improved due to the increase of surface energy and roughness. Costa *et al.* (2006) [9] analyzed the effect of the atmospheric composition in the plasma reactor on the surface of polyester fabric using mixture of oxygen, hydrogen, nitrogen and argon, and found an increase in the fabric wettability and roughness. However, they did not concluded that new functional groups (–OH and –NH_2) have been formed as it was expected.

In the present work was studied the surface modification of polyester fabric made from multifilament yarns using non-thermal plasma with different gaseous atmosphere. The study was based on hydrophilic and chemical changes of the polyester caused by mixtures of argon, nitrogen and oxygen in the plasma reactor.

2. Experimental

2.1. Materials and sample preparation

Multifilament yarns of polyester (1/318/95 or 1 cable with 318 Dtex and 95 filaments) commercially obtained were used to produce rectangular samples of fabric with 130 mm of length and 10 mm of width. Samples were washed with neutral soap for 30 min and kept in a bath of water for 8 hours.

2.2. Plasma Treatment

Non-thermal plasma was generated in a cylindrical chamber of 250 mm length and 180 mm internal diameter, made of borosilicate. The sample was fixed between two electrodes discs (anode and cathode). The scheme of the plasma system was shown in Figure 1. The ends of the chamber were closed using two stainless steel flanges. Thus, gases were introduced in by upper flange while the vacuum system, pressure sensor and thermocouples were held in the bottom flange.

Discharge plasma was produced at a vacuum pressure fixed in 1.35 mBar. The electric current and the time of tests were maintained in 0.08 A and 30 minutes, respectively.

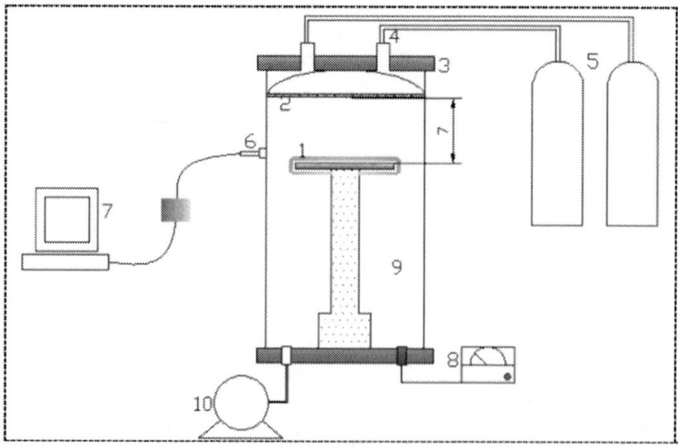

Figure 1. Schematic diagram of the plasma system. 1- cathode, 2- sample, 3- anode (superior flange), 4- gas entrance, 5- gas cylinder, 6- optical fibre, 7- computer, 8- pressure sensor, 9- vacuum chamber, 10- vacuum pump.

The fabric was placed perpendicularly to the cylindrical axis, between the parallel disc electrodes at 7 cm of distance from cathode. This distance ensured that there was no alteration thermal in the fabric due to the high temperatures reached in the cathode (Figure 1).

Journal of Physics: Conference Series **406** (2012) 012017 doi:10.1088/1742-6596/406/1/012017

Influence of the gaseous atmosphere on the surface modification of the polyester was analysed using nitrogen, oxygen and argon mixtures according to Table 1. Ocean Optics spectrometer was used to detect the species present in the reactor plasma during the treatment. Optical fibre cable had one end positioned to UV in the reactor and the other end linked to the optical emission spectrometer which was connected to the computer. During the plasma treatment a spectrum was recorded every 5 minutes.

Table 1. Gaseous atmosphere in the chamber of the plasma.

Gaseous	Proportion
Ar, N_2	4 : 2
N_2, O_2	2 : 0.6
Ar, N_2, O_2	4 : 2 : 2

2.3. Surface characterization

2.3.1. FTIR analysis
Fourier Transform infrared spectral analyses of the treated and untreated fabrics were recorded in the range of 4000 – 400 cm^{-1} using Nicolet-Avatar 360 with OMINIC 5.2 computerized software from Nicolet.

2.3.2. XPS analysis.
The chemical compositions of the sample surface were analyzed using an XPS model ESCALAB 200A (VG Scientific, UK) with data acquisition and analysis software PISCES. The X-ray source was Al Kα (1486.6 eV), operating at 300 W. The pressure within the XPS chamber was less than 10^{-6} Pa. The effect of the electric charge was corrected by the reference of the carbon peak (285 eV). The deconvolution of spectra was performed using the XPSPEAK41 program, in which an adjustment of the peaks was performed using peak fitting with Gaussian-Lorentzian peak shape and Shirley type background subtraction.

2.3.3. Wicking rate
A polyester fabric strip (130mm×10mm) was suspended vertically above the colored distilled water surface in a glass beaker in a way that the vertical bottom edge slightly touches the colored water. A spontaneous wicking occurs due to capillary force. The height of the liquid rise boundary is recorded every 20 s until 300 s.

3. Results and Discussion

3.1. Identification of species in the emission spectrum
The typical emission spectra obtained in UV region during the treatment are shown in the Figures 2, 3 and 4. Optical Emission Spectroscopy (OES) in UV is widely used for plasma diagnostics as non-invasive technique to identify the species present in the plasma. Thus, atomic and ionized species interacting with the samples can be identified.

Journal of Physics: Conference Series **406** (2012) 012017 doi:10.1088/1742-6596/406/1/012017

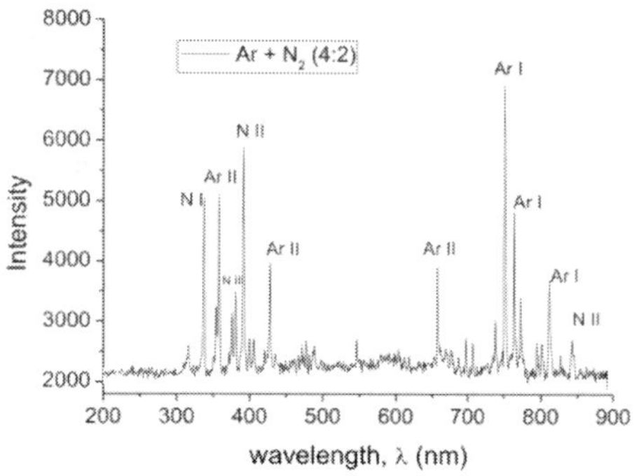

Figure 2. Emission spectrum in UV region during treatment with Ar and N_2.

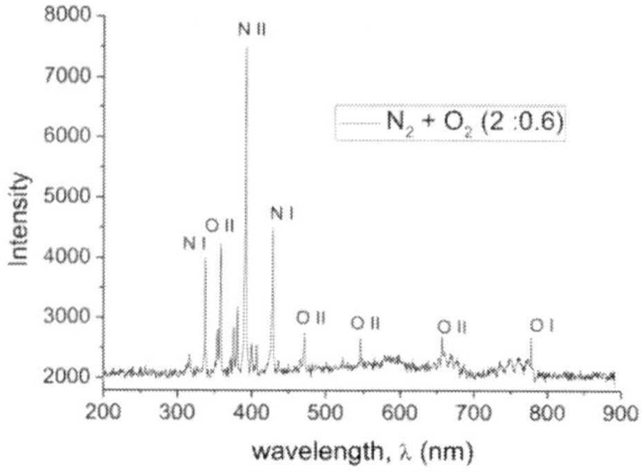

Figure 3. Emission spectrum in UV region during treatment with N_2 and O_2.

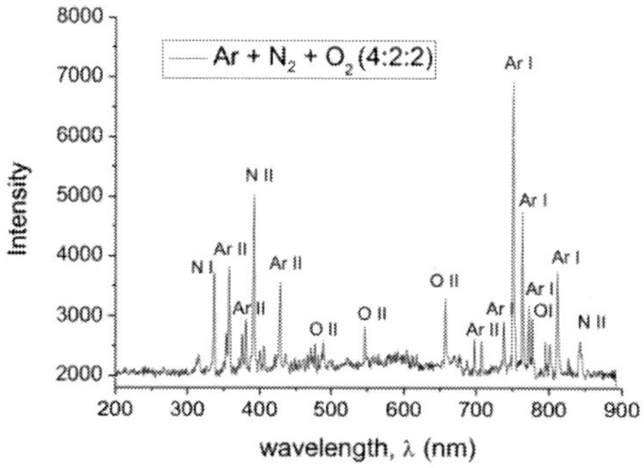

Figure 4. Emission spectrum in UV region during treatment with Ar, N_2 and O_2.

3.2. Surface Chemistry

The FTIR spectrum of polyester fabric before plasma treatment is shown in Figure 5. Analyzing the results, it is possible to observe that the spectrum is similar to polyester spectrum present in the literature (Li *et al.* 2010[10]; Yaman *et al.* 2009[4]). Principal peaks are presented in the Table 2 together with its functional groups. It was also identified a peak at 2360 cm^{-1} which is due to a different CO_2 concentration in the air in the IR cell.

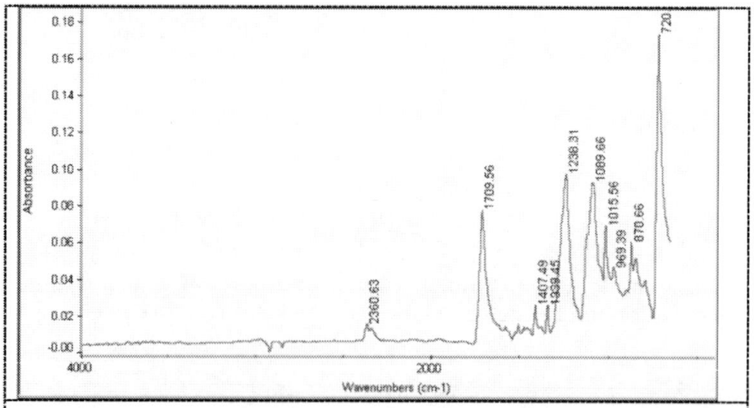

Figure 5. FTIR spectrum of polyester untreated.

Table 2. Functional groups associated with the peaks found in the FTIR spectrum of polyester untreated.

Functional group	Wavenumber (cm^{-1})
Carboxylic acid (C═O)	1709
Aromatic ring	1407
Alkane (CH_2)	1339
Carboxylic acid (C– O)	1238
Ester (O=C–O–C)	1089
Ester (O=C–O–C) or amine (C–N)	1015
Vinil (C–H)	969
Aromatic (C–H)	870
Aromatic (C–H)	720

Regarding the analysis of FTIR spectra of untreated and treated polyester fabric, it was not identified significant variations in the spectra before and after plasma treatment. Since plasma treatment works in a nanometer scale, it is impossible to be detected by the FTIR analysis that works in a micrometric scale.

XPS overall scan of untreated and plasma treated polyester fabric are shown in Figure 6. According with the results, it was detected four atoms: carbon, oxygen, nitrogen and silicon. It was also observed that after the plasma treatment with the three gaseous atmospheres ($Ar+N_2$, N_2+O_2 and $Ar+N_2+O_2$), the oxygen and nitrogen peak intensified while the carbon and silicon peak decreased. Table 3 shows chemical composition of the samples before and after the plasma treatment. It was also noted an indication of nitrogen presence (0.32 at %) in the untreated sample, and that these amount of nitrogen increased with the treatment, as can be seen in Table 3. Furthermore, Yaman *et al.* (2009)[4] found a small amount of nitrogen (0.4%) in untreated polyester using the XPS analysis and observed that this amount increases after plasma treatment in air atmosphere. Haverkamp *et al.* (2002) when studding aging of polyester powder coating also detected slight presence of nitrogen (0.3%) in the new polyester.

O/C and N/C ratio presented in the Table 3 facilitates the visibility of the increase of C and N concentration after the plasma treatment. Higher ratio O/C was obtained with N_2+O_2 atmosphere. After treatment with $Ar+N_2$ it was observed that O/C ratio increased from 0.365 to 0.453, which was unexpected since there was no oxygen in the atmosphere of the plasma reactor during the treatment. Charpentier *et al.* 2006[12] studying surface modification of medical grade polyester using nitrogen-plasma also detected an increase in the O/C ratio. They suggested that this increase in the oxygen occurred due to the reactive species presents on surface treated sample reacting with oxygen from the air immediately after removal of the samples from reactor. These results suggest that nitrogen and oxygen-containing polar groups were created on the polyester surface after the plasma treatment.

Silicon presence in the fabric samples is not a characteristic of the polyester which suggests that the material was contaminated during its manufacture. After the plasma treatment with $Ar+N_2$ the silicon groups almost disappeared of the polyester surface and decreased more of 50% when treated with N_2+O_2 and $Ar+N_2+O_2$ (Table 3).

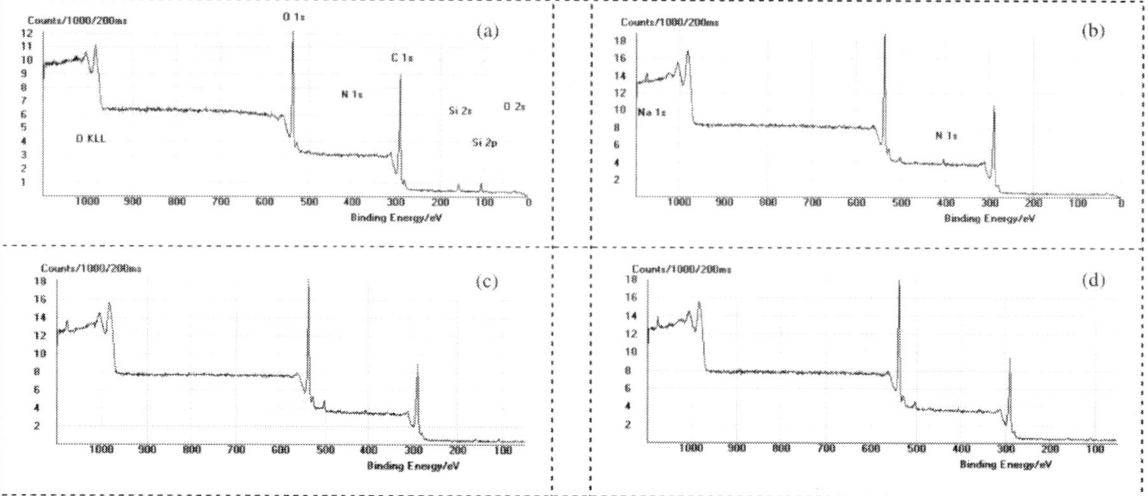

Figure 6. XPS overall scan of untreated (a) Ar and N_2 plasma treated (b) N_2 and O_2 plasma treated (c) and Ar, N_2 and O_2 plasma treated (d) polyester fabric.

Table 3. XPS data analysis.

Sample	Chemical composition (At%)				Ratio	
	Carbon	Oxygen	Nitrogen	Silicon	O/C	N/C
Untreated	68.94	25.19	0.32	5.55	0.356	0.004
Ar + N_2	66.99	30.36	1.77	0.88	0.453	0.026
N_2 + O_2	63.86	32.21	1.71	2.23	0.504	0.027
Ar + N_2 + O_2	64.92	31.59	1.12	2.36	0.486	0.017

Deconvulation analysis of C1s and N1s were realized to identify the chemical functional groups that were incorporated in the surface fabric after the plasma treatment (Figures 7 and 8). Figure 7 shows XPS spectra from the untreated and treated samples around the C1s energy region. The peak of the untreated polyester is composed of three main components: C–C or CH group which correspond at 285 eV and C–O and O–C=O groups referents at 286.5 eV and 289 eV, respectively. In the range of 286 and 286.5 eV can also be characterized the presence of the amine group (C–N) since it was detected nitrogen in the analysis.

After performing a treatment using three different gaseous atmospheres it is possible to observe that the peaks intensity regarding C–C and CH groups decreased while the peaks intensity referents to carboxyl and ester groups increased compared with those in the untreated sample. The variation of the relative area is presented in the Table 4 where can be noticed a reduction in the relative area of C–C and CH group after the treatment with Ar + N_2, N_2+O_2 and Ar+N_2+O_2, whilst the relative area of C–O and COO groups increased. These results evidence surface modification in some C–C and CH groups that have been oxidized forming C–O and O–C=O.

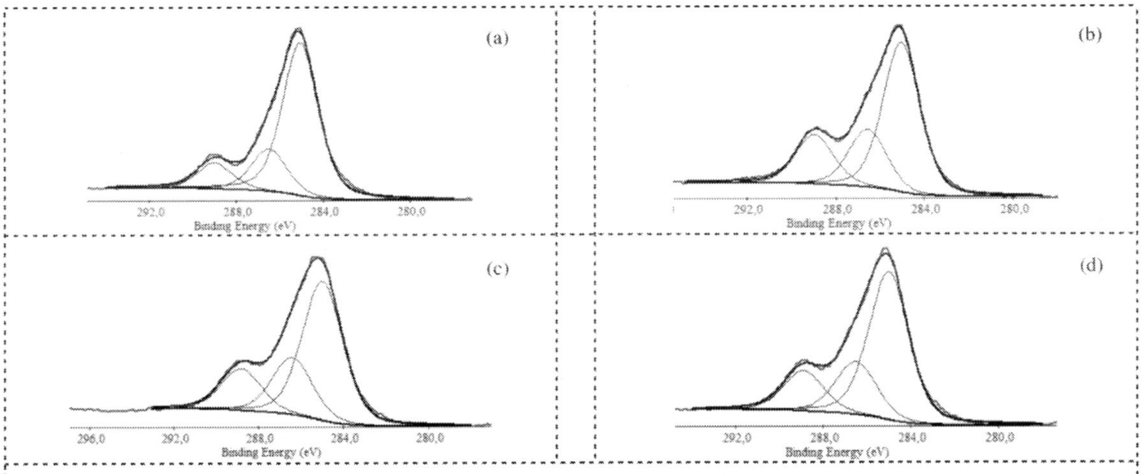

Figure 7. Deconvolution of XPS core level C 1s of fabric (a) untreated (b) $Ar+N_2$ (c) N_2+O_2 and (d) $Ar+N_2+O_2$.

Table 4. XPS data analysis of C 1s peak.

Sample	Relative area of different chemical bonds (%)		
	C – C and/or CH (285eV)	C–O and/or C–OH (286.5eV)[a]	O–C= O and/or COOH (289eV)
Untreated	69	12	19
$Ar + N_2$	58	19	23
$N_2 + O_2$	58	18	24
$Ar + N_2 + O_2$	61	17	22

[a] this value of energy also can refer to C–N

Observing Table 5 which shows the relative area of different chemical compositions around the O1s energy region, it can be observed two peaks: 532 and 533 eV that refers to C=O and O–C=O, respectively. Furthermore, significant modification was found to the sample treated with N_2+O_2, where can be noticed an increase of relative area referent at 533 eV peak that is associated to COO and NO_3^-.

Figure 8 shows XPS spectra from the untreated and treated samples around the N1s energy region and it shows a peak at 399.9 eV in the untreated sample (Figure 8(a)) which can indicate the presence of amine (–NH–) group in the polyester fabric as also identified by Cascalheira and Abrantes (2008)[13].

Analyzing Figure 8 it is also possible to observe that different functional groups appeared after the treatment, according to the different atmosphere gaseous. Thus, in the treatment with $Ar+N_2$ and $Ar+N_2+O_2$ two peaks appeared: one at 399.9 eV and the other at 400.6 eV (Figure 8(b) and (d)). The first peak which is referent to –NH– (399.9 eV) increased in area when compared with the untreated sample. This increase was 75% when treated with $Ar+N_2$ and 46% when treated with $Ar+N_2+O_2$. As long as the 400.6 eV peak is concerned, it may indicate N-C bonds formation. Regarding the treatment using N_2+O_2, two nitrogen peaks were observed, one at 400.07 eV and the other at 407.6 eV (Figure 8(c)). As Figure 8(c) shows, the first peak (400.07 eV) which concerns to –NH– had an area increase of 153% when compared with the untreated sample. Thus, the second peak (407.6 eV) refers to Nitrate (NO_3^-) as noticed by the high relative area value of 533 eV (57%).

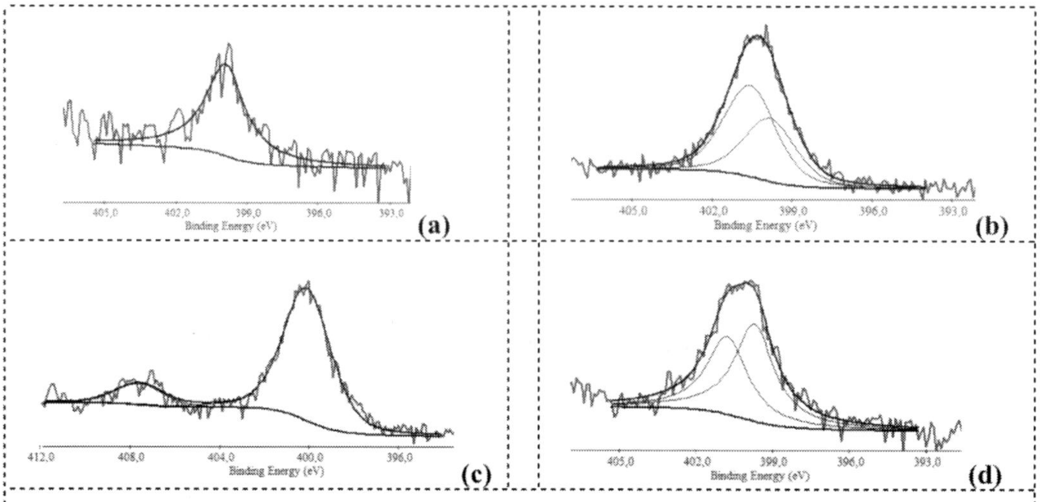

Figure 8. Deconvolution of XPS core level N 1s of fabric (a) untreated (b) Ar+N_2 (c) N_2+O_2 and (d) Ar+N_2+O_2

Table 5. XPS data analysis of O 1s peak.

Sample	Relative area of different chemical bonds (%)	
	C = O (~532 eV)	O – C= O (~533 eV)
Untreated	49	51
Ar + N_2	48	52
N_2 + O_2	43	57[a]
Ar + N_2 + O_2	50	50

[a]this value of energy can also refer to NO_3^-

3.3. Wicking rate – capillarity

As long as Figure 9 is concerned it shows the capillarity of untreated and treated samples. Thus, it is possible to observe that the untreated sample is totally hydrophobic since there was no vertical wicking during the test. However, after plasma treatment there was a significant increase in the samples capillarity. This behavior is due to the incorporation of polar functional groups as confirmed by XPS analysis.

During the first 25 seconds the wicking rate was the same to three treated samples. Although, after this time was noted a higher wicking rate to polyester fabric treated with Ar+N_2 which has reached the maximum length of the sample in 150s. This behavior can be associated to the higher concentration of nitrogen found in the fabric surface after this treatment, as shown in the Table 3. Furthermore, nitrogen can bind with hydrogen and carbon to form amines (NH and CN) that have high polarity which gives hydrophilicity to polyester. In this sample was also observed an almost complete reduction in silicon (Figure 6), which it may have contributed to higher capillarity.

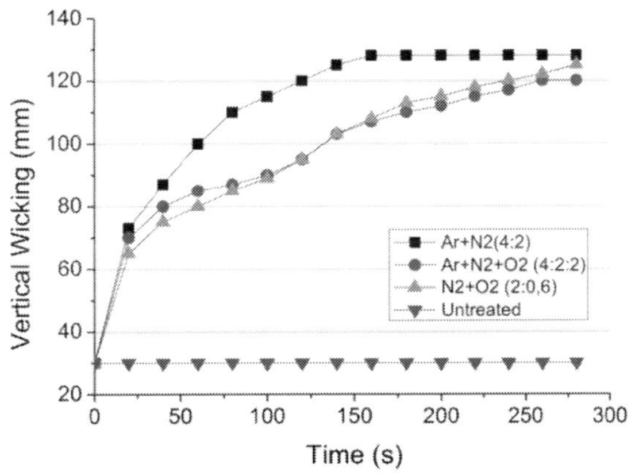

Figure 9. Vertical wicking *vs.* Time to untreated samples and treated with $Ar+N_2$; $Ar+N_2+O_2$ and N_2+O_2.

4. Conclusions

Polyester fabric was treated in non-thermal plasma with three different gaseous atmospheres ($Ar+N_2$, N_2+O_2, $Ar+N_2+O_2$). The samples were characterized by FTIR and XPS analysis and vertical wicking where was possible to conclude that:

1. The hydrophilicity of the polyester fabric was improved as found in the vertical wicking. This occurred due to the introduction of functional groups on the material surface as evidenced in the XPS analysis;
2. Combination of gaseous in the plasma atmosphere has influence on hydrophilicity results. Higher wicking rate was identified in the sample treated with $Ar+N_2$. In this sample was observed an increase in the –CO– and –COO– groups and a decrease in the C–C and CH groups as well as higher count of nitrogen and also an almost complete reduction of silicon. All this factors together contribute to a higher polyester fabric hydrophilicity that was treated with $Ar+N_2$.
3. Different compounds with nitrogen were found after the treatments. NH and N–C were identified in the samples treated with $Ar+N_2$ and $Ar+N_2+O_2$, while NO_3^- and NH were seen in the spectrum of the sample treated with N_2+O_2. These results underscore the importance of the gaseous atmosphere when pretending to carry out plasma treatment of polymers.

5. References

[1] Fang K and Zhang C 2009 *Applied Surface Science* **255** 7561–67
[2] Wang C X *et al.* 2008 *Surface & Coatings Technology* **202** 2775–82
[3] Barni R *et al.* 2005 *Plasma Processes and Polymers* 2 64 – 72
[4] Yaman N *et al. 2009* Improvement surface properties of polypropylene and polyester fabrics by glow discharge plasma system under atmospheric Condition. *Tekstil ve Konfeksiyon* **1**
[5] Morent R *et al.* 2008 *Surface & Coatings Technology* **202** 3427 – 49
[6] Sun S *et al.* 2011 *Applied Surface Science* **257** 2377–82
[7] Kale K H and Desai A N 2011 *Indian J. Fibre & Textile Research* **36** 289 - 299
[8] Navaneetha Pandiyaraj K and Selvarajan V 2008 journal *of materials processing technology* **199** 130–139
[9] Costa T H C *et al.* 2006 *Journal of Materials Processing Technology* **173** 40–43

[10] Li L, Frey M and Browning K J 2010 *Journal of Engineered Fibers and Fabric* **5** Issue 4
[11] Haverkamp R G, Siew D C W and Barton T F 2002 XPS study of the changes during the service life of polyester powder coatings. *Surf. Interface Anal* **33** 330–334
[12] Charpentier P A, Maguire A and Wan W 2006 Surface modification of polyester to produce a bacterial cellulose-based vascular prosthetic device *Applied Surface Science* **252** 6360–67
[13] Cascalheira A C and Abrantes L M 2008 *Corros. Prot. Mater.* **27** N.° 1

Acknowledgments
The authors are acknowledging to financial support supplied by CAPES – Brasília/Brazil through of the CAPES/IGC program – CGCI n° 055/2010.

Efficiency of inductively torch plasma operating at atmospheric pressure on destruction of chlorinated liquid wastes- A path to the treatment of radioactive organic halogen liquid wastes.

G Kamgang-Youbi[1,2]*, K Poizot[1] and F Lemont[1]

[1]French Atomic Commission-CEA, Marcoule- DTCD/SCDV/LPIC, BP 17171, 30207 Bagnols-Sur Cèze Cedex, France
[2] The University of Yaoundé I, Inorganic Chemistry Department, P.O Box 812 Yaoundé, Cameroon.

* To whom any correspondence should be addressed: kamyougeo@yahoo.fr

Abstract: The performance of a plasma reactor for the degradation of chlorinated hydrocarbon waste is reported. Chloroform was used as a target for a recently patented destruction process based using an inductive plasma torch. Liquid waste was directly injected axially into the argon plasma with a supplied power of ~4 kW in the presence of oxygen as oxidant and carrier gas. Decomposition was performed at $CHCl_3$ feed rates up to 400 $g \cdot h^{-1}$ with different oxygen/waste molar ratios, chloroform destruction was obtained with at least 99% efficiency and the energy efficiency reached 100 $g \cdot kWh^{-1}$. The conversion end products were identified and assayed by online FTIR spectroscopy (CO_2, HCl and H_2O) and redox titration (Cl_2). Considering phosgene as representative of toxic compounds, only very small quantities of toxics were released (< 1 $g \cdot h^{-1}$) even with high waste feed rates. The experimental results were very close to the equilibrium composition predicted by thermodynamic calculations. At the bottom of the reactor, the chlorinated acids were successfully trapped in a scrubber and transformed into mineral salts, hence, only CO_2 and H_2O have been found in the final off-gases composition.

1. Introduction

Organohalogens are often cited as responsible for climate change and occupational diseases. These compounds are used in many processes in the chemical, nuclear, and processing industries and the quantity of waste generated is a problem requiring a solution at industrial scale. Conventional thermal destruction is often used industrially to decompose, but if not sufficiently controlled it raises the risk of emissions that can be even more toxic than the initial waste. During combustion, chlorinated hydrocarbons are also known to interfere and modify the process, if only by the presence of chlorine which prevents the oxidation of CO to CO_2 [1] and favors the production of harmful products such as dioxins, furans, phosgene, etc. Organohalogens are also very poorly biodegradable, and their chemical stability requires destruction by energy-intensive processes.

Thermal plasma can be used; several studies have proposed the thermal plasma process for the destruction of organo-halogenated wastes [2-4]. In order to reduce the waste volumes, radioactive or

Journal of Physics: Conference Series **406** (2012) 012018 doi:10.1088/1742-6596/406/1/012018

not, the CEA recently patented an "Installation for the Destruction of Organo-Halogenated Liquids" (IDOHL: French patent FR 0450303) using inductively coupled thermal plasma. This type of plasma has several advantages as the geometry of the torch which allows the treatment not only of gases but also of liquids without prior evaporation (it is not always easy to place liquid directly in the plasma because of the viscosity of plasma medium), the absence of electrodes, the kinetics and the extended residence time. The initial tests with this device, at a higher power level (41–45 kW) than in the present work, demonstrated the chloroform destruction using water as the oxidant to minimize the production of toxic byproducts [5]. Pure liquid $CHCl_3$ was completely degraded with a calculated energy efficiency (waste mass decomposed per hour / torch power) of 19 $g.kWh^{-1}$ and converted to HCl and CO_2. Unfortunately, however, simultaneously supplying water and waste can lead to major overconsumption of the initial plasma energy because of the high energy necessary to heat and vaporize the water. This drawback can limit the waste input and thus affect the energy efficiency. In addition, not all waste compounds are fully miscible with water, and their destruction can occur under pyrolytic conditions, although the importance of destroying chloromethanes under oxidative rather than pyrolytic conditions is well known [6]. These limitations led us to choose a new procedure using oxygen instead of water. A smaller (~4 kW) plasma reactor was built to demonstrate the feasibility of processing radioactive liquid organochlorine waste. Chloroform was also selected as the organochlorine model. With its Cl:H ratio of 3:1, there is a strong possibility of forming toxic gases such as Cl_2 and $COCl_2$ during its destruction. Then, $CHCl_3$ represents a good model to study the complete and safe degradation. As the destruction of halocarbons is more likely to form toxic halogenated byproducts in the presence of O_2 [1,7,8] than in the presence of H_2O [9,10], the objective of the present work is not only to demonstrate the degradation and complete mineralization of chloroform to harmless CO_2 and H_2O by IDOHL process with oxygen as the oxidizing gas, but also to decompose without formation of toxic, thereby ensuring acceptable gas release as in the previous work [5]. Another objective is to improve the energy efficiency by substituting O_2 for H_2O. The influence of initial waste concentration and the oxidant/ratio molar ratio on the qualitative and quantitative nature of the off-gas will also be discussed.

2. Thermodynamic considerations

In order to predict the equilibrium composition, thermodynamic calculations based on minimizing the Gibbs free energy were performed with the software package, FactSage®. The calculations were also intended to predict the optimum quenching temperature for decomposition. The calculations assume the use of ideal gases and given the high temperatures of the plasma, equilibrium is assumed to have been reached. Calculations were performed for temperatures between 300 K and 4000 K at atmospheric pressure. For C-H-Cl-O system, more than 70 species were taken into account. The calculated equilibrium results for thermal destruction of chloroform are shown in **figure 1**.

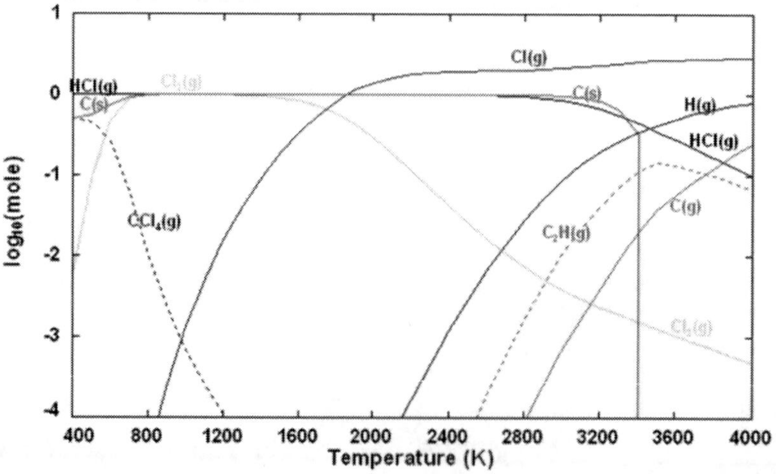

Figure 1. Equilibrium composition of system of 1 mol of $CHCl_3$ in Ar plasma ($n_{Ar}/n_{CHCl3} = 100$).
N.B: The graph of Ar is not represented because of the scale.

At 4000 K, $CHCl_3$ is entirely degraded; gaseous Cl, H, C radicals together with HCl are found in the system. C(s) appears below 3500 K. HCl, Cl, Cl_2 and C(s) are the main thermal decomposition products of chloroform between 2500 K and 3500 K. The apparition of CCl_4, which is formed in large amounts below 900 K, must be avoided. Indeed, it is preferable to avoid the formation of other haloalkanes (for example, the Global Warming Potential of CCl_4 is 1400 times greater than that of CO_2 and 10 times higher than that of $CHCl_3$). Moreover, C(s) must be converted to CO_2. For all these reasons, it is necessary to supply oxygenated specie to the system. **Figure 2** shows the equilibrium calculations in the presence of oxygen in equimolar quantity with the waste

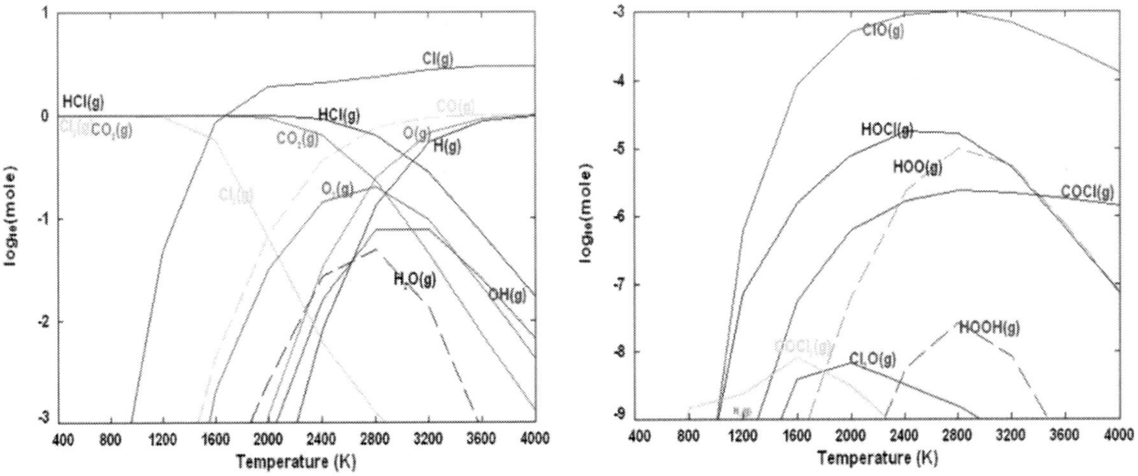

Figure 2. Equilibrium composition of $CHCl_3 + O_2 + Ar$ ($n_{O2} = n_{CHCl3} = 1$ mol, $n_{Ar}/n_{CHCl3} = 100$)
(Left) Major species; (Right) Minor species. N.B: The graph of Ar is not represented.

At high temperatures, the initial carbon is present in the form of CO, and soot is theoretically no longer formed. Between 2500 K and 3500 K the equilibrium composition is dominated by Cl, HCl, CO, CO_2, O_2 and H_2O. No new organic product with a molar fraction exceeding 10^{-3} is formed. The decomposition of chloroform becomes complete.

The oxidation kinetics of CO to CO_2 is known not to be fast when the temperature decreases; in practice, moreover, this oxidation is complicated by the presence of chlorine [1]. Excess O_2 is used to minimize these drawbacks.

The calculated results for a simulation with $O_2/CHCl_3$ molar ratio = 20:1are presented in **figure 3**.

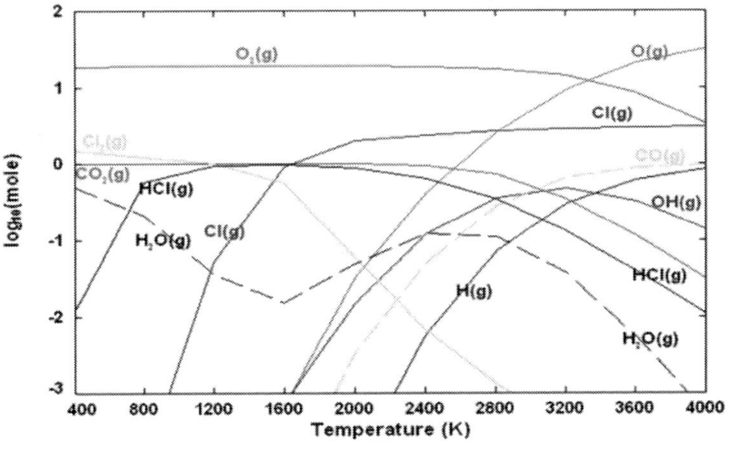

Figure 3. Equilibrium composition of $CHCl_3 + O_2 + Ar$ with excess of oxygen ($n_{CHCl3} = 1$ mol; $n_{O2}/n_{CHCl3} = 20$, $n_{Ar}/n_{CHCl3} = 100$).
N.B: The graph of Ar is not represented.

Compared with stoichiometric conditions (**figure 2**), **figure 3** shows that excess oxygen favors the formation of H_2O over HCl. This would result in increased quantities of molecular chlorine (Cl_2) in the cold off-gas stream. When dealing with chlorinated compounds it is preferable to favor the formation of HCl (more readily neutralizable and more soluble in aqueous solution) rather than Cl_2 which may also combine with CO to form toxic compounds such as phosgene. The peak of formation of HCl is around 1200 K. It is thus advisable to work with excess oxygen, but to ensure rapid quenching of the gas to prevent the formation of undesirable and enhance the formation of HCl, CO_2 and H_2O.
In order to suitably control the off-gas release, we also examined the formation of toxic. The phosgene was targeted in this study as a representative of the off-gas quality to limit its toxicity. Its formation, which has no economic advantage in this work, must be minimized. According to the literature, the formation of phosgene is relatively stable between 500 K and 2100 K during combustion of organohalogens [11-13].

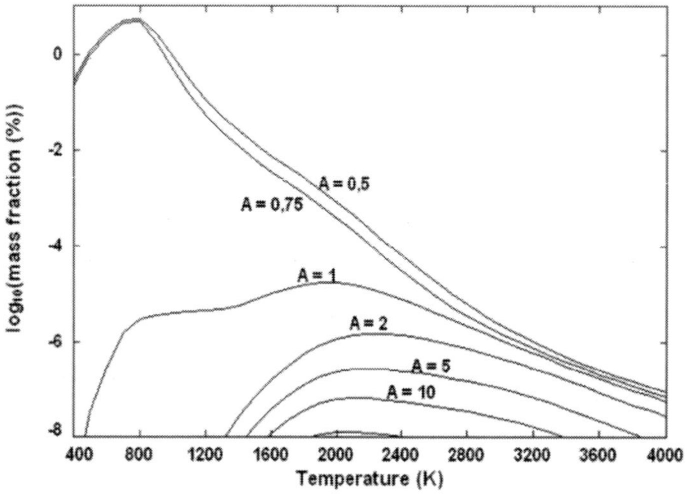

Figure 4. Phosgene formation as a function of oxygen quantity $CHCl_3$ + <A> O_2. (A<1: lean oxygen conditions, A = 1: stoichiometric mixture, A>1: excess of oxygen).

For a mixture with an O_2 deficiency the calculations predicted a large formation of phosgene with a peak near 780 K (**figure 4**, curves A = 0.5 and A= 0.75). Simply starting with a stoichiometric mixture or with excess oxygen has a beneficial effect on reducing the quantity of phosgene formed (**figure 4**, curves A = 1, A > 1). Excess O_2 appears to have a positive effect on reducing the quantities of $COCl_2$ liable to form. **Figure 4** shows that from O_2:$CHCl_3$ = 10:1, the $COCl_2$ concentration is < 1 ppm.
The thermodynamic analysis led to the following conclusions:

- oxygen must be added to the system in order to oxidize carbon to CO and CO_2, it must be in excess to accelerate the conversion to CO_2 and especially to minimize the risk of producing toxic gases such as phosgene and incomplete combustion products such as CO;
- the off-gas must be cooled rapidly after the reaction to prevent undesirable recombinations.

3. Experimental

3.1 Inductively coupled plasma and reactor

The experimental device is presented in **figure 5**.

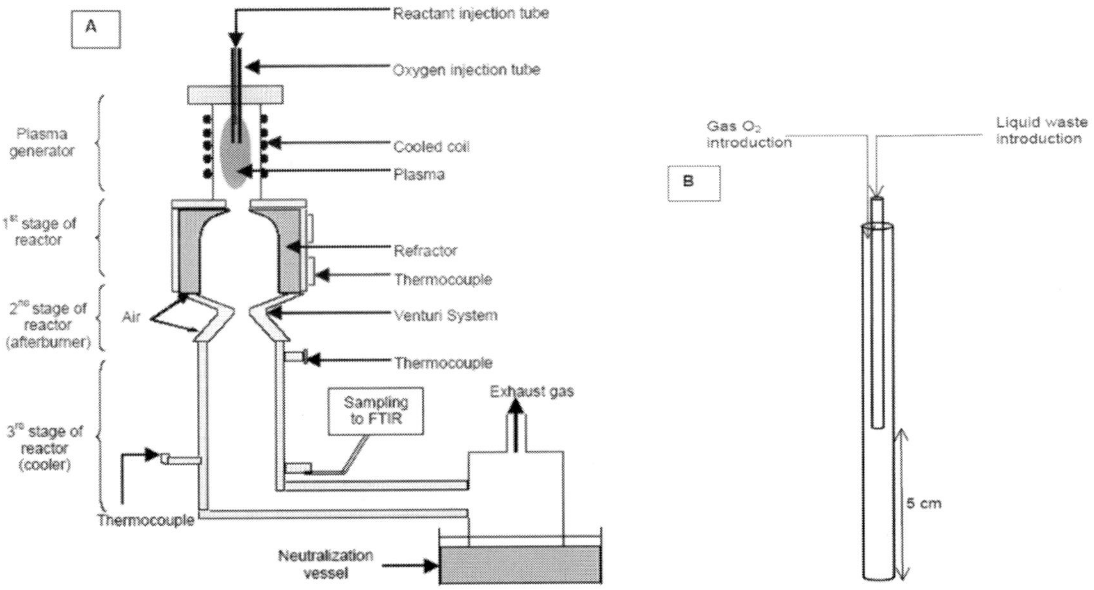

Figure 5. (A) Schematic diagram of IDOHL process. (B) Detailed setup of the injection system.

The plasma source is an induction plasma torch (Tekna PL-50) with high-purity argon as the plasma and sheath gas. It is supplied by a high-voltage power supply coupled with a high-frequency (64 MHz) transmitter supplying a 7-turn copper induction coil to generate the plasma. The maximum power rating is 4.5 kW. The plasma is confined in a quartz tube with an inside diameter of 35 mm. The torch is coupled with a reactor consisting of three main parts (**figure 5A**):

- The first (341 mm long, 43 mm inside diameter) presents an appropriate design to avoid cold areas, with a refractory wall. When only argon plasma is working, the reactor design allows gas temperatures of about 1200 K to be reached at the output of this stage.
- The second part is the afterburner stage consisting of a Venturi system. There is injection of airstreams to enrich the medium in oxygen in order to fully oxidize the residual carbon to CO_2 and hydrogen to H_2O, as well as to sharply reduce the temperature of the gas mixture from the first stage. A temperature of around 500 K is measured at the end of this zone.
- The third stage of the reactor is a stainless steel tube 300 mm long and 43 mm in inside diameter. The reaction products are further quenched in this column.

The entire process operates under slight negative pressure ($\leq$ 30 mbar) to prevent leakage.

3.2 Waste injection

The oxygen is injected in the reactor via an alumina tube (4 mm inside diameter). Then, liquid chloroform (> 99% purity, Sigma-Aldrich, France) is introduced via an alumina tube (1.5 mm inside diameter); the two tubes form a coaxial cylindrical injection system (**figure 5B**). As the chloroform passes through the tube already containing oxygen, the two reactants are mixed well before they enter into the plasma. The injection tube can be placed in several positions in the plasma. In this study the tube was positioned between the 5th and 6th coils from the top. As the system is hot, combustion begins in the lower part of the injection tube. We tested various flow rates for both $CHCl_3$ and O_2; **Table 1** indicates the test conditions.

Table 1. Operating conditions.

Parameters	Values
Supplied plasma power	4 – 4.2 kW
Pressure	1×10^5 Pa
Total argon flow	40 $NL{\cdot}min^{-1}$
Oxygen flow	7 – 28 $NL{\cdot}min^{-1}$
Chloroform	100 – 400 $g{\cdot}h^{-1}$
$O_2/CHCl_3$ molar ratio	15–30
Additional air flow	60 $NL{\cdot}min^{-1}$

3.3 Gas analysis

Analysis was performed at two locations: i) just at the bottom of the reactor to measure the efficiency of plasma-reactor alone, ii) after the scrubber, in order to have the performance of the entire process.

A gas sample is analyzed continuously by a Fourier transform infrared spectrometer (Gasmet™ DX4000, Finland). The spectrometer output signal is updated every 20 seconds. It is capable of simultaneously analyzing more than 30 gaseous species, notably the following in our study: $CHCl_3$, HCl, CO_2, H_2O, CO, $COCl_2$. The instrument was calibrated with certified chloroform concentrations ranging from 0 to 200 ppm. The same method was used to calibrate the other species analyzed.

Cl_2 was identified by bubbling a fraction of the reactor off-gas stream through a NaOH solution, followed by a reaction with KI. The iodine generated was titrated with an aqueous solution of sodium thiosulfate, $Na_2S_2O_3$.

Just at the bottom of the reactor, the off-gas is trapped in an aqueous NaOH solution. The main reason for caustic scrubbing is to absorb and thus neutralize the acidity due to the major chlorinated species (Cl_2 and HCl) generated. The scrubbing solution pH is automatically adjusted when it drops below 9.

4. Results and discussions

Throughout the tests, the argon flow rate was maintained constant at 40 $NL{\cdot}min^{-1}$. The plasma operating power was obtained by multiplying the instantaneous voltage and current readings; the power remained around 4.2 kW at all times (**figure 6**, curve A).

4.1 Gas temperature in the reactor

Gas temperature is the first parameter describes the good working of the process. **figure 6** shows typical operating conditions, with the temperatures variations measured in the first stage of the reactor (curve B) and those recorded after the injection of cooling air (curve C). **Table 2** describes the procedure. The air injection rapidly cooled the gas stream, as shown by a temperature difference of about 650°C between curves B and C.

After starting the plasma, the temperature rapidly increased as the electrical energy supplied to the plasma is converted into thermal energy. When thermal equilibrium is reached, the temperature at the bottom of the first stage is about 950°C (curve B, sequence 1). This temperature, measured relatively far from the plasma torch (29 cm from the tail of the plasma), is a good indicator of process performance.

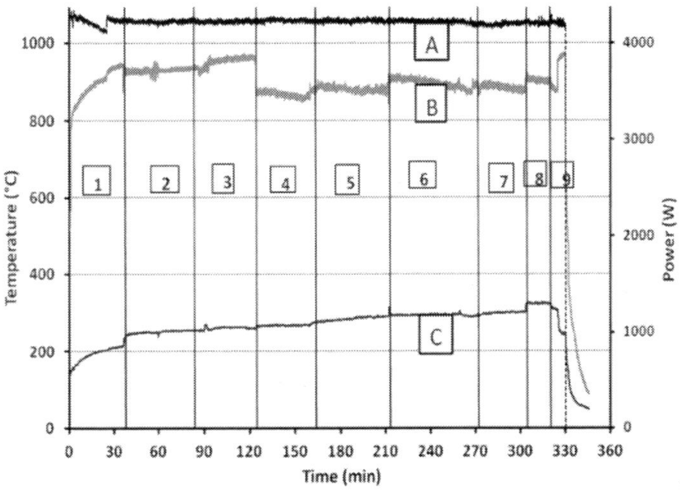

Figure 6. Plasma power (curve **A**) and gas temperatures (curve **B**: at the bottom of the first stage of reactor; Curve **C**: after air cooling) versus experiment time.

Table 2. Experiment description.

Sequence Number	Time (min)	Description
1	0-36	Heating with Ar plasma
2	36-81	1st injection of O_2 (7 NL·min^{-1})
3	81-124	Introduction of waste (100 g·h^{-1})
4	124-164	Increasing O_2 (10 NL·min^{-1})
5	164-212	Increasing waste (150 g·h^{-1})
6	212-273	Increasing waste (200 g·h^{-1})
7	273-304	Increasing O_2 (14 NL·min^{-1})
8	304-320	Increasing O_2 (20 NL·min^{-1})
9	320-350	Process shutdown

4.2 Decomposition in the presence of oxygen and effect of molar ratio

Decomposition is clearly illustrated in **figure 7**, which compares the FTIR spectra before and after the destruction of $CHCl_3$. After injection in the plasma-reactor torch, the spectral properties of $CHCl_3$ clearly disappear. Peaks are observed for CO_2, HCl and H_2O.

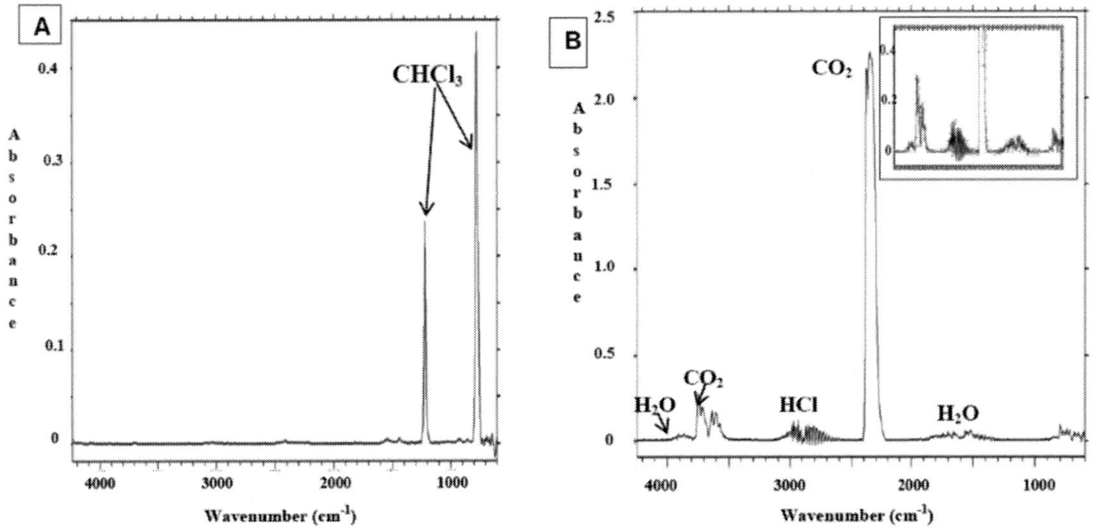

Figure 7. FTIR spectra of $CHCl_3$ (A) before and (B) after plasma processing (the figure inset represents an enlargement of the lower part of B). $CHCl_3$ feed rate: 100 g·h^{-1}, O_2: 10 NL·min^{-1}.

The decomposition efficiency is calculated by equation (1):

$$\text{Destruction of chloroform (\%)} = \frac{\text{moles of } CHCl_3 \text{ converted}}{\text{moles of initial } CHCl_3} \times 100 \quad \textbf{(1)}$$

The results obtained under the different test conditions are indicated in the following table (**table 3**).

Table 3. Conversion of $CHCl_3$ in inductively plasma reactor.

$CHCl_3$ feed rate ($g{\cdot}h^{-1}$)	O_2 ($NL{\cdot}min^{-1}$)	$CHCl_3$ destruction (%)	Conversion to CO_2 (%)	Initial Cl conversion to HCl (%)	Conversion to Cl_2 (%)
100	7	99.95	99.16	32.70	62.98
	10	99.96	99.34	31.55	64.04
150	7	99.09	88.59	31.14	ND
	10	99.71	99.67	29.44	66.75
	14	99.95	>99.99	28.72	69.14
200	14	99.70	98.45	27.40	64.50
	17	99.95	>99.99	28.44	70.00
	20	99.98	>99.99	26.38	73.35
300	20	99.96	97.20	24.88	ND
400	20	99.94	>99.99	24.32	ND
	24	99.98	96.18	25.74	ND
	28	99.98	94.33	26.30	ND

ND: Not determined

Near 100% destruction of the initial chloroform was obtained regardless of the chloroform injection rate (100–400 $g{\cdot}h^{-1}$). When sufficient oxygen is injected, complete destruction of the initial chloroform is guaranteed. The energy efficiency increases with the chloroform injection rate up to a relatively high value of 100 $g{\cdot}kWh^{-1}$ (i.e. 400 $g{\cdot}h^{-1}$ $CHCl_3$ entirely degraded in a 4 kW plasma). For the same degraded compound, the energy efficiency is significantly improved in comparison with the previous study with liquid H_2O as the oxidant where the calculated value was 19 $g.kWh^{-1}$ [5]. The energy efficiency obtained here is also higher than reported by Föglen et al. [3] for thermal destruction of $CHCl_3$ by inductively coupled plasma, or by Krawczyk et al. [14] and Song et al. [4] for gliding plasma arc decomposition of chloroform.

Another parameter investigated was the oxidation factor or CO_2 conversion, corresponding to the final step of oxidation of the initial carbon based on the equilibrium curves in excess oxygen (**figure 3**). Negligible CO concentrations were present in the reaction off-gas; CO_2 was thus the main carbon specie measured by FTIR. The measured CO_2 production can therefore be a suitable performance indicator for this process. The CO_2 conversion factor was calculated with the following relation:

$$\text{Fraction converted to } CO_2\ (\%) = \frac{\text{moles of } CO_2 \text{ produced}}{\text{moles of } CHCl_3 \text{ converted}} \times 100 \quad \textbf{(2)}$$

Column 4 in **table 3** shows very satisfactory conversion of the initial carbon. The conversion results are generally better than 90%, indicating that the initial carbon is effectively converted to CO_2. Although increasing the oxygen flow rate (and thus the molar ratio) reduces the temperature in the reactor under certain conditions (**figure 6**), the overall performance of the process was not affected by this temperature drop.

In addition to the conversion of the initial carbon to CO_2, it was very interesting to observe the fate of the initial chlorine. The fraction of initial Cl converted to HCl was calculated using the following relation:

$$\text{Fraction converted to } HCl\ (\%) = \frac{\text{moles of } HCl \text{ produced}}{3 \times \text{moles of } CHCl_3 \text{ converted}} \times 100 \quad \textbf{(3)}$$

Between 25% and 33% of the initial chlorine was converted to HCl (**table 3**, column 5). For each chloroform injection rate, the overall percentage of Cl converted to HCl tends to diminish when the oxygen flow rate increases (i.e. for increasing excess oxygen). The same behavior was reported by Huang et al. [8] for plasma decomposition of C_2HCl_3 in the presence of oxygen. Our findings are consistent with calculations at thermodynamic equilibrium (**figure 2 Left** and **figure 3**) showing that

excess oxygen is unfavorable to the formation of HCl: the oxygen atoms capture hydrogen atoms to form H_2O molecules instead. Most of the Cl atoms thus remain in the form of Cl_2 molecules.
To confirm this hypothesis, Cl_2 was measured. The test results are shown in the last column of **table 3**. Cl_2 is indeed formed; the chlorine balance shows that, regardless of the test conditions, more than 95% of the initial chlorine is finally present as HCl and Cl_2.

4.3 Influence of the initial chloroform concentration

To assess the possible influence of the initial waste concentration, the $O_2/CHCl_3$ molar ratio was maintained at 20, while the initial chloroform concentration in the initial mixture ($CHCl_3 + O_2 + Ar$) varied from 0.7 mol% to 2 mol% and the oxygen concentration was adjusted to maintain a constant molar ratio. The results are shown in **figure 8** for the decomposition rate of $CHCl_3$ and its conversion to CO_2 and HCl.

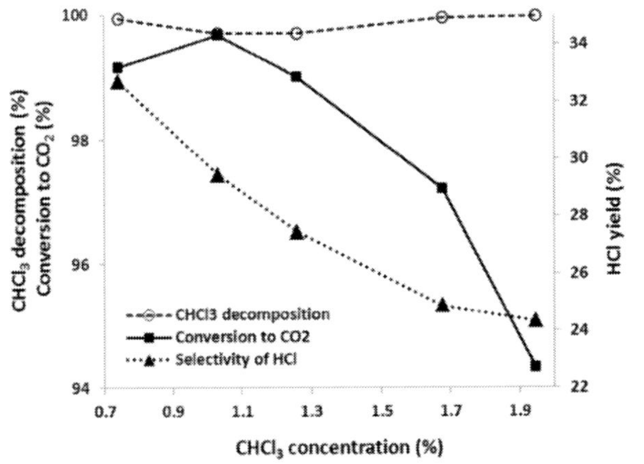

Figure 8. Effect of initial chloroform concentration on chloroform destruction, yields of CO_2 and HCl.

The overall decomposition of $CHCl_3$ exceeds 99% and varies only slightly in the range studied. The initial concentration has a minor effect on the conversion to CO_2. The curve decreases slightly as the initial $CHCl_3$ concentration increases. For example, the CO_2 conversion factor decreases from 99% to 94% when the concentration increases from 0.7 to 2%. Since the $O_2/CHCl_3$ molar ratio is constant, increasing the $CHCl_3$ concentration also increases the O_2 concentration; the argon concentration therefore diminishes in the initial mixture. Argon is not only the plasma gas, but also the energy transfer vector. Consequently, as the argon concentration increases, the amount of excited argon Ar* increases in the plasma and the degree of conversion of the reactants also increases due to the greater frequency of collisions. In addition, when the $CHCl_3$ concentration increases, the plasma energy density available for degradation diminishes. The lower available energy could account for the observed reduction in the conversion factor of $CHCl_3$ molecules to CO_2. Similarly, **figure 8** shows that the overall molar conversion of initial Cl to HCl diminishes when the initial $CHCl_3$ concentration increases. As the oxygen concentration also increases with the initial $CHCl_3$ concentration, and the global Cl conversion to ($HCl + Cl_2$) remain constant, the lower HCl conversion rate can be attributed to the Deacon reaction by which HCl is converted to Cl_2 in the presence of oxygen according to the following overall equation:

$$4HCl_{(g)} + O_{2\,(g)} = 2Cl_{2\,(g)} + 2H_2O_{(g)} \qquad \textbf{(4)}$$

4.4 Off-gas quality

Chloroform injected into the plasma is destroyed and converted mainly to CO_2, HCl, Cl_2, and H_2O. However, it is very important to check the level of oxy-chlorinated byproducts from decomposition. We present on the **table 4**, the maximum quantities of phosgene obtained for the various $CHCl_3$ injection rates are indicated.

Table 4. Maximum phosgene release per chloroform injected.

Initial composition		$COCl_2$ conc.	Mass of $COCl_2$
$CHCl_3$ ($g \cdot h^{-1}$)	O_2 ($NL \cdot min^{-1}$)	(ppm)	released ($g \cdot h^{-1}$)
100	7	11	0.30
150	14	12.5	0.37
200	20	30	0.93
300	20	22.5	0.70
400	20	28	0.93

The maximum release rate was almost 1 $g \cdot h^{-1}$ for waste injected at a $CHCl_3$ rate of 400 $g \cdot h^{-1}$. Current French legislation requires that the phosgene gaseous effluents maximum permissible hourly flow rate is 10 $g \cdot h^{-1}$, but if this value is exceeded, the release concentration limit is 1 ppm [15]. **Table 4** clearly shows that the maximum phosgene formation rate from plasma reactor was well below the limit.
Concerning the overall off-gas quality (outlet the scrubber), Cl_2 and HCl formed by decomposition of $CHCl_3$ must not be released untreated into the atmosphere. One technique for their simultaneous trapping is their reactions in excess NaOH to form mineral salts.
At the outlet of the scrubber, the FTIR spectrum shown in **figure 9** indicates that the final gas stream after destruction of $CHCl_3$ consists only of CO_2 and H_2O.

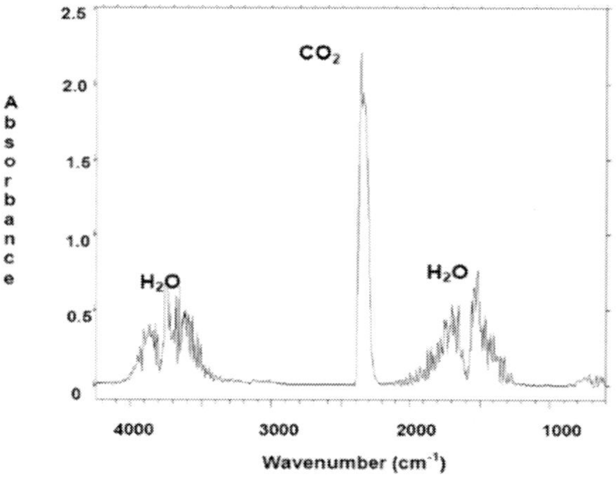

Figure 9. FTIR spectra of the final exhaust gas.

5. Conclusion

The effective destruction of chloroform -a model of liquid organochlorine- is proven using argon inductively plasma with oxygen as oxidant to obtain complete and safe destruction and minimize the energy losses observed in the original process when liquid water was used as the oxidant. Near-100% decomposition regardless of the test conditions with relatively high energy efficiency of 100 $g.kWh^{-1}$ and a satisfactory destruction rate (up to 400 $g \cdot h^{-1}$) were achieved. $CHCl_3$ was converted to CO_2, Cl_2 and HCl and the production of toxic such as $COCl_2$ was limited ($\leq 1\ g \cdot h^{-1}$). These experimental results are very near the values predicted by thermodynamic simulations. The chlorinated acidic produced were suitably trapped in an aqueous soda scrubbing column before release of the final off-gas into the atmosphere with [Cl] < 1 ppm.
The system presented in this work can be considered as a clean technology for the destruction of liquid organohalogens with the advantage of direct liquid waste injection into the plasma without any prior evaporation or treatment. These results are of considerable importance for industrial applications. The final objective of this study is to use this reactor for small-scale treatment of radioactive chlorinated organic liquid waste. The waste is contaminated by ^{14}C and ^{3}H. The encouraging results and destruction efficiency observed on a pure compound containing 89 wt% Cl allowed us to validate this small-scale unit. The process can subsequently be installed in a radioactive zone to test the destruction of actual radioactive waste. The transfer operations are currently in progress.

6. References

[1] Lou J C and Y S Chang 1997 *Combustion and Flame* **109** 188-197.
[2] Föglein K A *et al* 2005 *Plasma Chem. Plasma. Proc.* **25** 289-302.
[3] Murphy A B *et al* 2002 *Plasma Chem. Plasma. Proc.* **22** 371-385.
[4] Song H K *et al* 2005 *Toxic. Environ. Chem* **87** 509-519.
[5] Bournonville B and Meillot E 2007 *High Temp. Mat. Proc.* **11** 245-256.
[6] Lee W J *et al* 1993 *Environ. Sci. Technol.* **27** 949-960.
[7] Fitzsimmons C *et al* 2000 *J. Phys. Chem. A* **104** 6032-6038.
[8] Huang L *et al* 2006 *Chem. Eng. J.* **124** 81-87.
[9] Watanabe T and Tsuru T 2008 *Thin Solid Films* **516** 4391-4396.
[10] Amama P B *et al* 2001 *J. Molecular Catal. A* **176** 165-172.
[11] Lee W J *et al* 1996 *J. Hazard. Mater.* **48** 51-67.
[12] Tonics E J *et al* 2000 *J. Vac. Sci. Technol. A* **18** 393-400.
[13] Xu X P *et al* 2000 *J. Vac. Sci. Technol. A* **18** 213-231.
[14] Krawczyk K and Ulejczyk B 2003 *Plasma Chem. Plasma Proc.* **23** 265-281.
[15] Arrêté 2002-05-02 relatif à la pollution de l'air art. 2 *Journal Officiel de la République Française* 5 mai 2002.

Synthesis of single-walled carbon nanotubes using induction thermal plasma technology with different catalysts: thermodynamic and experimental studies

Y Alinejad[1,2], **A Shahverdi**[1,2], **N Faucheux**[1], **G Soucy**[1,*]

[1]Department of Chemical and Biotechnological Engineering, Université de Sherbrooke, 2500 blvd. de l'Université, Sherbrooke, QC, CANADA, J1K2R1

[2] Both authors had an equal contribution to this article

* Corresponding author, E-mail: Gervais.Soucy@USherbrooke.ca

Abstract. The effects of the type and quantity of three catalyst mixtures (i.e. Ni-Y_2O_3, Ni-Co-Y_2O_3, and Ni-Mo-Y_2O_3) on single-walled carbon nanotubes (SWCNT) synthesis by induction thermal plasma process have been examined in order to evaluate their individual influences on SWCNT production. Thermodynamic calculations, in gas and particularly in liquid solution phases, have also been performed to better understand the effects of the catalysts on the production of SWCNT. Characterization of the synthesized SWCNT by different techniques including Raman spectroscopy, high resolution scanning electron microscopy (HRSEM) imaging and thermogravimetric analysis (TGA) clearly indicated that the best quality of SWCNT was achieved using Ni-Co-Y_2O_3 catalyst mixture in the feedstock.

1. Introduction

Due to the very exceptional thermal, mechanical and electrical properties [1] of single-walled carbon nanotubes (SWCNT) discovered by Iijima in 1993 [2], much attention has been attracted to this novel nanostructured material. It has a broad range of applications in different fields such as sensors [3,4], polymers [5,6], microelectric devices [7], and biomedical engineering [8,9]. Different methods have been developed for the synthesis of SWCNT including arc discharge [2], laser ablation [10], chemical vapor deposition (CVD) [11], and induction thermal plasma [12]. These methods can be divided into two main categories depending on their catalytic or catalyst free synthesis. In catalytic synthesis methods, the presence of catalysts, mainly transition metals, in the feedstock or on the support plate is inevitable [13]. In laser ablation and arc discharge methods high temperature heating of the target, made from graphite-catalyst mixture plate, followed by a rapid cooling rate process of the generated vapor results in the formation of high quality SWCNT. However, in the induction thermal plasma method (ITP), the formation of SWCNT is based on the direct evaporation of feedstock materials containing carbon black (CB) and a mixture of catalysts in the high temperature plasma plume. For CVD methods, hydrocarbon decomposition takes place over the supported catalyst particles which indeed act as active sites for the nucleation of SWCNT [14]. Since different phenomena occur in various synthesis methods, it is somehow impossible to adopt a unique mechanism for the SWCNT formation. Among all suggested mechanisms so far, the nucleation and growth of SWCNT based on carbon-catalyst interaction [15] is more suited for the gas phase methods. Indeed, in the gas phase, SWCNT formation includes several steps (a) decomposition of catalysts and carbon source, (b) nucleation and growth of catalyst nanoparticles, (c) carbon diffusion or solubilization into the metal particles and (d) precipitation and solidification in the form of SWCNT. In this mechanism, diffusion and solubilization of carbon atoms into the catalyst liquid droplets is a key factor. The rate of dissolution is mainly dependent on the type of catalyst, temperature, and the presence of other solutes. For example, it has been shown that small amount of sulfur can reduce the surface tension of liquid metal which favors the higher solubility of carbon [16]. It has been shown experimentally that the solubility of carbon in different catalysts varies notably. For example, the solubility of carbon is high in Mo and Fe, while it is very low in Cu [13]. Therefore, it is expected that changes in the type of catalyst can distinctly affect the formation of SWCNT. Although the efficiency of different catalysts in production of SWCNT with the other methods of synthesis has been intensively studied, no systematic experimental study has been conducted on the effect of the type and quantity of catalysts on the SWCNT synthesis by ITP.

In the present work, since the solubility of carbon and the gas phase composition of plasma are expected to affect the SWCNT synthesis, a complete thermodynamic study on a wide temperature range (500-5 000 K) was carried out in order to fully understand the effect of the type and quantity of catalyst on the gas-phase during the synthesis of SWCNT by ITP process. The operating temperature window was limited to T< 5 000 K since at higher window temperature of plasma the system is noticeably far from the equilibrium state. It is expected that at lower temperatures considered in this work, thermodynamic studies will help to better understand the effect of different catalysts on the synthesis of SWCNT.

Following the thermodynamic study, the effect of three different catalyst mixtures on the SWCNT synthesis by ITP was experimentally studied. Along with Yttrium oxide (Y_2O_3), Ni, Co, and Mo have been selected due to their high efficiency in production of SWCNT using catalytic synthesis methods [13]. Moreover, since there is controversial data about the toxicity of metallic catalysts in the literature [17-19], it is important to find less harmful catalysts that are capable of producing high quality SWCNT.

2. Thermodynamic calculations

Theoretical thermodynamic study was carried out using FACTSAGE 6.2 software on the different carbon-catalyst mixture systems summarized in Table 1. To study the solution phase, particularly in liquid state the FSstel database was selected for the calculations. For the gas phase, the FACT53 database was considered without taking into account the inert gases (Ar, He). Moreover, the presence of sulfur in the CB was not neglected since it can alter the quality of SWCNT [20]. For these series of calculations, two group of mixtures were considered and their results were compared. In Group 1 (i.e. MIX 1, 2 and 3) three different mixtures of catalysts were considered with a constant total amount of catalysts (i.e., 5.2 wt. %). In fact, the Co content present in MIX 1 was replaced by the same amount of Ni in MIX 2 and by the same amount of Mo in MIX 3 so that the total metallic catalyst content remained constant. Group 2 which includes MIX 4, MIX 5 and MIX 6, contains mixtures with different ratios of Ni-Y_2O_3 and Ni-Mo-Y_2O_3.

3. Experimental procedure and set-up

SWCNT were synthesized at large-scale using ITP system (feed stock injection rate of 2 g/min), as shown in Figure 1. Detailed information about the experimental set-up can be found in a previous paper [21]. The physical properties of the feedstock materials are summarized in Table 2. The same mixture compositions used for thermodynamic calculations (Table 1) were used in the experimental studies.

Table 1 Feedstock material content in weight percent (wt. %) used for the synthesis of SWCNT

MIX No	Ni	Co	Mo	Y_2O_3	CB	S*
1[¥]	2.6	2.6	0	7.5	87.3	2
2[¥]	5.2	0	0	7.5	87.3	2
3[¥]	2.6	0	2.6	7.5	87.3	2
4[£]	12	0	0	3	85	2
5[£]	5.2	0	0.5	7.5	86.8	2
6[£]	1.3	0	3.9	7.5	87.3	2

* wt. % of sulfur in CB, [¥]Group 1, [£]Group 2

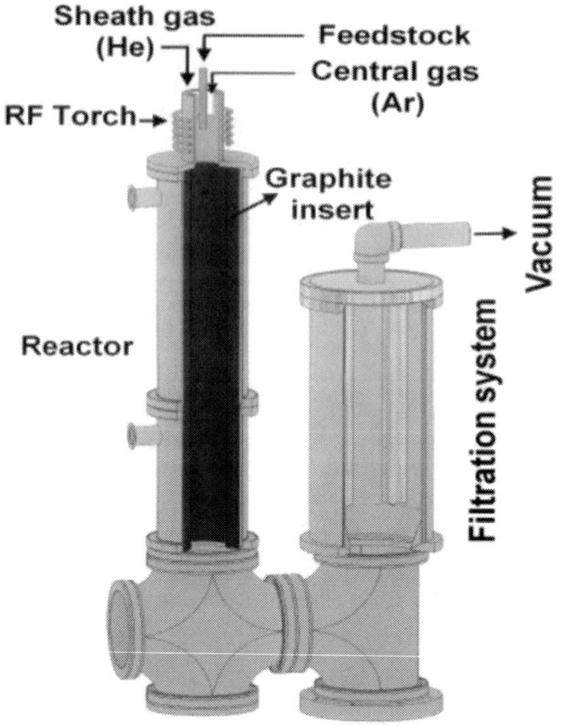

Figure 1 Schematic diagram of an ITP system designed for large-scale synthesis of high quality SWCNT

The synthesized SWCNT samples were characterized by means of Raman spectroscopy, X-ray diffraction (XRD, PANALYTICAL-X'Pret Pro MPD), thermogravimetric analysis (TGA, SETSYS 2400) and high resolution scanning electron microscopy (HRSEM, S4700, Hitachi). Raman spectroscopy was performed with Ar laser ($\lambda_{ext.}$ = 514 nm). TG was performed according to the protocol provided in [22].

Table 2 Properties of the materials used for the production of SWCNT

	Size	Purity (%)	$T_{melting}$ (°C)	$T_{vaporization}$ (°C)	Company
Ni	2μm	99.9	1455	2913	Cerac (CANADA)
Co	2μm	99.9	1495	2927	Sigma-Aldrich (CANADA)
Mo	2μm	99.9	2623	4639	Sigma-Aldrich (CANADA)
CB	45nm	98	N/A	3642	Grade: M280, Cabot Inc.(USA)
Y_2O_3	5μm	99.9	2425	4300	Hermann C.Starck (THAILAND)

4. Results and discussion

4.1. Thermodynamic study

Thermodynamic calculations were performed on different SWCNT feedstock mixtures summarized in Table 1. The results of thermodynamic calculations in gas phase and liquid solution phase on Group 1 are depicted in Figure 2.

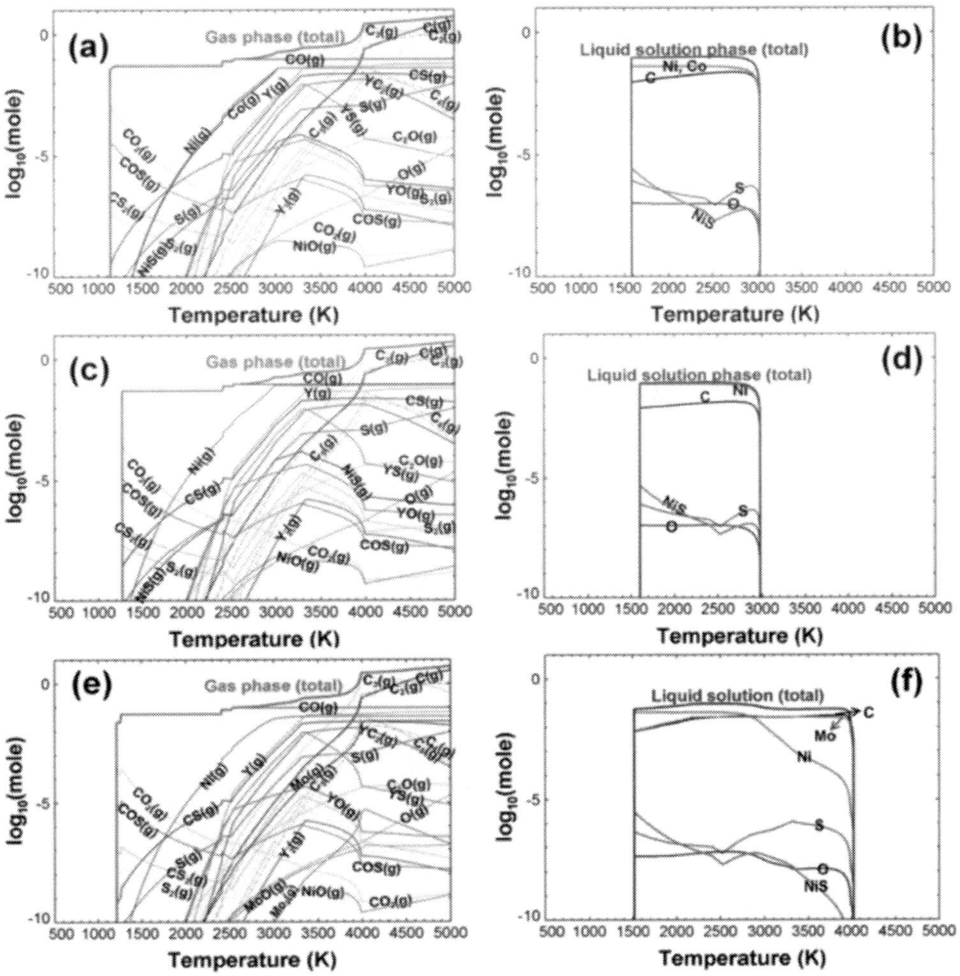

Figure 2 Thermodynamic equilibrium composition in gas phase (left-side) and liquid solution phase (right- side) of (a), (b) MIX 1, (c), (d) MIX 2 and (e), (f) MIX 3

The comparison of the total gas phase of each mixture revealed that the total gas phase was poorly affected by the type of metallic catalysts since it ended at almost the same temperature with the same pattern for all the mixtures. Nonetheless, slight differences were observed at the temperature where the total gas phase was terminated for the three mixtures. For example, a more extended gas phase at lower temperature was observed for the MIX 3, MIX 1, and MIX 2 respectively. Details of gaseous species are illustrated under the gas phase (total) curve (Figure 2, red curve). The concentration and evolution pattern of the most important precursors of SWCNT (i.e. C_i, i= 1 to 5) seemed not to be affected by the change in the catalyst mixtures. The exothermic reaction of Y-C along with other reactions such as Y-S, Y-O and Ni-S can be considered as a key phenomenon to improve the yield of SWCNT synthesis process since the released heat increases the temperature locally and enhances the supersaturation of carbon in liquid solution phase [23]. The heat released due to Y-C and Y-S reactions will be higher as their concentration in gas phase are higher compared to other reactions, as seen in Figure 2 (a), (c) and (e). Figure 2 (b), (d) and (f) show the liquid solution phase formed during the high temperature plasma process of feedstock materials in Group 1. The graphs show that Y did not participate in liquid solution in contrast with the metallic catalysts (i.e., Ni, Co and Mo). The liquid solution was formed abruptly for all mixtures as seen in Figure 2. However, for MIX 3, (Figure 2, f), Ni content gradually increased up to a point (around T= 2 800 K) where it became the main solvent, while Mo started to diminish at 2500 K due to solidification. The whole liquid solution phase solidified at a temperature lower than 1500 K. The presence of liquid solution phase containing metals in such a wide range of temperature compared to the pure liquid metals is due to the solubility of carbon which lowers the solidification temperature of metals [24,25]. The liquid solution phases depicted in Figure 2 showed also that along with the presence of liquid metals as solvent, and carbon as a solute, the other species including O, S and NiS exist in all liquid solutions. Compared to other species, a high degree of solubility for carbon in liquid metals favored the formation of SWCNT. Interestingly, Mo compared to other catalysts can markedly elongate the existence range of liquid solution phase about 1000 K.

The calculated liquid solution phase of MIX 1 and MIX 2 are depicted in Figure 3. The Mole-Temperature graphs of S and O for both mixtures were quite the same indicating that their solubility was the same in Ni and Co. However, three main differences between the liquid solution phase of MIX 1 and MIX 2 were observed (Figure 3). The first difference was the solubility pattern of C which started at a higher content for MIX 1 and ended at the same content as MIX 2. This observation firstly suggests that the solubility of C was enhanced by Co in the binary system of Ni-Co compared to unary Ni system. Moreover, taking into account the concentration gradient of C over temperature as a driving force for nucleation of SWCNT [26,27], the binary Ni-Co system, having a bigger $\Delta C/\Delta T$ in the same range of temperature than unary Ni system, eases the nucleation of SWCNT. The second difference was the range of temperature in which the liquid solution phase exists. As shown in Figure 3, this temperature range seemed to be slightly wider for MIX 1 than MIX 2. According to the very high temperature gradient and cooling rate along the induction thermal plasma reactor [23], the wider temperature range in which the liquid solution phase exists can increase the time of SWCNT growth leading to a higher quality final product. The third difference is that unlike Ni, Co did not react with S in liquid solution phase which consequently allowed the active catalysts sites to uptake more carbon atoms.

Figure 4 and Figure 5 show thermodynamic calculations for another mixture of Ni (MIX 4) and two other mixtures of Mo (MIX 5 and MIX 6). Figure 4 shows the gas phase and liquid solution phase of MIX 4 in which the amount of Ni and Y_2O_3 was changed. By reducing the amount of Y_2O_3 in the feedstock mixture, the oxygen content was also reduced. The gas phase composition of the feedstock mixture was consequently affected by this change in the oxygen content as observed in the thermodynamic calculation results (Figure 4, a). On the other hand, the liquid solution phase was more affected by the change in the Ni content. High content of Ni in the gas phase makes the formation of NiS thermodynamically more favorable which in turn may partially compensate the diminution in the heat released by the exothermic reaction of Y-C → YC_2. The shortage of O in the reaction system results in an increase in sulfur cluster (S_2) followed by the formation of another cluster (S_3) in the lower temperature regions (T< 3000 K) (Figure 4, a). The formation of the clusters of sulfur at temperatures where liquid solution phase exists can reduce the effectiveness of the process since these clusters can compete with carbon precursors in being solubilized into the liquid solution and occupy it and form the soluble NiS. This can be a reason why the amount of S and NiS in the liquid solution phase has been remarkably increased in Figure 4 (b).

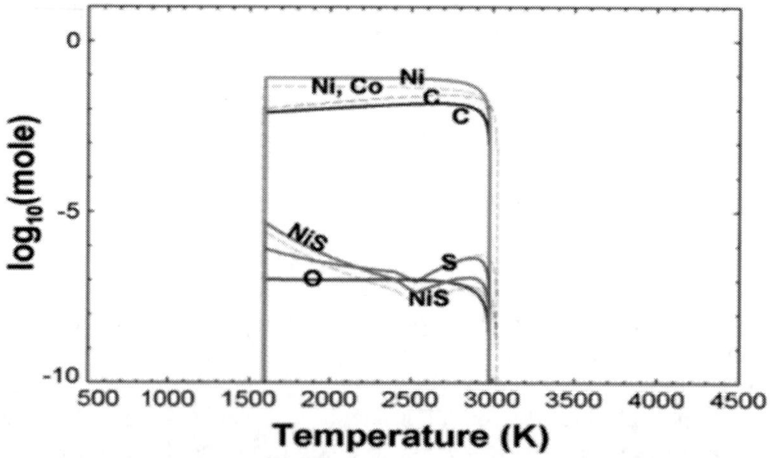

Figure 3 Superimposed liquid solution phase of MIX 1 (dashed-line) and MIX 2 (solid-line)

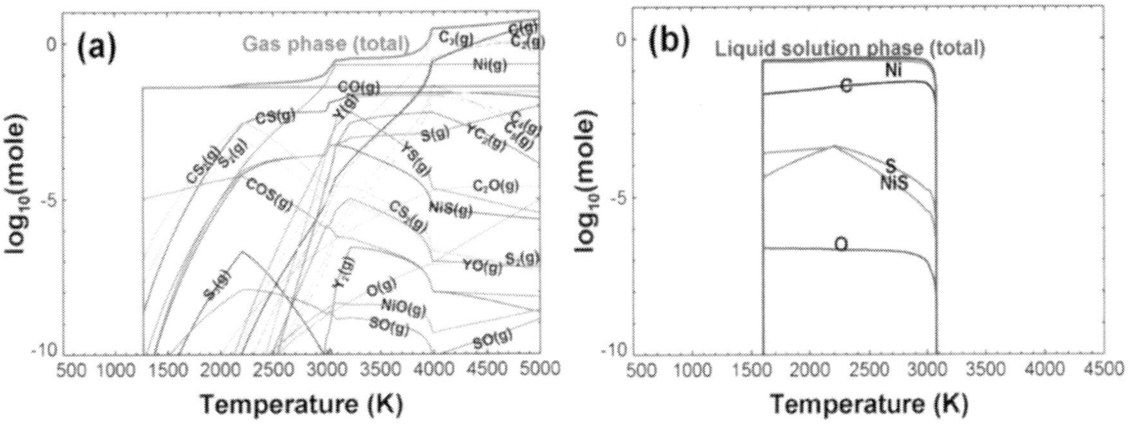

Figure 4 MIX 4 thermodynamic equilibrium compositions in (a) gas phase and (b) liquid solution phase

The total gas phase and also carbon precursors (C_1-C_5) seemed to be independent from the amount of metal catalysts and follow the same pattern, as shown in Figure 5. The total heat of dominant exothermic reactions, mentioned above, was poorly affected since the gas phase composition did not significantly change between MIX 5, MIX 6, and MIX 3 as shown in Figure 2 (c). The main changes were detectable from the liquid solution phases (Figure 5, b and d). By adding small amount of Mo (i.e., 0.5 wt. %) the liquid solution phase temperature window and the solubility of solute species were changed compared to MIX 2. By increasing the amount of Mo, the solubility of C was changed compared to MIX 2 which is in total agreement with the experimental observations where the solubility of carbon is very high in Mo [13].

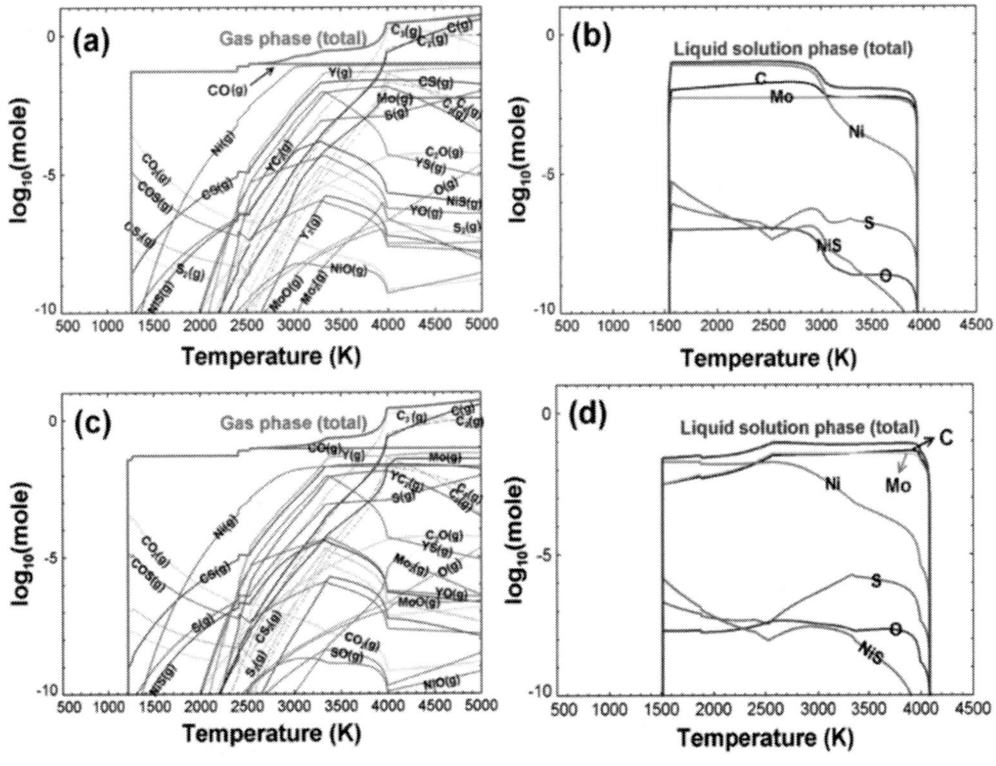

Figure 5 Thermodynamic equilibrium composition in gas phase (left side) and liquid solution phase (right side) of (a) MIX 5 and (b) MIX 6

4.2. Experimental results

HRSEM images of SWCNT samples synthesized with different feedstock mixtures are depicted in Figure 6. All samples represent the existence of SWCNT bundles entwined with impurities including graphitic particles, catalysts and amorphous carbon. HRSEM images demonstrated also the different morphology of SWCNT produced using different types and amounts of catalysts.

To compare the structural quality of SWCNT samples synthesized with different catalyst mixtures, Raman spectroscopy was performed and the results are summarized in Table 3. The intensity of tangential vibration mode of graphite at 1582 cm^{-1} (G-band), over the intensity of disordered sp^3 carbon at 1350 cm^{-1} (D-band), is an informative value for the structural quality assessment of SWCNT [28]. A higher G/D ratio is an indicator for a better quality of SWCNT. The Raman analysis showed that the highest quality of the SWCNT was achieved using MIX 1 in which Co was present. In contrast, the lowest quality of SWCNT was obtained using MIX 6 in which Mo had the highest content. Sample obtained with MIX 2, possessed SWCNT with a lower quality than those produced by MIX 1. Therefore, Ni-Co binary mixture would be a better choice than Ni alone for production of high structural quality SWCNT. Furthermore, Co can be considered as a promoter for the synthesis of SWCNT. The results of Raman analysis for MIX 1 and MIX 2 are in good agreement with the thermodynamic calculations that predicted a superior potential of Ni-Co mixture in the synthesis of higher quality SWCNT. For MIX 4, SWCNT quality seemed similar to that obtained using MIX 2. This observation is in good agreement with the thermodynamic results. Although the increase and decrease in the Ni and Y_2O_3 contents respectively, could affect the reaction system individually, their overall impact on the SWCNT synthesis seemed to be neutral.

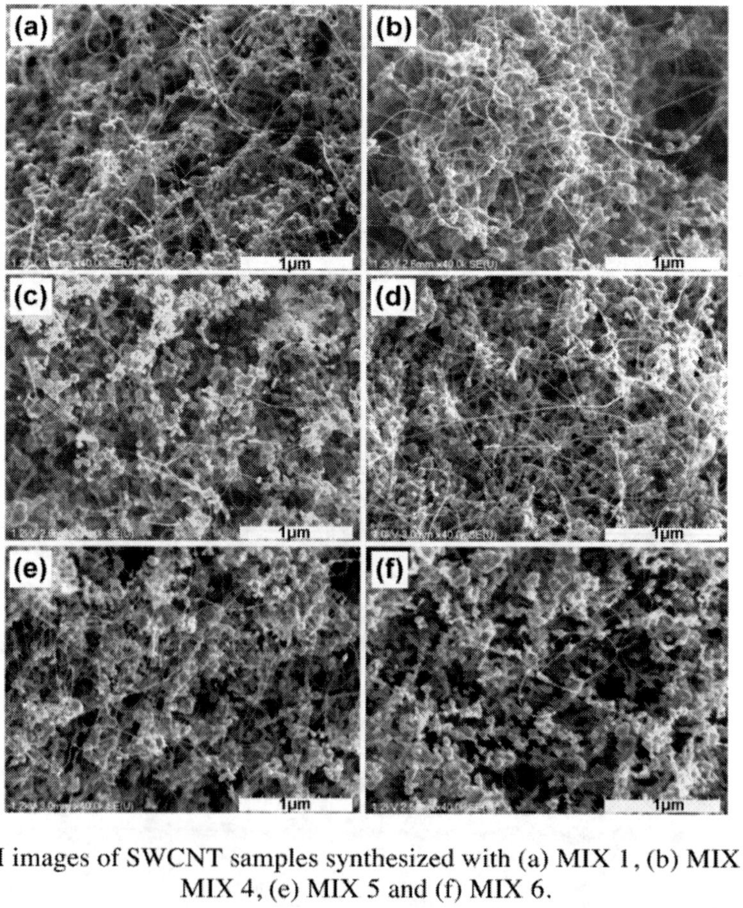

Figure 6 HRSEM images of SWCNT samples synthesized with (a) MIX 1, (b) MIX 2, (c) MIX 3, (d) MIX 4, (e) MIX 5 and (f) MIX 6.

Table 3 Quality assessments of SWCNT synthesized with different catalysts mixtures

Raman \ MIX No.	MIX 1	MIX 2	MIX 3	MIX 4	MIX 5	MIX 6
G/D	15.2	13.6	4	14	4.5	3.8

To closely examine the crystalline structure of SWCNT synthesized with different catalyst mixtures in Group 1, XRD analysis was performed and the results are shown in Figure 7.

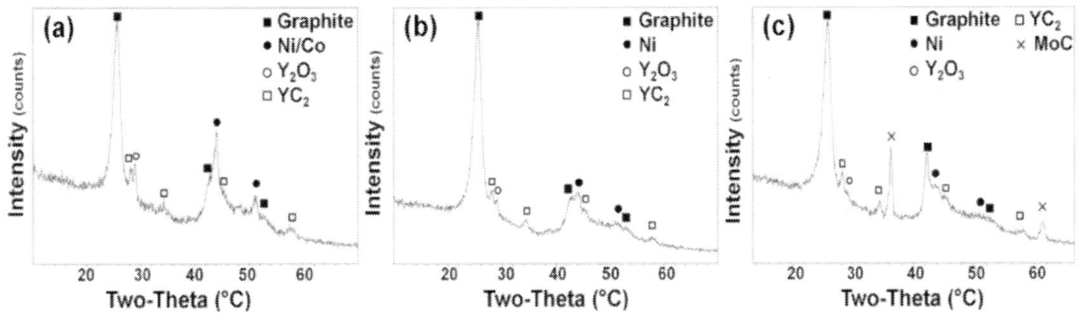

Figure 7 XRD patterns of SWCNT samples synthesized with (a) MIX 1, (b) MIX 2, and (c) MIX 3

The XRD analysis revealed that unlike pure Ni and Co, pure Mo did not exist in the final SWCNT product. However, the carbide form of Mo (MoC) was found. This observation suggested the higher tendency of Mo to form carbide than Ni and Co catalysts in the present ITP operating conditions. It should be mentioned that the carbide formation can reduce the yield of the SWCNT synthesis by putting away the active catalysts from the SWCNT nucleation sites. Therefore, the formation of MoC in high quantity during the ITP process is unfavorable and has to be minimized.

To further investigate discrepancies observed between experimental results and thermodynamic calculation for Mo, the temperature profile of the reactor along the central axis was plotted regarding to the numerical simulation of the ITP synthesis system (Figure 8) [29]. The eutectic temperature of Mo-C system ($T_{eutectic}$ = 2 500 K) [30] and Ni-C/Co-C ($T_{eutectic}$= 1 600 K) [29] are located at 0.6 and 0.75 m respectively from the torch exit (Figure 8). Two temperature zones are depicted in this graph: 3 000-2 000 K and 2 000-1 000 K. Since it has been experimentally proven that the optimal synthesis temperature lies 100-200 K below the $T_{eutectic}$ [25,31], the Zone 1 is expected to be a region in the ITP reactor which is more appropriate for the synthesis of SWCNT with Mo catalyst while the Zone 2 is suitable for using Ni and Co catalysts. As shown in Figure 8, the length of Zone 1 is only 13 cm in the reactor while the length of the Zone 2 is 30 cm. Therefore the nanotubes synthesized with Ni and Co have at least 2.3 times greater residence time in their suitable temperature zone than those synthesized with Mo.

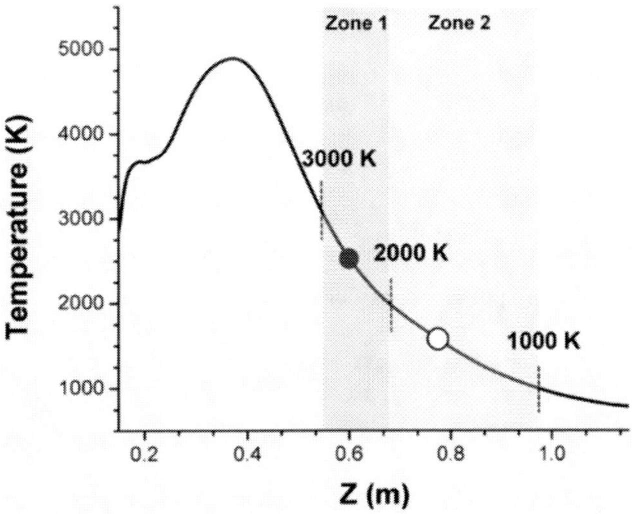

Figure 8 Axial profile of temperature along the centreline of the reactor. Black and white circle denote the eutectic temperature of Mo-C and Ni-C/Co-C systems, respectively.

Given that Co and Ni eutectic temperature is almost the same, this observation enables to explain why both Ni and Co produced SWCNT with similar quality but higher than that synthesized using Mo. Therefore, to overcome the problem observed with Mo it is necessary to optimize the ITP operating conditions to obtain similar temperature zone for Mo with the same effective length as Ni and Co.
To characterize the thermogravimetric behavior of each SWCNT sample produced with the feedstock mixtures in Group 1, TG analysis was performed (Figure 9). All SWCNT samples demonstrated a multistep weight loss in the temperature range of 200-800 °C indicating the presence of different carbonaceous structures including SWCNT. The mass loss became zero at temperatures above 800° C where the residual mass, attributed to the residual catalyst content, was calculated to be 15, 12 and 11 wt. % for the SWCNT samples synthesized with MIX 1, MIX 2, and MIX 3, respectively. The mass gain at temperature around 200 °C was due to the oxidation of metal catalysts in the samples which was then followed by a rapid mass loss at a temperature around 350 °C due to the fast reaction of amorphous carbon with oxygen [32]. The temperature at which SWCNT samples start to lose weight (T_{onset}) seemed to be identical. However, the multistep weight loss event (TG graph) was more identical for the samples synthesized with MIX 1 and MIX 3 than MIX 2. This observation suggested that the ternary mixture of catalysts with the identical catalyst oxide produced SWCNT samples with the same TG behavior while this behavior was slightly changed when the binary mixture was used.

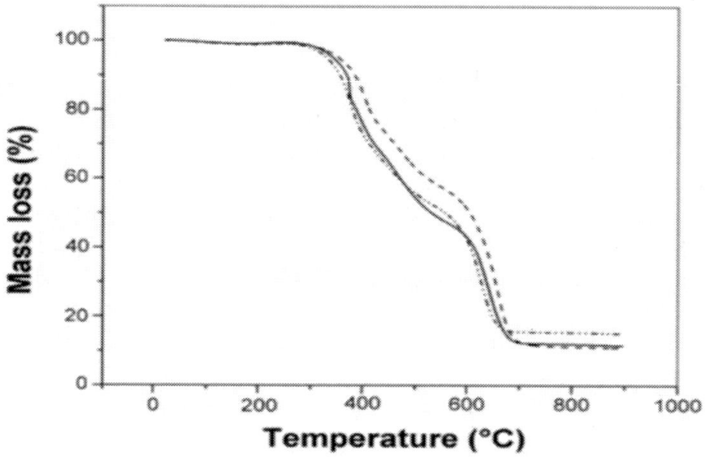

Figure 9 TG graphs of SWCNT samples synthesized using MIX 1 (dashed-dotted line), MIX 2 (dashed line) and MIX 3 (solid line)

5. Conclusion

The effect of metallic catalysts and their quantity in the feedstock material on the synthesis of SWCNT were thermodynamically and experimentally studied.
High quality SWCNT can be produced with a ternary catalyst mixture containing Ni, Co and Y_2O_3 or with a binary mixture of Ni and Y_2O_3. Thermodynamic calculations indicated that type and quantity of metal catalysts would affect more the liquid solution phase formed in the ITP reaction system during SWCNT production and less the gas phase, while the content of Y_2O_3 changed the gas phase with less impact on the liquid solution phase. Considering the eutectic temperature of the metallic catalyst-carbon system, the effective residence time for Ni/Co was found to be 2.3 times more than Mo in the ITP resulting in a higher quality Ni/Co-grown SWCNT. Thus, Mo potential for the synthesis of SWCNT can be improved through ITP process optimization by increasing the length of higher temperature zone (T > 2 000 K).

Acknowledgement
This study was supported financially by the Natural Sciences and Engineering Research Council of Canada (NSERC) and Research Chair of Canada (Laboratory of Cell-Biomaterials Biohybrid

Systems). We would like to thank our collaborators at National Research Council (NRC) for Raman spectroscopy measurements and Dr. Alireza Hekmat for the English proof reading of the manuscript.

References

[1] Baddour CE and Briens C 2005 *International Journal of Chemical Reactor Engineering* 3
[2] Iijima S and Ichihashi T 1993 *Nature* 363 603–603
[3] Cao Q and Rogers JA 2009 *Advanced Materials* 21 29–53
[4] Genest J, Kim KS, Sauve A, Boissy P, Soucy G et al. 2012 *J Appl Phys* 111 023502
[5] Ci L, Suhr J, Pushparaj V, Zhang X, and Ajayan PM 2008 *Nano Letters* 8 2762–2766
[6] Breuer O and Sundararaj U 2004 *Polymer Composites* 25 630–645
[7] LeMieux MC, Sok S, Roberts ME, Opatkiewicz JP, Liu D et al. 2009 *ACS Nano* 3 4089–4097
[8] Fadel TR, Look M, Staffier PA, Haller GL, Pfefferle LD et al. 2010 *Langmuir* 26 5645–5654
[9] Zanello LP, Zhao B, Hu H, and Haddon RC 2006 *Nano Letters* 6 562–567
[10] Guo T, Nikolaev P, Thess A, Colbert DT, and Smalley RE 1995 *Chemical Physics Letters* 243 49–54
[11] Dai H, Rinzler AG, Nikolaev P, Thess A, Colbert DT et al. 1996 *Chem Phys Lett* 260 471–475
[12] Kim KS, Cota-Sanchez G, Kingston CT, Imris M, Simard B et al. 2007 *Journal of Physics D: Applied Physics* 40 2375–2387
[13] Moisala A, Nasibulin AG, and Kauppinen EI 2003 *Journal of Physics Condensed Matter* 15 S3011– S3035
[14] Dai H, Rinzler AG, Nikolaev P, Thess A, Colbert DT et al. 1996 *Chemical Physics Letters* 260 471–475
[15] Celnik M, West R, Morgan N, Kraft M, Moisala A et al. 2008 *Carbon* 46 422–433
[16] Flint O 1965 *J Nucl Mater* 16 233–248
[17] Machado BI, Suro RM, Garza KM, and Murr LE 2011 *International Journal of Nanomedicine* 6 167–178
[18] Sakai T, Takeda S, and Nakamura M 2002 *Dental Materials Journal* 21 133–146
[19] Tessier DM and Pascal LE 2006 *Toxicology Letters* 167 114–121
[20] Alinejad Y, Shahverdi A, Kim KS, and Soucy G 2010 *High Temperature Material Processes* 14 101–117
[21] Kim KS, Cota-Sanchez G, Kingston CT, Imris M, Simard B et al. 2007 *Journal of Physics D: Applied Physics* 40 2375–2387
[22] Shahverdi A and Soucy G 2011 *Journal of Thermal Analysis and Calorimetry*
[23] Kim KS, Moradian A, Mostaghimi J, Alinejad Y, Shahverdi A et al. 2009 *Nano Research* 2 800–817
[24] Krivoruchko OP and Zaikovskii VI 1998 *Kinetics and Catalysis* 39 561–570
[25] Gorbunov A, Jost O, Pompe W, and Graff A 2002 *Carbon* 40 113–118
[26] Celnik M, West R, Morgan N, Kraft M, Moisala A et al. 2008 *Carbon* 46 422–433
[27] Wen JZ, Celnik M, Richter H, Treska M, Vander Sande JB et al. 2008 *Journal of Materials Chemistry* 18 1582–1591
[28] Dresselhaus MS, Dresselhaus G, Saito R, and Jorio A 2005 *Physics Reports* 409 47–99
[29] Kim KS, Moradian A, Mostaghimi J, Alinejad Y, Shahverdi A et al. 2009 *Nano Research* 2 800–817
[30] Hugosson HW, Jansson U, Johansson B, and Eriksson O 2001 *Chemical Physics Letters* 333 444–450
[31] Kataura H, Kumazawa Y, Maniwa Y, Ohtsuka Y, Sen R et al. 2000 *Carbon* 38 1691–1697
[32] Shahverdi A, Kim KS, Alinejad Y, and Soucy G 2012 *Journal of Nanoparticle Research* 14

Polymerization of 1-Naphthylamine by DC discharge

A Gilman[1], M Yablokov[2], M Augustyniak-Jablokov[3], K Tadyszak[4], and A Kuznetsov[5]
[1] Senior Researcher, Enikolopov Institute of Synthetic Polymer Materials, Russian Academy of Sciences, Moscow, Russia
[2] Senior Researcher, Enikolopov Institute of Synthetic Polymer Materials, Russian Academy of Sciences, Moscow, Russia
[3] Head of Laboratory, Institute of Molecular Physics, Polish Academy of Sciences, Poznan, Poland
[4] PhD Student, Institute of Molecular Physics, Polish Academy of Sciences, Poznan, Poland
[5] Head of Laboratory, Enikolopov Institute of Synthetic Polymer Materials, Russian Academy of Sciences, Moscow, Russia

[1] E-mail: plasma@ispm.ru

Abstract. Thin polymer films from 1-naphthylamine have been obtained for the first time by polymerization at the cathode and anode in DC discharge. The morphology of the films synthesized at the cathode was examined by SEM and AFM. The elemental composition, chemical structure, and thermal stability of the polymer have been studied by pyrolysis chromatography, IR and UV spectroscopy, and thermogravimetry, respectively. The surface contact properties of the films and their thickness have been determined with a microinterferometer. Polymer was also investigated by ESR method.

1. Introduction

It is well known that 1-naphthylamine holds promise for organic luminescent devices [1], and the plasma polymerization method is of undoubted interest in this regard, since it allows for the synthesis of polymers in the form of thin films and coatings [2, 3].

There are only chemical [4-6] and electrochemical [7-9] methods known for producing a polymer from 1-naphthylamine. For example, polymer from 1-naphthylamine (PNA) was synthesized chemically in methanol with $CuCl_2 \cdot 2H_2O$ as initiator in the presence of the surfactant *N*–cetyl–*N,N,N*–triethylammonium bromide [4].This PNA was soluble in organic solvents, such as ethanol, methanol, DMSO, DMF, MP, and tetrahydrofuran. The IR and UV studies of the polymer structure showed that it contained benzenoid and quinonoid moieties, and the polymer was a mixture of naphthylamine structures linked in the 1,4- and 1,5-positions one another. By the thermogravimetric technique, it has been shown that the polymer is quite stable up to ~ 180°C and its 50% mass loss is observed at ~ 420°C. It has also been reported [7] that electrochemical synthesis in a standard three-electrode cell using a saturated aqueous solution of 1-naphthylamine hydrochloride in the presence of KCl (at a potential of 0.9 V and an initial current of 0.08–0.1 mA, with a decrease to 0.01 mA) yields coloured films of oligomers. The structure of these oligomers was investigated by IR and UV spectroscopy, and 1,4-amine/imine bonds were found in the resulting oligomer products.

In this work we performed for the first time a plasma-chemical synthesis of a polymer from 1-naphthylamine in DC discharge and studied its structure and properties.

2. Experimental

1-Naphthylamine (NA) was purified by recrystallization from ethanol prior to use; its purity was checked by measuring the melting temperature (T_m=50°C) using differential scanning calorimetry.

The setup and procedure for DC discharge polymerization of compounds that are solid at room temperature are described by us elsewhere [10]. The process conditions were as follows: a temperature in the reaction chamber of 40°, an NA vapour pressure of 40 Pa, a discharge current of 5 mA, and a reaction time from 60 s to 420 s. A polymer from NA (PPNA) was obtained by DC discharge in the form of thin film on an aluminium foil, conducting single-crystalline silicon, or fused silica substrate.

The film thickness was measured by interferometry in white light using a Linnik microinterferometer.

The SEM images were taken with a Zeiss EVO 40 scanning electron microscope with SDD. The AFM study of the film was conducted with a Solver P47 microscope (NT-MDT, Russia) and software Gwyddion [11].

The elemental composition of PPNA was studied by pyrolysis chromatography using a Carlo-Erba IA 1108 CHNS analayzer.

To characterize the surface contact properties, the values of contact angle (θ) determined with a goniometer (error ±1°) using two test liquids, doubly distilled water and glycerol, were used. From the results, the work of adhesion, the total surface energy, and its polar and dispersion terms were calculated [12].

The chemical structure of the polymer was investigated by Fourier-transform IR and UV spectroscopy. The IR spectra were recorded with Bruker Equinox 50S FTIR spectrometer in CsI disks. The assignment of bands was carried out according to [13]. Electronic absorption spectra were obtained on a Shimadzu UV-2501 PC spectrometer, and luminescence and excitation spectra were recorded with scanning spectrofluorimeter ALS-1 (Russia).

The thermal stability of the polymer was studied by TGA in air and argon using a NETZSCH TG-209-F instrument and software package NETZSCH Proteus-Thermal Analysis, ver.4.8.4.

Polymer synthesized by the DC discharge was investigated by ESR method using spectrometer RADIOPAN SX (Poland) with cryostat Oxford CF935.

3. Results and discussion

The PPNA films neither dissolved nor swelled in ethanol, acetone, heptane, at all, but the films obtained at the cathode were easy detached from the aluminium foil substrate with ethanol.

Kinetics of the formation of polymer films obtained at the cathode has been investigated, and the data obtained are presented in Figure 1. It is seen that the growth rate of polymer films increases from 3 nm/s to 6 nm/s with the film thickness from 100 nm to 0.5–1 μm; then, the growth rate becomes constant.

The morphology of polymer films was examined by SEM and AFM, the results obtained by SEM are given in Figure 2. It was shown that the films have granular structure and a size of these granules is about 50 nm (Fig. 2a). The presence of micron-size agglomerates composed from granules was also established (Fig. 2b). The value of the surface roughness calculated from roughness profile obtained by AFM (Fig. 3) was increased from 4±1 nm for the film with the thickness of 90 nm to 35±12 nm for the film with the thickness of 1.5 μm. It is probably owing to the enhancement of the quantity of the agglomerates of globular particles in the structure of PPNA films.

Data on the elemental composition of PPNA films obtained at the cathode are shown in Table. It is evident that the relative amount of hydrogen and nitrogen in polymer is lower than that in NA, and oxygen appears as well. Calculation of these results of empirical formula for C_{10} shows that the amount of nitrogen in PPNA is reduced to 0.7 as compared to 1.0 in the initial NA.

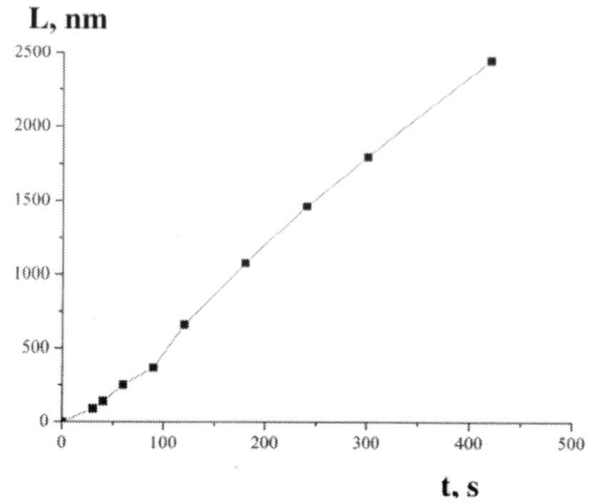

Fig. 1. Dependence of PPNA film thickness (L) on the time of polymerization (t).

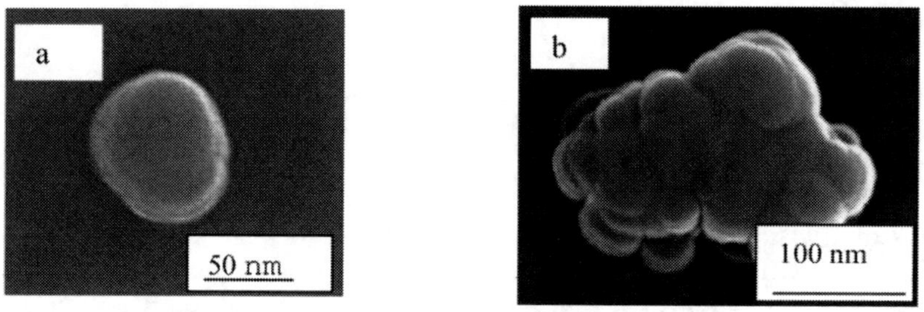

Fig. 2. SEM images of (a) globular particles and (b) the agglomerate of globular particles of a PPNA film synthesized at the cathode.

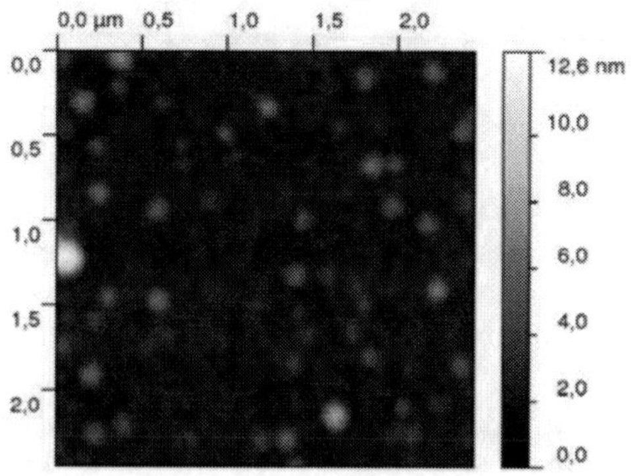

Fig. 3. AFM image of a PPNA film synthesized at the cathode.

Table. The elemental composition of 1-naphthylamine and the polymer (PPNA) synthesized at the cathode in DC discharge. Chemical formula of 1-naphthylamine is $C_{10}H_7NH_2$.

Sample	C, %	H, %	N, %	O,% (by difference)	Calculation in terms of C_{10}
1-naphthylamine (calculation)	83. 9	6.8	9.8	–	$C_{10}H_9N$
PPNA	81.6	3.5	6.8	8.1	$C_{10}H_{5.15}N_{0.71}O_{0.74}$

The results of the goniometric measurements of the contact angles (θ) on the surface of PPNA films deposited on an aluminium foil substrate at the cathode and anode are shown that the both films have the same values of θ by water (65°), the work of adhesion (103.6 mJ/m^2), the surface energy (38.7 mJ/m^2), and its polar (20 mJ/m^2) and dispersion (18.7 mJ/m^2) components.

The chemical structure of PPNA films was investigated by FTIR and UV spectroscopy. Figure 4 shows the IR absorption spectra of (*1*) the monomer and (*2*) the polymer synthesized at the cathode by DC discharge.

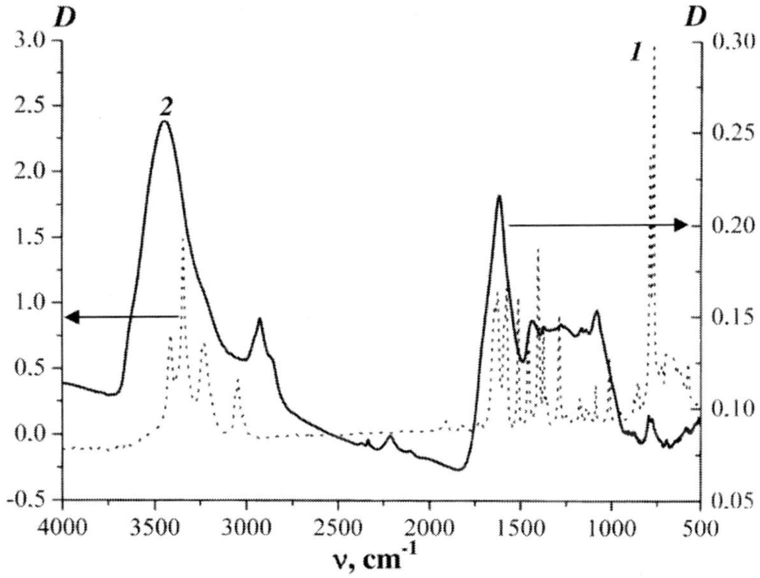

Fig. 4. IR absorption spectra of (*1*) parent 1-naphthylamine and (*2*) the polymer, synthesized at the cathode.

The broad band at 3370 cm^{-1} is associated with stretching vibrations of secondary amines [4]. The absorption bands at 1445 and 1520 cm^{-1} correspond to C=C skeletal vibrations of the aromatic rings, whereas the band at 1630 cm^{-1} is apparently due to the stretching vibrations of the imino groups [7]. The absorption bands at 1378 and 1260 cm^{-1} are associated with C–N stretching [14], and the absorption band peaked at 1141 cm^{-1} (C–N) indicates the presence of the imine and secondary amine groups, i.e. benzenoid and quinonoid structures [15]. Absorption in the region 700-900 cm^{-1} is attributed to CH bending vibrations for different kinds of attachment of the aromatic rings. The absorption bands at 720 and 879 cm^{-1} apparently indicate attachment in the 1,4-position [15].

However, the weak intensity of other absorption bands in this area and their low resolution do not permit accurate assignment concerning the addition position.

The luminescence spectra of (*1*) NA in ethanol and (*2*) PPNA obtained at the cathode are shown in Figure 5, as well as the NA (*3*) and polymer (*4*) UV absorption spectra. In the case of excitation at a wavelength of 325 nm, week luminescence of the polymer at 400–600 nm is observed, roughly coinciding with the monomer luminescence.

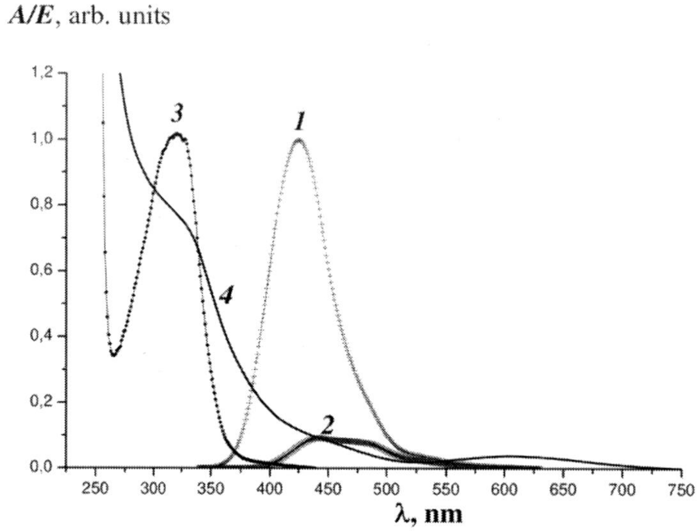

Fig. 5. Matching the (*1, 2*) luminescence and (*3, 4*) absorption spectra of (*1, 3*) parent 1-naphthylamine and (*2, 4*) the polymer, synthesized at the cathode.

A comparison of the absorption spectra of NA and PPNA leads to the conclusion that this may be naphthylamine nuclei luminescence strongly attenuated by the 440-nm polymer absorption band (inner filter effect). These data are shown, that the PPNA chain is apparently made of units with two different structures. In one case, the NA units are linked to one another by a spacer with broken π conjugation, as indicated by the presence of both the short-wavelength absorption, which coincides with the monomer absorption, and luminescence. In the other case, NA units are joined together without breaking the double bonds, as confirmed by the long-wavelength absorption (maximums at 440 and 615 nm) and the absence of luminescence.

Figure 6 shows TGA curves measured in (*1*) argon and (*2*) air atmosphere for the PPNA synthesized by DC discharge at the cathode. It is seen that PPNA loses less than 20% of mass in an argon atmosphere when heated to 400°C, whereas the polymer undergoes oxidative degradation in air: a 50% mass loss is observed upon heating to ~410°C and further heating leads to a rapid loss (up to 90% at ~470°C) and then slowly to the gradual complete degradation of the sample at 800°C. These data resemble those reported in [4] for the polymer synthesized by chemical method.

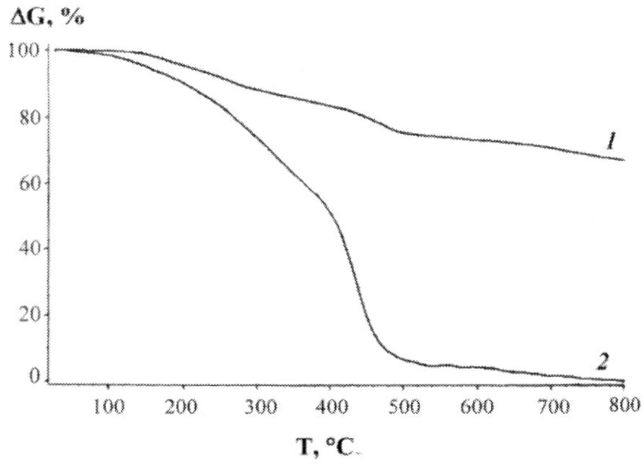

Fig. 6. TGA curves measured in (*1*) argon and (*2*) air for the PPNA synthesized at the cathode.

Polymer synthesized by DC discharge was investigated by ESR method. It was found that the spin concentration is equal to ~10^{19} spins/g, the line width at 300 K to ΔB=0.13 mT, and g-factor to 2.0020±0.0002, respectively. The (*1*) experimental ESR spectrum of PPNA and its resolution (***2, 3***) are shown in Figure 7.

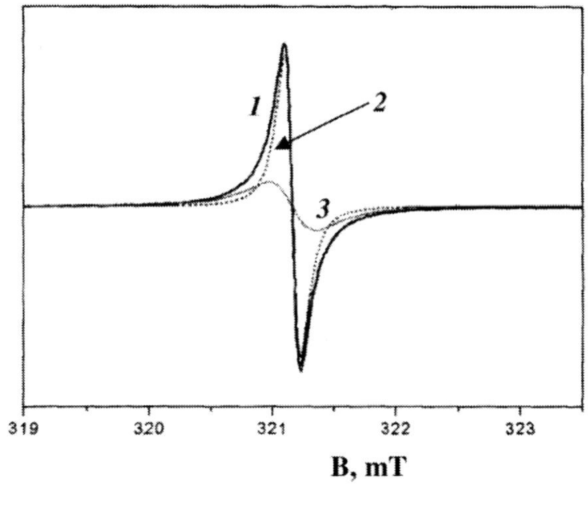

Fig. 7. ESR spectrum of PPNA synthesized at the cathode: ***1*** – experimental spectrum, ***2*** – signal resolution in terms of a line with small width and high amplitude ("long" chain), ***3*** – signal resolution in terms of a line with high width and short amplitude ("short" chain).

The data obtained confirm the presence of aromatic radicals localized near nitrogen atoms [16]. Previously it was shown that benzenoid-quinonoid moieties in PPNA polymer chain were also found by means of FTIR an UV spectroscopy.

4. Conclusion

Thus, polymer was obtained from 1-naphthylamine for the first time by plasmachemical polymerization at the cathode in DC discharge. Based on the elemental analysis, IR and UV spectroscopy and ESR data, we can conclude that the polymer structure is composed of benzenoid-quinonoid moieties to a considerable extent.

Acknowledgments

This work was financially supported by the Russian Foundation for Basic Research, project no. 10-03-00772a.

References

[1] Kim J-E *et al* 2009 *US Patent* 7604874
[2] Yasuda H 1985 *Plasma Polymerization* (Missouri: Academic Press)
[3] Plasma Deposition, Treatment and Etching of Polymers 1990 ed. R D'Agostino (N.Y.: Academic Press)
[4] Riaz U *et al* 2008 *J. Appl. Polym. Sci.* **108** 2604
[5] Moon D-K *et al* 1993 *Macromol.* **26** 6992
[6] Radhika S *et al* 2009 *J. Mater. Sci.* **44** 3542
[7] Schmitz BK and Euler WB 1995 *J. Electroanal. Chem.* **399** 47
[8] Chung C-Y *et al* 2001 *Mat. Chem. Phys.* **71** 148
[9] Lu B *et al* 2011 *Synth. Met.* **161** 188
[10] Drachev A *et al* 2004 *Synth. Met.* **112** 35
[11] http://gwyddion.net
[12] Wu S 1982 *Polymer Interfaces and Adhesion* (N.Y.: Marcel Dekker)
[13] *NIST IR Spectroscopy Database*, http://srdata.nist.gov
[14] Cox BM *et al* 1965 *Spectrochim. Acta, Part A* **21** 1663
[15] Marianovic GC *et al* 2002 *Serb.Chem. Soc.* **67** 867
[16] Gong X *et al* 1998 *J.Polym. Sci. Part A : Plym. Chem.* **36** 633

Journal of Physics: Conference Series **406** (2012) 012021 doi:10.1088/1742-6596/406/1/012021

TiN thin film deposition by cathodic cage discharge: effect of cage configuration and active species

Natália de Freitas Daudt[1]*, Júlio César Pereira Barbosa[2], Danilo Cavalcante Braz[1], Marcelo Barbalho Pereira[3] and Clodomiro Alves Junior[1]

[1] Universidade Federal do Rio Grande do Norte, Programa de Pós Graduação em Ciência e Engenharia dos Materiais, Natal, RN, Brazil 59072-970
[2] Universidade Federal Rural do Semi-Árido, Mossoró, RN, Brazil 59625-900
[3] Universidade Federal do Rio Grande do Sul, Porto Alegre, RS, Brazil 91501-970

E-mail daudt.natalia@gmail.com

Abstract. Plasma cathodic cage technique was developed recently in order to eliminate phenomena such as edge effects and overheating, which occur during conventional nitriding processes. In this work, the effect of plasma active species and cage configurations during thin film deposition of TiN were studied. This compound was chosen because its properties are very sensitive to slight variations in chemical composition and film thickness, becoming a good monitoring tool in fabrication process control. In order to verify the effect of cage geometry on the discharge and characteristics of the grown film, a cage made of titanium was used with different numbers and distribution of holes. Furthermore, different amounts of hydrogen were added to the Ar + N_2 plasma atmosphere. Flow rates of Ar and N_2 gas were fixed at 4 and 3 sccm, respectively and flow rates of H_2 gas was 0, 1 and 2 sccm. Plasma species, electrical discharge and physical characteristics of the grown film were analyzed by Optical Emission Spectroscopy (OES), Atomic Force Microscopy (AFM), X-Ray Diffraction. It was observed by OES that the luminous intensity associated to Hα species is not proportional to flow rate of H_2 gas. Electrical efficiency of the system, crystal structure and topography of the TiN film are strongly influenced by this behavior. For constant flow rate of H_2 gas, it was found that with more holes at the top of the cage, deposition rate, crystallinity and roughness are higher, if compared to cages with a small number of holes at the top of cage. On the other hand, the opposite behavior was observed when more holes were located at the sidewall of cage.

1. Introduction

Cathodic cage plasma technique was recently developed in order to eliminate edge effects and overheating, which occur during thermochemical treatment with plasma. These effects take place due to high superficial area/volume ratio, mainly in pieces with complex geometry [1-3].

The technique is used to protect electrically the workpiece during the process, resembling a Faraday cage device. An isolator is placed between the workpiece and the cathode, in order to keep the workpiece in a floating potential [4]. In this configuration, only the cage is in cathodic potential, producing a hollow cathode effect in each hole. Due to the higher density of ions formed in the holes of the cage; there is a higher sputtering rate in that region. The combinations of the gas species with the sputtered species present in plasma are directed towards the substrate surface, where they are

Journal of Physics: Conference Series **406** (2012) 012021 doi:10.1088/1742-6596/406/1/012021

deposited and diffused onto the sample [5]. Thus, it is possible to produce a hybrid process of deposition and diffusion, since the substrate temperature is high enough to promote diffusion.

Titanium nitride films are used in various industrial applications, for example, in metallurgical industry as coatings for high hardness and low friction [6] and in decorative coatings to replace gold, since different color tones may be achieved by variation of the Ti/N ratio [7]. The material is also used in optical applications as coatings of solar cells and solar control windows [8], in coatings of biocompatible alloys for biomaterials [9] and in coatings for microelectronic semiconductors [10].

Properties of titanium nitride films are very sensitive to small variations in chemical composition and thickness. Therefore, this sensibility is a useful tool to monitoring the fabrication process control.

Since this technique was developed recently, there are few studies about the influence of process parameters on the characteristics of the film grown. Therefore, this work proposes to investigate the efficiency of this technique to obtain TiN thin films. The effects of plasma active species and the influences of cage configuration on the deposition rate, on electrical parameters, on the microstructure and topography of the grown film on were investigated by Optical Emission Spectroscopy (OES), Atomic Force Microscopy (AFM) and X-Ray Diffraction.

2. Materials and Methods

Rectangular glass samples with 25 mm x 10 mm x 2 mm (length x width x thickness) were used in this work. The deposition was made using an ion nitriding reactor adapted to cathodic cage configuration. The samples were isolated of cathode by an alumina disk. The cages were made of commercially pure titanium (grade II) and have holes with 12 mm of diameter on the top and sidewall, covering all over the surface of the device. Figure 1 illustrates the reactor and a cage with its respective nomenclature.

Four different cages were used and denominated as L1T4, L1T8, L2T4 and L2T8. The L is referent to the number of lines of holes in the sidewall of cage and T is referent to the number of lines of holes on the top (covering) of the cage.

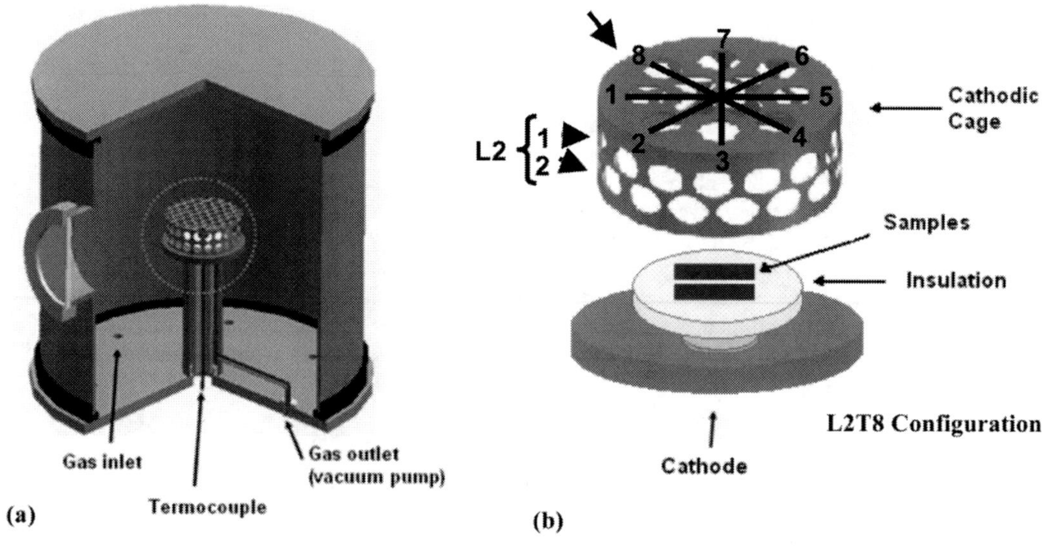

Figure 1. Schematic representation of (a) plasma reactor and (b) cathodic cage with respective nomenclature for different cage configurations.

Ar and N_2 gas flow rates were fixed at 4sccm and 3 sccm, respectively and for H_2 were used gas flow rates of 0, 1 and 2 sccm. Argon was used to increase titanium sputtering rate and to control nitriding rate of the cage [11]. Hydrogen was used to reduce the presence of superficial oxides and to increase process efficiency [12]. Pressure and temperature were fixed at 1.5 mbar and 450 °C, respectively. Table 1 shows the experimental conditions.

Table 1. Experimental conditions used in this work.

Sample	H_2 gas flow (sccm)	Pressure (mbar)	Current (A)	Voltage (V)	Power	Number of holes		Cage height
					W	Wall	Top	mm
L1T4	0	1.45	0.30	877	263	16	8	34
L1T8	0	1.46	0.33	868	286	16	12	34
L2T4	0	1.47	0,35	846	296	31	8	45
L2T8	0	1.45	0.35	890	311	31	12	45
L1T4-1	1	1.48	0.30	860	258	16	8	34
L1T8-1	1	1.49	0.31	820	254	16	12	34
L2T4-1	1	1.48	0.33	811	267	31	8	45
L2T8-1	1	1.50	0.36	878	316	31	12	45
L1T4-2	2	1.51	0.32	808	258	16	8	34
L1T8-2	2	1.49	0.31	820	254	16	12	34
L2T4-2	2	1.52	0.35	817	286	31	8	45
L2T8-2	2	1.52	0.36	828	298	31	12	45

The deposition process was monitored by Optical Emission Spectroscopy (OES), using an Ocean Optics USB 4000 spectrograph. The X-Ray Diffraction analyses were performed with a Shimadzu XRD-6000 diffractometer using Cu radiation Kα and an accessory for Grazing Incidence X-ray Diffraction (GIXRD) and angle of incidence fixed at 0.5°. The samples were also analyzed by Atomic Force Microscopy (AFM) in contact mode with a Shimadzu microscope model SPM 9600.

3. Results and discussion

Figure 2a shows a typical OES spectrum obtained during cathodic cage deposition with the L1T8 configuration, where some lines corresponding to Ar, H_2, N_2 and N_2^+ transitions are highlighted.

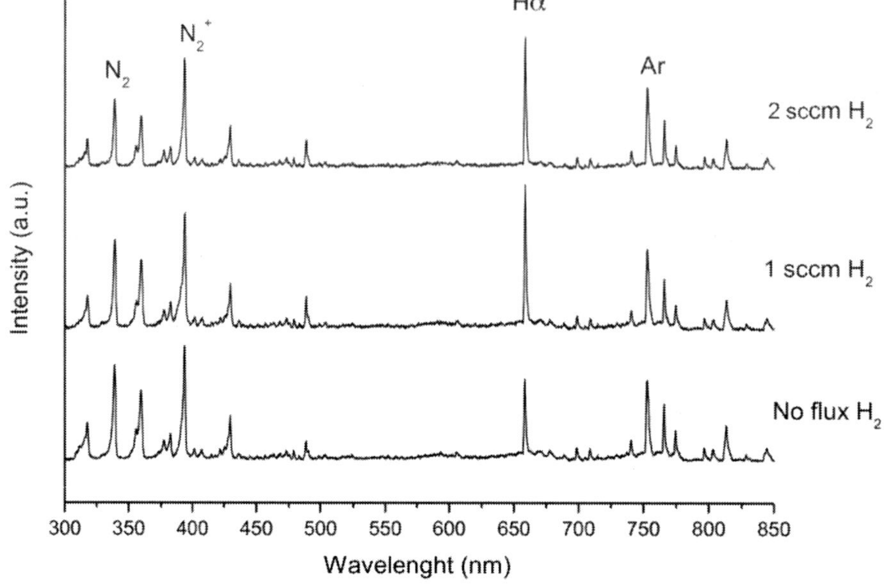

Figure 2. Plasma spectra obtained during deposition process in L1T8 configuration in different H_2 gas flow rates.

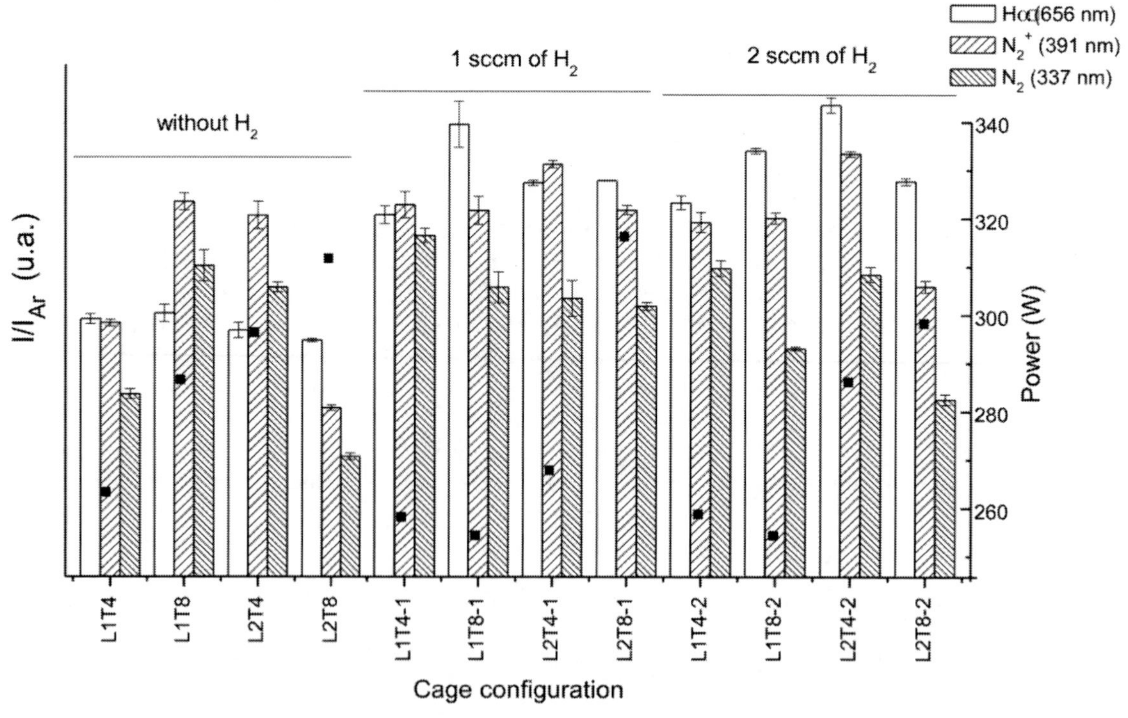

Figure 3. Species intensities for different configurations of plasma cage (bars) and values of the power supplied to keep the system at 450 ° C (black dots).

Figure 3 shows the intensities lines relative to argon (750.3 nm) of N_2^+, N_2 and Hα, with respective emissions lines at 391.4 nm, 337 nm and 656.3 nm.

The intensities of Hα lines are not proportional to the H_2 gas flow rate used in this work as shown in the Figure 2. The line at 656.3 nm corresponding to Hα has a maximal intensity at 1 sccm and it decreases in intensity when the gas flow rate is elevated to 2 sccm. Similar behavior is also observed for the other cage configurations (Figure 3). The intensities of N_2^+ lines are not related to the H_2 gas flow rate. Emission depends more on cage configuration than on H_2 gas flow rate (figure 3).

Plasma efficiency in the heat transfer can be estimated by the ratio between the power supplied in the system (electric power) and the average of the species energy (line or band intensities). The results indicated that the optimal configurations were the L1T8 and L1T4 cages.

A higher density of emitted species does not mean satisfactory energy efficiency to heat the cage, as can be observed in the configuration L2T4-2 (Figure 3), where the luminous intensity of species is maximal, but the power supplied to maintain the system at 450 °C is not the lowest one.

Figure 4 (a) shows the XRD patterns of deposited films with the L1T8 cage configuration using the different flow rates of H_2 gas (configuration that presented the highest TiN intensity). Moreover, Figure 4 (b) also shows a comparison between the XRD patterns of deposited films from L1T8 with the others cage configurations (all films deposited with 1 sccm flow rate of H_2 gas).

In general, the configuration with more holes at the covering exhibits higher efficiency. However, the highest efficiency was achieved when the lowest amount of holes is present at the sidewall. Apparently, this result is contradictory, since increasing the number of holes, it is expected that heating is promoted due to the increase in areas with hollow cathodes effect. This result is justified by the increase of the distance between the pieces and the top of the cage, when the number of holes in the sidewall is increased. Once that the heat from these holes is transferred to the chamber walls, to the inner and to the outer cage, there is an appropriate height at 34 mm which the maximum efficiency is promoted due to combination of heating by hollow cathode effect in covering and sidewall of cage.

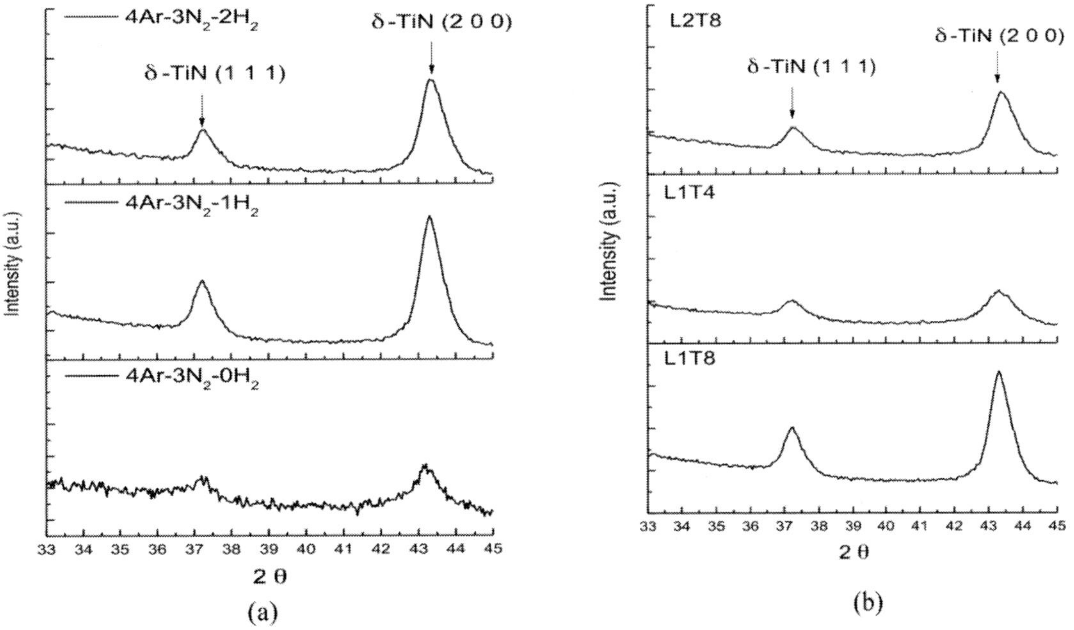

Figure 4. X-ray pattern with grazing incidence angle of 0.5 ° for films deposited with (a) the L1T8 cage configuration at different hydrogen flow rates and (b) different cage configurations at 1 sccm of H_2 gas flow rate.

Microstructural characteristics of the grown films obtained by GIXRD shown an abrupt increase in TiN intensity and decrease in the full width at half-maximum (FWHM) when flow rate of H_2 gas is added (Figure 4a). It has been observed by other authors [13-16] that hydrogen reduces both superficial oxide formed spontaneously on the cage and titanium nitride formed on the inner walls of cage holes during plasma treatment, which causes surface poisoning. In this work, it was observed that Hα species plays a key role in the deposition rate of TiN, because there is an increasing in deposition rate since there is a reduction in cage poising, promoting a formation of more stoichimonetry film.

The most intense TiN peak was observed with the cage configuration L1T8. On the other hand, the configuration L1T4 exhibited the lowest intense TiN peak due to low signal and high noise ratio. Furthermore, this configuration also produced lower luminous intensity associated to Hα species at 1 sccm of H_2 gas flow rate, as can be seen in Figure 3.

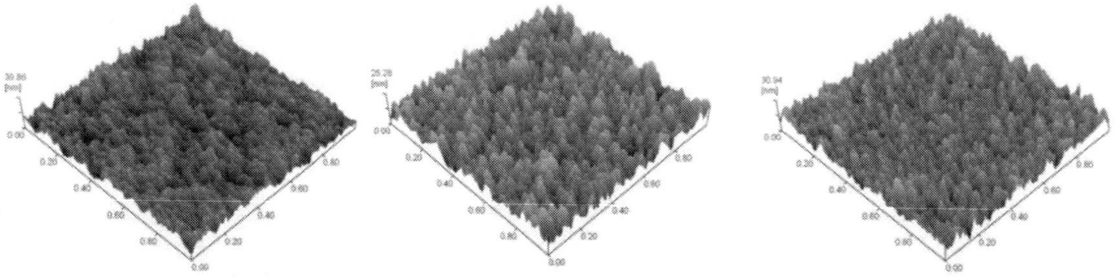

(a) (b) (c)

Figure 5. Nanotopography of the TiN films surfaces deposited by L1T8 cage configuration at different flow rates of H_2 gas (a) 0 sccm, (b) 1 sccm and (c) 2 sccm.

Table 2. Characteristics of the crystal structure and topography of films deposited at different H_2 gas flow rates and different cage configurations.

Cage configuration	Flow rate of H_2 gas (sccm)	$I_{H\alpha}/I_{Ar}$	Intensity δ-TiN (200)	FWHM	Roughness	
					Ra	Rms
L_1T_8	0	30267	264	2.75	4.49	3.60
L_1T_8	1	53643	5176	0.91	4.27	3.53
L_1T_8	2	48337	3158	1.05	4.32	3.49
L_2T_8	1	44900	2862	1.01	1.77	2.24
L_1T_4	1	41989	1520	1.57	3.66	2.95

Table 2 shows $I_{H\alpha}/I_{Ar}$ line intensity values obtained from OES spectra (Figure 3), FWHM values obtained from XRD patterns (Figure 4) and roughness values obtained from nanotopography (Figure 5). There is a correlation between the luminous intensity of Hα species and the grown film characteristics. The higher deposition rate, (demonstrated by the intensity of TiN (200) diffraction peak) and higher crystallinity (demonstrated by the FWHM) occurred when the $I_{H\alpha}/I_{Ar}$ ratio was higher. A relationship between roughness and Hα intensity was found, since roughness increases when the luminous intensity of Hα decreases. This effect occurred due to reduction of nitrogen and argon amount in plasma mixture, since nitrogen and argon have a higher atomic mass than hydrogen, they promote a higher sputtering rate in the grown film, resulting in a more roughness surface [11].

Furthermore it is possible to notice that the increase of the number of holes in the sidewall promotes a reduction of roughness. It is related to minor sputtering rate, since the increase of the distance between the pieces and the top of the cage decreases the probability of sputtered atoms collide with sample.

4. Conclusions

TiN films were obtained on glass substrates by cathodic cage discharge. It was found that deposition rates depend on the gas species density and cage configurations, which leads to obtain films with different roughness and microstructures.

Some conclusions can be made from these results:

1. The luminous intensity of the Hα species is not directly dependent on flow rate of H_2 gas. There is maximum line intensity at 1 sccm flow rate of H_2 gas and line intensity decreases when flow rate of H_2 gas is increased to 2 sccm.
2. L1T8 and L1T4 cage configurations, in this order, are the most efficient when the ratio between supplied power and average energy of the species is considered.
3. In general, configurations with more holes at the top of the cage exhibit higher efficiency. However, this is not observed when the holes at sidewall of the cage are more numerous. Actually, in this case, the efficiency is lower.
4. There is a correlation between the luminous intensity of Hα species and the characteristics of the grown film. Higher deposition rate and higher crystallinity were obtained with a higher $I_{H\alpha}/I_{Ar}$ ratio. Furthermore, there is a tendency of roughness increases when luminous intensity of Hα increases.

References

[1] Sousa R R M, Araújo F O, Ribeiro K J B, Costa J A P, Sousa R S and Alves Jr. C 2008 *Surface Engineering* **24** 313-318
[2] Sousa R R M, Araújo F O, Ribeiro K J B, Mendes M W D, Costa J A P and Alves Jr. C 2007 *Materials Science and Engineering A* **465** 223-227
[3] Ribeiro K J B, Sousa R R M, Araújo F O, Brito R A, Barbosa J C P and Alves Jr. C 2008 *Materials Science and Engineering A* **479**142-147.
[4] Alves Jr. C, Araújo F O, Ribeiro K J B, Costa J A P, Sousa R R M and Sousa R S 2006 *Surface and Coatings Technology* **201** 2450-2454.
[5] Gallo S C and Dong H 2010 *Vacuum* **84** 321-325
[6] Peng Z, Miao H, QI L, Yang S and Liu C 2003 *Acta Materialia* **51** 3085-3094
[7] Roquiny Ph, Bodart F and Terwagne G 1999 *Surface and Coatings Technology* **116-119** 278-283
[8] Smith G B, Ben-David A and Swift P D 2001 *Renewable Energy* **22** 79-84
[9] Cheng Y and Zheng Y F 2006 *Materials Science and Engineering A* **434** 99-104
[10] Wittmer M, Noser J R and Melchior H 1983 *Journal of Applied Physical* **54** 1423-1428
[11] Sankar M B, Pal A R, Bailung H and Cuthia J 2008 *Applied Surface Science* **254** 5760-5765
[12] Tamaki M, Tomii Y and Yamoto M 2000 *Plasma and Ions* **3** 33-39
[13] Safi I 2000 *Surface and Coating Technology* **127** 203-217
[14] Figueroa C A and Alvarez F 2006 *Applied Surface Science* **253** 1806-1809
[15] Priest J M, Baldwin M J and Fewel M P 2001 *Surface and Coatings Technology* **145** 152-163
[16] Sharma M K, Saikia B K, Phukan A and Ganguli B 2006 *Surface and Coatings Technology* **201** 2407-2413

Journal of Physics: Conference Series **406** (2012) 012022 doi:10.1088/1742-6596/406/1/012022

In-flight particle measurement of glass raw materials in hybrid heating of twelve-phase AC arc with oxygen burner

Y Liu, M Tanaka, T Ikeba, S Choi and T Watanabe
Department of Environmental Chemistry and Engineering, Tokyo Institute of Technology, Yokohama, Kanagawa 226-8502, Japan

E-mail: watanabe@chemenv.titech.ac.jp

Abstract. The high temperature provided by a 12-phase AC arc plasma is beneficial to finish vitrification reaction in milliseconds. Another heating method called "hybrid plasma" combines multi-phase AC arc and oxygen burner are expected to improve glass quality and increase productivity with minimum energy consumption. In this study, recent works on the development of in-flight particle measurement in hybrid plasma system are presented. Two-colour pyrometry offers considerable advantages for measuring particle temperatures in flight. A high-speed camera equipped with a band-pass filter system was applied to measure the in-flight temperatures of glass particles. The intensity recorded by the camera was calibrated using a tungsten halogen lamp. This technique also allows evaluating the fluctuation of the average particle temperature within millisecond in plasma region.

1. Introduction

Current glass melting technology is based on the typical Siemens-type melter fired in air with heavy oil or natural gas as the fuel. This type of melter has been used over 140 years because of its good large-scale performance and continuous melting system with minimal financial or technological risk. Fine-grained ingredients, closely controlled for quality, are mixed to make a batch, flows into the furnace that is heated to 1600 °C. Several processes - melting, refining, homogenising - take place simultaneous in the 2,000 tonnes of molten glass in the furnace, lasting as long as 50 hours. The melting and refining process is key to glass quality but it is energy-intensive and time-consuming.

Many innovations in industrial glassmaking have been explored during the second half of the 20th century as glassmakers have sought to solve critical industry problems. Few of these innovations have been commercialized. The British Glass Industry firstly used plasma in place of traditional electrical firing in glass melting in 1994. The high melting efficiency of soda-lime-silica glass was demonstrated [1]. Similarly, a two-year program began in 2003 by Plasmelt Glass Technologies, LLC, used a small skull melter with all electrodes above the glass batch [2]. However, both of the two programmes have the principle limitations that torch materials and design is not yet quite satisfactory. Recent years, our research group has successfully developed an innovative "in-flight melting" technology to reduce the vitrification reaction of glass batches. The glass raw materials are injected above the furnace and penetrated though the plasma region to be melted and vitrified within milliseconds. This totally changes the traditional radiation heating and improves the thermal efficiency considerably.

A stable 12-phase AC arc plasma furnace has been successfully developed to melt granulated glass raw materials. The particle diagnostics during the in-flight treatment was also studied in order to understand the plasma-particle energy exchange dynamics [3]. Furthermore, a hybrid heating combined the oxy-firing with the multi-phase AC arc was under research. This new kind of heat source was expected to enhance the energy efficiency and the in-flight glass productivity. However, the in-situ measurement of in-flight particles by hybrid plasma has not been studied yet. Thermal spray particle temperature is most often measured by two-colour pyrometry, also spectroscopic techniques have also been used [4]. An advantage of a two-colour pyrometry is in circumventing the need to know the emissivity of the body in order to determine its temperature. Greybody conditions, however, are often an unwarranted assumption. This paper present the works on the development of particle diagnostics based on the two-colour pyrometry principle.

2. Experimental

2.1. Raw Material

The raw materials of alkali-free glass were supplied by Asahi Glass Co., Ltd., Japan. Reagents of H_3BO_3, Al_2O_3, $BaCO_3$, and SiO_2 (quartz) were mixed in appropriate mass ratio and dispersed in water to form slurry. The slurry was sprayed from an atomizer into a heated air chamber at 200 °C to form spherical powders. The average grain size of the raw materials was 135 μm. The target composition of alkali-free glass is $15B_2O_3$-$10Al_2O_3$-$25BaO$-$49SiO_2$ in wt%.

2.2. Plasma System Description

The schematic illustration of the hybrid plasma apparatus is shown in figure 1. It combines a multi-phase AC arc plasma reactor with an oxygen burner installed over it. The plasma was generated by 12 electrodes, which are symmetrically arranged rotating on the axis and divided into two layers to increase the plasma volume. The angle between the upper six inclined electrodes and the six lower horizontal electrodes is 30°. The electrodes are made of 2%-thoriated tungsten with 3.2 mm in diameter. They are cooled by city water at a flow rate of 3 L/min, and argon shield gas with purity of 99.9% is injected around each electrode at a flow rate of 5 L/min. The AC power supply provides the total current of 320 A and the total voltage of 200 V. The chamber was made of stainless steel and cooled by water. The oxygen burner contains gas supply, powder feeding and cooling system. Propane (C_3H_8) was used as the fuel at a flow rate of 6 L/min. The primary oxygen and the secondary oxygen were 6 and 24 L/min, respectively. The combustion energy of oxygen burner was 9 kW. The distance from burner nozzle to arc was 300 mm and diameter of arc was 80 mm.

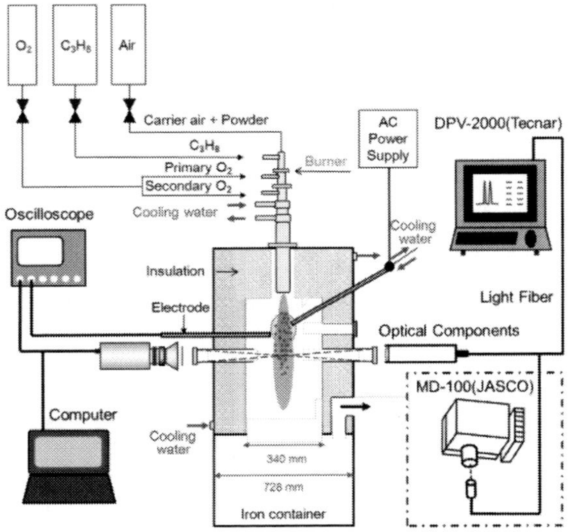

Figure 1. Schematic of hybrid plasma apparatus with diagnostic systems.

2.3. Particle Diagnostics

In-flight particle diagnostics were performed at an axial distance of 60 mm down from the lower electrode level and focused on the centre of the plasma plume. According to Figure 1, two diagnostic systems were used for in-situ measurement. One is the optical system DPV-2000 (Tecnar, Canada) which enables to measure single particle velocity and temperature [5]. The spectral regions chosen to provide high sensitivity and low interference from non-thermal background light are $\lambda_1 = 787 \pm 25$ and $\lambda_2 = 995 \pm 25$ nm, respectively. When use DPV-2000 system, it is inevitable that several non-thermal signals will also be collected by optical lens. The intensity recorded by the sensor was corrected by spectroscopic measurement. The emission spectrum separate the continuum signal emitted by particles and other non-thermal radiations from the vaporized materials and scattered plasma emission [3].

Another implementation of two-colour pyrometry for particle measurement involves the use of a high-speed camera (Photron, FASTCAM SA WTI, Japan). As shown in figure 1, an oscilloscope (Yokogawa, ScopeCorder DL850, Japan) synchronized with high-speed camera measured the current and voltage of the 12 electrodes. The choice of spectral bandwidths used to measure the two-colour particle intensity signal should avoid the direct and scattered plasma light. As the glass particles are low emissivity ceramics, only a limited amount of radiation is available to measure. Figure 2 shows the line emission spectra of the plasma gas species and vaporized particle material (Barium) considered here. For a good sensitivity and to minimum the plasma radiation signal, the camera used a short wavelength band covering 785 ± 5 and 880 ± 5 nm. The optical system observes the plasma plume and splits the single object into two identical images simultaneously, as shown in the figure 3. Histograms of particle properties can be displayed in real time (>10^4 Hz). The technique also indicates how the average particle temperatures change with time due to the multi-phase arc fluctuation.

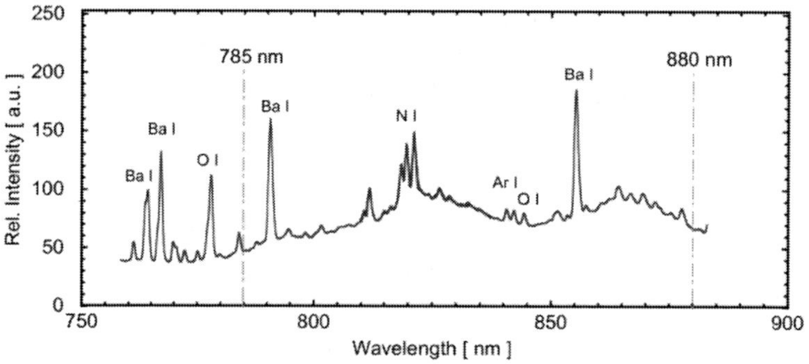

Figure 2. Emission line spectrum of hybrid plasma.

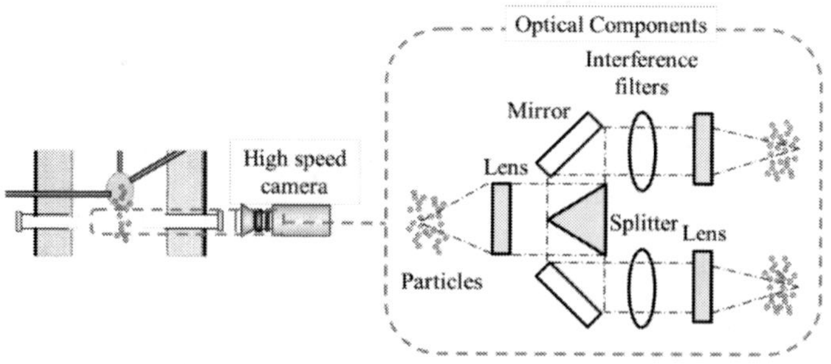

Figure 3. Conceptual diagram of temperature measurement by high-speed camera.

Journal of Physics: Conference Series **406** (2012) 012022 doi:10.1088/1742-6596/406/1/012022

3. Results and Discussion

3.1. Particle Temperature Calculation

Figure 4 gives a sample (dual) image of glass particles streaks. Although narrow band pass filters were used, the fluctuation of the plasma flame can be seen periodically. If the carrier gas flow rate is higher than 30 L/min, the particles fly quickly through the measurement volume and it is impossible to separate the particles one by one. The sensor described here uses a commercially available 12-bit 1024×1024 pixel CCD camera. In a single measurement, up to 1051 images of particle streaks can be recorded at a rate of 5000 frames/s with the exposure time of 200 μs. The camera focused on the centerline at the same height with DPV-2000 but the field of view was much wider. The calculated range of the image was 200×200 pixels which corresponding to the real size of 11×11 mm^2. Analysis software scans each image to identify individual particle streaks and compute velocity and temperature based on the Planck's law as in equation (3.1):

$$T = \frac{K_2(\lambda_1 - \lambda_2)}{\lambda_1 \cdot \lambda_2} \cdot \left[\ln \frac{TR_{\lambda 1}}{TR_{\lambda 2}} + 5 \ln \frac{\lambda_1}{\lambda_2} \right]^{-1} \quad (3.1)$$

where K_2 is the second radiation constant in Planck's law which value is 1.439×10^{-2} m · K, $TR_{\lambda1}$ and $TR_{\lambda2}$ are the thermal radiations emitted by a particle at λ_1 and λ_2, which represents the integrating pixel intensities in the video images. The sensor's spectral sensitivity was calibrated by a tungsten-halogen lamp.

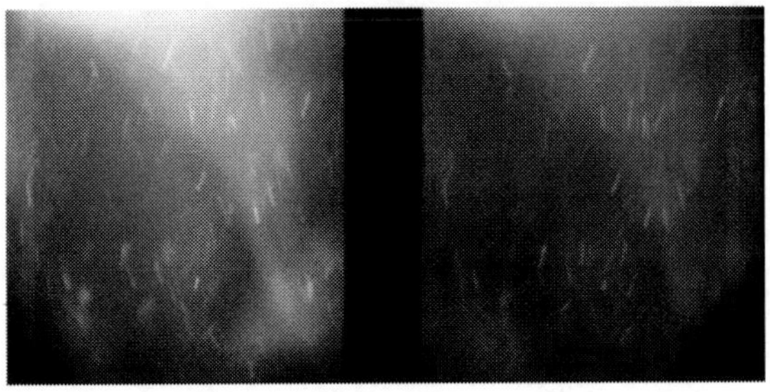

Figure 4. Image of glass particle streaks obtained by two different band-pass filters, 785 ± 5 nm (Left) and 880 ± 5 nm (Right).

3.2. Time-Resolved Particle Temperatures

It is already know that the fluctuation from the 12-phase AC arc causes the temporal non-uniformity of the whole temperature distribution. Considering for the particle velocity and an appropriate estimation, every 5 images (represent for 1 ms) were calculated to obtain the average temperature value of particles. Since the fluctuation period of multi-phase AC arc is evaluated as about 20 ms, the total calculated time for particle temperature was over 5 periodic cycles of 100 ms.

Figure 5 presents the frame-averaged particle temperatures as a function of time under different carrier gas flow rate. Each point was obtained from calculation of all the particles over 5 images. For example in Figure 5(a), the counted particles have a temperature distribution at time of 33 ms. The effect of carrier gas flow rate on the average temperature and particle velocity are summarized in figure 6. The particle velocity is mostly dependent on the carrier gas flow rate. The temperature of particles decreased due to shorter residence time in plasma region. Moreover, a gradual decrease in the temperature fluctuation as a function of carrier gas flow rate can be understood by considering their Fourier amplitude spectra, which is shown in figure 7. The peak around 100 Hz is clearly visible, indicates the periodicity of the temperature evolution. The amplitude decreased with carrier gas flow rate increased, which can be explained by the higher gas flow rate decrease the effect of the arc fluctuation on the whole temperature field.

Note that the energy transfer to a particle depends on the input power and its dwell time in the plasma. The particle experienced non-uniform temperature field when it pass through the plasma region. The hybrid plasma decreases the arc fluctuation due to the use of oxygen burner. Moreover, the particles experienced low temperature heating and short residence time in the oxygen flame before they enter the high temperature plasma zone. The particle surface was melted so that the initial void in the granulated raw powders were decreased, thus the diffusivity of the B_2O_3 vapor can become smaller. Then, the short residence time in the high temperature plasma region can finish the vitrification reaction and inhibits the volatilization.

(a)

(b)

(c)

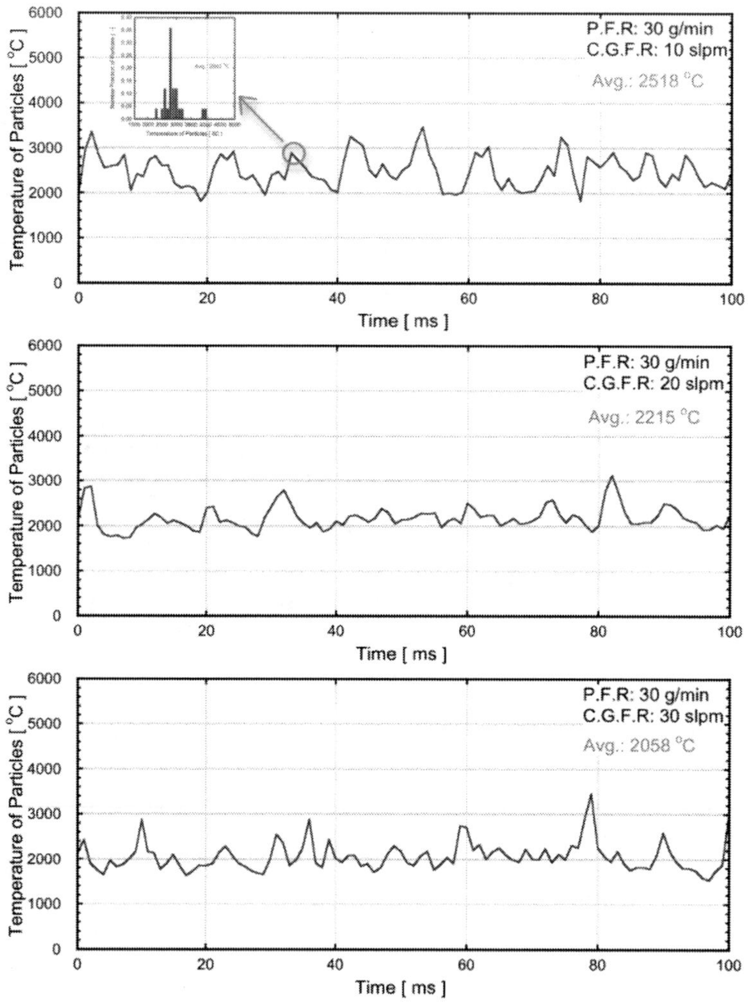

Figure 5. Fluctuation of the frame-averaged temperature of particles under different carrier gas flow rate: (a)10 L/min, (b)20 L/min, (c) 30L/min.

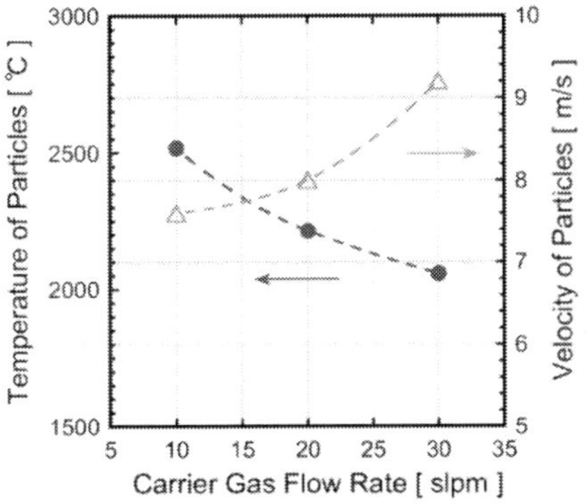

Figure 6. Effect of carrier gas flow rate on the particle average temperature and velocity.

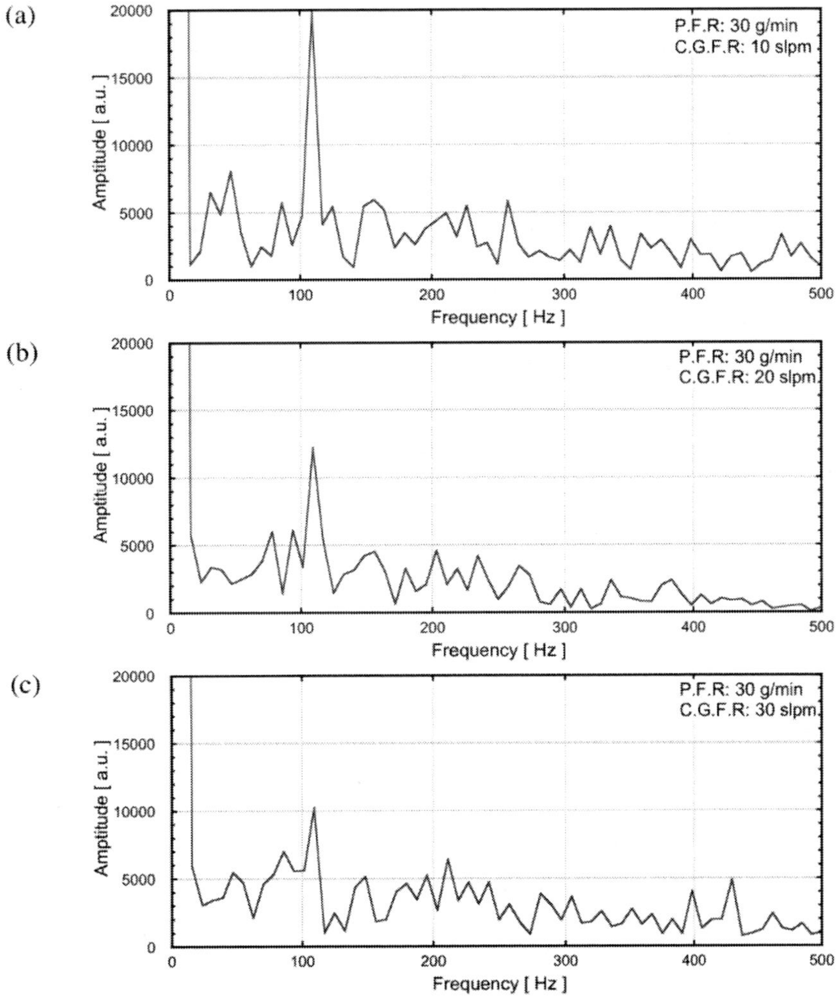

Figure 7. Frequency spectra of fluctuation of temperatures under different carrier gas flow rate: (a)10 L/min, (b)20 L/min, (c) 30L/min.

3.3. Comparison Between DPV-2000 and High-Speed Camera

In our experiment, the lens of DPV-2000 was placed at a constant position so it only focused on the small measurement volume (about 1 mm^3) of the center. On the contrary, the calculated range of the high-speed camera was about 11×11 mm^2. Figure 8 presents particle surface temperature distribution obtained from DPV-2000 and high-speed camera at carrier gas flow rate of 10 L/min. The comparison of the results is found to be in good agreement. This also confirms the measurement accuracy of high-speed camera. However, increase the carrier gas flow rate cause more smaller-size particles penetrate into the center plasma region, thus the DPV-2000 measured more higher temperature particles. More experiments should be done with DPV system in order to get reasonable data.

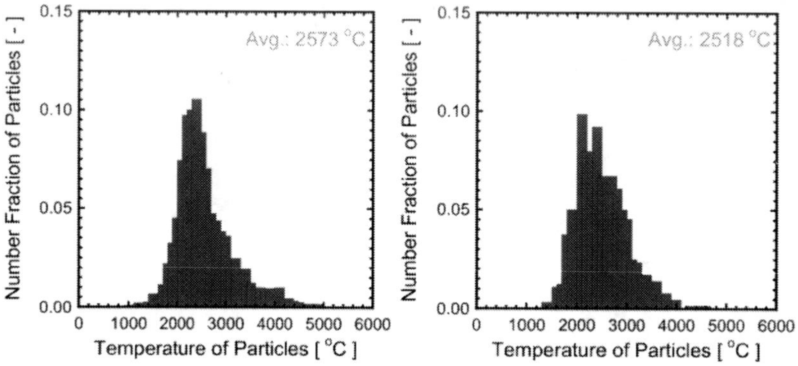

Figure 8. Particle surface temperature distribution estimated by (a) DPV-2000 and (b) high-speed camera.

4. Conclusion

In-flight glass particle measurement in the hybrid plasma heating was firstly introduced in this paper. The influence of the plasma fluctuations on the in-flight particle temperature was evaluated by a high-speed camera with two-colour pyrometry system. Time-dependent particle temperature due to arc fluctuations was observed. Amplitude of the variation in the in-flight particle state decrease with increases the carrier gas flow rate. The particle surface temperature distribution obtained by DPV-2000 analysis shows good agreement with high-speed camera. Using two-colour pyrometry method will help to successfully apply the hybrid heating technique.

Acknowledgment

The financial support provided by the Strategic Development of Energy Conservation Technology Project of NEDO (New Energy and Industry Technology Development Organization, Japan) is gratefully acknowledged.

References

[1] Ross C P, Tincher G L and Rasmussen M 2010 Glass Melting Innovations Glass Melting Technology: A Technical and Economic Assessment *KANCH* **3** 5-13

[2] High Intensity Plasma Glass Melter Project, Final DOE, USA Department of Energy, Report, 2006

[3] Liu Y, Manabu T, Choi S and Watanabe T 2012 In-Flight Particle Measurement of Alkali-Free Glass Raw Materials in 12-Phase AC Arc Plasma *J. Therm. Spray Technol.* 21 **5** 863-872

[4] Hollis K and Neiser R 1998 Particle Temperature and Flux Measurement Utilizing a Nonthermal Signal Correction Process *J. Therm. Spray Technol.* **7** 392-402

[5] Salhi Z, Gougeon P, Klein D and Coddet C 2005 Influence of Plasma Light Scattered by In-Flight Particle on the Measured Temperature by High Speed Pyrometry *Infrared Phys. Technol.* **46** 394-9

Journal of Physics: Conference Series **406** (2012) 012023 doi:10.1088/1742-6596/406/1/012023

Equilibrium analysis of hydrogen production using the steam-plasma gasification process of the used car tires

V A Kuznetsov, I I Kumkova, A S Lerner, V E Popov

Institute for Electrophysics and Electric Power of RAS

18 Dvortsovaya emb., 191186 St.-Petersburg, Russia

E-mail: lab10@iperas.nw.ru

Abstract. The paper deals with the treatment of used car tires. The method of used tires plasma gasification is proposed. The investigation of the syngas composition was carried out according to the temperature and plasma flow rate variation. The method of the steam catalytic conversion of CO, which is a part of the syngas, and CaO usage are suggested. The results of the calculation modeling at various temperatures, pressures, and steam flow rates are presented.

1. Introduction

Over 1 billion car tires [1] are produced annually. Thus, their demand constantly increases due to the improved living standards. The used tires practically do not dissolve in the native habitat. The main ways of their processing are: combustion, recapping, recycling and landfilling. According to the research data [2] about 65 % of the used tires are processed in the developed countries. Another part is put on dumps, which is environmentally unfriendly. Therefore a cost effective method of the used tires processing is an important problem.

The used tires generally consist of carbon – 83-89 % and hydrogen – 6-8 %, their ash content makes up 4.4-11.1 %, low calorific value – 32-39 MJ/kg [3]. Tire processing using the gasification technology is promising. In this case the organic part of tires turns into syngas, and inorganic transforms into slag. Syngas is the mixed gas mainly consisting of H_2 and CO. The gasification process divides to allothermal and autothermal by means of a heat supply. Autothermal process is realized due to the partial raw material combustion. Heat is supplied with the external heat carrier in the allothermal method. The low temperature plasma is the most perspective heat carrier [4,5]. It also can be used as an oxidizer, thus high temperatures of the process are reached and the end product yield increases. The plasma usage allows controlling the plasma forming gas flow rate in wide range. The use of steam as a plasma forming gas is considered to be the most effective, as that allows obtaining the syngas which almost does not contain minor components (N_2, CO_2)[6-8].

Syngas can be used for hydrogen production. Thus it is expedient to carry out steam catalytic conversion of CO, which is a part of syngas, on the reaction with CaO [9, 10]:

$$CO_{(g)} + H_2O_{(g)} + CaO_{(s)} \rightarrow H_{2(g)} + CaCO_{3(s)} \qquad (1)$$

It is necessary to carry out the syngas cleaning [11] to prevent the poisoning of the steam conversion catalyst, since the used tires contain sulfur-containing components which generally turn into H_2S at the gasification process. It is expedient to use ethanolamine (or diethanolamine, or methyl diethanolamine) as an absorber of sulfur-containing components of the syngas, so that CO_2 also is absorbed [12].

2. The analysis of the process

The composition calculation of gaseous products under the condition of the thermodynamic equilibrium at flow rates of the steam plasma of 0.2-5 kg/kg of tires, temperatures 1150-1500 K and atmospheric pressure for modeling the process of used tires plasma gasification was settled. Chemical WorkBench ver.3.5, Kinetic Technologies, which is based on the principle of entropy maximum, was used for the calculations. Thus it was accepted that the process is adiabatic, and the inorganic part of the used tires does not participate in chemical transformations. The following tire composition was used in calculation: C – 81.48, H – 8.00, O – 3.38, N – 0.47, S – 1.80, ashes – 4.88 % of mass (on working mass). The low calorific value is 36.55 MJ/kg.

The calculation of H_2 concentration in a range of temperatures 650-1000 K, steam flow rate 0.8-3.1 kg/kg of gas and pressure 1-7 atm was carried out for determining the optimum process conditions of steam catalytic conversion of CO in the presence of CaO.

The preliminary calculation of the hydrogenous gas composition was carried out at the mole stoichiometric and excess ratios under the condition of the thermodynamic equilibrium for determining the CaO:C ratio during the steam catalytic conversion of CO. It showed that the excess content of CaO slightly influences on the hydrogenous gas composition. Therefore the stoichiometric mole ratio of CaO:C was used in further calculations.

3. Results and their discussion

Figure 1 shows the calculated H_2+CO content in dependence on the specific (on 1 kg of tires) steam plasma flow rate.

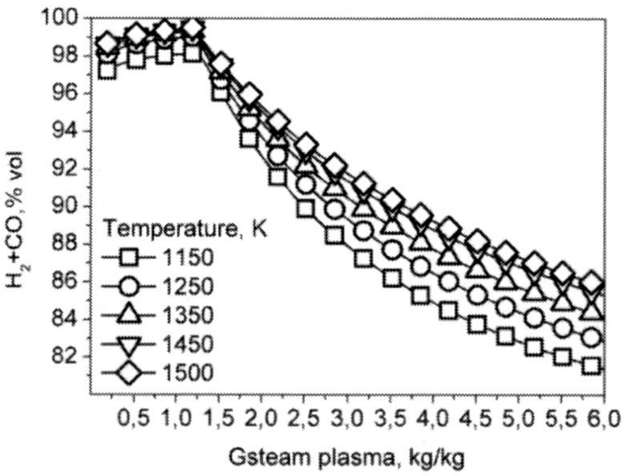

Figure1. The dependence of the total volume density H_2+CO on the specific steam plasma flow rate.

According to the calculations, the formation of the graphite is taking place at the steam plasma flow rate less than ~1.2 kg/kg of tires. The maximum concentration of CO+H_2 is observed at the flow rate close to stoichiometric (~1.2 kg/kg of tires). The stoichiometric flow rate is an oxidizer flow rate when all carbon and oxygen of the fuel and the oxidizer turns into CO, and fuel hydrogen – in H_2. The further increase in the steam plasma flow rate leads to interaction between CO and H_2O with H_2 formation and to syngas yield increasing. Thus the total fuel component quantity does not change. It leads to decrease in their concentration in syngas.

Apparently from fig. 1 the temperature increase leads to increase in H_2+CO content in the syngas. However at a mode close to stoichiometric at temperatures above 1350 K H_2+CO concentration practically does not change.

The specific energy consumption was calculated from the energy balance:

$$E_{in} = H_{sg}\left(1350\right)\cdot G_{sg} - H_{tires}(298) - H_{H_2O}\left(298\right)\cdot G_{plasma} \quad (2)$$

where E_{in} – energy consumption (MJ/kg), $H_{sg}(1350)$ – gasification products enthalpy at temperature 1350 K (MJ/kg), $H_{H2O}(298)$ – enthalpy of formation of water (MJ/kg), $H_{tires}(298)$ – enthalpy of formation of tires (MJ/kg), G_{sg} – gasification products yield (kg/kg), G_{plasma} – steam plasma consumption (kg/kg).

If syngas thermal energy is used to decrease the energy consumption, the steam plasma flow rate in dependence of the energy consumption at temperature 1350 K will be as follows (fig. 2):

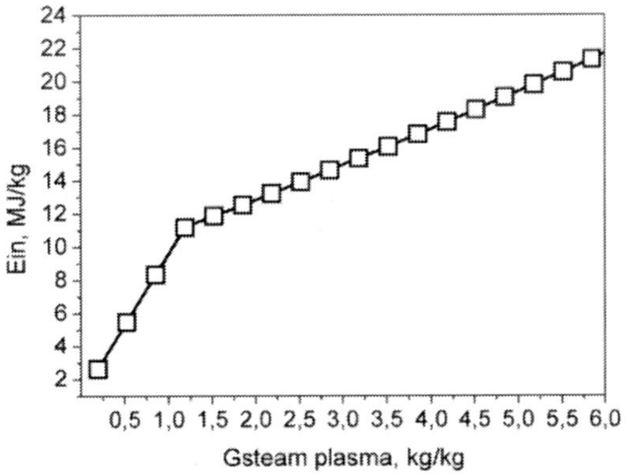

Figure 2. The dependence of the energy consumption on the steam plasma flow rate.

It can be seen that when the steam plasma flow rate increases, the energy consumption increases significantly. Thus the H_2+CO yield also raises (fig. 3), at the energy consumption more than ~11.2 MJ/kg (steam plasma flow rate is ~1.2 kg/kg of tires and more) the maximum total yield of CO+H_2 is observed.

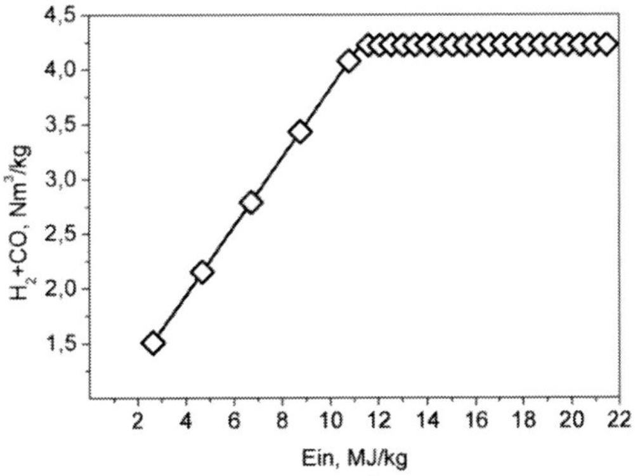

Figure 3. The dependence of total H_2+CO yield on energy consumption.

Therefore, the gasification process is expedient to carry out at 1350 K and at steam plasma flow rate of ~1.2 kg/kg of tires. It is possible to produce 4.2Нм^3 syngas, consisting of H_2 – 60.3 % and CO – 39.0 %, from 1 kg used automobile tires at such process organization. Thus it is necessary to obtain steam plasma with ~14 MJ/kg enthalpy, taking into account the heat of vaporization. The energy consumption on the steam plasma gasification of 1 kg used tires is 11.2 MJ/kg of tires.

Figure 4 shows the calculated hydrogen content, which is produced at steam conversion of CO, in dependence of the temperature at steam flow rate 0.8-3.1 kg/kg syngas.

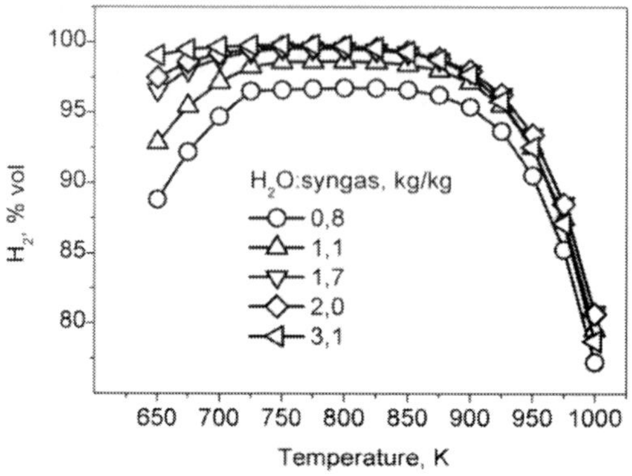

Figure 4. The dependence of hydrogen content on the temperature at various steam flow rates in the atmospheric pressure.

Findings in Fig. 4 show that the steam plasma flow rate growth promotes more complete process progress. However, the H_2 concentration reaches maximum and practically does not change at flow rates higher than ~2.0 kg/kg of tires and at temperatures ~725-850 K. Low H_2 concentration at temperatures less than ~725 K is connected with reactions course with the formation of CH_4. The efficiency of CO_2 absorption considerably decreases at temperatures more than ~850 K that also leads to the H_2 content decrease in hydrogenous gas.

Figure 5 shows the dependence of H_2 yield on temperature and pressure at the steam flow rate 2 kg/kg syngas.

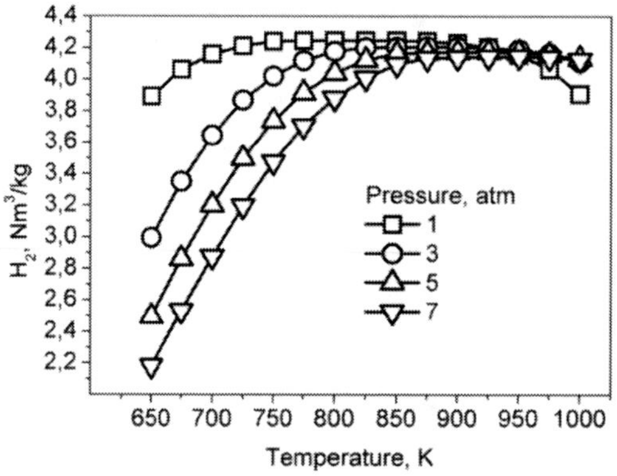

Figure 5. The dependence of H_2 yield on the temperature at steam flow rate 2 kg/kg syngas for various pressures.

Pressure growth promotes the reactions course with the methane formation that leads to considerable decrease in the H_2 yield (fig. 5) at temperatures lower than ~850 K.

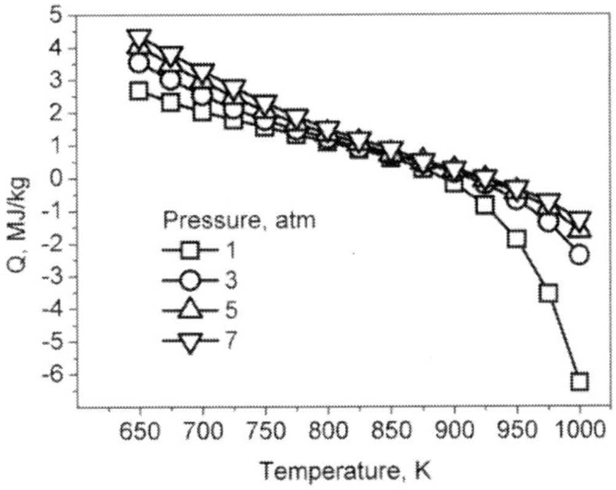

Figure 6. The dependence of the specific energy (Q), released in the process of the CO conversion, on temperature for various pressures.

As was said, the process of CO steam conversion is exothermic. The released energy amount considerably decreases with the temperature growth and rises with the pressure increase (fig. 6) at CO conversion.

Thus, it is most convenient to carry out the process of steam catalytic conversion of CO at steam flow rate ~2 kg/kg syngas, at temperature ~725-750 K and at atmospheric pressure. The energy released at the stage of steam catalytic conversion of CO in this process mode is ~1.6-1.8 MJ/kg syngas. It is possible to produce ~4.2 Nm^3 of hydrogenous gas from 1 kg of the used tires under such conditions. The gas composition will be the following: H_2 – 99.5-99.7 %, CH_4 – 0.2-0.4 % and N_2 – 0.1 %. The conversion level of CO according to the calculation reaches ~99.99 %.

It is possible to use the Fe-Cr-Cu- oxide catalyst as the catalyst of the CO steam conversion at such process organization [13].

The total energy consumption on 1 Nm^3 of H_2 production without calculation of expenditures on the absorbent reactivation and the syngas cleaning is ~0.6 kW.

4. Conclusion

The calculation shows that it is possible to produce ~4.2 Nm^3 of the syngas, consisting of H_2 – 60.3 and CO – 39.0 %, when processing 1 kg of the used tires by the steam plasma gasification process at the temperature 1350 K and the steam plasma flow rate of ~1.2 kg/kg of tires.

It is possible to produce ~4.2 Nm^3 of the hydrogenous gas, consisting of H_2 – 99.5-99.7, CH_4 – 0.2-0.4 and N_2 – 0.1 %, with the subsequent steam conversion of CO, which is a part of syngas, at the process temperature of ~725-750 K, at the steam flow rate of ~2 kg/kg syngas and at atmospheric pressure.

The energy consumption quantity on 1 Nm^3 of H_2 production without calculation of expenditures on the absorbent reactivation and the syngas cleaning is ~0.6 kW.

References

[1] Ho M-H 2010 One Dimensional compressibility characteristics of clay stabilised with cement-rubber chips *IJSCET* **1** **(2)** 91-104

[2] "Tire recycling industry: a global view" – http://www.irevna.com/pdf/Industry%20report.pdf (accessed: 21.06.2012)

[3] Lopez F 2011 Distillation of granulated scrap tires in a pilot plant *J. of Hazard. Mater.* **190** 285-92A reference

[4] Bratsev A N, Popov V E, Shtengel S V and Rutberg A Ph 2006 Some aspects of development

and creation of plasma technology for solid waste gasification *High Temp. Mater. Processes* **10(4)** 549–556

[5] Bratsev A N, Kuznetsov V A, Popov V E, Rutberg A Ph, Ufimtsev A A and Shtengel S V 2009 Experimental development of methods on plasma gasification of coal as the basis for creation of liquid fuel technology *High Temp. Mater. Processes* **13(2)** 147–154

[6] Popov V E, Bratsev A N, Kuznetsov V A, Shtengel S V and Ufimtsev A A 2011 Plasma gasification of waste as a method of energy saving *JPCS* **275(1)** 012015

[7] Bratsev A N, Kuznetsov V A, Popov V E, Ufimtsev A A and Shtengel S V 2009 Estimation of perspectivity of steam-plasma methane conversion *High Temp. Mater. Processes* **13(2)** 241–246

[8] Rutberg Ph G, Bratsev A N, Kuznetsov V A, Popov V E, Ufimtsev A A and Shtengel' S V 2011 On efficiency of plasma gasification of wood residues *Biomass and Bioenergy* **35(1)** 495–504

[9] Iulianelli A, Manzolini G, De Falco M, Campanari S, Longo T, Liguori S and Basile A 2010 H_2 production by low pressure methane steam reforming in a Pd-Ag membrane reactor over a Ni-based catalyst: Experimental and modeling *Int. J. Hydrogen Energy* **35** 11514–11524

[10] Li Zh, Liu Y, Cai N 2012 Effect of CaO hydration and carbonation on the hydrogen production from sorption enhanced water gas shift reaction *Int. J. Hydrogen Energy* **37(15)** 11227–11236

[11] Zhang L 2009 Deactivation characteristics of Fe–Al–Cu water-gas shift catalysts in the presence of H_2S *J. Mol. Catal. A: Chem.* **309(1–2)** 63–70 (10.1016/j.molcata.2009.04.016)

[12] Hagewiesche D P, Ashour S S, Al-Ghawas H A and Sandall O C 1995 Absorption of carbon dioxide into aqueous blends of monoethanolamine and n-methyldiethanolamine *Chem. Eng. Sci.*, **50(7)** 1071-1079

[13] Baronskaya N A, Minyukova T P, Khassin A A, T M Yurieva, V N Parmon, 2010 Enhancement of water–gas shift reaction efficiency: catalysts and the catalyst bed arrangement *RUSS CHEM REV* **79(11)** 1027–1046

Plasma gasification of organic containing substances as a promising way of development of alternative renewable power engineering

Ph G Rutberg, A N Bratsev, V A Kuznetsov, I I Kumkova, V E Popov, A V Surov

Institute for Electrophysics and Electric Power of RAS

18 Dvortsovaya emb., 191186 St.-Petersburg, Russia

E-mail: lab10@iperas.nw.ru

Abstract. The paper deals with perspectives of large-scale implementation of the plasma gasification process of solid organic-containing substances as a source of renewable energy. First of all, such substances as wood waste, agriculture waste, solid household waste are considered. Thanks to the process of the plasma high-temperature gasification the energy of their combustion can be completely converted into the energy of the synthesis gas combustion, which use as a fuel for the combined cycle allows electricity generation with efficiency of ~60 %. Thus, if the psychogenesis production wastes are considered, this technology enables avoiding additional emission of carbon dioxide into biosphere as for production of biomass from biosphere it is extracted the same amount of carbon dioxide as is emitted at its combustion. The report represents the realized and developing designs of plasma gasification, their advantages and deficiencies.

1. Introduction

World consumption of primary energy in 2011 has made more than 12.274 billion tons in an oil equivalent [1]. This magnitude steadily grows with increase in rates of growth. So from 1990 to 2000 the consumption growth is ~16 %, and in the next decade ~28 %. First of all, it is connected with the economic growth and improvement in living standards of the population. Oil, gas and coal constitute ~87 %, and renewed sources make up hardly more than 1 %. The oil and gas will last for ~50 years, and coal for ~100 years at such rates of consumption of the proven reserves. The back of this process is escalating streams of gaseous, liquid and solid wastes mainly remaining in the environment causing ecological problems and global warming. So, for example, throughout last decades the concentration of the carbon dioxide steadily grows in the atmosphere, and in 2010 the global volume of carbon dioxide emissions has grown on ~5.9 % and for the first time has exceeded $9{\cdot}10^{15}$ g in a year [2]. Volumes of solid household waste also increase. It is quite apparent that energy needs will only increase, therefore the replacement of fossil fuel by alternative energy sources and use energy in more efficient ways in general is a pressing issue.

Plasma gasification as an alternative energy

Usage of biomass and waste energy is one of the solutions of the problems mentioned above. Biomass has a lower heating value than fossil fuels, nevertheless, it can be also used for production of electric and thermal energy. Besides volumes of biomass annual formation are great enough to cover an essential share of mankind needs in energy already now, and under forecasts [3] in 2050 the global

Journal of Physics: Conference Series **406** (2012) 012024 doi:10.1088/1742-6596/406/1/012024

potential of biomass energy will be 1135-1300 EJ (without use of algae as biomass) while world consumption on the average, under different scenarios, will reach the order of 826 EJ. One more essential advantage of the biomass usage for energy generation is the absence of the additional emission of carbon in biosphere as the biomass consumes this carbon from biosphere for its formation.

Table 1 tabulates the major types of biomass and wastes and some of their properties.

Table 1. **The major types of biomass and wastes and their properties.**

Parameter: Name		Units of measurement	Wood waste	Switch-grass	Cow manure	Chicken manure	Sludge	RDF	Peat	Used car tires	Plastic waste
Composition of organic part of feedstock	C	% mass	50.25	49.16	50.39	53.96	48.60	49.96	57.56	86.13	82.91
	H		6.09	6.36	5.77	5.60	6.57	6.69	6.10	6.93	12.29
	N		0.20	0.63	3.94	7.92	5.15	0.96	2.17	0.57	0.71
	S		0.10	0.13	1.31	0.96	1.71	0.11	0.24	1.73	0.14
	O		43.35	43.73	38.58	31.56	37.97	42.28	33.88	3.47	2.54
	Cl		0.00	0.00	0.00	0.00	0.00	0.00	0.04	1.16	1.41
Feedstock moisture			20.00	6.29	36.60	20.20	6.10	7.83	9.40	1.14	1.00
Ash content of feedstock			0.80	8.51	25.20	21.23	36.06	8.71	16.87	13.35	28.20
Fuel LHV	published	MJ/kg	13.90	15.82	6.49	12.09	12.92	15.01	16.46	33.57	28.68
	calculated		14.26	15.45	5.8	10.84	10.15	15.73	15.77	31.11	28.89
Adiabatic temperature of combustion in dry air		K	2123	2304	1815	2210	2562	2174	2337	2621	2421
Oxygen consumption needed for complete gasification		g/kg	9.2	129.5	0	56.8	100.7	133.0	232.0	941.2	755.0
Oxygen consumption needed for complete combustion		g/kg	1101	1174	546	923	841	1202	1240	2416	2235
Maximal yield of chemical energy at complete gasification		MJ/kg	17.84	17.20	8.51	14.05	11.93	17.56	16.64	24.47	24.36
References			[4]	[5]	[6]	[7]	[8]	[9]	[10]	[11]	[12]

Wood versus the fossil fuels, mainly does not contain sulphur and other ecologically harmful elements, however its use in a wood-working industry on the economic efficiency considerably surpasses energy use. Therefore, it is expedient to process the remains formed during timber cutting and the municipal waste mainly consisting of wood materials.

Energy crops are called plants, raised for energy applications. They are the major kind of biofuels. Generally, they represent grassy or wood fast-growing plants, for example switchgrass [13] and willow [14]. Algae is another one promising kind of biofuels. Algae do not require fertile lands for their cultivation and they grow practically in any water [15].

Waste of cattle and poultry raising (manure) also are biomass and, accordingly, can be used as the renewed energy source. Their mishandling leads to the bacteriological infection of ground waters [16]. The manure composition, even produced by the same group of animals can vary greatly depending on technologies used for its disposal.

Sewage sludge represents a solid-state component of sewage separated by mechanical methods and sedimentation at sewage purification plants. Depending on sources of formation they can contain a considerable quantity of mineral inclusions and possess high moisture. Thus a mineral component, as a rule, is difficult to separate from deposits. The moisture containing in pores of sludge particles, can be separated by traditional methods (dewatering, evaporation, and other drying procedures) [17]. Sewage treatment also includes stages of sanitation and recovery of sludge demanding separate investments.

The household waste polluting environment can become the energy source. However, their direct energy treatment is not always possible because of high variability of morphological composition and mechanical properties. Therefore the household waste is preliminary processed into RDF. RDF possesses lower humidity and ash content and higher combustion heat, in comparison with the primary waste.

Peat is a product of decomposition of plants and it is formed on marshes. By various estimations peat world's reserves are 270-6000 billion t [18]. Due to rather low calorific value this kind of fuel is not widely used for energy production. Currently peat attracts more interest as it is considered as the renewed energy source.

Energy properties of used car tires are close to coals. They have high calorific value and carbon containing and low oxygen content. Nevertheless the efficiency of car tires combustion is not high enough because of a low reactive capacity. Average world's rubber consumption is ~3.8 kg/man×year, approximately half of this amount is car tires [19].

Plastic waste is a part of a household waste and RDF. It can be separated into individual fraction. Besides, huge garbage islands mainly comprising plastic debris have accumulated in the world ocean. This kind of waste also can be considered as the renewed energy source. The table also tabulates the calorific value of presented kinds of waste and biomass calculated on its composition [20], as far as the published data do not always well agree with each other.

Wood is used as a fuel by millennia and is one of the most simple in use energy resources. Nevertheless, the efficiency of electricity production by wood combustion is 18-24 % [21]. The efficiency of electricity production can be increased up to ~29 % using the integrated gasification combined cycle (IGCC), and at usage of plasma gasification — to ~35 % [22]. Usage of plasma blasting allows increasing the temperature of the process, rate of physical and chemical transformations, conversion level of fuel, residence time of gases in the reactionary volume, hydrogen and carbon monoxide content in syngas, thermal stress of the reactionary volume and specific output (in ~7.8 times at increase in temperature from 1500 K to 1900K). Development of power engineering on the basis of plasma gasification is one of the most promising directions.

2. Potentially achievable energy parameters

The numerical modeling of the plasma gasification allows evaluation of the energy potential of biomass and waste using the following assumptions:

- Thermodynamically equilibrium composition of a mixture of plasma and ash-free weight of wood waste calculated at temperature 1500 K and atmospheric pressure (101.325 kPa) is assumed for ash-free composition of products of plasma gasification of biomass and waste;
- The inorganic component does not participate in chemical reactions, and power inputs on its heating from 298 K to 1500 K make up ~1.367 MJ/kg;
- There is no change in the composition of gasification products at their cooling to 298 K;
- Mass composition of air: N_2 – 75.52; O_2 – 23.14; Ar – 1.29; CO_2 – 0.05;
- Value of the specific flow rate of oxidizing agent and level of corresponding specific power inputs were selected from a condition of achievement of adiabatic combustion temperature of produced gas 2000 K;
- The specific yield of the electric power, using the produced syngas as a fuel for the combined cycle, was evaluated on the basis of dependence of the combined cycle on the temperature range 1400-2000 K [23].

The data obtained as a result of calculations (Table 2) show, that use of plasma in the course of biomass and waste gasification allows increasing the efficiency of fuel usage for electric power production in 1.2-1.6 times (in comparison with efficiency 29% for IGCC).

Table 2. Potentially achievable energy parameters.

Parameter (Name)		Units of measurement	Wood waste	Switch-grass	Cow manure	Chicken manure	Sewage water sludge	RDF	Peat	Used car tires	Plastic waste
Air flow rate		kg/kg	1.637	1.784	0.673	1.371	1.247	1.826	1.954	4.064	3.770
Specific power inputs for LHV fuel	Published	MJ/kg	2.691	1.571	2.970	1.955	-0.099	2.762	1.557	-0.941	1.752
	Calculated	MJ/kg	2.331	1.941	3.66	3.205	2.671	2.042	2.247	1.519	1.542
Specific yield of syngas		Nm^3/kg	2.385	2.516	1.184	2.005	1.786	2.571	2.648	4.954	4.667
Syngas composition	CO	% vol.	26.10	27.63	21.86	25.56	26.51	27.00	27.83	30.25	24.41
	H_2	% vol.	20.10	18.53	27.51	20.02	19.50	19.26	17.50	14.54	20.19
	N_2	% vol.	45.32	46.95	38.61	47.11	47.49	47.11	49.19	54.18	53.37
	CO_2	% vol.	7.91	6.29	11.26	6.56	5.54	6.05	4.84	0.01	1.23
Yield of steam		kg/kg	0.278	0.203	0.321	0.197	0.139	0.212	0.154	0.001	0.091
Syngas LHV		MJ/Nm^3	5.012	5.035	5.315	4.980	5.081	5.033	4.960	4.988	4.823
Specific yield of chemical energy		MJ/kg	11.95	12.67	6.293	9.984	9.08	12.94	13.13	24.71	22.51
Specific yield of electricity minus its own internal power requirements		MJ/kg	4.928	6.505	1.041	4.409	5.79	5.486	6.814	15.75	12.59

The exception is the cow manure, that is connected with its increased (in relation to other kinds of considered raw materials) humidity. Probably the same refers to sewage sludge as the specified humidity at level of 6.1 % for the given kind of raw materials can be achieved at the expense of essential energy consumption. Nevertheless, it is impossible to lose sight of these kinds of raw materials, as they are formed everywhere in enough considerable quantities, and additional drying can be organized at the expense of the excesses of heat formed at various stages of the process. It is also necessary to note, that not all specified kinds of raw materials are easily treated even by the combustion process and plasma gasification is actually a unique and simple way of energy extraction.

3. Survey of existing designs

A lot of companies around the world promote plasma gasification to the market. AlterNRG [24], Integrated Environmental Technology LLC [25], Advanced Plasma Power [26], Plasco Energy Group [27], Pyrogenesys Canada Inc. [28] and others achieved the better results. These companies created pilot installations of various scales. Currently, some of them are at the stage of implementation of large-scale commercial projects. The most widespread are gasification processes using molten material layer (Integrated Environmental Technology LLC), with the subsequent plasma conversion of the crude syngas (Pyrogenesys), traditional gasification in the updraft process with plasma conversion of the crude syngas (Plasco Energy Group, Advanced Plasma Power), or the updraft process of steam-oxygen gasification with plasma use for liquid slag discharge (AlterNRG).

The metallurgical furnaces with Joule heating of molten material layer are used in case of the gasification on molten layer. Wastes are fed on a melt surface. In this area either the electric arc is ignited, or the plasma jet from the plasma torch is injected. In some designs the plasma torch is mounted in a special chamber at the outlet, providing mixture of off-gases with high-temperature oxidizing agent for conversion of tars containing in crude synthesis gas supplied from the melting chamber. Creation of large-scale installations is the advantage of the process. The second advantage is capability to remove noncombustible waste components as melt with its separation on metal and non-metal fractions. The essential disadvantage is the increase of energy consumption for melt maintenance.

The AlterNRG updraft gasification uses supply of waste product simultaneously with metallurgic coke and limestone. Plasma torches are mounted at the bottom part of the gasifier and serves for heating-up and maintenance of melt in a liquid state (the lime stone is fed to decrease the melting temperature). Oxygen is fed into the gasification zone through the special tuyeres. The produced gas is removed from the top of the gasifier. Syngas, produced in the gasification zone passing through colder upper layers of waste is polluted by tars, forming in the pyrolysis zones, evaporating water does not participate in the conversion due to low temperatures in the zones of its release. It is the essential deficiency of the updraft process. The produced gas also needs additional conversion or tar cleaning.

The common essential disadvantage of these technologies is application of free-burning arcs or DC plasma torches. Very low efficiency of energy transfer from the arc to gas ~30 % is a feature of free-burning arcs. The efficiency of DC plasma torches is about 60 % due to high losses in the power supplies.

We consider AC plasma torches to be the most efficient plasma generating systems. Plasma torches developed in IEE RAS provide the efficiency of electricity conversion into thermal plasma energy of 90-94 %. A large-scale experimental installation for plasma gasification has been created on the basis of these plasma torches in IEE RAS [29-35].

References

[1] 2012 BP Statistical Review of World Energy 2012 BP p.l.c. p 45

[2] Peters G P, Marland G, Le Quéré C, Boden T, Canadell J G and Raupach M R 2012 Rapid growth in CO_2 emissions after the 2008–2009 global financial crisis. *Nature Climate Change* **2**, 2-4 (doi:10.1038/nclimate1332)

[3] Ladanai S and Vinterbäck J 2009 *Global Potential of Sustainable Biomass for Energy* (Uppsala: Swedish University of Agricultural Sciences)

[4] Green D W, Perry R H 2007 Perry's Chemical Engineers' Handbook, Eighth Edition p 2400 (McGraw-Hill, United States)

[5] Weiland N T, Means N C and Morreale B D 2012 Product distributions from isothermal co-pyrolysis of coal and biomass *Fuel* **94** 563-570 (doi:10.1016/j.fuel.2011.10.046)

[6] Santoianni D A, Bingham M F, Woodard D M and Kinnell J C 2008 Power from Animal Waste—Economic, Technical, and Regulatory Landscape in the United States *Journal of EUEC* **2**

[7] Giuntoli J, de Jong W, Arvelakis S, Spliethoff H and Verkooijen A H M 2009 Quantitative and kinetic TG-FTIR study of biomass residue pyrolysis: Dry distiller's grains with solubles (DDGS) and chicken manure Journal of Anal. and Appl. Pyrolysis **85** 301-312 (doi:10.1016/j.jaap.2008.12.007)

[8] Thipkhunthod P, Meeyoo V, Rangsunvigit P, Kitiyanan B, Siemanond K and Rirksomboon T 2005 Predicting the heating value of sewage sludges in Thailand from proximate and ultimate analyses *Fuel* **84(7-8)** 849-857

[9] Raghunathan K and Bruce K R 1997 Control of Emissions from Cofiring of Coal and RDF p 25 URL: http://www.nrel.gov/docs/legosti/fy97/26036.pdf (access date: 22.06.2012)

[10] Raniere F D, Combs L P and Falk A Y Experimental investigation of peat hydrogasification URL: http://www.anl.gov/PCS/acsfuel/preprint%20archive/Files/24_3_WASHINGTON_09-79_0064.pdf (access date: 22.06.2012)

[11] 1995 Environmental Factors of Waste Tire Pyrolysis, Gasification, and Liquefaction. Report № 1364 (California: CalRecovery, Inc.)

URL: http://www.calrecycle.ca.gov/publications/Tires/62095001.pdf (access date: 06.08.2010)

[12] Adrados A, de Marco I, Caballero B M, A. Lopez, Laresgoiti M F and Torres A 2012 Pyrolysis of plastic packaging waste: A comparison of plastic residuals from material recovery facilities with simulated plastic waste *Waste Manage.* **32(5)** (doi:10.1016/j.wasman.2011.06.016)

[13] Barney J N and DiTomaso J M 2010 Bioclimatic predictions of habitat suitability for the biofuel switchgrass in North America under current and future climate scenarios *Biomass and Bioenergy* **34(1)** 124–133 (doi:10.1016/j.biombioe.2009.10.009)

[14] Keoleian G A and Volk T A 2005 Renewable Energy from Willow Biomass Crops: Life Cycle Energy, Environmental and Economic Performance *Crit. Rev. Plant Sci.* **24** 385–406 (doi:10.1080/07352680500316334)

[15] Demirbas M F 2011 Biofuels from algae for sustainable development *Appl. Energy* **88** 3473–3480 (doi:10.1016/j.apenergy.2011.01.059)

[16]] Ribaudo M, Gollehon N, Aillery M, Kaplan J, Johansson R, Agapoff J, Christensen L, Breneman V and Peters M 2003 *Manure Management forWater Quality: Costs to Animal Feeding Operations of Applying Manure Nutrients to Land* (U.S. Department of Agriculture, Economic Research Service, Resource Economics Division. Agricultural Economic Report 824. Washington)

[17] Sattho T and Yamsaengsung R 2005 Vacuum drying of rubberwood *PSU-UNS International Conference on Engineering and Environment.* Novi Sad, Serbia **T11-3.4**. p 1-5

[18] 2009 Mineral commodity summaries p 195
URL: http://minerals.usgs.gov/minerals/pubs/mcs/2009/mcs2009.pdf (access date: 22.06.2012)

[19] 2008 World Rubber & Tire, Industry Study with Forecasts for 2011 & 2016. Study #2282
URL: http://www.freedoniagroup.com/brochure/22xx/2282smwe.pdf (access date: 22.06.2012)

[20] Reed T B, Das A 1988 *Handbook of Biomass Downdraft Gasifier Engine Systems* v(Colorado, US, Golden: Solar Energy Research Institute) p 140

[21] Wood Biomass for Energy. URL: http://www.fpl.fs.fed.us (access date: Feb 16 2012)

[22] Rutberg Ph G, Bratsev A N, Kuznetsov V A, Popov V E, Ufimtsev A A and Shtengel' S V 2011 On efficiency of plasma gasification of wood residues *Biomass and Bioenergy* **35(1)** 495-504 (doi:10.1016/j.biombioe.2010.09.010)

[23] Ishikawa M, Terauchi M, Komori T and Yasuraoka J 2008 Development of High Efficiency Gas Turbine Combined Cycle Power Plant Mitsubishi Heavy Industries Technical Review **45(1)** 15-17

[24] Alter NRG. http://www.alternrg.com/ (access date: 21 June 2012)

[25] InEnTec http://www.inentec.com/ (access date: 21 June 2012)

[26] Advanced Plasma Power. http://www.advancedplasmapower.com/ (access date: 21 June 2012)

[27] Plasco Energy Group http://www.plascoenergygroup.com/ (access date: 21 June 2012)

[28] PyroGenesis Canada Inc http://www.pyrogenesis.com/ (access date: 21 June 2012)

[29] Bratsev A N, Popov V E, S V Shtengel and A Ph Rutberg 2006 Some aspects of development and creation of plasma technology for solid waste gasification *High Temp. Mat. Processes. An International Quarterly of High-Technology Plasma Processes* **10(4)** 549-556.

[30] Kuznetsov V A, Bratsev A N, Kovshechnikov V B, Kumkova I I, Popov V E, Shtengel S V and Ufimtsev A A 2007 Distinctive features of biomass gasification using ac plasma generators working on air *2007 IEEE Pulsed Power Conference (PPPS-2007)* (Omnipress, Madison WI USA) p 1223-1226 (doi:10.1109/PPPS.2007.4652407)

[31] Bratsev A N, Glezin I L, Kovshechnikov V B, Kumkova I I, Kuznetsov V A, Popov V E, Shtengel S V and Ufimtsev A A 2007 Experimental Research of Air Gasification of Waste. The first results *XXVIII International Conference on Phenomena in Ionized Gases (ICPIG*

2007) (Czech Republic: Institute of Plasma Physics AS CR, Prague, July 15-20, 2007) p 1848-1851

[32] Bratsev A N, Kuznetsov V A, Popov V E, Rutberg A Ph, Ufimtsev A A and Shtengel S V 2009 Experimental development of methods on plasma gasification of coal as the basis for creation of liquid fuel technology *High Temp. Mater. Processes* **13 (2)** 147-154

[33] Popov V E et al 2011 *J. Phys.: Conf. Ser.* 275 012015

[34] Brattsev A N, Kuznetsov V A, Popov V E and Ufimtsev A A 2011 Arc gasification of biomass: Example of wood residue *High Temperature* **49(2)** 244-248 (doi:10.1134/S0018151X11010020)

[35] Bratsev A N, Kumkova I I, Kuznetsov V A, Popov V E, Shtengel' S V, Ufimtsev A A 2011 Air plasma gasification of RDF as a prospective method for reduction of carbon dioxide emission *IOP Conf. Series: Materials Science and Engineering* **19** 012004 (doi:10.1088/1757-899X/19/1/012004)

Degradation of *m*-cresol in aqueous solution by dielectric barrier discharge

B Jaramillo-Sierra[2], A Mercado-Cabrera[1], R López-Callejas[1,2], R Peña-Eguiluz[1], S R Barocio[1]†, R Valencia-Alvarado[1], B Rodríguez-Méndez[1], A Muñoz-Castro[1], A De la Piedad-Benitez[2]

[1] Instituto Nacional de Investigaciones Nucleares, Plasma Physics Laboratory AP 18-1027, 11801 México D. F., México.
[2] Instituto Tecnológico de Toluca, AP 890, Toluca, México
† In Memoriam

E-mail: antonio.mercado@inin.gob.mx

Abstract. It was carried out a theoretical and experimental study of the *m*-cresol degradation in aqueous solution using a non-thermal plasma induced by dielectric barrier discharge. For the experimental setup a coaxial reactor vertically placed was used. Wherein a liquid solution flowing inside the internal electrode was impelled by a peristaltic pump and falling over the external surface of the internal electrode. Working gas was applied in a parallel direction respect to the surface of the liquid and inside of the quartz tube. Non-thermal plasma was generated at the gas-liquid interface in Ar-O_2 gas mixtures with a high-voltage power supply system. The electric power applied ranged from 10-60W at a 3.0 kHz frequency. The initial concentration of *m*-cresol was of 5×10^{-3} mol/L, and the removal efficiency up to 97.3% was obtained after 1 h of treatment. For the theoretical study a simplified model of the chemical kinetics was developed where the temporary evolution of the compounds generated in the process of degradation of the *m*-cresol was analyzed. Byproducts as oxalic acid and CO_2 were found.

1. Introduction

Cresols are organic compounds obtained from petroleum, wood tar or coal tar and they have a great variety of applications in chemical, petrochemical, cosmetic and pharmaceutical industries [1,2]. Conventionally treatments as biological and physical methods are used for their degradation. Nevertheless, because cresols are chemically stable, they are not totally eliminated thus noxious byproducts as chlorocresols can be generated. For this reason they represent a source of pollution of the water. The non-thermal plasma is an alternative for the degradation of pollutants in the wastewater and its effectiveness has already been demonstrated for dyes [3], organic compounds as phenol [4], as well as bacteria deactivation.

In this work non-thermal plasma induced by dielectric barrier discharge was intended to carry out the *m*-cresol degradation. For that, an experimental development was made, where several variables of the process like gas composition and released electric power were analyzed, with the objective of deciding the best operating conditions for the degradation as well as the determination of generated byproducts. A simple theoretical kinetics model was developed. It included only the liquid phase in order to determine in function of the time the final products of the degradation of the *m*-cresol.

2. Experimental setup and theoretical model

2.1. Experimental procedure

The experimental device shown in Figure 1 is composed by (1) plasmagen gases, which are supplied to the reactor according to the selected gas mixture, (2) a quasi-sinusoidal high-voltage power supply working at a repetition frequency of 3.0 kHz and released power ranged from 10 to 60 W, (3) a reactor of treatment with coaxial configuration is constituted by a stainless steel hollow tube inner electrode with external diameter of 12.0 mm, a quartz tube as a dielectric material and a fine mesh of copper working as outer electrode, (4) for the analysis of the collected samples a gas chromatography/mass spectrometry (GC/MS) system from Agilent Technologies was used, it meets the GC 5973N and the MSD 6890, which uses a column HP-5MS (30 m $\times$ 0.25 mm $\times$ 0.25μm).

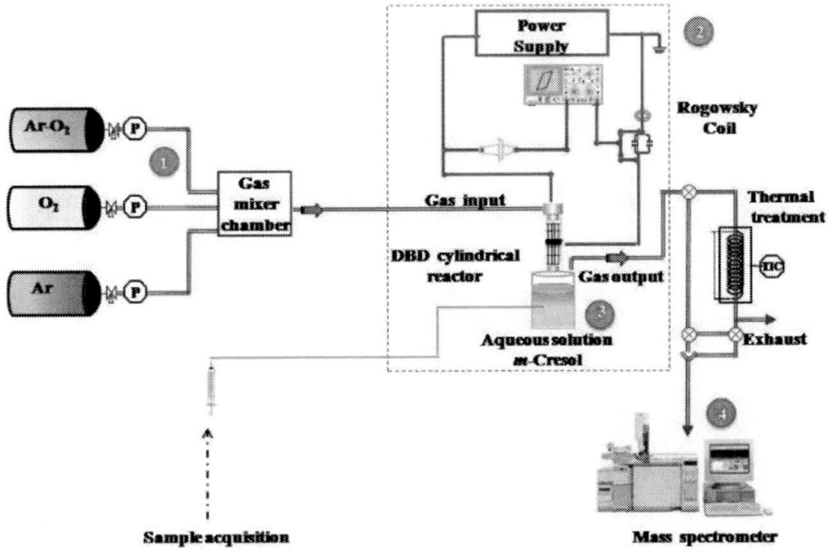

Figure 1. Experimental apparatus

The experimental procedure consisted on the treatment of synthetic aqueous solutions of *m*-cresol with an initial concentration of 5.0×10^{-3} mol/L. The treatment time was 1h for 500 mL of solution. The cylindrical reactor was placed vertically and the solution was impelled inside of the internal electrode, made of a hollow tube of stainless steel, using a submersible pump; when the liquid arrives to the top overflows on the external wall of the internal electrode and forms a thin layer of liquid establishing an gas-liquid interface as is showed in Figure 2. The residence time of the liquid was estimated of approx. 1.47 seg. Non-thermal plasma was generated in the gas phase using a gas mixture of Ar-O_2. During the treatment (1h), every five minutes solution samples were collected and analyzed by gas chromatography/mass spectrometry (GC/MS) system. The power supply working at a constant frequency of 3.0 kHz and a constant difference of potential was imposed during all treatment time. This procedure was repited imposing a different electric potential scanning from 15 kV up to 23 kV.

Journal of Physics: Conference Series **406** (2012) 012025 doi:10.1088/1742-6596/406/1/012025

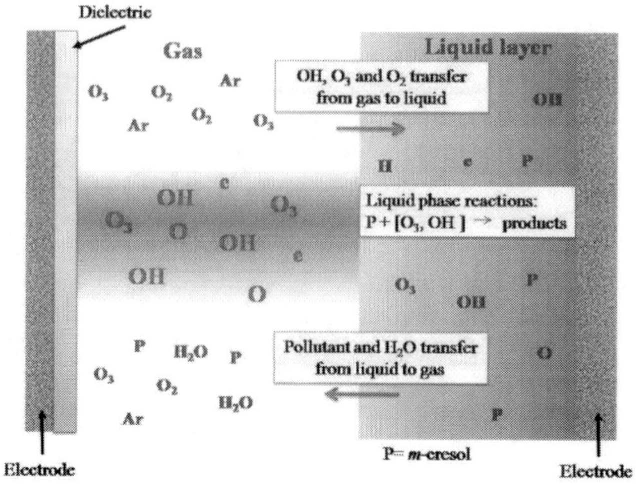

Figure 2. Gas-liquid interface

2.2. Kinetics Model

The theoretical model is constituted by 33 reactions and 23 chemical species only for the liquid phase [5]-[11], and was developed from the equation of continuity of the species. It was neither considered the diffusion effect nor the convection effect. In this way the equation of continuity can be simplified (1), where n_i is the density of the i species, S_i represents the generation and decomposition of the i species.

$$\frac{dn_i}{dt} = S_i \tag{1}$$

3. Results and discussion

3.1. Theoretical study

In Figure 3 the theoretical degradation in liquid phase of the m-cresol is shown, the simulations were carried out at the atmospheric pressure with an initial concentration of 5×10^{-3} mol m-cresol/L. The m-cresol degradation is carried out by combined chemical ways in gas and liquid phases. In gas phase electrons generated during the electric discharge react with O_2, Ar and vapour of water, producing active species like O_3, OH and O (2)-(5). Moreover, mass transfer mechanism is stablished in both phases, when some of the species in gas phase transfer to the liquid phase and vice versa [12] by diffusion process, Figure 2.

$$e + Ar \rightarrow Ar^* + e \tag{2}$$

$$e + O_2 \rightarrow e + O^* + O \tag{3}$$

$$e + H_2O \rightarrow OH + H \tag{4}$$

$$O + O_2 + O_2 \rightarrow O_3 + O_2 \tag{5}$$

$$Ar^* + Ar^* \rightarrow Ar^+ + Ar^{**} + e \tag{6}$$

$$Ar^{**} + O_2 \rightarrow Ar + O + O \tag{7}$$

The elimination mechanism starts when active species diffused in water O, H_2O_2, OH and other generated in liquide phase due to the dissociation of H_2O (7)-(11), react with the m-cresol generating primary products (12)-(13).

$$H_2O \rightarrow H + OH \tag{8}$$

$$H_2O + H_2O \rightarrow H_2O_2 + H_2 \tag{9}$$

$$H_2O_2 \rightarrow OH + OH \tag{10}$$

$$O_3 + H_2O \rightarrow H_2O_2 + O_2 \tag{11}$$

$$m - cresol + 2OH \rightarrow 2 - methyl - hydroquinone + 2H_2O \tag{12}$$

$$2 - methyl - hydroquinone + 2OH \rightarrow 2 - methyl - p - benzoquinone + 2H_2O \tag{13}$$

Later on, the m-cresol and his derived react with each other and OH in the solution producing smaller molecules such as oxalic acid and acetic acid. Finally, these species are oxidized to CO_2 and H_2O following the way showen in [11].

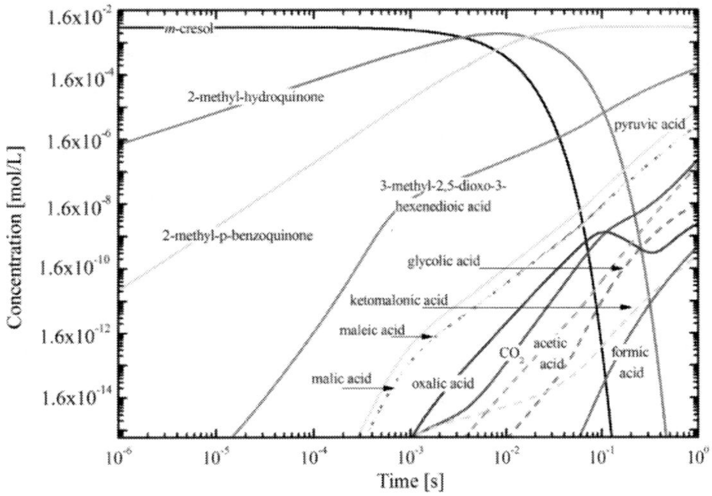

Figure 3. Theoretical degradation of m-cresol

3.2. Experimental Results

a) Gas mixture: [*90% Ar - 10%* O_2]

Figure 4 makes evident that the increments of the applied electric power and the time of treatment had a positive effect on m-cresol degradation. In liquid phase, chemical reactions are carried out generating active species like OH and H_2O_2 (8)-(11), (14)-(16). Radical OH reacts with the m-cresol and starts its oxidation, based on the mechanism proposed by *Da Silva Pimentel* [11], the primary by-products are generated, as they are chemically unstable break down generating intermediate species as oxalic acid and acetic acid, which react successively in the solution producing CO_2 and H_2O like final products.

$$H_2O + O_3 \rightarrow 2OH + O_2 \tag{14}$$

$$H + O + O \rightarrow HO_2 \tag{15}$$

$$H + HO_2 \rightarrow OH + OH \tag{16}$$

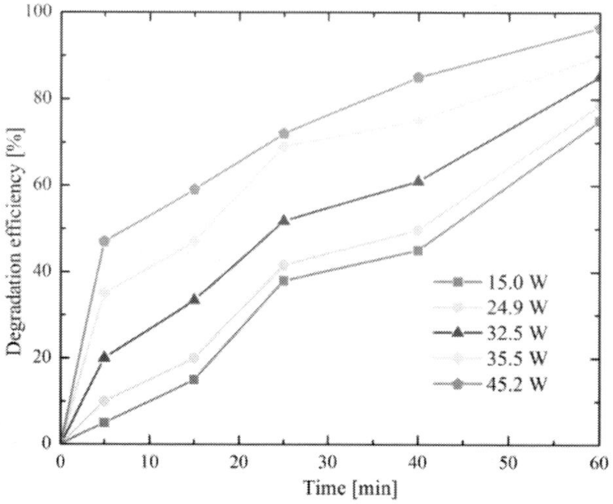

Figure 4. Efficiency of *m*-cresol degradation (*90% Ar - 10%* O_2)

b) Gas mixture: [*80% Ar - 20%* O_2]
In Figure 5 the effect of the time of treatment and electric power supplied are shown. The tendency is the same as in the previous case (Fig. 4). However from powers beyond of 23.7 W a significant influence of the power is not clearly observed, efficiencies were obtained of up to 92.2% after 60 minutes of time of treatment. Comparing Fig. 5 with Fig. 4 a lower degradation efficiency is observed, two causes were possible. The first one attributed to an increase of the O_2 quantity. Indeed, at *10%* O_2, the released power was consumed mainly by *Ar*. At *80% Ar - 20%* O_2, the energy should be shared with O_2. Therefore *Ar* does not participate efficiently in the dissociation of O_2 (7), and the power consumption is increased as it is displayed on legend curves in Figures 4, 5 and 6. If the oxygen quantity increase the electron attachement and the power consumtion too. The second one may comes from the power consumption, as it becomes higher reactor and aqueous solution temperatures increase. This augment of temperature (results not shown) was measured at the end of the experiment and a value up to 64°C was observed. Temperature increasing rises humidity in the gas phase and therefore instabilities in the discharge were observed affecting the efficiency. As a consequence, the elimination process finishes prematurely, indeed when free radicals as *OH* are generated in excess, they associate with each other finishing the reaction path (17)-(18) [13].

$$H + OH \rightarrow H_2O \tag{17}$$

$$OH + OH \rightarrow H_2O_2 \tag{18}$$

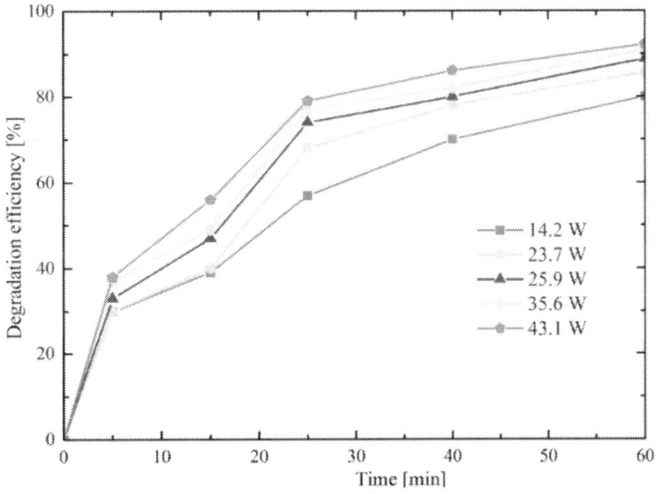

Figure 5. Efficiency of *m*-cresol degradation (*80% Ar - 20%* O_2)

c) Gas mixture: [*70% Ar - 30%* O_2]
Treatment time and applied power effects are shown in Figure 6. Here, efficiencies were obtained up to 91.0% after 60 minutes of time of treatment. Comparing the three last figures, one can observe a decrease tendency of the degradation efficiency.

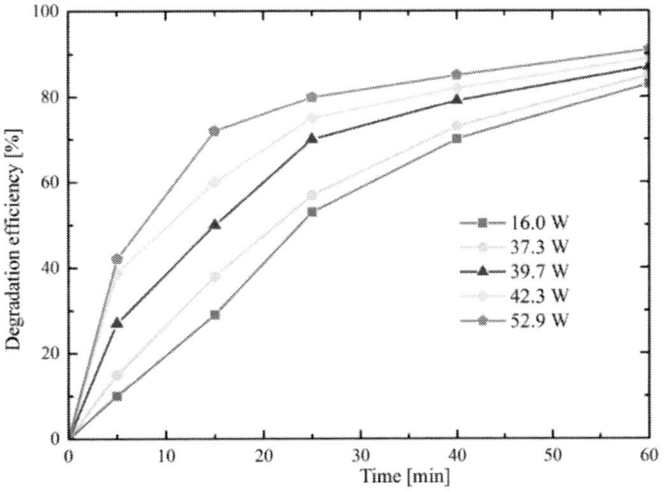

Figure 6. Efficiency of *m*-cresol degradation (*70% Ar - 30%* O_2)

d) Gas mixture: [*0% Ar - 100%* O_2]
In Figure 7 is depicted, as the applied power was increased the efficiency of *m*-cresol degradation was augmented as in the previous cases. For this gas mixture the efficiencies were obtained up to 97.3%.
Due that the gas mixture was constituted totally by O_2, the whole energy was consumed by O_2 and competition was not exist with *Ar*. The presence of O_2 was propitious to the generation of active species as *O* and O_3, which contribute to the formation of radical *OH*, (to see reactions

5, 10, 11, 14 and 15) and therefore in the *m*-cresol degradation.

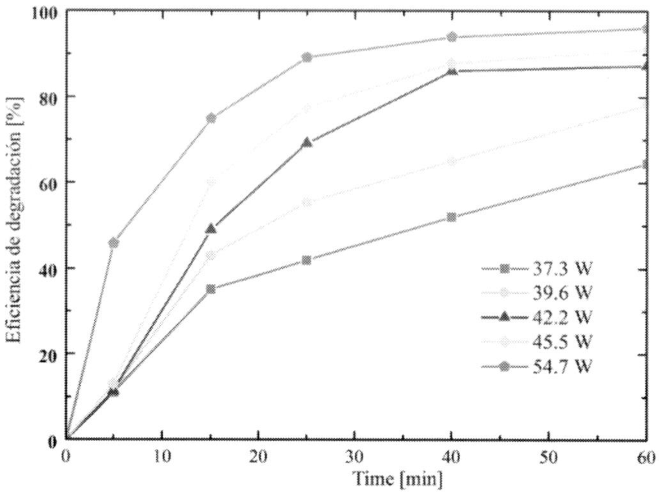

Figure 7. Efficiency of *m*-cresol degradation (*0% Ar - 100%* O_2)

3.3. *Identification of by-products*

In the liquid phase were identified oxalic acid, glyoxal, 4-hydroxy-2-pentanone, methyl alcohol, ethanol-2-methoxy-acetate, formic acid-methyl ester and ethanol-2-methoxy-acetate using gas chromatography and mass detection. Figure 8(a) shows the variation in time of oxalic acid, which can be explained by the formation of CO_2.

Furthermore, exhausted gases were also analyzed, the main compound found was CO_2. *m*-cresol in the exhausted gases was not detected. Figure 8(b) shows the variation of CO_2 during one hour of treatment. It can be seen in the figure that CO_2 generation raises as the treatment time is increased. CO_2, according to [11] comes from the oxidation of oxalic acid and formic acid.

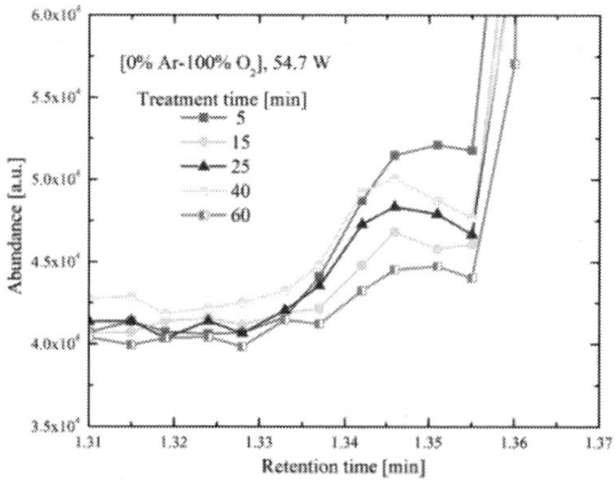

a) Identification of oxalic acid

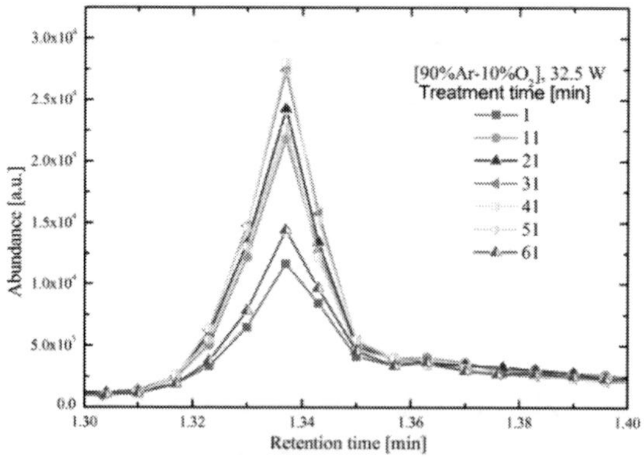

b) Identification of CO_2

Figure 8. Identification of by-products

Furthermore, color alteration of the residual liquid was special noticed to be function of the treatment time. It can be observed that the sample without treatment (showed as 0, Figure 9) after some minutes of treatment becomes sepia color and in some cases it turns to a light yellow color. According to some authors [14]-[15] the color change is attributed to the presence of acids as for example: oxalic acid and acetic acid generated during the process of degradation of *m*-cresol. When the residual liquid presents a light yellow color, it indicates a higher mineralization and when it has practically a light color the by-products have been transformed almost completely in CO_2 and H_2O. This treatment time evolution is described in the Figure 9.

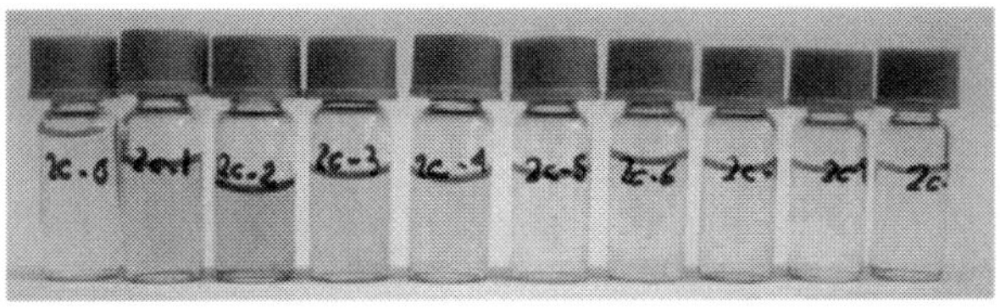

Figure 9. Changes of color of the residual liquid [*0% Ar - 100%* O_2], 54.7 W

The pH variation as a function of applied power and time of treatment is shown in Figure 10. As an exemple, in the case of 100% O_2 initial pH measured was 7.5. It was observed an augmentation in acidity as the treatment time was increased. The acidity measured is explained by the carboxylic acids generated after the oxidation of *m*-cresol. Inversely a pH value increasing indicate that some carboxylic acids are oxidized to CO_2 and water, as it is shown in Figure 10 when applied power was 54.7 W.

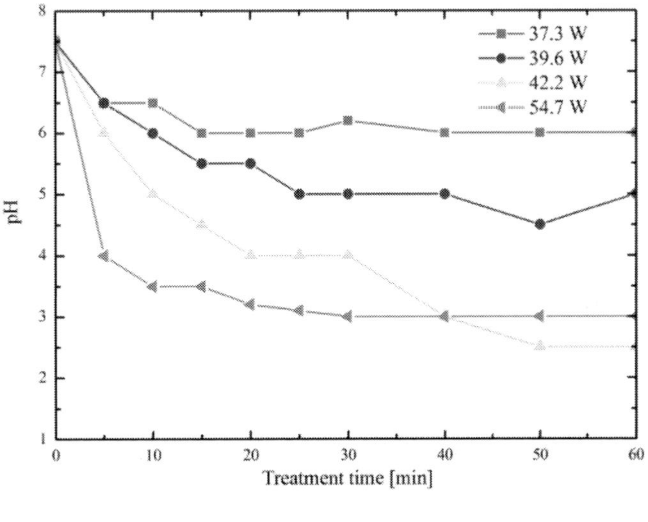

Figure 10. pH variation

4. Conclusion

Theoretically, it was possible to determine the main products of oxidation of *m*-cresol, the water and dioxide of carbon are among the final products. Another intermediate product as oxalic acid was also observed theoretically. Experimentally, during oxidation process as the oxygen proportion in the gas mixture was increased the efficiency of removal of *m*-cresol was diminished, a higher power was consumed and an increasing of temperature of the aqueous solution was measured. Nevertheless, a 97.3 % of efficiency of *m*-cresol removal was obtained using 100% O_2. It shows the effectiveness of the non-thermal plasma for the degradation of compound organic in water and the dielectric barrier discharge process was able to oxidize compound organic in aqueous solution. Oxalic acid and carbon dioxide were determined in liquid and gas phase respectively. Measurements of pH had confirmed the organics acids formation.

Acknowledgments

This work received financial support from CONACYT, Mexico, under Contract N°. 288097/229210. The authors appreciate the technical support received from P. Angeles Espinosa, M. T. Torres Martinez, and I. Contreras Villa

References

[1] Duu-Jong L, Kuo-Ling H and Yu-You C 2011 *Appl. Microbiol. Biotechnol.* **89** 209-15
[2] Abdollahi Y, Abdullah A H, Zainal Z and Yusof N A 2011 *Int. J. Chem.* **3** 31-43
[3] Tang Q, Jiang W, Zhang Y, Wei W and Lim T M 2009 *Plasma Chem. Plasma P.* **29** 291-305
[4] Jaramillo-Sierra B, Mercado-Cabrera A, López-Callejas R, López-Fernández J A, Peña-Eguiluz R, Barocio S R, Valencia-Alvarado R, Rodríguez-Méndez B, Muñoz-Castro A and de la Piedad-Beneitez A 2011 *Eur. Phys. J. Appl. Phys.* **56** 24026
[5] Hoigné J and Bader H 1976 *Water Res.* **10** 377-86
[6] Hoigné J and Bader H 1983 *Water Res.* **17** 173-83
[7] Hoigné J and Bader H 1983 *Water Res.* **17** 185-94
[8] Devlin H R and Harris L J 1984 *Ind. Eng. Chem. Fund.* **23** 387-92
[9] Dors M, Nichipor G V and Mizeraczyk J 2005 *IEEE International Conference on Dielectric Liquids* 95-98
[10] Grymonpré D R, Finney W C and Locke B R 1999 *Chem. Eng. Sci.* **54** 3095-105
[11] Da Silva Pimentel M A 2008 PhD. dissertation Université Paris-Est Marne-La-Vallée
[12] Ognier S, Iya-sou D, Fourmond C and Cavadias S 2009*Plasma Chem. Plasma P.* **29** 261-73
[13] Joshi A A, Locke B R, Arce P and Finney W C 1995 *J. Hazard. Mater.* **41** 3-30
[14] Tomizawa S and Tezuka M 2006 *Plasma Chem. Plasma P.* **26** 43-52
[15] Kavitha V and Palanivelu K 2005 *Water Res.* **39** 3062-72

Syngas production by plasma treatments of alcohols, bio-oils and wood

K. Arabi, O. Aubry, A. Khacef, J-M Cormier

GREMI, CNRS-Université d'Orléans, 14 Rue d'Issoudun, 45067 Orléans, France

E-mail: jean-marie.cormier@univ-orleans.fr

Abstract. Exploitation of forest resources for energy production includes various methods of biomass processing. Gasification is one of the ways to recover energy from biomass. The Syngas produced from biomass can be used to power internal combustion engines, or, after purification, to supply fuel cells. The paper is summarizing results obtained through a non thermal arc plasma reactor at laboratory scale. A stationary discharge (I = 150mA) is used to perform physical diagnostics and also chemical analysis. The arc is formed between two electrodes made of graphite. We first present results on plasma-steam reforming of alcohols and bio-oils mixed in water. The outlet gas compositions are given from various alcohols and-bio-oils obtained at different experimental conditions. The second part of the paper is dedicated to a direct plasma treatment of wood (beech) at laboratory scale. One of the electrodes is surrounded by wood. The final part of the paper is a general discussion about efficiencies and comparisons of plasma treatments presented. The results obtained are discussed by considering the steam reforming reactions and the water gas shift reaction.

1. Introduction

Biomass is considered as a renewable, storable and transportable energy source which is available in various forms such as wood, agricultural and forest residues.

Pyrolysis and Gasification are efficient ways for biomass utilization [1-4] and specially Syngas production. The thermochemical and biological conversion of biomass provide a great variety of products: oils, alcohols and gases. After treatment, these products can be used for various energy applications.

Reduction of greenhouse gas emission can be improved using syngas and hydrogen obtained from biomass. These requirements present a necessity to develop new types of method for the hydrogen production and Syngas production [4, 5].

Some authors have performed the possibility to use plasma reactor for hydrogen production from water and hydrocarbons, alcohol or biomass with high H_2 yields with limited costs [6-9]. Steam reforming using non thermal plasma processes is studied many years ago [6,7].

This paper is devoted to a comparative study of Syngas (H_2+CO) production from alcohol (methanol, ethanol, phenol) and others hydrogenated liquids and also wood.

2. Experimental techniques

The experimental plasma reactor used in this work was described previously [8]. Laboratory-scale experiments were carried out with a liquid mixture injected at atmospheric pressure into a non-thermal plasma reactor (Figure 1).

The discharge is called Non Thermal Arc (NTA) [9] although the voltage drop at the electrodes is significantly greater than voltage drop of conventionnal arcs.
The plasma reactor includes a quartz tube (400 mm length and 30 mm inner diameter) containing two electrodes made of graphite. The conical extremities of the two electrodes are placed opposite each other. The electrode gap is 10 mm. The discharge column looks like a plasma string with a diameter of 2 mm diameter and length of 10 mm.
Liquid mixtures are injected with a syringe pump, which permits to adjust the total liquid flow rate at the inlet of the reactor from about 0.5 to 2.1 sccm.

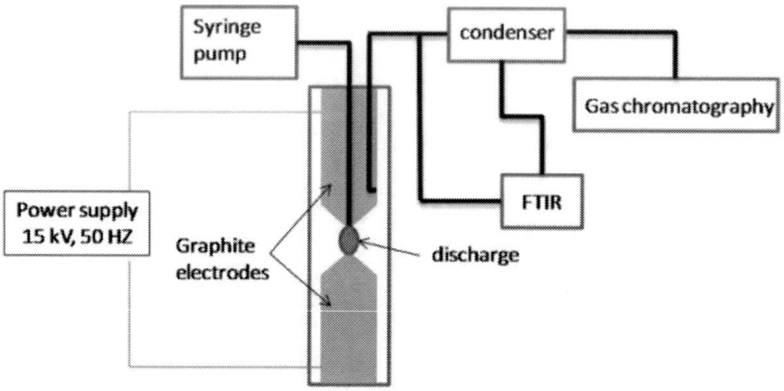

Figure 1. Experimental set up.

The inlet alcohol/water liquid mole ratios studied are from 0.07 to 0.19 and from 0.1 to 0.27 for the ethanol and methanol mixtures, respectively. The liquids are vaporized during their flow in the upper electrode before the discharge injection. Graphite electrodes are heated by plasma.
The gas discharge is powered by a 50 Hz high-voltage transformer with leakage flux (primary 230V, secondary 20kV, I_2= 155 mA). The current and voltage waveforms are measured using a current and a differential voltage probes respectively. The signals are recorded on a digital oscilloscope. From the current and voltage data, the power can be calculated and the energetic cost of production H_2 can be estimated. Plots of voltage and current are given on figure 2. At a maximum of current (about 230 mA), a quasi constant voltage is observed (900V). Periodic high voltages peak are observed at the zero current transition of the alternating current.

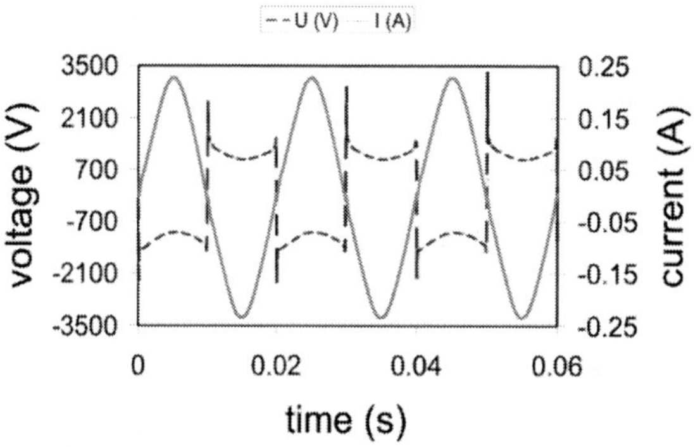

Figure 2. Variations of the voltage and current as a function of the time, inlet liquid volume ethanol/water ratio = 0.23.

At the outlet reactor, analyses are performed on exhaust gas to measure species concentrations in wet gas or in dry exhaust gas. For this latter, the outlet gas is condensed using a cryogenic trap (-20°C) and analyzed using a gas phase chromatography analyzer (GC-Varian CP 3800). H_2, CO, CO_2, CH_4, C_2H_2, C_2H_4 and C_2H_6 species can be easily quantified.

3. Plasma steam reforming of methanol and ethanol

3.1Effects of the total flow rate and inlet alcohol concentration

Figures 3 display the concentration of the main produced species as a function of the total flow rate. Two ethanol/water and methanol/water inlet liquid volume ratios are studied. Main species are H_2, CO, CO_2 and CH_4. C_2 species are also produced but their concentrations are remaining lower than 1.5%.

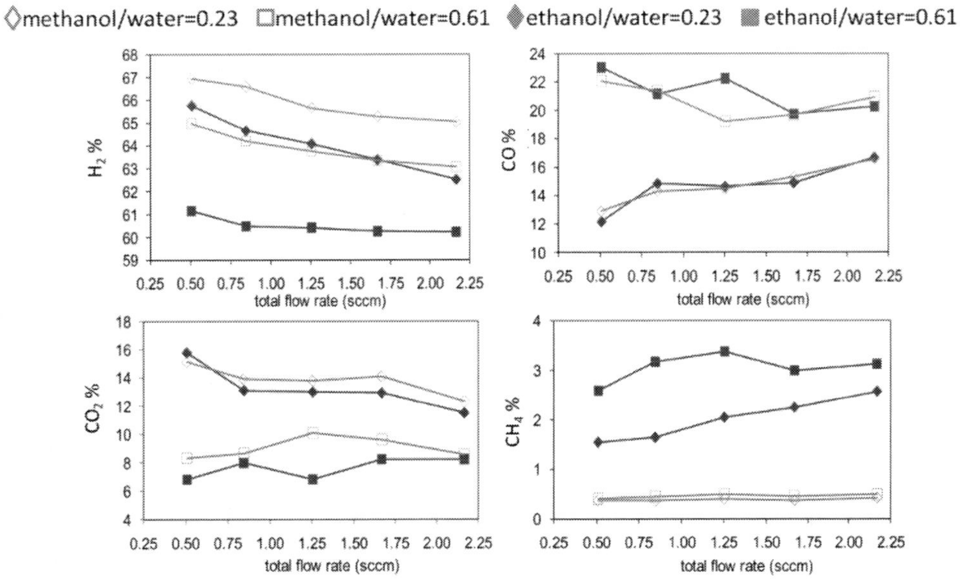

Figure 3. H_2, CO, CO_2 and CH_4 concentrations vs. the inlet total flow rate. Open symbols: methanol+water mixtures; full symbols: ethanol+water mixtures. Alcohol/water volume ratios: square symbols: 0.23; diamond symbols: 0.61.

Increase of flow rate generates a decrease in hydrogen concentration probably due to the decrease of residence time in the reactor. A rise of the alcohol/water ratio leads to a decrease of H_2 production. The lowest CO concentrations corresponding to the highest CO_2 concentrations are observed for highest dilutions of alcohols in water. CH_4 is the main hydrocarbon produced; its maximum concentration is about 3% in the ethanol mixtures. CH_4 is promoted when alcohol/water ratios and/or the total flow rate increase. In the ethanol mixtures, hydrocarbon species seem to be produced more easily corresponding to an increase of the inlet carbon number. Similar results are obtained for C_2 species, which are not presented here because their concentrations remain low.

From these results it can be proposed the following reactions involved in the process for methanol and ethanol.

$CH_3OH \rightarrow CO+2H_2$ (1)

$C_2H_5OH + H_2O \rightarrow 2\,CO + 4\,H_2$ (2)

$H_2O + CO \rightarrow CO_2+H_2$ (3)

Journal of Physics: Conference Series **406** (2012) 012026 doi:10.1088/1742-6596/406/1/012026

Main part of hydrogen is produced by cracking (1) or steam reforming (2) and the water gas shift (3) reaction(WGS) is acting simultaneously to produce CO_2 and hydrogen enrichment.
In this case, given the results, we can assume that the WGS reaction is more efficient in excess water. This may explain higher hydrogen concentrations at low concentrations of alcohols. A more detailed interpretation of these results, based on this model is underway.

3.2 Energy balance

From the results, losses energy in the reactor have been estimated. To evaluate the losses, vaporization energy (to vaporize inlet liquid mixtures) and reaction energy are take into account. Energy losses, E_l, are calculated from the next expression:

$$E_l = Et - Er - Evap \qquad (4)$$

Where, Et is the total energy injected in the reactor. Et is calculated from injected power by integration of P over time. P is define in equation 5. It should be noted that the only external source of power is the electrical generator.

$$P=\frac{1}{T}\int_0^T u(t).i(t)\,dt \qquad (5)$$

Er takes into account steam-reforming and water gas shift reactions energies, Evap is the consumed energy to vaporise the inlet liquid mixture. In table 1, we present the various energy for ethanol steam reforming.

X C_2H_5OH	Time (h)	Et (Wh)	Evap (Wh)	Er (Wh)	E_l (Wh)
0.07	0.012	1.77	0.94	0.121	0.709
0.22	0.006	1.01	0.48	0.115	0.415
0.49	0.004	0.76	0.32	0.112	0.328

Table 1. Energies implied in the process.

One can observe that a rise of the inlet alcohol mole fraction leads to a decrease of the energy losses, because Et highly decreases in these mixtures. This decrease is due to a decrease of the conductance because of less water vapor is in the plasma. Same trends are observed for other alcohols treatments.

About half of the total energy injected in the reactor is lost. It is possible to reduce the energy costs by increasing the electrode gap. The efficiency of the process can be improved by inceasing the length of the plasma column that reduce the relative power dissipated at the electrodes.

4. Comparison of results obtained from three alcohols: methanol, ethanol and phenol

To summarize the results obtained with different alcohols concentrations, main species are represented, according to the atomic ratios H/C and H/O of alcohol-water mixtures injected at the reactor. Inlet ratios are calculated from the inlet mixtures composition by considering that all the injected liquid is vaporized through upper electrode. Thus, liquid moles and vapor moles of ethanol and water are the same. An increase of the inlet alcohol concentration leads to a decrease of H/C and a rise of H/O values. . We report, figures 4, H_2, CO and CO_2 mol% as functions of the inlet H/C and H/O ratios.

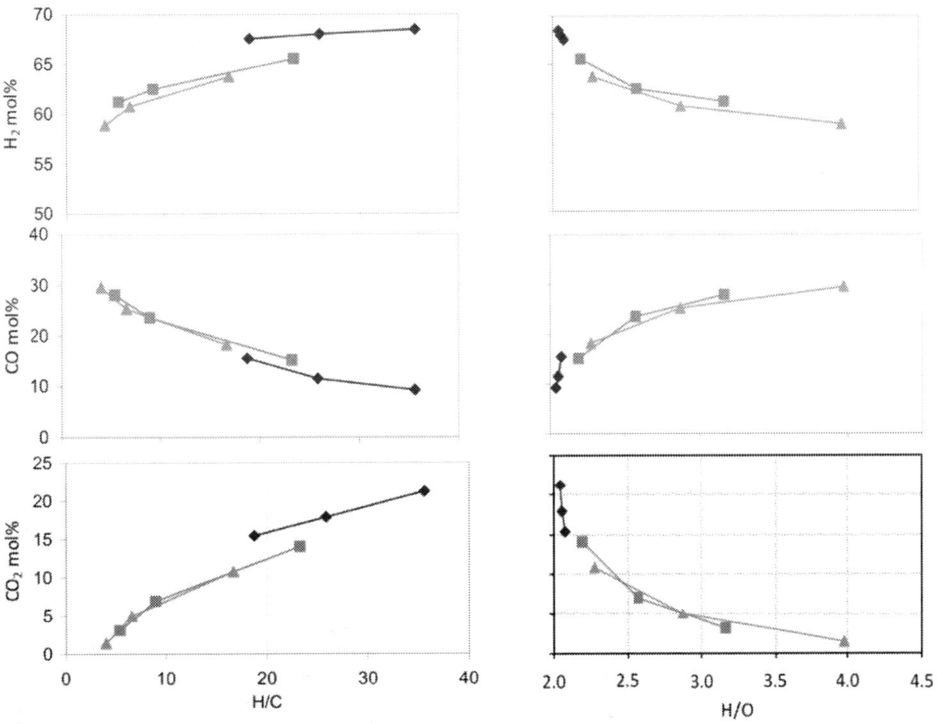

Figure 4. .H_2, CO, CO_2 mol% vs. inlet H/C and H/O ratios. ▲ methanol+H_2O; ■ ethanol+H_2O; ◆ phenol+H_2O.

H_2 and CO_2 concentrations increase when H/C increases or H/O decreases. CO concentrations decrease for H/C increasing or H/O decreasing. Alcohol nature, at given H/C and H/O values, have not significant effects on the species concentrations. Changes in different concentrations depending on H/C and H/O can be approximately described by continuous functions independent of alcohols nature.

5. Experiments with other reactants

Experiments with others reactants have been also performed with ammonia and aBio- oil diluted or not in water. In Table 2, we report the inlet mixtures parameters and main dry gases produced and their concentrations.

Table 2. Produced species concentrations in dry gas from ammonia and Bio-oil mixtures.

Inlet mixtures	H/O	H/C	flow rate (sccm)	H_2 (%)	CO (%)	CO_2 (%)	CH_4 (%)
13% vol. $NH_3 + H_2O$	**2.45**	**-**	**1.25**	**55**	**0.6**	**-**	**-**
Pure Bio oil	**2.25**	**1.5**	**0.026**	**63.5**	**14.5**	**21.5**	**0.38**
10% vol. Bio oil + H_2O	**2.1**	**28.3**	**0.026**	**59.6**	**27.7**	**17.1**	**0.4**

This table shows that oils obtained from various process can be easily treated by plasma to produce Syngas.In comparison with classical gassification process, high values of H_2/CO (>1) can be obtained.

Journal of Physics: Conference Series **406** (2012) 012026 doi:10.1088/1742-6596/406/1/012026

Ammonia could be an interesting hydrogen source by reforming in-situ because NH_3 is easily transportable and storable. H_2 concentration obtained from the ammonia mixture is sligthly lower than in alcohol ones for the same total flow rate and inlet H/O: about 64-66% in alcohols mixtures against 55% with NH_3.

6. Direct plasma steam reforming of wood

6.1 The experimental system

The device is practically the same (Figure 5) as presented above excepted for the top electrode. This electrode is made of a very fine graphite rod surrounded by wood (beech).
The water is introduced through the lower electrode and is vaporized in plasma region. At start-up, the plasma column is developed only between the electrodes. Subsequently, due to the wood pyrolysis, the discharge can move on the surface of the carbon layer appeared on the wood. The figure 6 shows an example of the discharge obtained from wood treatment.

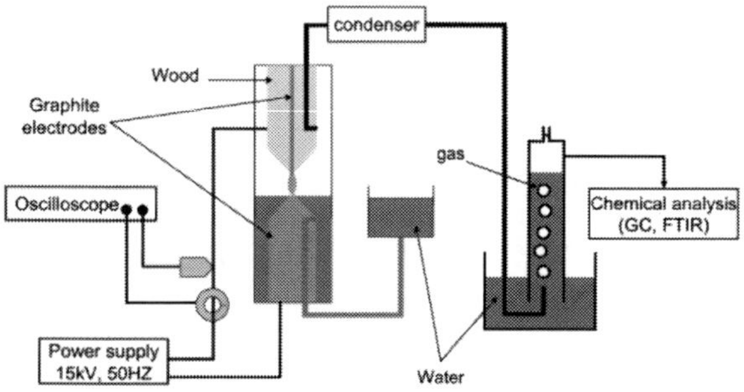

Figure 5: Schematic of experimental reactor

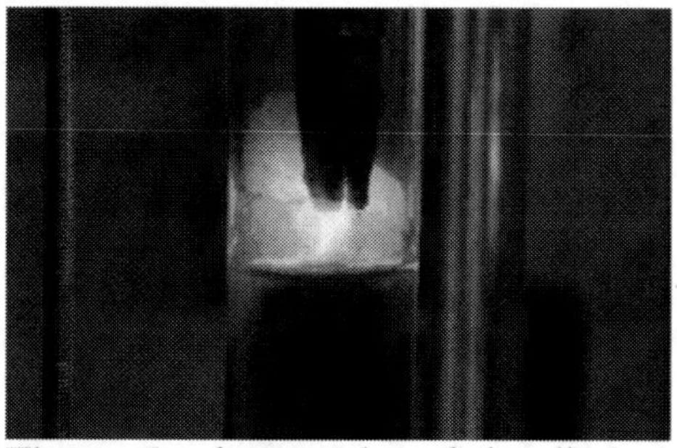
Figure 6: photography of the discharge between electrodes.

6.2 Results

The equation (4) of the biomass treatment (beech wood) is written as:
$C_6H_9O_4(s) + 5\ H_2O\ (g) \rightarrow 3CO\ (g) + 3CO_2(g) + 9.5\ H_2(g)$ (6)

In table 3, the results obtained with biomass treatment are reported for two values of the electrodes gap (D_{int} = 1cm and D_{int} = 2.5cm).
The observed species are mainly H_2, CO, CO_2 and CH_4. C_2 hydrocarbons species are less than 0.5 %

Table 3 : Species concentration in dry gas from wood and gas flow rate (L/h) as a function of electrodes gap

D int (cm)	Analysis GC dry gas				Produced gas flow rate (L/h)
	H_2	CO_2	CO	CH_4	
1	57.8	17.9	22.5	1.2	5.6
2.5	58.3	19.3	19.7	2.3	13.6

The concentrations of the produced species are of the same order for the two distances between electrodes.
Moreover, the flow of gas production increases (from 5.6L/h to 13.6L/h) when the length of the plasma column increases.

The efficiency of plasma increases with the length of the plasma column.
Theoretically, the concentration of H_2, CO and CO_2 are calculated from stoichiometric coefficient using reaction (4) for the biomass treatment. The following mixture composition of dry gas is found as:
H_2 = 61%
CO = 19%
CO_2 = 19 %
We observe that H_2, CO and CO_2 concentrations are similar to those obtained from experiment.
Therefore, the reaction (4) of biomass treatment well describes the experimental results and it is the dominant process.

6.3 Energy interest of wood plasma treatment
The energy produced from the Syngas combustion is compared to the energy obtain from combustion of wood consumed. The initial mass of electrode surrounded by beech wood is: m_i = 4.0g; the total duration of the experiment is 45min and the consumed mass of wood is equal 2.5g.
The heating value of wood is 18.10^3 kJ/mol, so the combustion energy of this mass equal to 46 kJ.
The total volume of gas produced, corresponding to 0.75h, is equal to 7.2L. The composition of gas mixture is: 58% H_2 (46 kJ) and 21% CO (20 kJ).
Taking into account the heating values of H_2 and CO (241 kJ/mol, 283 kJ/mol respectively), we calculate the combustion energy of the gas mixture contained in 7.2L. The energy production is equal to 66 kJ.
The energy combustion of the wood consumed is less compared to that obtained with the combustion of Syngas.

7. Discussion
To show a synthetic view of these experiments a global interpretation is given by estimating the maximum and minimum concentrations of hydrogen obtained from cracking, steam reforming, and water gas shift reactions.
The first set of equations (table 4) describes a hypothetical behaviour without CO_2 production (without WGS).
The second ones correspond to a maximum H_2 production with a total conversion of CO by water gas shift reaction.
Estimates are shown in the table 4; one column is dedicated to theoretical value of H_2 (% vol) the other one to an arbitrary calculation taking into account 10% of other molecules in the exhaust gas mixture.

Table 4 : Theoretical and simplified analysis		
Global reactions	**H_2 % (vol p_{atm})**	**H_2 % (vol p_{atm})**
Without WGS	**Theoretical**	**+10%mol total**
$CH_3OH \rightarrow CO + 2\ H_2$	66	60.6
$C_2H_5OH + H_2O \rightarrow 2\ CO + 4\ H_2$	66	60.6
$C_6H_5OH + 5\ H_2O \rightarrow 6\ CO + 8\ H_2$	57	52
$C_6H_9O_4 + 2\ H_2O \rightarrow 6.5\ H_2 + 6\ CO$	52	47
Full WGS		
$CH_3OH + H_2O \rightarrow CO_2 + 3H_2$	75	68
$C_2H_5OH + 2H_2O \rightarrow 2\ CO_2 + 6\ H_2$	75	65
$C_6H_5OH + 11\ H_2O \rightarrow 6\ CO_2\ (g) + 14\ H_2$	70	63.6
$C_6H_9O_4 + 8\ H_2O \rightarrow 12{,}5\ H_2 + 6\ CO_2$	68	61

A quick overview of this table shows similar orders of magnitude for hydrogen concentrations regardless of biomass type considered (wood or alcohol). Global results well agree with Figure (3).

WGS reaction is an interesting point because this reaction reduces the amount of CO while increasing the amount of hydrogen.
This reaction takes place differently in various plasma conditions. The above study is of great interest for scientific and also from industrial point of view.

8. Conclusion

Alcohol steam-reforming from a non-thermal arc is an interesting pathway to produce H_2 and Syngas. The CO and CO_2 concentrations are in accordance with usual thermo-chemical analysis. The nature of the inlet alcohol does not play a role on the mol% composition of the dry gas at outlet of the reactor.
The results obtained in this study present the first development of the wood treatment using a non-thermal plasma reactor. The flow rate of the hydrogen production is increasing with the electrodes gap. Efficiency of plasma process is increasing with the length of the plasma column. So "Gliding Discharges" can be used to produce large volumes of plasma. The energy produced by the combustion of Syngas is higher than the energy combustion of wood consumed to obtain the Syngas.
This work shows that the biomass treatment using non thermal plasma could be a promising technique for Syngas production. A complete and a large study remain to be developed to improve the chemical and physical efficiency of the plasma reactor.

9. References

[1] L. Shiguang, X. Shaoping, L. Shuqin, Y. Chen, L. Qinghua. 2004 *Fuel Process Technology* 1201
[2] G. Chen, J. Andries, H. Spliethoff, M. Fang, P. J. van de Enden. 2004 *Solar Energy* 76 345
[3] H. Qinglan, W. Chang, L. Dingqiang, W. Yao, L. Dan, L. Guiju. 2010 *International Journal of Hydrogen Energy* 35 8884
[4] K. Aasberg-Petersen, J.-H. Bak Hansen, T.S. Christensen, I. Dybkjaer, P. Seier Christensen, C. Stub Nielsen, S.E.L. Winter Madsen, J.R. Rostrup-Nielsen. 2001 *Applied Catalysis A: General* 221 379
[5] J.D. Holladay, J. Hu, D.L. King, Y. Wang. 2009 *Catalysis Today* 139 244
[6] I. Rusu, J-M. Cormier 2003. *Chemical Engineering Journal* 91, 23,.
[7] I. Rusu, J-M. Cormier. 2003 *International Journal of Hydrogen Energy* 28, 1039,
[8] O. Aubry, C. Met, A. Khacef, J.M. Cormier. 2005 *Chemical Engineering Journal* 106, 241,.
[9] J.D. Rollier, J. Gonzalez-Aguilar, G. Petitpas, A. Darmon, L. Fulcheri, R. Metkemeije. 2008 *Energy & Fuels* 22, 556

Net emission coefficient of CO_2-Cu thermal plasmas: role of copper and molecules.

T Billoux[1], Y Cressault[1], V F Boretskij[2], A N Veklich[2] and A Gleizes[1]
[1] *Université de Toulouse; UPS, INPT; LAPLACE (Laboratoire Plasma et Conversion d'Energie), 118 route de Narbonne,F-31062 Toulouse cedex 9, France*
[2] *Taras Shevchenko Kiev National University, Radio Phys. Dept., 64, Volodymirs'ka Str., Kiev, 0133, Ukraine*

E-mail: yann.cressault@laplace.univ-tlse.fr

Abstract
The present work was conducted to calculate the radiative transfer in CO_2–Cu mixtures at atmospheric pressure. The plasma is supposed to be in local thermodynamic equilibrium (LTE) for temperatures between 500 and 30000 K. In the first part of the work, we briefly present the plasma composition. The second part is devoted to the various mechanisms responsible for the radiation (atomic continuum, molecular continuum, atomic lines and molecular bands). A particular attention is paid to the radiation of molecular bands. These radiative properties are then introduced in the net emission coefficient (NEC) which is used to estimate the radiation in an isothermal and homogeneous medium. Finally, some results are proposed for various concentrations of copper and different plasma sizes. For optically thin plasmas, the radiation of molecular bands and metallic vapors strongly increase the NEC at low temperature. For higher plasma sizes, the copper lines do not only have high intensities but are also strongly self-absorbed contrarily to the radiation of the molecular bands which becomes the major part, indicating that we must not neglect their contribution in the total radiation.

1. Introduction

In a previous work [1], the plasma parameters of an electric arc discharge in a gas flow between copper electrodes were experimentally investigated. The electrical and thermal conductivities of Cu-Air and Cu-CO_2 plasma were calculated. Experimental temperature values were low (3000 K<T<6000 K) for a discharge at current 3.5A. To complete this experimental study and develop further numerical models, this paper deals with the radiation of CO_2-Cu plasmas. Some values can be found in the literature concerning pure CO_2 [2] and pure copper or mixtures with copper as Ar-Cu [3-6], N_2-Cu [4], SF_6-Cu [3,4,7], Air-Cu [8,9] and Ar-H_2-Cu [6]. Few data are available for this mixture. We then estimated the continuum radiation (molecular continuum, bremsstrahlung, radiative attachment) and the atomic line radiation spectra neglecting the atomic lines overlapping. According to the measured temperatures, a particular attention was focused on the emission of the molecular bands of CO, C_2, O_2 and CO_2 (with overlapping of molecular lines) since no data were available in the literature for the CuO species. Finally, we estimated the net emission at the centre of an isothermal and homogenous sphere, by applying the method of the Net Emission Coefficient [10].

2. Composition of the plasma

For thermal plasmas, the radiation strongly depends on the species present in the medium. In the present work, the computation of the equilibrium compositions of CO_2 plasmas with copper metallic vapors in the temperature range 500K – 30kK and at atmospheric pressure was realized with a method based upon the mass action law and the chemical base concept described by *Godin and Trépanier* [11]. We have considered 34 gaseous species: e^-, C, O, Cu, C^-, O^-, Cu^-, C_2^-, O_2^-, C_3^-, CO_2^-,

C^+, O^+, Cu^+, C_2^+, O_2^+, CO^+, C^{2+}, O^{2+}, Cu^{2+}, C^{3+}, O^{3+}, Cu^{3+}, C_2, O_2, Cu_2, CO, C_3, O_3, CO_2, C_2O, CuO, C_4, C_3O_2. Internal partition functions (IPF) of atoms and their positive ions were calculated with electronic level energies and degeneracies taken from the NIST database [12]; these IPF were systematically confronted with the Drawin and Felenbok's IPF compilation [13]. IPF were assumed to be equal to the degeneracy of the ground state for negative atomic ions and they were calculated using the Morse potential minimization method for diatomic species. The spectroscopic data (Dunham coefficients, vibrational frequencies and degeneracies, rotational constants, moments of inertia and symmetry numbers) essential to the calculation of the internal partition functions of molecules were taken from *Chase et al* [14] and *Huber and Herzberg* [15]. First order corrections corresponding to deviations to the ideal gas law were included in the computation: the Virial correction [16] associated to interactions between neutral chemical species at low temperature and the Debye-Hückel correction [17] due to Coulombian interactions between charged particles at high temperature. An example with 99%CO_2-1%Cu (in mass proportions) is given in *Figure 1*.

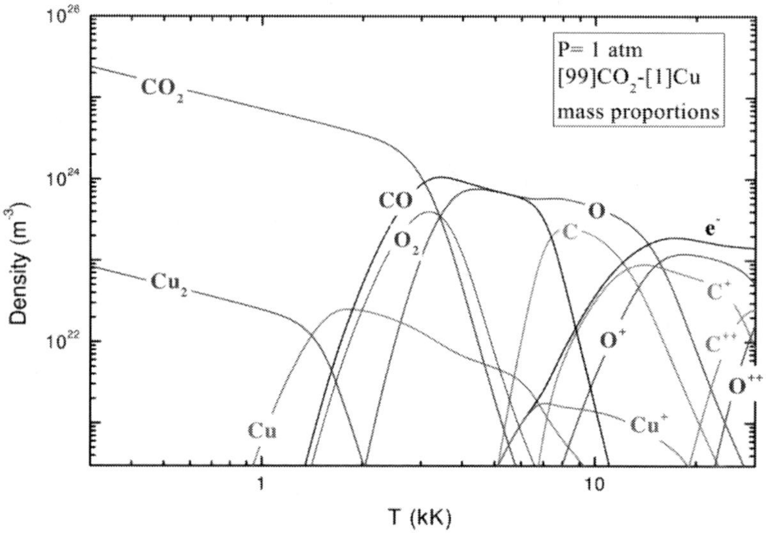

Figure 1. Composition for a 99%CO_2-1%Cu plasma at atmospheric pressure.

3. Total absorption coefficient

The computation of the NEC requires the total absorption coefficient which was obtained after a preliminary study of the radiation resulting from the atomic continuum, the molecular continuum (6 molecules C_2, CO, CO_2, O_2 and O_3), the atomic lines and the molecular bands (Schumann-Runge system of O_2, 4th positive and IR systems of CO, Swan system of C_2 and IR bands of CO_2).

3.1. Atomic continuum

For the atomic continuum radiation, we have to take into account three important phenomena: the radiative attachment, the radiative recombination and the Bremsstrahlung. Even if the radiative attachment is often negligible on the total continuum absorption coefficient, we nevertheless considered it because of the O^-, C^- and Cu^- species in the plasma. The monochromatic emission coefficient is given by *Gleizes et al* [4] and requires the negative ion population number density and the photo-detachment cross-sections which were issued from *Robinson and Geltman* for oxygen [18], *Yachkov* for carbon [19] and *Schultz-Gulde E and Gilde* for copper [20].

The radiative recombination is often essential in the continuum emission of thermal plasmas and was calculated according to the method of the Scaled Thomas-Fermi potential [21]. By taking into consideration LTE assumption, the monochromatic emission coefficient is given for a temperature T and a wavelength λ by *Gleizes et al* [4] and requires the electron and ion number densities, the ion charge, the partition function and the ion ground state statistical weight and the Biberman-Schlüter factor taking into account the difference between the photoionization cross section of the energy levels and the corresponding hydrogen-like cross section [22,23]. This factor was calculated by *Hofsaess* for oxygen and carbon species [24]. As the Biberman-Schlüter factor is not available in the literature for the copper species, the photoionization cross sections is obtained according to the hydrogenoid-like atoms approximation developed by *Okuda et al* [25]. For this last method, the energy levels and principal quantum numbers of the considered species are issued from the tables of *Moore* [26], corrected by the recent data of the *NIST* [27] and completed if necessary by those of *Kurucz and Peytremann* [28].

The Bremsstrahlung coefficient due to the interaction between electrons and atoms is given by *Gleizes et al* [4] knowing the Gaunt factor introduced to correct the variation from the hydrogen-like approximation, and calculated from the total elastic cross section. We used the total cross sections issued from *Neynaber et al* and *Robinson and Geltman* [18,29] for oxygen and carbon, and from [18,30-32] for copper. The Bremsstrahlung coefficient due to the interaction between electrons and ions is obtained according to *Gleizes et al* [4] with the use of the Gaunt factor.

3.2. Molecular continuum

A special attention has been devoted to the description of the molecular continuum spectrum too since it represents one of the major contributions to the total emission at low temperature. We considered the C_2, CO, CO_2, O_2 and O_3 molecules assuming that the photoabsorption cross-section did not depend on temperature. Consequently, each molecular continuum could be obtained by the multiplication of this section by the corresponding molecular density. More than 50 references were used to reconstruct the photoabsorption cross-sections. *Figure 2* shows the molecular continuum obtained at 300K, for 5 molecules and the atmospheric pressure, the number density for O_3 being negligible.

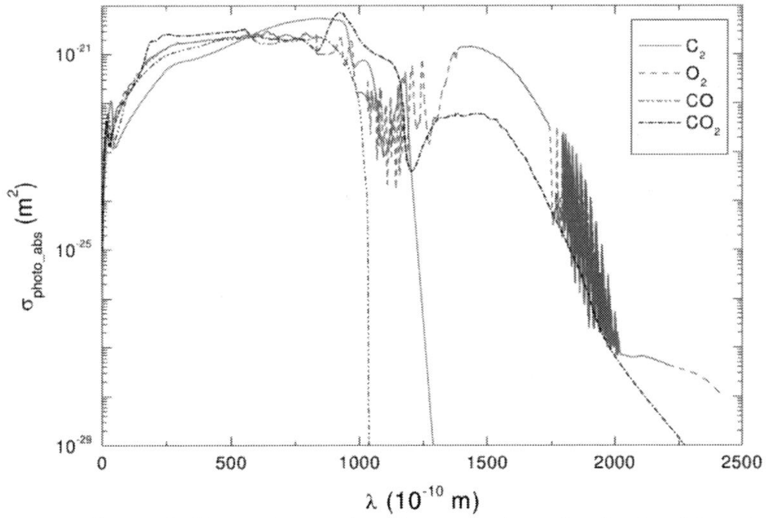

Figure 2. Photoabsorption Cross-sections for C_2, O_2, CO and CO_2.

3.3. Atomic lines

To calculate the emission of the atomic lines, we elaborated a complete data bank including the properties of 18069 lines (6672 lines for carbon species, 6217 lines for oxygen species and 5180 lines for copper species) whose characteristics (energy levels, quantum numbers, transition probabilities, oscillator strengths…) were issued from [26-28]. The radiation transfer requiring the knowledge of the line profiles, we took into account the Doppler broadening, the pressure effects (resonance and Van der Waals broadenings) and the Stark broadening [33]. The FWHM was obtained from the relation given by *Laux* [34] for the Doppler effect, from *Ali and Griem* [35] for the resonance broadening, from *Walkup et al* [36] for the Van der Waals broadening and from *Griem* [37] for the Stark broadening including the corrections of *Griem* [33,37,38] to take into account the influence of the ion. Due to the numerous lines, we assumed that the lines' overlapping had a negligible influence on the radiation allowing us to treat each line by means of an escape factor Λ_{line} introduced by *Drawin and Emard* [39]. This factor allowed us to simplify the treatment of the spectra and to spare calculation time. In the case of an isothermal plasma, this escape factor Λ depends on the line profile and thus, of all the mechanisms of line broadening. It can be defined by the ratio between the radiative flux escaping an isothermal plasma of thickness R_p with the consideration of the absorption and the radiative flux without absorption (optically thin plasma). Therefore, the resonance line escape factor lies between 0 and 1, and its value is equal to 1 for the other lines. The consequences are that the use of the escape factor tends to over-estimate the radiation whose values are however acceptable assuming the precision of the method. With the definition of the escape factor, the line radiation is written as:

$$\varepsilon_\lambda^{line}(T,R_p) = \left(\frac{e^2}{4\pi\varepsilon_o}\right)\cdot\left(\frac{\pi}{m_e c}\right)\cdot L_\lambda^o(T)\cdot f_{\ell u}\cdot \overline{n}_\ell(T)\cdot \Lambda_{line}\cdot\left(1-\exp\frac{-hc}{\lambda k_b T}\right) \quad (1)$$

where $L_\lambda^o(T)$ is the Planck function at the centre of the line, e is the electron charge, λ is the wavelength corresponding to the transition between the upper energy level E_u and the lower energy level E_l, m_e is the mass of the electron, $\overline{n}_\ell$ is the average number density of the lower energy level, f_{lu} is the oscillator strength of this transition and Λ_{line} is the escape factor of the line.

3.4. Molecular bands

Finally, the most important work of this paper is the consideration of the radiation issued from the molecular bands. In a previous work focus on the radiation of air plasmas, 9 molecular electronic systems corresponding to 1788 vibration bands and more than 10^6 rotation lines [40] were considered in the visible and UV parts of the spectrum (Schumann-Runge system for O_2, first and second positive systems for N_2, β, γ, ε and δ systems for NO, first negative and Meinel systems for N_2^+). This work had enabled to highlight the fundamental role of molecules in the emission of an air plasma for temperatures under 8000 K. Unfortunately, in the case of CO_2 plasmas, our team only has data for O_2 which are limited to temperatures inferior to 10000K, for an optically thin plasma and the atmospheric pressure. Therefore, it was necessary to develop new studies in order to consider the radiation of molecular bands of O_2, CO, CO_2 and C_2 species whatever the pressure, the temperature and with or without absorption. Molecular emission and absorption spectrums were described using 4.470.000 points covering the spectral range [30-4500] nm. All molecular systems were treated in the coupling Hund's case b which is valid for Σ states and which tends to be valid for Π states when the rotational level increases. In the Hund's case b, the quantum numbers which permit to describe the system are Λ, S, N and J. The overlapping of molecular bands was considered by summation in each wavelength point of all molecular systems spectrum before the evaluation of the self-absorption.

3.4.1. Diatomic systems

The main diatomic molecular systems contributing to CO_2 mixture radiation [41] (IR and 4th positive systems of CO, Swan system of C_2 and Schumann-Runge system of O_2) were studied. To describe the radiation spectrums of molecular bands, it was necessary to calculate for each wavelength the spectral emission coefficient by the relation:

$$\varepsilon_\lambda = \frac{hc\left(E_{up}^{e'v'N'J'} - E_{down}^{e"v"N"J"}\right)}{4\pi}\,\Phi\left(\frac{N(T)\exp\left(-\frac{hcE_{up}^{e'v'N'J'}}{kT}\right)}{Q_{\text{int}}^{Z}(T)}\right)A_{e"v"}^{e'v'}S_{N"J"}^{N'J'}P_\lambda^{Voigt} \quad (2)$$

where the energies E^{evNJ} are expressed in cm^{-1}, N(T) is the total density of the molecule in m^{-3}, $Q_{\text{int}}^{Z}(T)$ is the internal partition function, Φ is the nuclear spin degeneracy of the emitting state, $A_{e"v"}^{e'v'}$ is the vibrational transition probability in s^{-1}, $S_{N"J'}^{N'J'}$ the Hönl-London factor corresponding to the rotational transition involved and P_λ^{Voigt} the Voigt line profile in m^{-1}. In the formalism of hund's case b, the numbers e, v, N and J corresponds respectively to electronic, vibrational, rotational and sub-rotational levels (spin splitting). Spectrums were computed using a line by line simulation spectrum method which consists in determining for each rotational transition involved the emitting density population, the line position, its integrated intensity and the line shape.

The line position was related (in cm^{-1}) to the energy difference between two rotational levels, which were calculated in the Klein-Dunham formalism (equilibrium spectroscopic constants for each electronic state). In the Born-Oppenheimer approximation, the expression of energy levels is given as follows (3):

$$E_{e,v,N,J} = \sum_{i,j} Y_{ij} (v + 1/2)^i [F(N)]^j \tag{3}$$

F(N)=N(N+1) being the rotational term dependency in the case of singlet transition. In the case of multiple transitions, a more complex expression must be found to describe spin multiplicity [42]. Lambda-doubling have been neglected for the Π states. All spectroscopic constants for each electronic states used for this work are referenced in *Table 1*.

Table 1. Molecular systems and spectroscopic constants used for each electronic state

Mol	States	Band system	Transitions	Spectroscopic constants	Bands (v',v'')	Number of rotational lines
O_2	$X^3\Sigma_g^-$ $B^3\Sigma_u^-$	Schumann-Runge	$B^3\Sigma_u^- - X^3\Sigma_g^-$	[34] [43]	(19,21)	200.000
CO	$X^1\Sigma^+$ $A^1\Pi$	Infrared 4th positive	$X^1\Sigma^+ - X^1\Sigma^+$ $A^1\Pi - X^1\Sigma^+$	[44] [45]	(30,29) (20,20)	200.000 110.000
C_2	$a^3\Pi_u$ $d^3\Pi_g$	Swan	$d^3\Pi_g - a^3\Pi_u$	[46] [47]	(20,20)	250.000

The integrated line intensity is proportional to the product of the vibrational transition probability (Einstein coefficients $A_{e''v''}^{e'v'}$) and the Hönl-London factors $S_{N''J''}^{N'J'}$ which define the relative rotational intensity between transitions allowed by the selection rules. Hönl-London factors used for this work are referenced in *Table 2*. All are normalized with the Whiting convention:

$$\sum_{J''} S_{N''J''}^{N'J'} = (2 - \delta_{0,\Lambda'}\delta_{0,\Lambda''})(2S'+1)(2N'+1) \tag{4}$$

Some Einstein coefficients and absorption oscillator strengths are available in the literature and were deduced from calculations or experiment measurements. Unfortunately, the comparison of the references showed some discrepancies at rather high vibrational levels, except for the case of C_2 Swan system. Consequently, we decided to use the published values for C2 Swan system and to set up the classical method of calculation vibrational transition probabilities for the electronic systems IR and 4th positive of CO, and for the Schumann-Runge system of O_2. We used the well-known RKR method [47] in the formalism of Zare [48], which consists in the reconstruction of the internuclear potentials by inversion of the equilibrium spectroscopic constants for each electronic state. Out of the experimental range of internuclear distance validity, we extrapolated the RKR potential with a repulsive potential $V_{rep}(r) = ar^{-b}$ at low distance, and with a Hulburt-Hirschfelder (HH) at large distance. Then, this potential was injected in the radial Schrödinger equation to calculate the electronic wave functions using a Numerov iteration algorithm (two-step) by a method based on the works of Cooley [49]. This calculation was achieved for all couple (v',v'') up to the limit of dissociation (a cut off could be done when the high vibrational levels have no significant interest for this work). Knowing the rotationless vibrational wave functions for all upper vibrational levels $\psi_{v'}(r)$ and all lower $\psi_{v''}(r)$ vibrational, the vibrational transition probability between two vibrational states is obtained by the formula:

$$A_{e''v''}^{e'v'} = 2.026.10^{-6}(\sigma_{v'v''})^3 \left(\frac{2-\delta_{0,\Lambda'+\Lambda''}}{2-\delta_{0,\Lambda'}}\right)\left[\int_0^{\infty}\psi_{v'}(r)R_e(r)\psi_{v''}(r)dr\right]^2 \quad (5)$$

The electronic transition dipole moments $R_e(r)$ used in the calculation of each system are referenced in *Table 2*. Einstein coefficients obtained were validated by comparison with experimental lifetimes or partial Einstein coefficients of absorption oscillator strength published in the literature. Furthermore, we obtained very accurate results for vibrational energy levels eigenvalues by our method compared to the experimental energy obtained with the Klein-Dunham formalism, which tends to valid the wave functions and the RKR potentials obtained.

Finally, ***the line shape*** of each line was calculated by a Voigt profile, taking into account the Doppler and the collisional broadenings. Doppler broadening for molecules obeys to the same formula as the atomic case [34]. For collisional broadening and species involved in this work, we did not find information in the literature with a sufficient pressure and temperature validity range. Consequently, we used systematically, for all the diatomic molecular bands, the semi-empirical approximation law for molecular collisional broadening HWHM (Half width at Half Maximum) $\gamma_c(T,P) = \gamma_{Traf} \cdot P(T_{ref}/T)^n$ extended to be pressure dependant. We used in a first approximation the order of magnitude defined by Breene [50] ($\gamma_{Tref} = 0.1$ and n=0.7) for the broadening parameters of all molecules.

Table 2. Molecular systems

Molecule	Band system	Einstein coefficient	Hönl-London factors	Dipole moment function Re(r)
O_2	Schumann-runge	This work	[52]	[55]
CO	Infrared	This work	[53]	[56]
	4th positive	This work	[53]	[57]
C_2	Swan	[51]	[54]	-

3.4.2. Linear triatomic contribution: CO_2

The emission of CO_2 infrared bands was obtained using two databases: CDSD-1000 for the temperature range 500-2000K and CDSD-4000 for the temperature range 2000-5000K [58]. Those databases give us the required parameters to compute the emission and the absorption spectra of CO_2. It contains for each rovibrational line: the band position, its integrated intensity (in cm^{-1} / ($molecule.cm^{-2}$) at 296 K), the rovibrational transition probability in s^{-1}, the energy of the lower state, and the broadenings parameters. The CDSD-4000 is not convenient to be used in routine because it contains 628.324.454 integrated rovibrational line entries, which require at least a thousand billions of points to compute. Therefore, we decided to set up partial databases by steps of 500K to keep only lines contributing significantly to the radiation in this range of temperature. It can be reduced to around 15.000.000 integrated lines for each temperature, leading to a difference of more than 0.1% of the total emission.

4. The Net Emission Coefficient (NEC)

The plasma's radiation is obtained according to the Net Emission Coefficient (NEC) coefficient. It consists in calculating the net emission at the center of an isothermal and homogenous sphere of radius R_p:

$$\varepsilon_N(T,R_p) = \int_0^{\infty} L_\lambda^o(T)\cdot K_\lambda^{'}(T)\cdot \exp\left(-K_\lambda^{'}(T)\cdot R_p\right)\cdot d\lambda \quad (7)$$

where λ (m) is the wavelength, T and P the local temperature and pressure respectively, L_λ^0 ($W/m^2/sr/m$) the Planck function, R_p is the size of the sphere assimilated to the plasma's size and $K_\lambda^{'}$ (m^{-1}) the monochromatic absorption coefficient corrected by the induced emission and correlated with the local emission coefficient by the Kirchhoff law. Moreover, due to the exponential term, self-absorption of radiation in the emitting regions is rather well estimated. In numerical modelling, the NEC is a good method to characterize the radiation losses in hot regions while taking into account the absorption. Our last studies also showed that this coefficient depends not only on the temperature, the pressure and the plasma size but also on the presence of the molecules or metallic vapours for low temperatures. Indeed, *Figure 3* shows the NEC and the different contributions obtained in the case of pure CO_2 at atmospheric pressure and two plasma's sizes: optically thin plasma and 1mm of thickness.

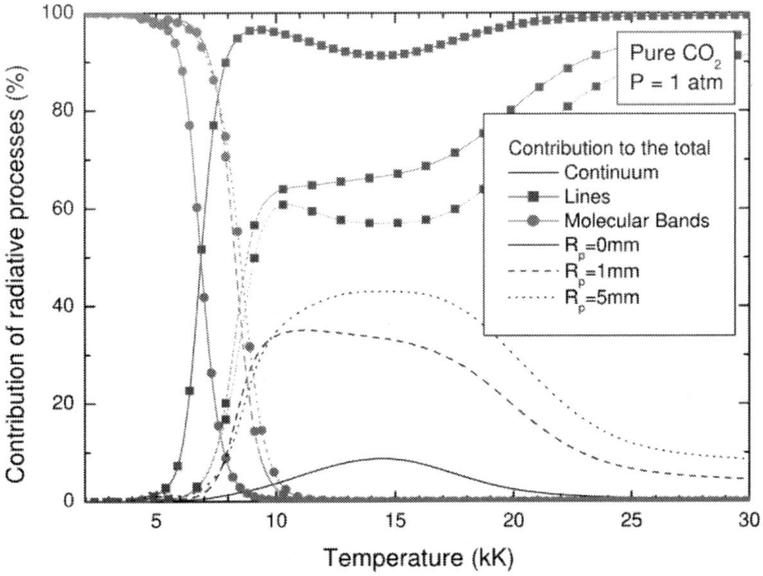

Figure 3. Contribution of the different radiative processes to the NEC for pure CO_2, three values of R_p and the atmospheric pressure.

For optically thin plasmas (R_p = 0mm), we observe that the radiation of the molecular bands represents the only contribution up to 5kK, and shares the radiation with the atomic lines up to 10kK with more than 50% at 6.9kK. For this size of plasma, the continuum is negligible. In the case of $R_p \neq 0$mm , the lines are strongly self-absorbed in the first millimeter which can explain the reason why their contributions decrease with only 60% at 10kK against 97% for R_p=0mm. The consequence of this absorption is that the contribution of the molecular bands, which are less absorbed, always plays an important role at 7.5kK and that the contribution of the continuum can represent up to 40% of the NEC between 10kK and 15kK.

Always in the case of pure CO_2, we compare on *Figure 4* the contribution of the 4 molecular systems (O_2, CO, CO_2 and C_2 molecules) to the total radiation of the molecular bands for pure CO_2 and two values of R_p. We observe that the radiation of the molecular bands is mainly due to the CO_2 system for temperatures lower than 3.5kK-4kK, CO IR for temperatures between 3.5kK-5kK, and 4th CO for higher temperatures. However, if the plasma's size increases, the contributions of CO_IR and 4th CO seem to decrease to the detriment of O_2 (SR) and CO_2. Nevertheless, if the plasma's size is inferior to 5mm, the molecular systems of C_2 (Swan) and O_2 (SR) do not play an important part in pure CO_2 radiation.

Finally, *Figure 5* compares the NEC of pure CO_2 plasma with the NEC of mixtures containing various concentrations of copper for R_p=5mm. We observe that a small concentration of metallic vapours (few percents) increases the radiation of the plasma for temperatures ranging from 5000K to 10000K. This increase is more visible for low sizes of plasma since the absorption minimizes the influence of the resonance lines of the metallic vapours species. At lower temperature, the increase of the plasma's size tends to diminish the important role of the metallic lines to the detriment of the molecular bands radiation which is less absorbed.

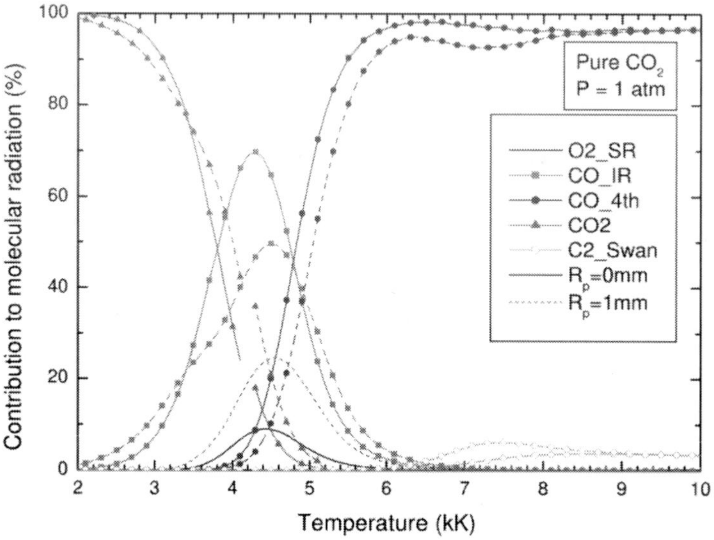

Figure 4. Contribution of the molecular systems to the total radiation of the molecular bands for pure CO_2, the atmospheric pressure and two values of R_p.

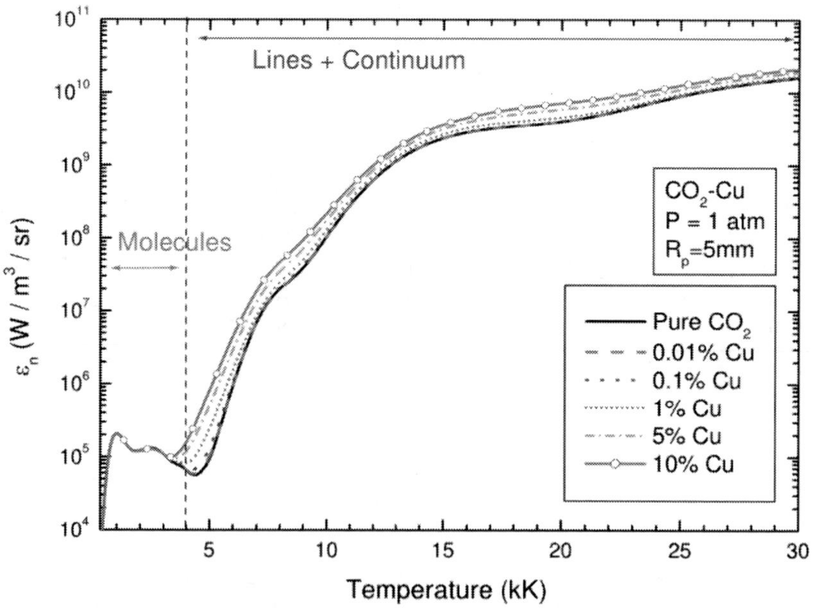

Figure 5. NEC for different concentration of metallic vapors in CO_2-Cu mixtures at R_p=5mm and atmospheric pressure.

5. Conclusions

This paper proposed a study of the radiative properties for CO_2-Cu thermal plasmas calculated according to the method of the Net Emission Coefficient, for temperatures between 500K and 30000K, and at the atmospheric pressure. The results showed that the NEC was mainly due to the emission of the molecular bands which should be considered in the NEC calculation for temperatures inferior to 4000K and all plasma sizes, then due to the emission of the atomic lines for higher temperatures and optically thin plasmas, and finally due to the emissions of the atomic lines and of the continuum for higher temperatures and higher plasma's sizes.

In pure CO_2, we also showed that C_2 emission was negligible whatever the plasma size and that CO and CO_2 emissions were the main contributions to the molecular bands emission for optically thin plasmas. For higher thickness of plasma R_p, the emission of CO became less important to the detriment of the O_2_SR emission. In CO_2-Cu mixtures, we highlighted that the continuum could represent up to 10% of the NEC for concentrations of vapours inferior to 10% and optically thin plasmas. We also put in evidence the decrease of the molecular bands emission with the increase of the metal vapours in the plasma. When the thickness of plasma increases, the results showed that the molecular bands and the continuum contributed more significantly to the NEC due to a strong self-absorption of the emission of the resonance atomic lines. In this case, the continuum could represent up to 45% of the NEC for the intermediate temperatures.

6.References

[1] Boretskij V F, Cressault Y, Veklich A N, Teulet Ph 2011 Plasma of electric arc discharge in carbon dioxide with copper vapours, *XIXth Symposium on Physics of Switching Arc,* Brno (Czech Republic)

[2] Cressault Y, Teulet Ph, Gonzalez J J, Gleizes A, Robin-Jouan Ph 2005 Transport and radiative properties of CO_2 arc plasma: application for circuit-breaker modelling, *XVIth Symposium on Physics of Switching Arc*, Brno (Czech Republic), 1, 46

[3] Gleizes A, Gonzalez J J, Liani B and Rahmani B 1990 Calculation of the net emission coefficient in Ar-Cu and SF_6-Cu thermal plasmas, *Colloque de physique* **C5 Suppl.18** tome 51 C5-213

[4] Gleizes A, Gonzalez J J, Liani B and Raynal G 1993 *J. Phys. D: Appl. Phys.* **26** 1921

[5] Menart J, Heberlein J and Pfender E 1996 *Plasma Chem. Plasma Proc.* **16** Suppl.1 S245

[6] Cressault Y 2001 *Propriétés des plasmas thermiques dans des mélanges Argon-Hydrogène-Cuivre*, Thesis, University Paul Sabatier, Toulouse (France)

[7] Raynal G, Vergne P J and Gleizes A 1995 *J. Phys. D: Appl. Phys.* **28** 508

[8] Cressault Y, Hannachi R, Teulet Ph, Gleizes A, Gonnet J P and Battandier J Y 2008 *Plasma Sources Sci. Technol.* **17** 035016

[9] Aubrecht V, Bartlova M and Coufal O 2010 *J. Phys. D: Appl. Phys.* **43** 434007

[10] Lowke J J 1974 *J.Q.S.R.T,* **14**, 2, 111

[11] Godin D and Trépanier J Y 2004 *Plasma Chem. Plasma Proc.* **24** 447

[12] NIST http://physics.nist.gov/PhysRefData/ASD/levels form.html

[13] Drawin H W and Felenbok P 1965 *Data for plasma in Local Thermodynamic Equilibrium* Ed. Gauthier-Villars, Paris

[14] Chase Jr M W, Davies C A, Downey Jr J R, Frurip D J, McDonald R A and Syverud A N 1985 *J. Phys. Chem. Ref. Data, JANAF Thermochemical Tables*, 3rd Edition, vol.**14**, supplement n°1

[15] Huber K P and Herzberg G 1978 *Molecular spectra and molecular structure, IV. Constants of diatomic molecules* (Van Nostrand Reinhold, New-York)

[16] Hirschfelder J O, Curtis C.F and Bird RB 1964 *Molecular theory of gases and liquids* (2nd Edition, John Wiley and Sons, New York)

[17] Boulos M I, Fauchais P and Pfender E 1994 *Thermal Plasmas,Fundamentals and Applications* **Vol.1** (New York, Plenum press)

[18] Robinson E J, Geltman S 1967 *Physical Review,* **153(1)**, 4

[19] D'Yachkov L G, Golubev O A, Kobzev G A, Vargin A N 1978 *J.Q.S.R.T.* **20** 175

[20] Schultz-Gulde E, Gilde J 1980 *J. Phys. D: Appl. Phys.* **13** 793

[21] Stewart J C and Rotenberg M 1965 *Physical Review* **140**(5A) 1508

[22] Biberman L M and Norman G E 1960 *Optics and Spectroscopy* **8** 230
[23] Biberman L M, Norman G E and Ulyanov K N 1961 *Optics and Spectroscopy* **10**(5) 297
[24] Hofsaess D 1978 *J.Q.S.R.T* **19** 339
[25] Okuda S, Ueda Y, Murai Y, Miyamoto T, Doi Y and Uenosono C 1980 *IEEE Transactions on Plasma Science* **PS-8**(4) 395
[26] Moore C E 1949 *Atomic energy levels, Vol I.* (Washington, Circular of the National Bureau of Standards)
[27] Ralchenko Y, Kramida A E, Reader J and NIST ASD team (2008) 2009 *NIST Atomic spectra database* (version 3.1.5), available online at http://physics.nist.gov/asd3, National Institute of Standards and Technology, Gaithersburg, MD
[28] Kurucz R L and Peytremann E 1995 *Atomic line data* (Kurucz R L and Bell B) Kurucz CD-ROM N0.23, Cambridge, Mass: Smithsonian Astrophysical Observatory, available online at http://cfa-www.harvard.edu/amp/ampdata/kurucz23/sekur.html
[29] Neynaber R H, Marino L L, Rothe E W and Trujillo S M 1961 *Phys. Rev.* **124** 1 135
[30] Trajmar S, Williams W and Srivastava S K 1977 *Journal of Physics B: Atomic, Molecular and Optical Physics* **10** 3323
[31] Mayol R and Salvat F 1997 *Atomic Data and Nuclear Data Tables* **65** 56
[32] Msezane A Z and Henry R J 1986 Electron-impact excitation of atomic copper, *Phys. Rev. A* **33**(3) 1631
[33] Traving G 1968 *Interpretation of line broadening and line shift in Plasma diagnostics* (North-Holland, W. Lochte-Holtgreven, 66-134)
[34] Laux C O 1993 *Optical diagnostics and radiative emission of air plasmas* (Ph.D Thesis, HTGL report T-288, Stanford University, California)
[35] Ali A W and Griem H R 1965 *Physical Review* **140** 4A 1044
[36] Walkup R, Stewart B and Pritchard D E 1984 *Physical Review A* **29** 1 169
[37] Griem H R 1974 *Spectral line broadening by Plasmas* (New York, London, Academic press)
[38] Griem H R 1964 *Plasma spectroscopy* (New York, McGraw-Hill)
[39] Drawin H W and Emard F 1973 *Beitrag Plasmaphysik* **13** (3) 143
[40] Naghisadeh-Kashani Y, Cressault Y and Gleizes A 2002 *J. Phys. D: Appl. Phys.* **35** *2925*
[41] Babou Y, Rivière Ph, Perrin M Y and Soufiani A 2009 *J.Q.S.R.T.* **13** 3 143
[42] Rouille G, Millot G, Saint-Loup R and Berger H 1992 *J. of Molecular Spectroscopy* **154** 372
[43] Cheung A S C, Yoshino K, Parkinson W H and Freeman D E 1986 *J. Mol. Spectrosc.* **119** 1
[44] Farrenq R, Guelachvili G, Sauval J, Grevesse N and Farmer C B 1991 *J. Mol. Spectrosc.* **149** 375
[45] Simmons J D, Bass A M and Tilford S G 1969 *The Astrophysical Journal* **155** 345
[46] Prasad C V V and Bernath P F 1994 *The Astrophysical Journal* **426** 812
[47] Rydberg R 1931 *Ann. Physik* **73** 376, Klein O *Z. Physik* **76** 226, Rees A L G 1947 *Proc. Phys. Soc.* **A59** 998
[48] Zare R N 1964 *J. Chem. Phys.* **40** 7 1934
[49] Cooley J W 1961 *Math. Computation* **15** 76 363
[50] Breene R G 1967 Applied Optics **6** 141
[51] Lino Da Silva M and Dudeck M 2006 *J.Q.S.R.T* **102** 348
[52] Tatum 1966 *Canadian Journal of Physics* **44** 11 2944
[53] Hertzberg G 1950 *Molecular spectra and molecular structure I. Spectra of diatomic molecules* (New York, Van-Nostrand Company)
[54] Schadee A 1964 *Communication from the Observatory at Utrecht* B.A.N. **17** 5 331
[55] Friedman R S 1989 *JQSRT* **43** 3
[56] Langhoff S R and Bauschlicher C W 1995 *J. Chem. Phys.* **102** 13 5220
[57] Kirby K W and Cooper D L 1989 *J. Chem. Phys.* **90** 9
[58] Tashkun S A, Prevalov V I, Teffo J L, Bykov A D and Lavrentieva N N 2003 CDSD-1000, *the high temperature carbon dioxide spectroscopic databank, J.Q.S.R.T.* **82** (1-4) 165-198; CDSD-4000, *J.Q.S.R.T.* **112** (9) 1403–1410

Journal of Physics: Conference Series **406** (2012) 012028 doi:10.1088/1742-6596/406/1/012028

The investigation of an electric arc in the long cylindrical channel of the powerful high-voltage AC plasma torch

Ph G Rutberg[1], S D Popov [1,2], A V Surov[1,2], E O Serba[1], Gh V Nakonechny[1], V A Spodobin[1], A V Pavlov[1], A V Surov[1]

[1] Institute for Electrophysics and Electric Power Russian Academy of Sciences, 18 Dvortsovaya emb., 191186 St.-Petersburg, Russia

[2] St.Petersburg State Polytechnical University, St.Petersburg, Polytechnicheskaya, 29., 195251, St.-Petersburg, Russia

E-mail: sergey_popov1973@mail.ru, rc@iperas.nw.ru, alex_surov@mail.ru

Abstract. The comparison of conductivity obtained in experiments with calculated values is made in this paper. Powerful stationary plasma torches with prolonged period of continuous work are popular for modern plasmachemical applications. The maximum electrode lifetime with the minimum erosion can be reached while working on rather low currents. Meanwhile it is required to provide voltage arc drop for the high power achievement. Electric field strength in the arc column of the high-voltage plasma torch, using air as a plasma-forming gas, does not exceed 15 V/cm. It is possible to obtain the high voltage drop in the long arc stabilized in the channel by the intensive gas flow under given conditions. Models of high voltage plasma torches with rod electrodes with power up to 50 kW have been developed and investigated. The plasma torch arcs are burning in cylindrical channels. Present investigations are directed at studying the possibility of developing long arc plasma torches with higher power. The advantage of AC power supplies usage is the possibility of the loss minimization due to the reactive power compensation. The theoretical maximum of voltage arc drop for power supplies with inductive current limitations is about 50 % of the no-load voltage for a single-phase circuit and about 30 % for the three-phase circuit. Burning of intensively blown arcs in the long cylindrical channel using the AC power supply with 10 kV no-load voltage is experimentally investigated in the work. Voltage drops close to the maximum possible had been reached in the examined arcs in single-phase and three-phase modes. Operating parameters for single-phase mode were: current - 30 A, voltage drop - 5 kV, air flow rate 35 g/s; for three-phase mode: current (40- 85) A, voltage drop (2.5-3.2) kV, air flow rate (60-100) g/s. Arc length in the installations exceeded 2 m.

1. Introduction

Global population and economic growth is accompanied by growth of energy consumption and waste cumulative weight. Therefore, there is an urgent need to develop new waste treatment technologies for electricity generation.

Promising methods for the solution of this challenge are plasma gasification and conversion [1-4] used for syngas production from organic, woodworking industry and agricultural wastes. Construction of powerful electric arc plasma torches with low operational costs is required for the development of mentioned above technologies. Usage of a multiphase (three-phase) AC system comprising AC arcs used as a basis of an electric arc plasma torch enables creation of a simple, low-cost and reliable

Journal of Physics: Conference Series **406** (2012) 012028 doi:10.1088/1742-6596/406/1/012028

power supply system. Air usage as a plasma forming gas dramatically decreases operational costs of the electric arc plasma system.

IEE RAS (Institute for Electrophysics and Electric Power Russian Academy of Sciences) has an extensive experience in development of powerful multiphase electric arc AC plasma torches. Powerful single-chamber electric arc plasma torches with rail electrodes and power up to 500 kW have been created (figure 1) [5, 6].

Figure 1. Single-chamber electric arc plasma torches with rail electrodes Power 100 – 500 kW, supply voltage 480 V, arc current 500 – 1500 A, air flow rate 15 – 70 g/s.

The power increase of such plasma torches is attained by increase of arc current at sufficiently low voltage drop (about 200 V). However, the lifetime of electrode units of these plasma torches is 100 hours that is not enough for some plasmachemical applications. Power of the plasma torch can be increased without its resource indexes degradation by increasing of the arc voltage drop instead of the arc current. AC plasma torches with arcs burning in long cylindrical channels have been developed in IEE RAS (figure 2) [7-12].

Figure 2. High-voltage electric arc AC plasma torch with rod electrodes. Supply voltage 6 – 10 kV, arc voltage drop 900-1200 V, arc current up to 35 A, plasma forming gas: air, argon, carbon dioxide, gas flow rate from 1 to 30 g/s.

In electric arc plasma torches of such type the power variation is controlled not by current change but by arc voltage drop change, which in turn depends on the change of plasma forming gas flow rate (see figure 3).

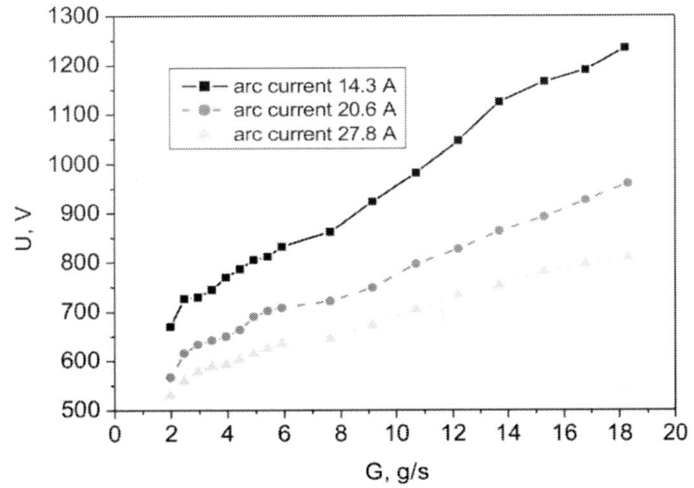

Figure 3. Dependence of variation of arc voltage drop versus plasma forming gas flow rate for various values of power supply source current. Plasma gas is air.

Thus, it is necessary to search for achievement of theoretically possible values of voltage drop for the used power supply. The possible arc voltage drop is 53 % from the idle-running power supply for a single-phase AC arc, and for a three-phase electric arc system the theoretically possible voltage drop is 33 - 35 %. What at the idle-running power supply of 10000 V can exceed 3000 V. At operation of the plasma torch on air at atmospheric pressure the strength in the arc lays in the range 10 - 15 V/cm, and the arc length, voltage drop in which is about 3000 V, should exceed 200 cm.

At present IEE RAS has achieved stable operation of the experimental three-phase AC plasma torch with arcs burning in long cylindrical channels. Figure 4 shows arc voltage drop versus flow rate of the plasma forming gas.

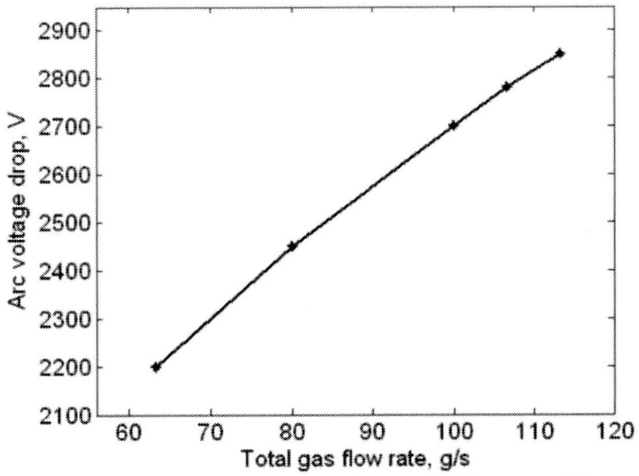

Figure 4. Dependence of voltage drop from gas flow rate. Plasma gas is air.

Thus the length of an arc remains constant and accordingly cannot influence the change of the electrostatic strength. The electric field strength is connected with the current density and conductivity by equation (1).

$$\sigma = \frac{j}{E} = \frac{e^2 n_e}{2\sqrt{3m_e kT}\left(\sum_i n_i s_i + n_{ion} s_{ion}\right)} \tag{1}$$

where, σ – conductivity, j – current density (A/m^2), E – electrostatic strength (V/m), n_e – electron concentration (1/m^3), e and m_e – electron charge and mass (C) and (kg) respectively, k – Boltzmann (J/K), n_i and s_i – concentration and cross-section of neutral plasma components scattering (1/m^3) and (m^2) respectively, s_{ion} – cross-section of Coulomb scattering (m^2), n_{ion} – ion concentration (1/m^3).

2. Results of numerical simulation

Model is considered one of two channels of high-voltage AC plasma torch (one of three channels for three-phase plasma torch). Topology of the plasma torch is such that the separated channels have the same direction at a slight angle to each other. Each channel includes a hollow cylindrical electrode, loops of the tangential feeding of the plasma forming gas and the outlet nozzle. The gas flow is the same in all channels.

As the arc length remains approximately constant it is possible to draw a conclusion that with change of flow rate ratios of the arc conductivity or current density varies. Hence, the arc voltage drop changes owing to the change of thermo-and gas dynamic processes (regimes) of gas flow in the channel. It is required to investigate the change of a gas flow in the channel depending on magnitude of a total gas flow rate. A cold blasting can be considered for this purpose as a first approximation and computing experiments are carried out. The investigation of gas flow in the channel was solved in three-dimension stationary problem statement. The approach developed for high-speed viscous compressible flows was chosen. Besides Navier-Stokes and continuity equations, the system is supplemented with the equation of energy and a set of the empirical constants accompanying connection of a turbulence model (k-ε). The fields of thermo and gasdynamic parameters distribution have been derived during computing experiments.

The grid containing $2{\cdot}10^6$ meshes has been built in the independent grid generator for the solution of this problem. It has been thickened in the injection area for the correct solution.

It is necessary to carry out preliminary theoretical calculation of gas-dynamic parameters to setup parameters at the input rated operating conditions. Figure 5 shows the set of cross-sections transverse to the channel axis where visualization of flow was performed.

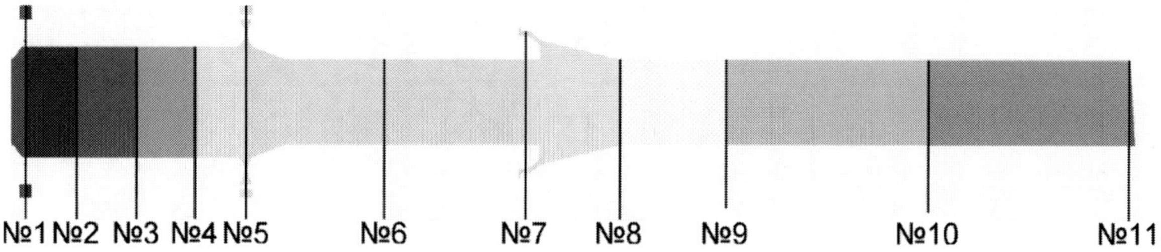

Figure 5. Layout of cross-sections.

The computing experiment has confirmed presence of a wall ring zone with the decreased to speed values (~10 m/s), which size and position depends on magnitude of gas flow rate and its distributions between supply loops. Figure 6(a) shows the field of module of the rotational component of the velocity vector for the total flow rate of supplied gas 68 g/s. The lines of gas flow are twirled on a spiral in the vicinity of the axis of symmetry, values of modules of a speed vector and its rotational components are close to zero. Figure 7 shows that with increase in the total magnitude of the flow rate of the supplied gas this area of the decreased values is constricted to the channel axis of symmetry. Thus the wall current speeds in the vicinity of the cross-sections noted by numbers 6-11 in figure 5, increase. The process is accompanied by rise of disturbances transferred by a current upwards the stream from the third injection orifice (figure 6(h)) and significant change of the flow pattern in the nozzle area.

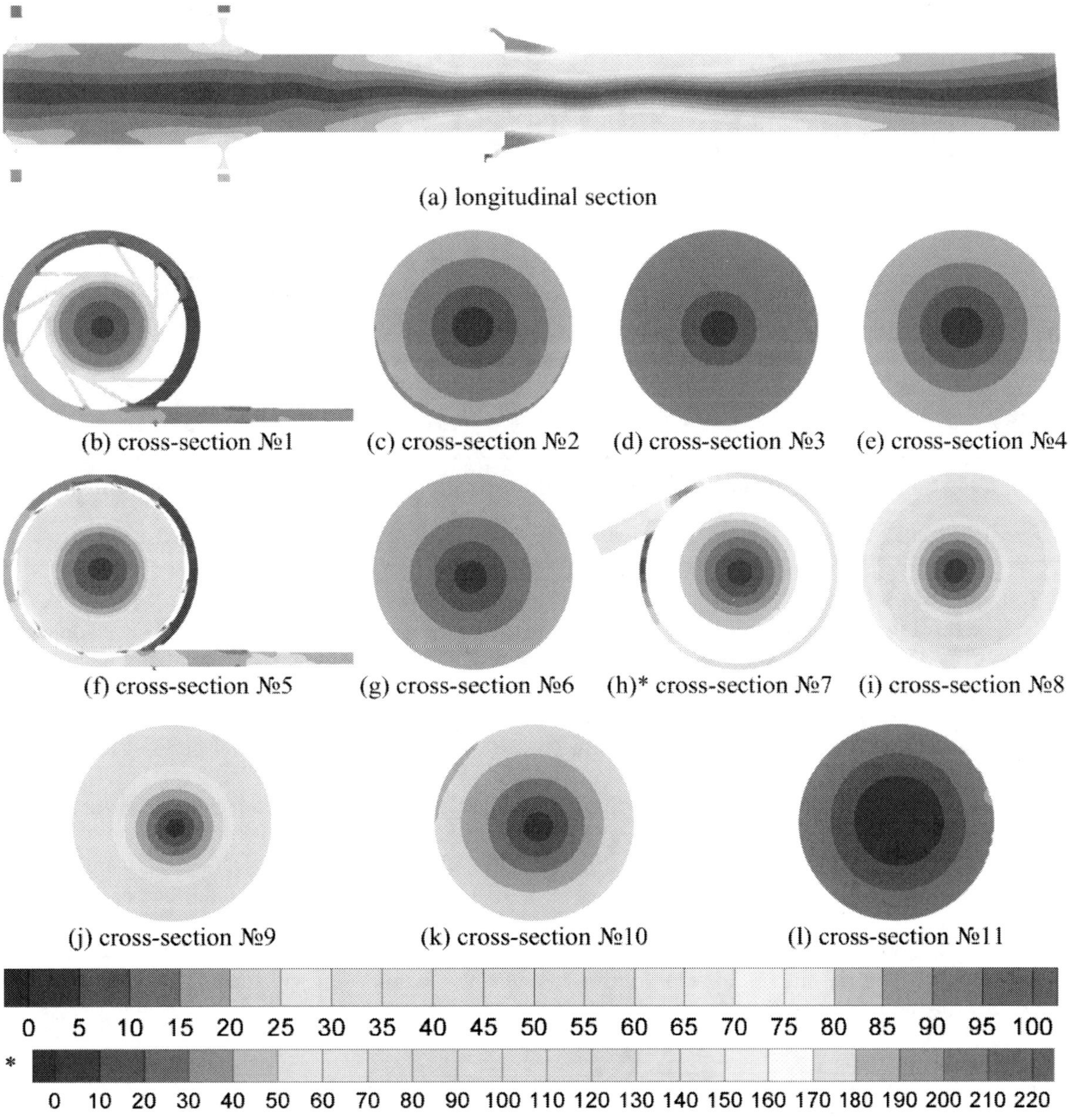

Figure 6. Field of module of the rotational component of the velocity vector for air flow rate 68 g/s.

In the cross-section area noted by number seven (figure 5) the flow acceleration up to speeds varying on the module over the range ~200÷350 m/s depending on magnitude of gas flow rate is observed.

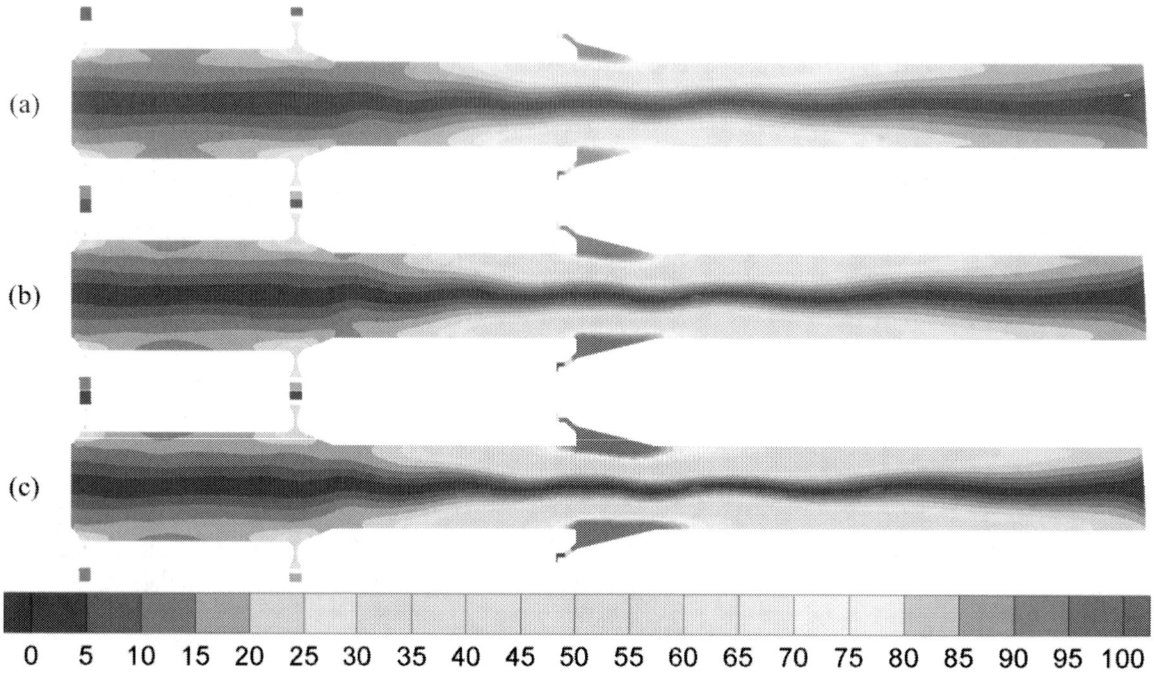

Figure 7. Comparison of field of module of the rotational component of the velocity vector for air flow rates (a) 68 g/s, (b) 86 g/s, (c) 107.5 g/s.

3. Conclusion

The third injection orifice has the greatest influence on change of the stream speed. The sonic flow caused by the channel geometry is realized in its area, at high values of the total flow rate of the supplied gas. In this area there is an essential reduction of temperature and pressure due to flow acceleration. Analogue of such flow is a classical gasdynamic flow in the confuser. The greatest gradients of gas- and thermodynamic parameters are observed in the channel cross-sections in the areas immediately adjacent to the injection orifices.

The increase of wall velocity at growth of a total gas flow rate complicates the electric arc closing on the case, and also promotes its stabilization along the symmetry axis.

The calculations show that at high flow rates the flow is more compressed and arc-gas heat exchange is more intensive. As a consequence, it first of all can influence upon increase in current density.

The obtained results represent the first stage of creation of a virtual model of the electric arc interaction with a gas stream. In what follows it is planned to enter "energy supply" into this model. At present the mathematical model of an arc and its interaction with a stream is under development.

4. Acnowledgements

The work is supported by Russian Academy of Sciences and also RFBR grants 11-08-01022-a and 12-08-00916-a.

References

[1] Bratsev A N, Kuznetsov V A, Popov V E, Rutberg A Ph, Ufimtsev A A and Shtengel S V 2009 *High temperature material processes: An international journal* **13(2)** 147-54

[2] Rutberg Ph G, Bratsev A N, Kuznetsov V A, Popov V E, Ufimtsev A A and Shtengel' S V 2011 *Biomass and Bioenergy* **35(1)** 495-504

[3] Bratsev A N, Kuznetsov V A, Popov V E and Ufimtsev A A 2011 *High Temperature* **49(2)** 244-8

[4] Bratsev A N, Kuznetsov V A, Popov V E, Ufimtsev A A and Shtengel 2009 *High temperature material processes: An international journal* **13(2)**, 241-6

[5] Rutberg Ph G, Safronov A A, Popov S D, Surov AV and Nakonechny Gh V 2005 *Plasma Phys. Control. Fusion* **47** 1681-96

[6] Popov S D, Rutberg A F and Safronov A A 2007 *High Temperature* **45(1)** 1-6

[7] Rutberg Ph G, Gorbunov V P, Nakonechny Gh V, Popov S D, Safronov A A, Serba E O and Surov A V 2006 *V International Conference Plasma Physics and Plasma Technology (PPPT-5), contributed papers* (Minsk, Belarus, 18-22 September) vol. 1 pp. 110-113

[8] Rutberg Ph G, Kumkova I I, Kuznetsov V E, Popov S D, Rutberg A P, Safronov A A, Shiryaev V N and Surov A V 2007 *IEEE Pulsed Power Conference (PPPS-2007) Digest of Technical Papers 1976-2007* (Albuquerque, New Mexico, USA, 17-20 June) Omnipress, Madison p. 1556-1559

[9] Rutberg Ph G, Gorbunov V P, Kuschev S A, Lukyanov S A, Nakonechny Gh V, Popov S D, Popov V E, Spodobin V A, and Serba E O 2007 *IEEE Pulsed Power Conference (PPPS-2007) Digest of Technical Papers 1976-2007* (Albuquerque, New Mexico, USA, 17-20 June) Omnipress, Madison p. 1560-1564

[10] Rutberg Ph G, Popov S D, Safronov A A, Serba E O and Nakonechny Gh V 2007 *XXVIII Internatuional Conference on Phenomena in Ionized Gases (ICPIG 2007) Proceedings* (Prague, Czech Republic, 15-20 July) Published by Institute of Plasma Physics AS CR, v.v.i. pp. 1762-1765

[11] Rutberg Ph G, Safronov A A, Surov A V and Popov S D 2007 *XXVIII Internatuional Conference on Phenomena in Ionized Gases (ICPIG 2007) Proceedings* (Prague, Czech Republic, 15-20 July) Published by Institute of Plasma Physics AS CR, v.v.i. pp. 1774-1777

[12] Rutberg Ph G, Lukyanov S A, Kiselev A A, Kuschev S A, Nakonechny Gh V, Nikonov A V, Popov S D, Serba E O, Spodobin V A and Surov A V 2011 *Journal of Physics: Conference Series* **275** 012006

A novel 2.45 GHz/200 W Microwave Plasma Jet for High Temperature Applications above 3600 K

Christoph Schopp[1], Frederik Nachtrodt[2,3], Holger Heuermann[1], Ulrich W. Scherer[1], Domiziano Mostacci[3], Torsten Finger[1], Wolfgang Tietsch[2]

[1]FH Aachen – University of Applied Sciences
[2]Westinghouse Electric Germany GmbH
[3]University of Bologna

christoph.schopp@gmail.com

Abstract. State of the art atmosphere plasma sources are operated with frequencies in kHz/MHz regions and all high power plasma jets make use of tungsten electrodes. A microwave plasma torch has been developed at FH Aachen for the application in various fields. The advantages over other plasma jet technologies are the high efficiency combined with a maintenance-free compact design and non-tungsten electrodes. In this paper the development of a 200 W torch is described. Argon is used as the primary plasma gas and a second gas can be applied for additional purposes. For the plasma generation a microwave at 2.45 GHz is sent through the torch. The special internal topology causes a high electric field that ignites the plasma at the tip and leads to the ionization of the passing Argon atoms which are emitted as a jet. By designing the copper electrode as a cannula it is possible to gain plasma temperatures higher than the electrode's melting point. The electric field simulations are made with Ansoft HFSS. Experiments were carried out to verify the simulations. The upcoming steps in the development will be the scale-up to higher power levels of several kW with a magnetron as power source.

1. Introduction

State of the art AC plasma sources are operated with frequencies in kHz/MHz regions and mostly contain tungsten electrodes [1]. By increasing the frequency the minimum voltage for ignition is reduced.

In this paper a novel microwave plasma torch is described which makes use of a bi-static matching in addition to an inner topology that allows a single frequency ignition and operation. This makes it possible to apply fixed frequency power supplies, such as magnetrons, which is an important step to achieve high power levels.

2. Development of the microwave plasma jet

For the development of a plasma jet the two different states, ignition and operation (bi-static), need to be considered.

The ignition state requires a maximum electric field strength at the tip of the electrode exceeding the avalanche voltage of the process gas [2]. In contrast to this the operation state demands getting as much power as possible into the plasma after it has been ignited.

2.1. First plasma jets

The first plasma jet designed enabled the tuning of both states for their optimal performance [3]. The torch was able to generate a plasma with a constant input power of 200 W. The drawback of this concept was the presence of two different frequencies for the respective states. Therefore this plasma jet could not be used with a magnetron. To amend this, a new topology is developed which is shown in Figure 1.

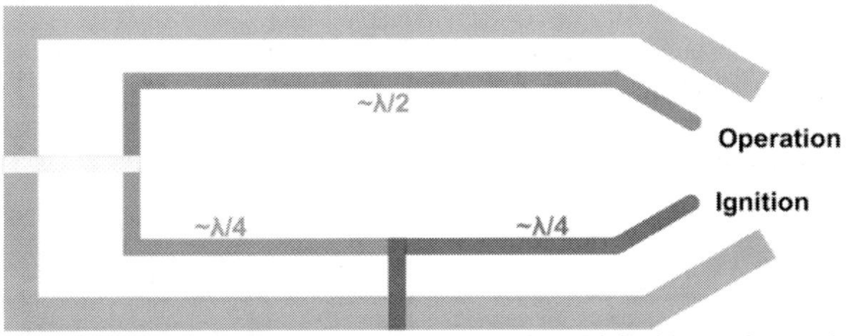

Figure 1. Inner topology of the new plasma jet with ignition and operation path

Figure 1 displays the inner structure of the so-called two-way plasma jet whose development can be found in [3].

The two-way plasma jet includes two separate electrodes for ignition and operation, which are ideally independent. Argon is flooded into the setting. Inner conductors are solid brass wire.

Ignition is realized by the help of the electrode creating a high electric field at the tip of the torch. After ignition the electrical properties are changed due to the presence of the plasma stream, which forms a conducting body itself. This behavior can be explained by comparing the plasma with a transistor switching from the off to the on state.

First tests showed successful ignition and operation at a single frequency so that proof of concept was achieved.

2.2. Cannula-electrode plasma jet

To improve the system performance the operating electrode was replaced with a cannula carrying a process gas, allowing a better cooling and a centric exit of the gas flow. In Figure 2 the new configuration for the head can be seen.

A second stream can be applied to flood the torch with another gas to streamline the plasma and add additional cooling to the case.

3. Electric field simulations

The electrical properties of the plasma jet can be simulated and described in detail [4]. Simulations are performed with Agilent ADS by designing the equivalent circuit that gives a principle idea of the torch's dimensions. These dimensions serve as the starting point for an EM-simulation in Ansoft HFSS and the final simulation results are illustrated in the next figure with an input power of 50 W.

Journal of Physics: Conference Series **406** (2012) 012029 doi:10.1088/1742-6596/406/1/012029

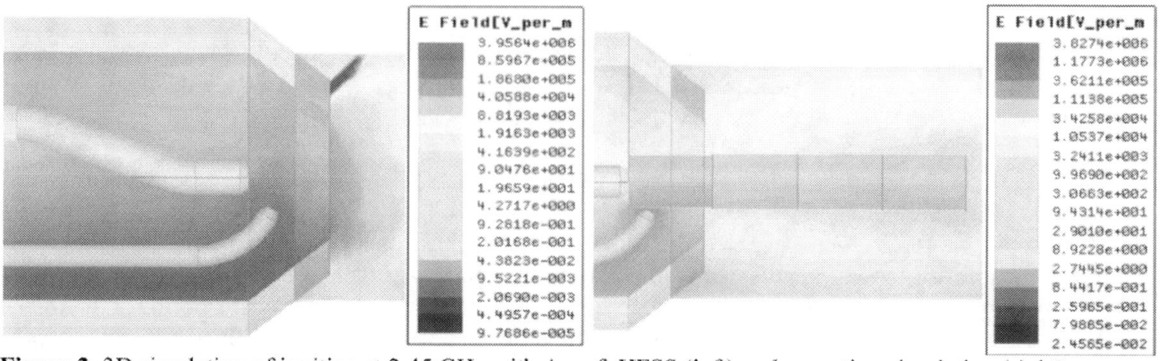

Figure 2. 3D simulation of ignition at 2.45 GHz with Ansoft HFSS (left) and operation simulation (right)

Figure 2 shows the 3D-EM-simulation of the new configuration. On the left the ignition state with a maximum electric field strength of about 1e6 V/m is illustrated. The right part of Figure 2 presents the operation state including an approximated plasma jet which is represented by four cylindrical bodies with different electrical conductivity. In the next figures the results of the HFSS simulation are plotted.

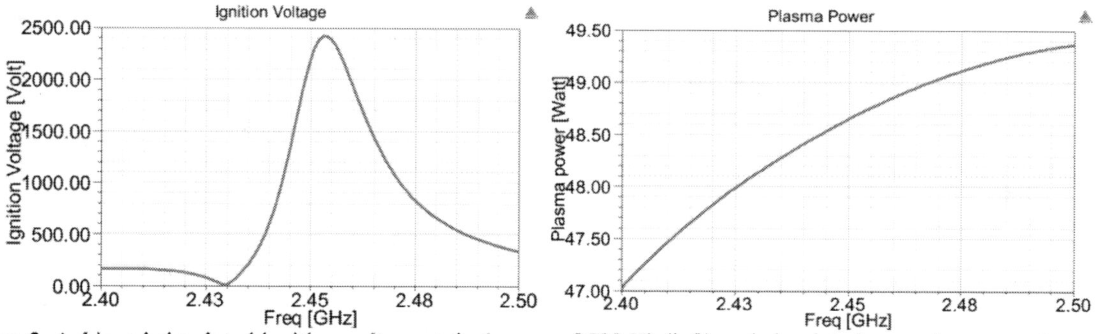

Figure 3. Achieved simulated ignition voltage at tip (approx. 2500 V) (left) and simulated operation power in plasma (min. power efficiency 93%) (right)

On the left side of Figure 3 the resulting ignition voltage of the bottom electrode against the outer case is presented with a maximum value of 2500 V. The right plot shows the operation power with a minimum efficiency of about 93%. This value is retrieved by calculating the volume density loss inside the cylindrical bodies.

The characterization of the plasma in terms of ion and electron temperature and density is being performed. Temperatures must be above 3700 K, since it is possible to melt tungsten wire (melting point 3695 K) which is shown in Figure 4. Located in the center of the plasma is the peak temperature with a high gradient. The plasma works in an atmospheric environment and runs with an Argon flow.

Figure 4. Cutting tungsten wire with a 200 W plasma torch

4. Applications

Figure 5. Self-ignited Argon plasma of first prototype (200 W input power), ignition electrode does not influence the plasma stream, experiments confirm simulation results

In Figure 5 a prototype with a self-ignited Argon plasma is shown with 200 W CW input power. An Argon stream of 0.1 l/min is run through the cannula and the torch also has an outer air stream of 1 l/min. The shown plasma consists only of the ionized Argon atoms and is additionally focused by the second gas. Experiments show a plasma ignition at about 60 W of input power.

Figure 6. Melting of a metal plate

Tests are performed on raw surfaces with different results and effects on different materials. Input power is yet too low to melt down larger metal bulks, however thin foils of metal can be treated as intended. Organic materials combust instantly and non-organics melt. In Figure 6 the melting of a metal plate can be seen. In addition melting of non-metal solids like various salt crystals was possible, where the thermal conductivity is sufficiently low to prevent heat dissipation. A local melting of the material on the treatment spot is the result.

An oven is set-up with a 200W plasma torch as combustion source for demonstrating the principal possibility of treating low-level radioactive wastes [5]. Typical waste components and compositions are treated and results show an intended combined combustion and melting of organics and non-organics, respectively. One result is that the power is still too low to melt larger specimen, but the temperature is sufficient to treat all kinds of material. When applying the plasma on a mixture of typical low-level waste and glass particles, all substances melt together to form an amorphous substance, proving the possibility to vitrify material with this plasma application.

Journal of Physics: Conference Series **406** (2012) 012029 doi:10.1088/1742-6596/406/1/012029

5. Outlook

Figure 7. First prototype with a magnetron as power source and air plasma at 850 W continuous wave

A second and bigger prototype is built and connected to a magnetron.

Air is used as process gas and the resulting plasma with 850 W is shown in Figure 7. This proofs the ability of the new topology to endure high power and supports the intention to further scale up the input power to several kW allowing improvements and further applications like welding, cutting, air cleaning, surface treatment, and also melting.

6. Summary

In this paper the development of a 2.45 GHz microwave plasma torch is described. Changes in topology for single-frequency operation, simulation and test results of the first prototype are presented. The 200 W torch is successfully tested by melting and burning of different substances. Various applications needing a high temperature in an atmospheric environment can be carried out. A second, magnetron-driven prototype for high power applications was successfully built and tested.

References

[1] Tendero C, Tixier C, Tristant P, Desmaison J, Leprince P, 2006 Atmospheric pressure plasmas: A review *Spectrochimica Acta B* 61, 2-30

[2] Holtrup S, Heuermann H, 2009, Fundamentals and ignition of a microplasma at 2.45 GHz, *European Microwave Conference, Proceedings* pp.1607-1609, Sept. 29 -Oct. 1 2009

[3] Finger T 2011, Entwicklung einer neuartigen bistatischen 2.45 GHz Transformationsschaltung zur Plasmageneration, *Master's Thesis*, FH Aachen

[4] Schopp C 2012, Entwicklung eines Kanuelenplasmastrahlers bei 2,45 GHz mit CW-Bi-Static-Netzwerk, *Master's Thesis*, FH Aachen

[5] Nachtrodt F, Tietsch W, Mostacci D, Scherer U W 2011, Plasma torch treatment of low level waste, *Proceedings of the Waste Management, Decommissioning and Environmental Restoration for Canada's Nuclear Activities,* Sep 11 - Sep 14 2011, Toronto

Radiation of transient high-current arcs: energy measurements in the optical range

J M Bauchire[1], **D Hong**[1], **H Rabat**[1], **G Riquel**[2]

[1] GREMI, UMR 7344 CNRS/Université d'Orléans, 14 rue d'Issoudun, BP 6744, 45067 Orléans Cedex 2, France
[2] EDF R&D, avenue des Renardières – Ecuelles, 77818 Moret-sur-Loing Cedex, France

E-mail: jean-marc.bauchire@univ-orleans.fr

Abstract. When no protection is used, the radiation emitted by a high-power electric arc can be dangerous for the eyes and the skin of a person. To ensure effective protection, it is first necessary to know the energy emitted by such arcs. The aim of our work was to experimentally determine the energy emitted by high-current (from 4 to 40 kA) transient arcs, for two different (10 cm and 2 m) lengths and for electrodes in copper or steel. These experiments enabled the radiative energy of the arcs to be quantified and also showed the influence of metal vapors in the spectral distribution of the radiation.

1. Introduction

One of the main features of electrical arcs is their radiation: the higher the current, the higher the radiation emitted. During welding or cutting operations for example, operators are efficiently protected against this radiation by protective clothing. However, undesirable transient arcs can also occur, as in maintenance operations on the electrical supply network.

To be effectively protected against radiation, it is necessary to know the precise amount of energy radiated by such arcs [1-3]. As the arcs do not radiate simply as black bodies, the radiated energy needs to be measured in several spectral bands.

The objective of this study is to go some way to answering this question by measuring the radiated energy, by spectral bands, in two arc configurations, in the field of optical radiation ranging from ultraviolet to infrared via visible radiation. However, an electrical arc creates a cloud of smoke, usually from erosion of the electrodes and of the surrounding materials, and this cloud changes the radiative transfer, so accurate measurement is very difficult to obtain.

2. Experimental setup

This section describes the two arc configurations studied. They correspond respectively to a short 10 cm arc and a long 2 m arc. The power supply used in the creation of each of these arcs is also different.

2.1. *Short arc*

The experimental setup consists of two opposed round copper rod electrodes 18 mm in diameter (figure 1). The distance between the electrodes is 10 cm. The electrodes are initially connected by a wire 0.2 mm in diameter in order to ignite an arc in the air at the beginning of the electric current discharge when the discharge is higher than a threshold value. The power source supplies a sinusoidal positive half-cycle current of close to 10 ms, with a 10 kA peak. The discharge current and voltage were recorded for each run, and their product was integrated on the total arc duration to estimate the input electrical power of the discharge. The total electrical energy dissipated by the arc is about 15 kJ.

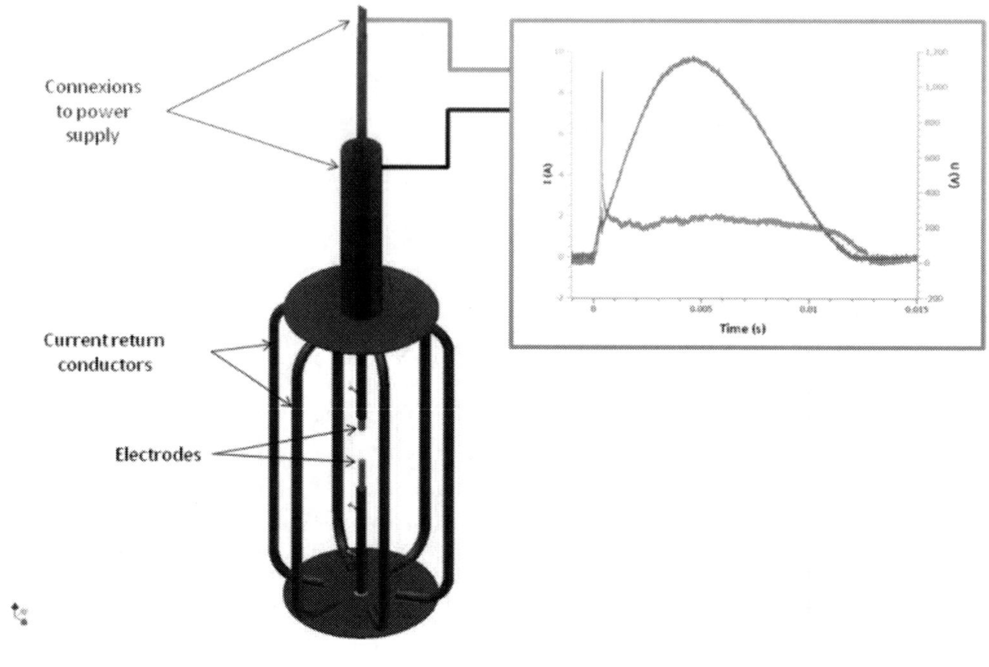

Figure 1. Device for the creation and study of the short arc. Typical electric current (green) and voltage (red) curves are shown.

2.2. *Long arc*

The device for the creation of the 2 m arc is more or less the same as the previous one (opposite electrodes, current return, etc.) but with larger dimensions. The power supply is different, since it allows the creation of alternating currents during 5 periods of 50 Hz and intensities ranging from 4kA to 40 kA rms.

The electrical energies are respectively about 0.75, 0.7, 3.1 and 5.5 MJ. In comparison with the previous configuration, the electrical energies are much higher, because of the duration of the discharge, the current intensities and the size of the arc.

2.3. *Energy measurements*

The radiated energy was measured with two joulemeters XPLF12, from Gentec Electro-Optics®.

In the short arc configuration, they were positioned 178 cm from the axis of the electrodes on the horizontal plane defined by the middle of the inter-electrode gap, and directed towards the center of the arc. In the second configuration, the joulemeters were positioned 9.4 m from the arc.

To obtain the energy distribution of spectral radiation, these detectors were equipped with filters in order to determine radiation in the UV, visible and IR bands for the two configurations. For the short arc configuration, it was possible to obtain more precise information on 6 spectral bands (2 UV, visible and 3 IR). Table 1 lists the standard boundaries of the spectral bands in the optical field of electromagnetic waves.

Journal of Physics: Conference Series **406** (2012) 012030 doi:10.1088/1742-6596/406/1/012030

Table 1. The 7 standard spectral bands of the optical range.

Spectral band	Lower wavelength (nm)	Upper wavelength (nm)
UVC	100	280
UVB	280	315
UVA	315	400
Visible	380	780
IRA	780	1400
IRB	1400	3000
IRC	3000	10^6

Note that the radiation below 190 nm cannot reach the sensor and that the spectral sensitivity of the sensor above 20000 nm is negligible. Thus in the results the spectral bands UVC and IRC are not completely measured.

3. Results and discussion

The contribution of each spectral UV, visible and IR band is shown in figure 2. UV is in the range 190-400 nm, visible in the range of 400 (in our case) to 780 nm, and the IR from 780 to 20000 nm. Whatever the nature of the electrodes, the largest proportion of radiation is UV and the lowest is visible radiation. As the arcs are created in air with fuse wire of the same nature as the electrodes, this radiation seems governed by metallic vapors coming from the electrodes rather than by the ambient gas.

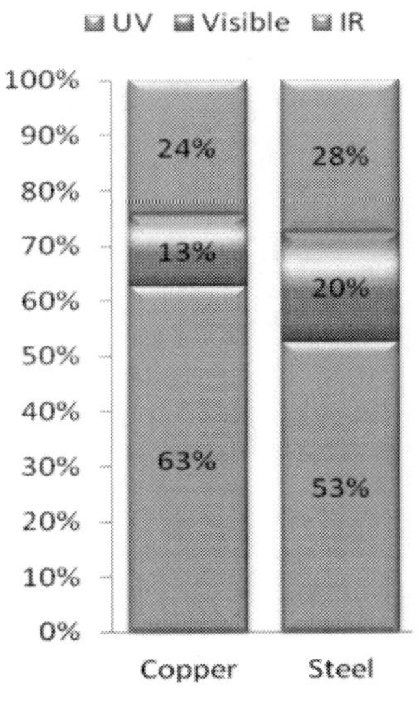

Figure 2. Spectral distribution of the measured radiated energy vs the type of electrodes, for the short arc.

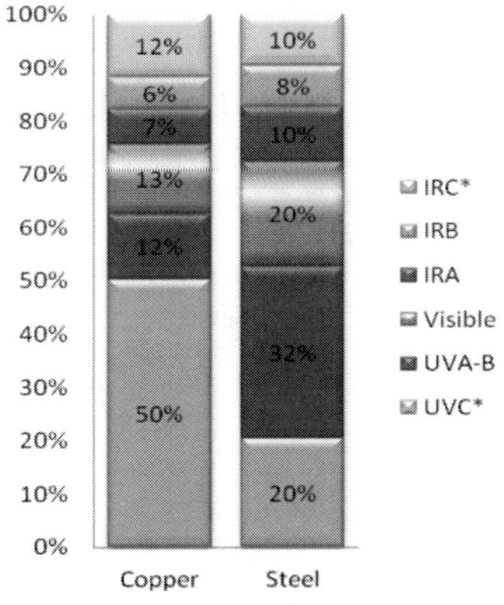

Figure 3. Detailed spectral distribution of the measured radiated energy vs the type of electrodes, for the short arc

Calculations [4] have shown that radiation from a plasma in air is mostly in the visible and the IRA spectral bands for a fairly wide temperature range. This confirms the important role of the metal vapors in our results.

The steel electrodes consist more than 97% of iron. In figure 3, which gives a more precise spectral distribution of radiation from the arc, we see that the essential difference between the two radiations is in the UVA-B and UVC* bands. * Remember that the 100-190 nm part of the UVC band is not measured, because this radiation is very quickly absorbed in air. The observed difference between the two types of plasma (copper and iron) comes from the spectral characteristics of the two elements, but also depends on their concentration in the plasma and on its temperature. The two parameters are linked since the plasma temperature tends to decrease when the proportion of metallic fumes increases.

It is admitted that the radiation of metallic vapors in a plasma is mostly in the UV range and that the contribution of UVC is substantial since the plasma is hot. This is in good agreement with our measurements in the case of copper electrodes. In the case of the steel electrodes the low proportion of UV and especially UVC is surprising. We can observe these trends provided that the proportion of iron in the plasma is significant (several %) and that the temperature is low enough (between 10 and 15 kK). However in this case the proportion of visible and IR should be lower. We observed [5] that the arc creates a cloud of smoke, composed of metallic vapor and droplets of molten metal. This cloud is dense and changes the radiative transfer around the arc. It is possible that part of the UV radiation is absorbed by this cloud. As the temperature of the latter does not exceed 1000 or 2000 K, its radiation will be rather in the near-infrared, which might explain our results.

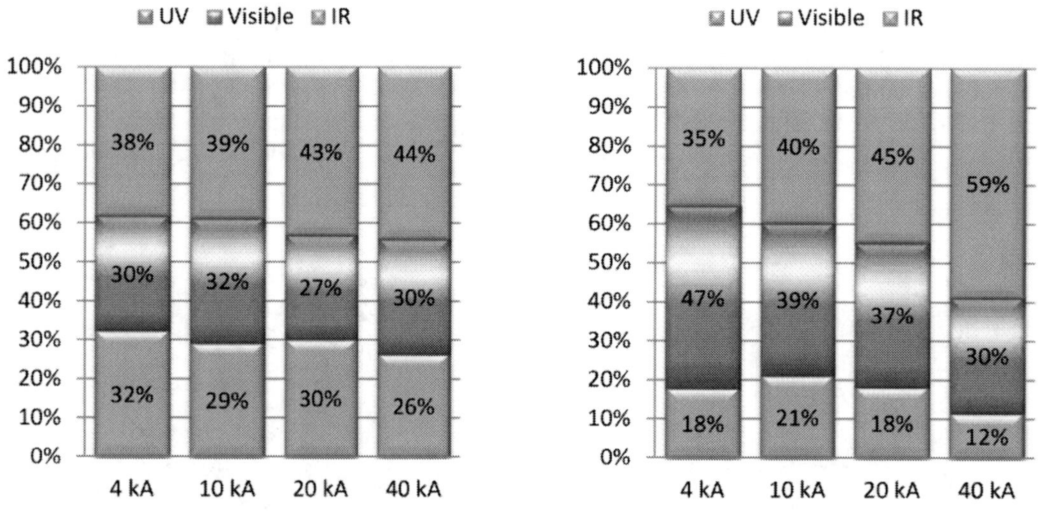

(a) Copper (b) Steel

Figure 4. Spectral distribution of the measured radiated energy vs the type of electrodes and the current, for the long arc.

Figures 4 shows the same kind of results as figure 2, but for the long arc configuration. The first trend that emerges is the high proportion of visible and infrared radiation. These proportions are more characteristic of air. One might think that in this configuration, the proportion of metal vapors in the plasma would be lower than in the short arc configuration. Given the high currents, erosion is greater but the volume provided by the plasma is significantly broader.

It can also be seen that UV radiation is greater for copper and visible radiation for iron. As the current increases, the UV radiation decreases while the visible and infrared ranges increase. When the current increases, the electrode erosion also increases, and with it, the proportion of metallic vapors. Radiative losses due to these fumes tend to cool the plasma so that the temperature of the plasma does not necessarily increase with the current. Here also the cloud of smoke certainly plays an important

Journal of Physics: Conference Series **406** (2012) 012030 doi:10.1088/1742-6596/406/1/012030

role, especially in this configuration. Indeed, the arc lasts 10 times longer than in the short arc configuration. The expansion of the cloud is greater; it occupies a larger volume, thus promoting the absorption of part of the arc radiation and the retransmission in a spectral domain closer to the visible and infrared spectral ranges.

4. Conclusion

In this study, we measured the energy emitted by strong current transitional arcs in two different configurations: one for a short arc and short duration, one for a long arc with longer-lasting and stronger currents, for two electrode materials: copper and steel.

We have demonstrated that this radiation does not depend only on the nature of the initial gas, namely air. Metallic vapors generated by the erosion of electrodes play an important role in the radiative transfer, both when it comes from metallic atoms in the plasma, and from the fact that the droplets form a cloud that modifies the radiative transfer around the arc.

As the erosion of these electrodes differs depending on the material used, we cannot ignore a dynamic effect of this cloud, which is more or less hot depending on the case, and is evacuated differently. This feature was verified using fast videos. We also plan to conduct these experiments for aluminum electrodes. They will not only confirm the hypotheses advanced, but will also be of practical interest, as arcs are increasingly created with aluminum electrodes.

References

[1] Décret n°2010-750 2010 *Protection des travailleurs contre les risques dus aux rayonnements optiques artificiels* (France: JO Ministry of employment)

[2] Directive 2006/25/CE 2006 (IEC: European parliament and council)

[3] Technical Report 1999 *Part 9 - Compilation of maximum permissible exposure to incoherent radiation* (60825-9)

[4] Cressault Y, Gleizes A, Riquel G 2011 Calculation of the radiation emitted by isothermal arc plasmas in air and air-metal mixtures *Proc. 20th International Symposium on Plasma Chemistry (Philadelphia,* USA, *25-29 July 2011)* paper 115

[5] Bauchire J M, Hong D, Rabat H, Riquel G 2011 Radiative energy measurements of pulsed high-current arcs *Proc. 20th International Symposium on Plasma Chemistry (Philadelphia,* USA, *25-29 July 2011)* paper 526

Simulation of charged and excited particle transport in the low-current discharge in argon-mercury mixture

G G Bondarenko[1], M R Fisher[2] and V I Kristya[2,3]

[1] Moscow Institute of Electronics and Mathematics of National Research University Higher School of Economics, 3/12 B. Trekhsviatitelskiy st., Moscow, 109028, Russia

[2] Kaluga Branch of Bauman Moscow State Technical University, 2 Bazhenov st., Kaluga, 248000, Russia

E-mail: kristya@bmstu-kaluga.ru

Abstract. Simulation of the electron, ion and metastable excited atom transport in the argon-mercury mixture low-current discharge is fulfilled. Distributions of the particle densities along the discharge gap under different mixture temperatures are obtained and it is demonstrated that the principal mechanism of mercury ion generation is the Penning ionization of mercury atoms by argon metastables, which contribution grows sharply with the mixture temperature due to mercury density increase. Calculations show that the mercury and argon ion flow densities near the cathode are of the same order already under the relative mercury content of about 10^{-4} corresponding at the argon pressure 10^3 Pa to the mixture temperature 30 C. Therefore, at the room temperature the electrodes of mercury illuminating lamps at the stage of their ignition are sputtered predominantly by mercury ions.

1. Introduction

Mixtures of argon with mercury vapor are widely used in gas-discharge illuminating lamps [1, 2]. After lamp turning on, the gas breakdown occurs under the applied voltage and the low-current (Townsend) discharge is initiated between the electrodes, which then transits to the glow and arc modes [3, 4]. Electrons are accelerated by the electric field from the cathode to the anode, ions are accelerated in the opposite direction and excited atoms diffuse in the gas in the direction of their density reduction. As a result of the cathode bombardment by ions, electrons necessary to maintain the discharge are emitted from its surface and the cathode material sputtering takes place, which terminates the electrode lifetime [1, 5, 6]. In the argon-mercury mixture, besides of direct ionization of gas atoms by electrons, ionization of mercury atoms in collisions with metastable excited argon atoms (the Penning reaction) proceeds as well [1, 2]. Therefore, the charged and excited particle densities and flows in the interelectrode gap depend substantially on the mixture temperature, as the density of saturated mercury vapor grows with its increase. But the influence of the temperature on the transport of charged and excited particles in the argon-mercury mixture low-current discharge plasma under conditions typical for high pressure illuminating lamps at the stage of ignition is studied insufficiently. In particular, the gas ionization coefficient and contributions of the main types of particle interactions

[3] To whom any correspondence should be addressed.

in it under different relative mercury contents in the mixture and electric field strength values were found in [7], where the electron density was obtained by fitting the results of calculations to experimental data of Ref. [8] and the densities of ions and excited atoms were estimated from their balance equations averaged over the discharge volume. The charged particle distributions in the lamp tube were computed in [9] under specific discharge conditions with using of the macroscopic description of their motion.

In this work, simulation of electron, ion and metastable excited atom motion and interactions in a low-current discharge between the flat electrodes in argon-mercury mixture is fulfilled. Computation of electron trajectories is carried out by the Monte Carlo method [3-5], whereas ion and metastable excited atom motion is described on the basis of their transport equations. An influence of the mixture temperature on the particle densities and flows in the interelectrode gap is studied.

2. Mathematical model

Let the low-current discharge be burning in the argon-mercury mixture between the flat parallel electrodes operating as the cathode and the anode, which dimensions substantially exceed the distance d between them. Suppose that axis z of the coordinate system is directed perpendicularly to the electrodes, and the cathode and the anode coincide with planes $z=0$ and $z=d$, respectively. In the low-current mode, which can exist at quite low discharge current density j values, the space charge in the interelectrode gap is negligible and does not distort the external electric field in it, which is directed along z - axis and equals to $E=U/d$ in each point, where U is the discharge voltage [5].

Simulation of electron motion in the gas is carried out with the Monte Carlo method. Calculation of the trajectories of electrons emitted from the cathode (the primary electrons) is executed at each time step, which value is taken small enough to ensure the length of the trajectory section passing by an electron during this time being much less than its mean path length between collisions with gas atoms. The probability of an electron-atom collision at the trajectory section, its type, as well as the electron direction of motion and energy after the collision, are obtained from the corresponding expressions of the collision theory with using of random numbers [10-14]. The trajectory of an electron emitted from the cathode is calculated until it reaches the anode or returns to the cathode. Motion of the secondary electrons generated in the interelectrode gap due to ionization of atoms by electrons and metastables is simulated in a similar way. As a result, the electron velocity distribution function $f_e(z_i, v, v_z)$ is found in each of s subintervals of the length $\Delta z = d/s$, into which the interelectrode distance is divided, where z_i is the coordinate of the central point of the i - subinterval, v and v_z are the electron velocity and its z - component.

Main types of heavy charged particles in the low-current discharge in argon-mercury mixture under the pressure $10^3 - 10^4$ Pa at the room temperature, typical for high-pressure mercury lamps, are atomic ions of the both gases Ar^+ and Hg^+, whereas the density of metastable argon atoms Ar^* and Hg^* exceeds considerably the density of atoms excited at the resonant levels ($Ar^{*'}$ и $Hg^{*'}$) [7, 9].

The following particle interactions are taken into account:

- electron elastic scattering on an atom;
- electron ionization of a ground-state atom

$$Ar + e \rightarrow Ar^+ + 2e, \tag{1}$$

$$Hg + e \rightarrow Hg^+ + 2e; \tag{2}$$

- electron excitation of a ground-state atom at the metastable and resonant levels

$$Ar + e \rightarrow Ar^* + e, \tag{3}$$

$$Ar + e \rightarrow Ar^{*'} + e, \tag{4}$$

$$Hg + e \rightarrow Hg^* + e, \tag{5}$$

$$Hg + e \rightarrow Hg^{*'} + e; \tag{6}$$

- ionization of a metastable atom

$$Ar^* + Ar^* \rightarrow Ar^+ + Ar + e\,, \quad (7)$$

$$Hg^* + Hg^* \rightarrow Hg^+ + Hg + e\,; \quad (8)$$

- Penning ionization

$$Ar^* + Hg \rightarrow Ar + Hg^+ + e\,; \quad (9)$$

- metastable quenching

$$Ar^* + Ar \rightarrow 2Ar\,, \quad (10)$$

$$Hg^* + Ar \rightarrow Hg + Ar \quad (11)$$

$$Hg^* + Hg \rightarrow 2Hg\,. \quad (12)$$

At high electric field strength values typical for the low-current discharge, the charged particles move in the gas in the drift mode [5], and the main mechanism of excited atom motion is diffusion in the direction of their density reduction. Therefore, their transport equations can be presented as follows:

$$\frac{d}{dz}\left(n_{Ar^+}\mu_{Ar^+}E\right) + \Delta n_{Ar^+} + k_1 n_{Ar^*}^2 = 0\,, \quad (11)$$

$$\frac{d}{dz}\left(n_{Hg^+}\mu_{Hg^+}E\right) + \Delta n_{Hg^+} + k_2 n_{Hg^*}^2 + k_3 n_{Ar^*} n_{Hg} = 0\,, \quad (12)$$

$$D_{Ar^*}\frac{d^2 n_{Ar^*}}{dz^2} + \Delta n_{Ar^*} - 2k_1 n_{Ar^*}^2 - k_3 n_{Ar^*} n_{Hg} - k_4 n_{Ar^*} n_{Ar} = 0\,, \quad (13)$$

$$D_{Hg^*}\frac{d^2 n_{Hg^*}}{dz^2} + \Delta n_{Hg^*} - 2k_2 n_{Hg^*}^2 - k_5 n_{Hg^*} n_{Ar} - k_6 n_{Hg^*} n_{Hg} = 0\,, \quad (14)$$

where n_{Ar^+}, n_{Hg^+}, n_{Ar^*} and n_{Hg^*} are the densities of argon ions, mercury ions, argon metastables and mercury metastables, μ_{Ar^+} and μ_{Hg^+} are the argon and mercury ion mobilities, D_{Ar^*} and D_{Hg^*} are the argon and mercury metastable diffusion coefficients, $k_1 - k_6$ are the constants of interactions (7) – (12). The number of p-type particles (Ar^+, Hg^+, Ar^* Hg^*) generated in a unit volume in a unit time in collisions of electrons with atoms of the mixture l-component (Ar, Hg) equals to $\Delta n_p = n_l \iint \sigma_l^p(v) f_e(z, v, v_z) v\,dv\,dv_z$, where $\sigma_l^p(v)$ is the cross-section of generation of a p- type particle in collision of an electron having energy v with an atom of the mixture l- component having density n_l (reactions (1) – (6)). Boundary conditions for the equations are zero values of the ion densities at the anode and the metastable atom densities at the both electrodes:

$$n_{Ar^+}(d) = n_{Hg^+}(d) = 0, \quad n_{Ar^*}(0) = n_{Ar^*}(d) = n_{Hg^*}(0) = n_{Hg^*}(d) = 0\,. \quad (15)$$

Solution of the system (11) - (14) with the conditions (15) is found numerically by the finite difference method. Then the number of electrons generated in a unit volume in a unit time by argon metastables in reactions (7) - (9) is calculated

$$\Delta n_e = k_1 n_{Ar^*}^2 + k_2 n_{Hg^*}^2 + k_3 n_{Ar^*} n_{Hg}\,, \quad (16)$$

and the corresponding number of the secondary electrons, which must be added to the primary electrons in each subinterval of the length Δz, is found.

After that, simulation of the electron velocity distribution function is carried out again taking into account the secondary electrons and the ion and metastable transport is computed. This cycle is repeated until the relative differences of values obtained in two successive iterations become sufficiently small.

3. Results and Discussion

Calculations have been performed for the discharge in the interelectrode gap of the length $d = 1.5\times10^{-3}$ m under the sustaining voltage $U = 255$ V in the mixture of argon with the constant density $n_{\mathrm{Ar}} = 6.58\times10^{23}$ m^{-3} corresponding to its pressure 2660 Pa at the temperature 293 K and saturated mercury vapor, which relative density $n_{\mathrm{Hg}}/n_{\mathrm{Ar}}$ takes values in the interval 10^{-7} - 10^{-3} under temperature variation from -30 to 60 C [15] (this corresponds to the typical situation when the lamp bulb is filled with argon with an addition of liquid mercury and is sealed off indoor at the room temperature, and then the lamp operates outdoor at different temperatures). The current density value $j \sim 10^{-5}$ A m^{-2} is supposed, ensuring the low-current discharge mode under the indicated conditions [5].

In the process of electron motion simulation with the Monte Carlo method, the interelectrode distance and the intervals of v and v_z variation are divided into 100 subintervals, the number of primary electrons is 10^4, their initial energies are considered as uniformly distributed in the range from 0 to 4 eV and the initial direction of motion is assumed to be isotropic. The cross-sections of the main types of electron collisions with argon and mercury atoms presented in [16-18] are used, as well as the constants $k_1 - k_6$ of the heavy particle interactions (7) – (12), the ion mobilities and the metastable atom diffusion coefficients taken from [9, 13, 19-24]. The calculated dependences of the gas ionization coefficient and the electron drift velocity on the electric field strength in pure argon and mercury, as well as the dependence of the ionization coefficient on the relative content of mercury in the mixture are in an agreement with experimental data [14].

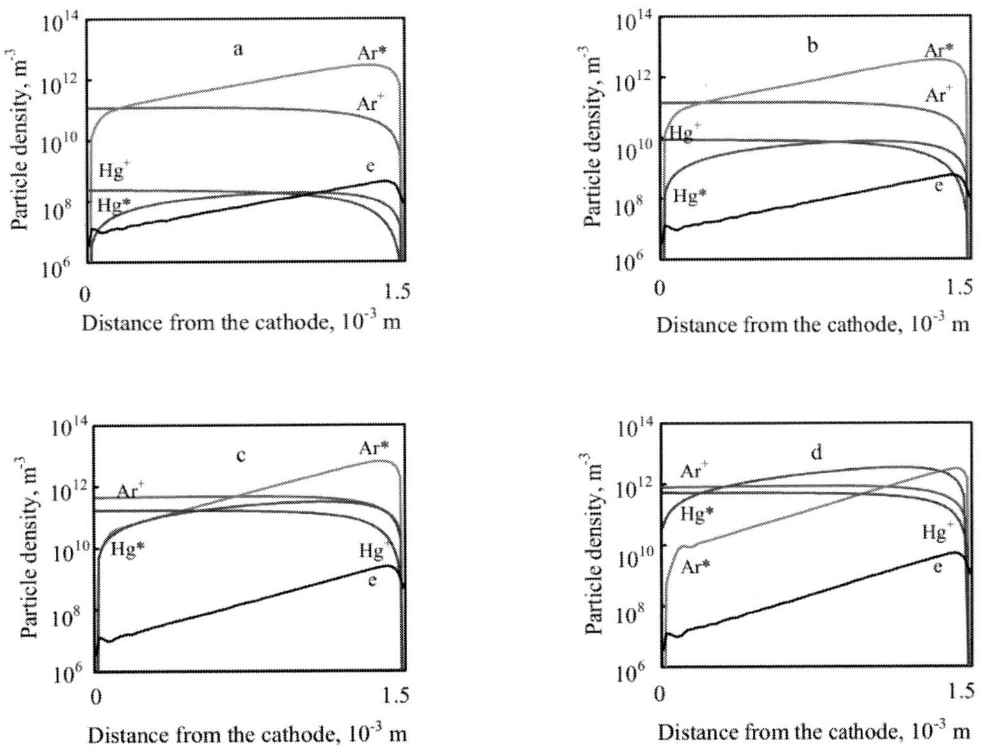

Figure 1. Distributions of the electron, ion and metastable densities in the discharge gap at different mixture temperature values: -30 C (a), 0 C (b), 30 C (c) and 60 C (d).

Journal of Physics: Conference Series **406** (2012) 012031 doi:10.1088/1742-6596/406/1/012031

In figure 1 the obtained distributions of the charged and excited particle densities between the electrodes at different temperature values are presented. It follows from it that the density of electrons is exponentially increased from the cathode to the anode as a result of generation of secondary electrons in ionizing collisions (1) – (2) and (7) – (9), and its value grows substantially with the temperature because of more intensive Penning ionization at higher mercury vapor densities (this fact can be ascertained by comparison of figures 1c and 2, in which distributions of the particle densities at the temperature 30 C, calculated with and without taking into account the Penning ionization (9), are shown). Argon metastables arise under excitation of argon atoms by electrons (in reaction (3)) and are destroyed mainly in the Penning and quenching reactions, whereas diffusion is of no importance for them. Therefore, distribution of their density is similar to that of electrons and its value is varied insignificantly with temperature increase. Mercury metastables appear under electron excitation of mercury atoms and are lost due to diffusion to the electrodes, therefore, their density distribution in the central area of the discharge gap is quite uniform. Argon ions are produced under ionization of argon atoms by electrons, whereas the dominant mechanism of mercury ion generation is the Penning ionization of mercury atoms by argon metastables, and ions of the both types are lost mainly due to their drift to the cathode under the influence of the electric field. Therefore, the mercury ion density, which at the temperature -30 C is by a factor of a hundred lower than the argon ion density, grows considerably with the temperature and at 60 C becomes comparable with the argon ion density.

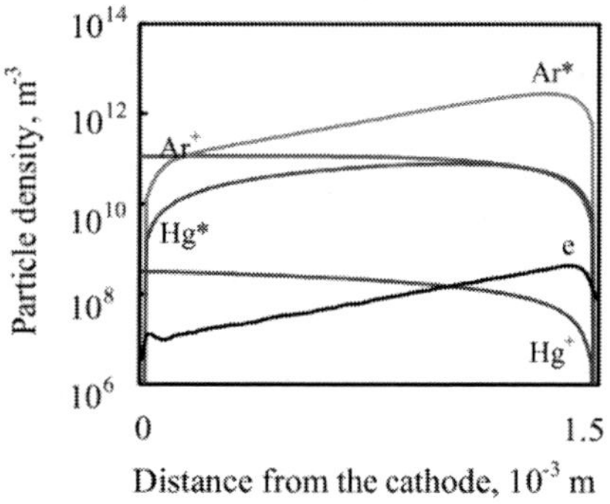

Figure 2. Distributions of the electron, ion and metastable densities in the discharge gap at the mixture temperature 30 C calculated without taking into account the Penning ionization (under k_3=0).

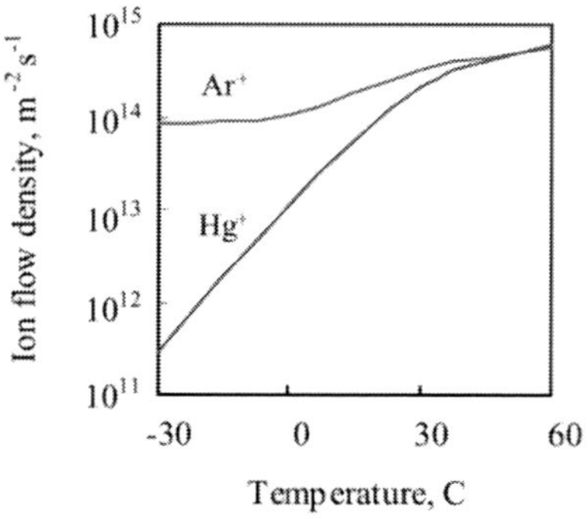

Figure 3. Temperature dependence of the ion flow densities at the cathode surface.

As a result, the mercury and argon ion flow densities at the cathode are of the same order already under the relative mercury content in the mixture of about 10^{-4}, corresponding to the mixture temperature 30 C (see figure 3). Because the mean path length of a mercury ion in the mixture between the resonant charge exchanges on parent gas atoms is much more than that of an argon ion, the energies of mercury ions exceed considerably the energies of argon ions, and they make the main contribution to the cathode sputtering, which reduces the lamp service time [1].

4. Conclusions

In this work, simulation of the electron, ion and metastable excited atom transfer in a low-current discharge in the argon-mercury mixture between the flat electrodes is carried out. Calculation of electron trajectories is performed with the Monte Carlo method, and motion of ions and metastable excited atoms is described on the basis of their transport equations, which are solved by the finite difference method. Distributions of the particle densities along the discharge gap are calculated under

Journal of Physics: Conference Series **406** (2012) 012031 doi:10.1088/1742-6596/406/1/012031

different mixture temperature values, and the main mechanisms of their generation and loss are determined. It is shown that the principal mechanism of mercury ion production is the Penning ionization of mercury atoms by argon metastable excited atoms, and its contribution grows sharply with the mixture temperature. It follows from the obtained results that the mercury and argon ion flow densities near the cathode are of the same order already under the relative mercury content of about 10^{-4}, corresponding at the argon pressure 10^3 Pa to the mixture temperature 30 C. Therefore, under the room temperature the electrodes of mercury illuminating lamps at the stage of their ignition are sputtered predominantly by mercury ions.

References

[1] Waymouth J F 1971 *Electric-Discharge Lamps* (Cambridge, MA: MIT Press)
[2] Lister G G, Lawler J E, Lapatovich W P and Godyak V A 2004 *Rev. Mod. Phys.* **76** 541
[3] Luijks G M J F and van Vliet J A J M 1988 *Lighting Res. Technol.* **20** 87
[4] Byszewski W W, Li Y M, Budinger A B and Gregor P D 1996 *Plasma Sources Sci. Technol.* **5** 720
[5] Raizer Y P 1991 *Gas discharge physics* (Berlin: Springer)
[6] Lieberman M A and Lichtenberg A J 2005 *Principles of Plasma Discharges and Materials Processing* (New York: Wiley-Interscience)
[7] Sawada S, Sakai Y and Tagashira H 1989 *J. Phys. D: Appl. Phys.* **22** 282
[8] Burgmans A L and Smeeth A H 1983 *J. Phys. D: Appl. Phys.* **16** 755
[9] Lay B, Moss R S, Rauf S and Kushner M J 2003 *Plasma Sources Sci. Technol.* **12** 8
[10] Boeuf J P and Marode E 1982 *J. Phys. D: Appl. Phys.* **15** 2169
[11] Sun J, Gong Y and Wang D 1993 *J. Phys. D: Appl. Phys.* **26** 436
[12] Bogaerts A, van Straaten M and Gijbels R 1995 *Spectrochim. Acta B* **50** 179
[13] Bogaerts A, Gijbels R and Goedheer W J 1995 *J. Appl. Phys.* **78** 2233
[14] Kristya V I and Fisher M R 2010 *Bull. Rus. Acad. Sci.: Phys.* **74** 277
[15] Kikoin I K 1976 *Tables of Physical Quantities* (Moscow: Atomizdat)
[16] ftp://jila.colorado.edu/collision_data
[17] Rockwood S D 1973 *Phys. Rev. A* **8** 2348
[18] Mason N J and Newell W R 1987 *J. Phys. B: At. Mol. Phys.* **20** 1357
[19] Biondi M A 1953 *Phys. Rev.* **90** 730
[20] Chanin L M and Biondi M A 1957 *Phys. Rev.* **107** 1219
[21] Krjukov N A, Penkin N P and Redjko T P 1977 *Optik. Spectrosc.* **42** 33
[22] Lyberopoulos D P and Economou D J 1993 *J. Appl. Phys.* **73** 3668
[23] Bashlov N, Zissis G, Charrada K, Stambouli M, Milenin V and Timofeev N 1994 *J. Phys. D: Appl. Phys.* **27** 494
[24] Bogaerts A and Gijbels R 1995 *Phys. Rev.* A **52** 3743

Fluid model of dc glow discharge with nonlocal ionization source term

I R Rafatov§, E A Bogdanov†, A A Kudryavtsev†

§ Middle East Technical University, Ankara, Turkey
† Saint Petersburg State University, St.Peresburg, Russia

E-mail: rafatov@metu.edu.tr

Abstract. We developed and tested a simple hybrid model for a glow discharge, which incorporates nonlocal ionization by fast electrons into the fluid framework. Calculations have been performed for an argon gas. Comparison with the experimental data as well as with the hybrid (particle) and fluid modelling results demonstated good applicability of the proposed model.

1. Introduction

Different approaches that are used for the glow discharge modelling can by classified as fluid models, kinetic (particle) models, and their combinations known as hybrid models (see, e.g., [1]). All the discharge models have their advantages and disadvantages. Advantages of the fluid models are their relative simplicity and computational efficiency. However, these models are very approximate, mainly due to the local field approximation (LFA), in which ionization is assumed to be dependent on the local value of the electric field [2]. Nonlocal ionization is approximately taken into account in the "extended fluid models", in which the electrons are described in terms of the fluid dynamics but their transport and kinetic coefficients are calculated as functions of the electron temperature, through the solution of the local Boltzmann equation (see, e.g., Refs. [3, 4, 5, 6]).

Exact description of the electron behavior can be obtained by solving the kinetic Boltzmann equation [7, 8, 9, 10]. However, this approach is mathematically very complicated, especially if more than one dimension needs to be simulated. Method known as PIC/MCC (Particle in Cell/Monte Carlo Collisions [11, 12]) which couples MC simulations for the behavior of electrons and ions to the Poisson equation for the self-consistent electric field, is also time consuming, and has not come into wide use in the glow discharge plasma modelling.

A compromise between computationally effective but very approximate fluid models and accurate but time consuming particle and kinetic models is represented by hybrid models [5, 11, 13, 14, 15, 16, 17]. These methods are based on separating the electron ensemble into different, independently behaving groups, which cannot be described within the fluid model, which deals with variables averaged over the whole ensemble. Basically, two main groups of electrons are identified. The first group is formed by high energetic (fast) electrons, which emerge as secondary electrons at the cathode surface or are created in the ionization processes in the cathode-fall region. Second group consists of low energetic (slow) bulk electrons, which mainly locate in the quasineutral region of the discharge.

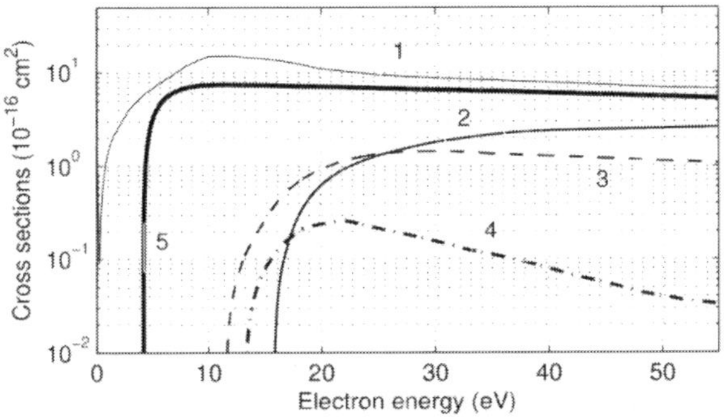

Figure 1. Electron cross sections for (1) elastic, (2) direct ionization, (3) excitation, (4) excitation, and (5) stepwise ionization collisions in argon, used in the model. Curve labels correspond to indices of corresponding processes in Table 1.

Within this approach, slow electrons are responsible mainly for the transport of electron current and provide the balance of plasma density and temperature (mean energy) over the electron ensemble, so they can be described in the framework of the fluid model (simple or extended). Concerning the fast electrons, these are treated as contributing to the ionization processes only and their dynamics are usually described by the Monte Carlo method. This approach is mathematically complicated, and its accuracy is severely limited by poorly known collision cross-sections. At the same time, what is really needed to be known for the discharge description, is such an integral characteristic of the fast electrons as a spatial distribution of the ionization sources, which is then substituted into the balance equations of the ions and plasma electrons. However, in order to find the ionization sources, there is no need to engage in labor-intensive calculations of EDF of fast electrons, accuracy of which is unknown a priori. This characteristic can be approximated analytically and tabulated on the basis of the measured Townsend ionization coefficient or alternatively obtained from the separate MC simulations, and then integrated into the fluid model.

In the context of this approach, in this work, we incorporated effect of fast electrons into the "simple" and "extended" fluid models of glow discharge. Nonlocal ionization source has been formulated in the way first suggested as early as in Ref. [18], and then used in Refs. [2, 19, 20]. We carried out test calculations for the discharge in argon gas, which results were compared with the data obtained from measurements as well as from the "extended fluid" model and hybrid (particle) simulations. Calculations performed showed that the proposed model, which can be called as a simple hybrid model, allows to overcome the shortcomings of the "extended fluid" models, since it takes properly into account the nonlocal ionization in the negative glow. At the same time it is computationally much more efficient compared to the models involving Monte Carlo simulations.

2. Model

2.1. *"Simple" fluid model*

"Simple" fluid model for the gas-discharge includes continuity equations for the charged and excited particles,

$$\frac{\partial n_k}{\partial t} + \nabla \cdot \mathbf{\Gamma}_k = S_k, \tag{1}$$

which are completed by the Poisson's equation for the electrostatic field,

$$\epsilon_0 \nabla \cdot \mathbf{E} = \sum_k q_k n_k, \quad \mathbf{E} = -\nabla \phi. \tag{2}$$

In these equations, subscript k indicates the kth species (we will also use subscripts i, e, m and g for the ions, electrons, metastable atoms and background gas, respectively), n stands for the number density, S denotes the particle creation rate, $\mathbf{E}$ and ϕ are the electric field and potential, q is the charge, and ϵ_0 is the dielectric constant. $\mathbf{\Gamma}$ denotes the particle flux, which, with parameters μ and D denoting the mobility and diffusion coefficients, is given in the drift-diffusion approximation,

$$\mathbf{\Gamma}_k = sgn(q_k) n_k \mu_k \mathbf{E} - D_k \nabla n_k. \tag{3}$$

Within this model, particle creation rate is determined as function of reduced electric field E/p ("local field approximation"), which, in general, is an unacceptable approximation (see the discussion in Section 1). In order to remedy the drawbacks associated with the LFA and include, to a certain extent, the nonlocal transport of electrons, "extended" fluid model has been suggested in [3] (which different versions have been applied in Refs. [4, 5, 6]).

2.2. "Extended" fluid model

Within the "extended fluid" model for a glow discharge, the electron transport coefficients (diffusion, D_e, and mobility, μ_e) as well as the rates S of the electron-induced plasma-chemical reactions are calculated as functions of the mean electron energy, $\bar{\varepsilon}$, rather than the magnitude of the reduced electric field, through the electron energy equation

$$\frac{\partial n_\varepsilon}{\partial t} + \nabla \cdot \mathbf{\Gamma}_\varepsilon = S_\varepsilon,$$

which is added to the system (1)-(3), and LUT's ("look-up tables"), obtained from the solution of the local electron Boltzmann equation (see, e.g., Ref. [4]). Here, $n_\varepsilon = n_e \bar{\varepsilon}$ denotes the electron energy density, $\bar{\varepsilon} = 3/2 k_B T_e$ is the mean electron energy, and density of the electron energy flux is

$$\mathbf{\Gamma}_\varepsilon = -D_\varepsilon \nabla n_\varepsilon - \mu_\varepsilon \mathbf{E} n_\varepsilon, \tag{4}$$

where energy transport coefficients are related to particle transport coefficients via $\mu_\varepsilon = (5/3)\,\mu_e$ and $D_\varepsilon = (5/3)\,D_e$ [4].

Source function for the electron energy equation has a form $S_\varepsilon = P_{heat} + P_{elastic} + P_{inelastic}$, where the first term describes the Joule heating (or cooling) of electrons in the electric field, $J_{heat} = -e\Gamma_e \cdot E$, the second term expresses the elastic losses,

$$P_{elastic} = -\frac{3}{2}\,\delta\,\nu_{ea} n_e k_B \left(T_e - T_g\right),$$

and the last term is the energy loss in inelastic collisions, $P_{inelastic} = -\sum_j \Delta E_j R_j$. In these equations, ν_{ea} denotes the electron-atomic elastic collision frequency, m is the particle mass, $\delta = 2m_e/m_g$, background gas temperature $T_g = 300$ K, ΔE_j and R_j are the energy loss (or gain) due to inelastic collision and corresponding reaction rate.

Mobility and diffusion coefficients of heavy particles are approximated by constant parameters, which depends on background gas density, and related by the expression $D_i/\mu_i = k_B T_i/e$ with $T_i = T_g$. Electron mobility, μ_e, and diffusion, D_e, are computed from

$$\mu_e = -\frac{1}{n_e}\frac{e}{m_e}\int_0^\infty D_r \sqrt{\varepsilon}\frac{\partial}{\partial \varepsilon} f_0\left(\varepsilon\right) d\varepsilon, \tag{5}$$

Table 1. Elementary reactions considered in this study. Label Boltz. indicates that constant was calculated from local Boltzmann equation.

Index	Reaction	Type	$\Delta E\,(eV)$	Constant
1	$e + Ar \rightarrow e + Ar$	Elastic collision	0	Boltz.
2	$e + Ar \rightarrow 2e + Ar^+$	Direct ionization	15.8	Boltz.
3	$e + Ar \leftrightarrow e + Ar^*$	Excitation	11.4	Boltz.
4	$e + Ar \rightarrow e + Ar$	Excitation	13.1	Boltz.
5	$e + Ar^* \rightarrow 2e + Ar^+$	Stepwise ionization	4.4	Boltz.
6	$2Ar^* \rightarrow e + Ar^+ + Ar$	Penning ionization	–	$6.2 \times 10^{-10}\,\mathrm{cm^3\,s^{-1}}$
7	$Ar^* \rightarrow h\nu + Ar$	Radiation (including trapping)	–	$1.0 \times 10^7\,\mathrm{s^{-1}}$

$$D_e = \frac{1}{n_e}\int_0^\infty D_r\sqrt{\varepsilon} f_0(\varepsilon)\,d\varepsilon, \tag{6}$$

where $\varepsilon = mv^2/2e$ is the electron kinetic energy (in eV units), v is the electron velocity, $D_r = 2\varepsilon/3m_e\nu_{ea}$ is the space diffusion coefficient, and $f_0(\varepsilon)$ is the EEDF obtained from the solution of the local Boltzmann equation.

We used the version of "extended fluid" model applied in [6], where the calculations were performed for an argon gas, and three plasma species, namely, electrons, positive ions, and metastable atoms were taken into account. The set of reactions is given in Table 1, and collision cross-sections are shown in Fig. 1.

Balance of charged particles, in the absence of recombination, is determined by the direct, stepwise, and Penning ionization processes,

$$S_e = S_i = K_2 n_0 n_e + K_5 n_m n_e + K_6 n_m^2. \tag{7}$$

Here, K denotes the constant of the corresponding reaction, n_0 is the concentration of neutral atoms. Reaction constants are numbered according to the list of processes in Table 1. Balance of excited atoms is determined by the reactions of excitation, de-excitation, stepwise ionization, Penning ionization and radiation,

$$S_m = K_3 n_0 n_e - K_3' n_m n_e - K_5 n_m n_e - 2K_6 n_m^2 - K_7 n_m, \tag{8}$$

where K_3' is the rate constant of superelastic electron collisions, which correspond to the left-directed arrow for the process 3 in Table 1. Cross-section corrsponding to this process is determined by the detailed balance relationship.

For the electron-induced reactions (processes 1–5 in the Table 1), rate constants K are calculated by convolving the EEDF, obtained from the solution of the local Boltzmann kinetic equation, with the corresponding cross-sections,

$$K_R = \int_0^\infty \sigma_R(\varepsilon)\,v(\varepsilon)\sqrt{\varepsilon}\,f_0(\varepsilon)\,d\varepsilon.$$

2.3. Models with a nonlocal ionization source.

As we discussed in Section 1, the fast electrons contribution in the hybrid models of glow discharge is limited only to developing the spatial distribution of the ionization (and excitation) rate. Therefore, their effect can be approximated by the analytical formulation of the ionization (and excitation) source function S_{fast}, and then integrating it into the fluid model.

In the context of this approach, we formulated the nonlocal ionization source in the way suggested in Refs. [2, 19, 20]. The discharge gap is divided into two regions, namely, the

Journal of Physics: Conference Series **406** (2012) 012032 doi:10.1088/1742-6596/406/1/012032

cathode sheath region where the electric field is strong, and the plasma region, where the field is weak. These regions are separated by a point $x = d$, the sheath boundary, which is determined as a distance x from the cathode such that the equality $n_e(x) = 0.5\, n_i(x)$ holds. (Due to sharp increase in the electron density in the vicinity of this point and quasineutrality in the plasma, replacing the factor 0.5 in this equation with a number from the range between 0.4 – 0.9 has little effect on the value of d.) Since the ionization source term $S_{\text{fast}}(x)$ decays exponentially with distance from the sheath boundary, it can be approximated as

$$S_{\text{fast}}(x) \propto \mathrm{e}^{-(x-d)/\lambda} \quad (x \geq d)$$

with a decay constant λ [2, 20]. Taking into account that the maximum value of the ionization source is

$$S_{\text{fast}}(d) = \Gamma_e(0)\, \alpha\, \mathrm{e}^{\alpha d},$$

where $\Gamma_e(0)$ is the electron flux density at the cathode and α is the Townsend ionization coefficient, function S takes the form

$$S_{\text{fast}}(x) = \begin{cases} \Gamma_e(0)\, \alpha\, \mathrm{e}^{\alpha x} & \text{for} \quad x < d, \\ \Gamma_e(0)\, \alpha\, \mathrm{e}^{\alpha d}\, \mathrm{e}^{-(x-d)/\lambda} & \text{for} \quad x \geq d. \end{cases} \tag{9}$$

In Ref. [2], authors used the conventional approximation for the Townsend ionization coefficient of the form $\alpha = A p \exp(-\mathrm{B}\, \mathrm{p}/\mathcal{E})$, with the effective field $\mathcal{E}$ calculated as average field over the cathode sheath, $\mathcal{E} = \varphi(d)/d$. In this work, we approximated α by a more accurate estimation from Ref. [21]. As a decay constant λ, we used an estimation proposed in [2],

$$\lambda = \frac{\phi(d)/(pB) - d}{\alpha d}, \tag{10}$$

which makes it possible to approximate the decay scale λ in (9), given thickness d of the cathode sheath and voltage $\phi(d)$. In this expression, parameter B is from the formula for the Townsend ionization coefficient α and is equal to 180 V/(cm · Torr) for an argon gas [22].

We incorporated ionization function S_{fast} defined by (9) into

(i) "simple" fluid model of glow discharge (section 2.1), where the plasma is composed of two species only, namely, the positive ions and electrons, and the ionization source function is

$$S_e = S_i = S_{\text{fast}},$$

(ii) "extended" fluid model of glow discharge (section 2.2), where we have added S_{fast} term in the right hand of equation (7) and the excitation term $0.5\, S_{\text{fast}}$ in in the right hand of (8), respectively:

$$S_e = S_i = S_{\text{fast}} + K_2 n_0 n_e + K_5 n_m n_e + K_6 n_m^2,$$
$$S_m = 0.5\, S_{\text{fast}} + K_3 n_0 n_e - K_3' n_m n_e - K_5 n_m n_e - 2K_6 n_m^2 - K_7 n_m.$$

In the model (i), slow electron mobility and diffusion, μ_e and D_e, have been taken to be constants corresponding to those of the model (ii) corresponding to the temperature $T_e = 1$ eV.

In the near-cathode plasma of NG and FDS, the electric field is weak (and even may change a sign), so that the Joule heating in this region is also weak. Heating of the plasma (bulk) electrons is carried out by the Coulomb collisions with fast electrons. The corresponding term in the equation of electron energy balance can be estimated (see Refs. [23, 24] for details) by

$$P_{\text{fast}} = \frac{1}{2}\, \varepsilon^*\, S_{\text{fast}}\, \frac{\nu_{ee}}{\nu_{ee} + \delta\nu_{ea} + \nu_{df}},$$

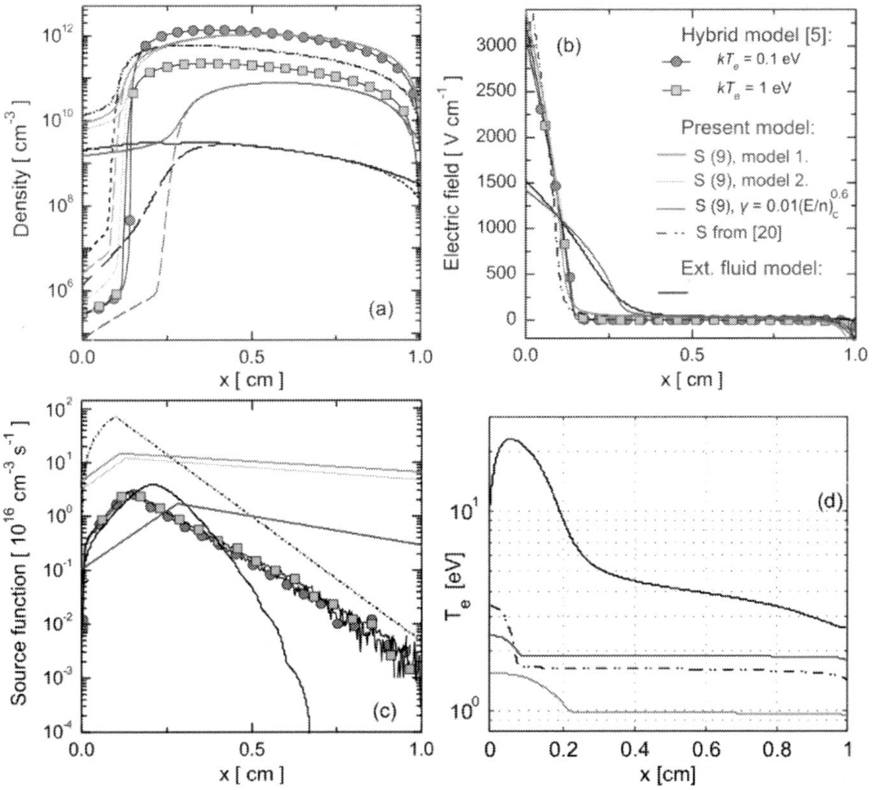

Figure 2. Comparison with the hybrid model results from [5], for the set of conditions $p = 133$ Pa, $V = 250$ V, $L = 1$ cm, $\gamma = 0.06$. Spatial distributions of (a) electron and ion density, (b) electric field, (c) ionization source function, (d) electron temperature.

where $\nu_{df} = 1/\tau_{df} = \Lambda^2/D_{ef}$, Λ is the characteric diffusive scale ($R/2.4$ for a cylinder), D_{ef} is the free diffusion coefficient of electrons. It is assumed that the electrons coming down from the inelastic region of EDF, uniformly fill in an energy range from 0 to ε^* (ε^* denotes the threshold energy for inelastic collisions) so that their average energy is $1/2\,\varepsilon^*$. Thus, $D_{ef}, \nu_{\mathrm{ea}}$, and ν_{ee} are calculated for this value of energy (or temperature) of electrons [23, 24].

2.4. Boundary Conditions

Boundary condition for the positive ions, metastable atoms, and electron energy density at the anode and cathode are given as follows,

$$\hat{\mathbf{n}} \cdot \mathbf{\Gamma}_{\mathrm{i}} = 1/4 v_{\mathrm{i}} n_{\mathrm{i}} + \alpha n_{\mathrm{i}} \mu_{\mathrm{i}} \left(\hat{\mathbf{n}} \cdot \mathbf{E}\right), \tag{11}$$

$$\hat{\mathbf{n}} \cdot \mathbf{\Gamma}_{\mathrm{m}} = 1/4 v_{\mathrm{m}} n_{\mathrm{m}}, \tag{12}$$

$$\hat{\mathbf{n}} \cdot \mathbf{\Gamma}_{\varepsilon} = 1/3 v_{\mathrm{e}} n_{\varepsilon}. \tag{13}$$

Here, $v_j = \sqrt{8k_BT_j/\pi m_j}$ ($j = e, i, m$) denotes the thermal velocity, the flux density $\mathbf{\Gamma}$ is described by the equations (3) and (4), $\hat{\mathbf{n}}$ is the normal unit vector pointing towards the surface, and α is a switching function (either 0 or 1) depending on positive ion drift direction at the surface: $\alpha = 1$ if $(\hat{\mathbf{n}} \cdot \mathbf{E}) > 0$, and $\alpha = 0$ otherwise.

Boundary conditions for the electron density at the anode is

$$\hat{\mathbf{n}} \cdot \mathbf{\Gamma}_{\mathrm{e}} = 1/4 v_{\mathrm{e}} n_{\mathrm{e}}, \tag{14}$$

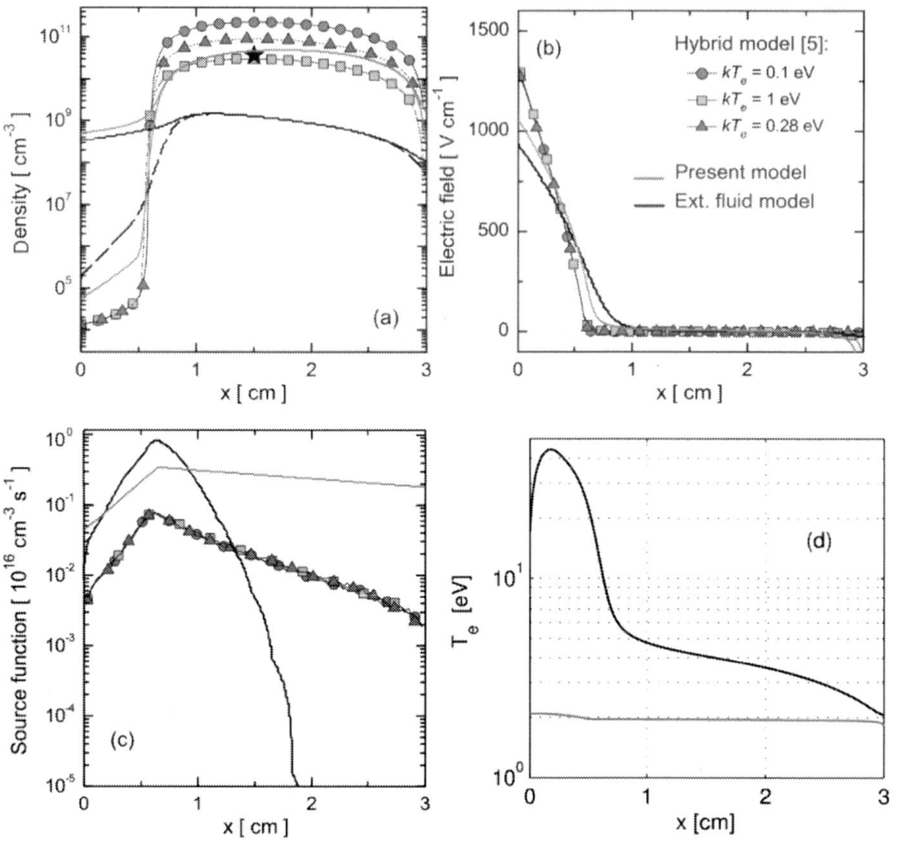

Figure 3. Comparison with the hybrid model results from [5], for the set of conditions $p = 40$ Pa, $V = 441$ V, $L = 3$ cm, $\gamma = 0.033$. Spatial distributions of (a) electron and ion density, (b) electric field, (c) ionization source function, (d) electron temperature. "Star" in panel (a) indicates the measured electron density from [5].

and at the cathode

$$\hat{\mathbf{n}} \cdot \mathbf{\Gamma}_{\rm e} = 1/4 v_{\rm e} n_{\rm e} - \gamma\, \hat{\mathbf{n}} \cdot \mathbf{\Gamma}_{\rm i}, \tag{15}$$

where γ is is the secondary electron emission coefficient.

For the electric potential, we set $\phi = U_d$ at the anode and $\phi = 0$ at the cathode.

3. Modelling results

In order to demonstrate the performance of the model, first we carried out test simulations for the discharge conditions from [5]. Figures 2 and 3 compare spatial distributions of the discharge parameters obtained by the hybrid model from Ref. [5] with those obtained from the present model, for the two sets of conditions. The first set comprises $p = 133$ Pa, $V = 250$ V, $L = 1$ cm, $\gamma = 0.06$. For the second set, the parameters are $p = 40$ Pa, $V = 441$ V, $L = 3$ cm, $\gamma = 0.033$. Figures 2 and 3 contain (a) the distributions of the electron and ion densities, $n_{\rm e}$ and $n_{\rm i}$, (b) electric field magnitude E, (c) ionization source functions S, and (d) electron temperature, $T_{\rm e}$. These figures illustrate also numerical results obtained by the "extended fluid" model (described in section 2.2), and the results obtained with the ionization function from Ref. [20] integrated into the "extended" fluid model.

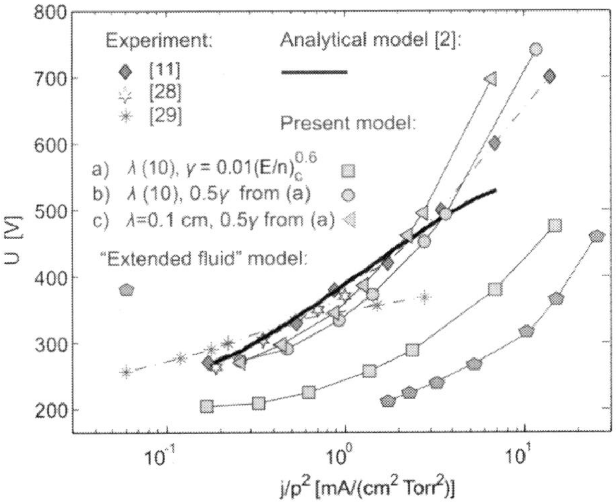

Figure 4. Voltage as a function of the reduced current density, for $pL = 0.5$ cmTorr. Measured data are from [11, 28, 29]. Present model results with different approximations for the parameters γ and λ as indicated in the legend.

As can be seen from Figs. 2 and 3, proposed model provides much better performance compared to the "extended fluid" model. Magnitudes and shapes of the charged particle densities (panels (a)) as well as the electric field (panels (b)) are closed to those obtained from the hybrid model. Also, the electron temperature is not overestimated as in the case of the "extended fluid" model (panels (d)). Recall the fact that the strong overestimation of T_e (and correspondingly the underestimation of n_e) is one of the main disadvantages of the fluid model, and attempts to improve it by taking into account non-Maxwellian EDF could lead to even worse results [25]. This is associated with the specific features of fluid description (more preciese, the "extended" fluid description), in which the kinetic characteristics, specifically, the ionization rates are determined by the temperature (mean energy) of the electron ensemble. Within such a model, the electron gas in the plasma is heated by the heat flux $\mathbf{\Gamma}_\varepsilon$ from the cathode layer due to heat conduction rather than by a small group of fast electrons escaping from the layer. In a real situation, plasma electrons are heated by collisions with fast electrons, relative density of which is extremely low. Other sources of electron heating (such as Penning ionization, dissociative ionization in collisions of two excited particles, deexcitation by electron impact, etc.) here are also ineffective. Therefore, temperature of plasma electrons is uniformly distributed and low (no higher than several fractions of an electronvolt [26]).

As can be seen from Fig. 2, models with nonlocal ionization source, derived from the "simple" and "extended" fluid models of sections 2.1 and 2.2, practically coincide under the studied (short) discharge conditions, because the nonlocal ionization term dominates over other ionization processes involved, namely stepwise and Penning ionizations.

As always in models for the glow discharge, secondary emission coefficient γ is one of the main sources of uncertainty. In order to demonstrate the effect of this parameter, Fig. 2 contains also numerical result obtained using the parameter γ approximated by [27]

$$\gamma = 0.01\,(E/n)_c^{0.6}\,, \tag{16}$$

where the reduced electric field is evaluated at the cathode, and given in units of kTd ($1\,\mathrm{kTd} = 10^{-20}\ \mathrm{V\,cm^2}$). For the second reference conditions set, γ determined by (16) appears to be very close to $\gamma = 0.033$ such that the numerical results obtained with these parameters practically coincide (in Fig. 3 we presented results for $\gamma = 0.033$ only).

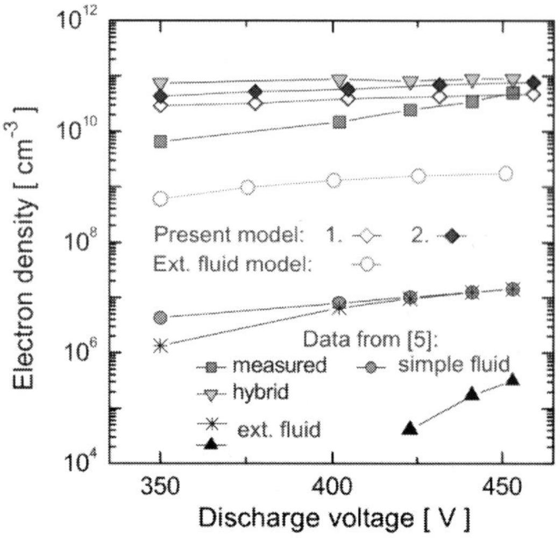

Figure 5. Electron density at $x = 1.5$ cm from the cathode as a function of discharge voltage. $L = 3$ cm, $p = 40$ Pa, $\gamma = 0.033$. Figure also contains measured and computed data from [5].

Figure 4 compares measured volt-ampere characteristics (VAC) of Refs. [11, 28, 29] and VAC from the semi-analytical calculations of Ref. [2] with those obtained from the present model for $pL = 0.5$ cmTorr, with parameter γ approximated by the equation (16) and one half of this γ, and with parameter λ determined by (10) and by constant $\lambda = 0.1$ cm. It is seen that results of the proposed model are close to those obtained from the measurements and it provides much better performance compared to the "extended fluid" model.

Figure 5 plots the electron density at distance $x = 1.5$ cm from the cathode as function of the discharge voltage, for the discharge with $L = 3$ cm, $p = 40$ Pa, and $\gamma = 0.033$. This figure compares the present model results with measured data and data obtained from the hybrid model as well as from different fluid models of Ref. [5]. As can be seen from Fig. 5, present model more adequately predicts the charged particle densities than fluid models. It should be also noticed that version of the "extended fluid" model applied in this work demonstrates better performance than those from Ref. [5].

4. Conclusions

We proposed a simple hybrid model for a glow discharge, which incorporates nonlocal ionization by fast electrons into the fluid framework, and thereby overcomes the fundamental shortcomings of the fluid model. At the same time, proposed model is computationally much more efficient compared to the models involving Monte Carlo simulations. Calculations have been performed for an argon gas. Comparison with the experimental data as well as hybrid (particle) and fluid modelling results exhibits good applicability of the proposed model.

Acknowledgments

The work was supported by the joint research grant from the Scientific and Technical Research Council of Turkey (TUBITAK) 210T072 and Russian Foundation for Basic Research (RFBR).

References

[1] *Electron Kinetics and Applications of Glow Discharges* 2002 (NATO Advanced Study Institute, Series B: Physics vol 367) ed U Kortshagen and L D Tsendin (New York: Kluwer Academic Publishers)
[2] Kudryavtsev A A, Morin A V and Tsendin L D 2008 *Tech. Phys.* **53** 1029

[3] Boeuf J P and Pitchford L C 1995 *Phys. Rev.* E **51** 1376
[4] Hagelaar G J M and Pitchford L C 2005 *Plasma Sources Sci. Technol.* **14** 722
[5] Derzsi A, Hartmann A, Korolov I, Karácsony J, Bánó G and Donkó Z 2009 *J. Phys. D: Appl. Phys.* **42** 225204
[6] Rafatov I, Bogdanov E A and Kudryavtsev A A 2012 *Phys. Plasmas* **19** 033502
[7] Winkler R, Arndt S, Loffhangen D, Sigeneger F and Uhrlandt D 2004 *Contrib. Plasma Phys.* **44** 437
[8] Kolobov V and Arslanbekov R 2006 *IEEE Trans. Plasma Sci.* **34** 895
[9] Robson R E, White R D and Petrović Z Lj 2005 *Rev. Mod. Phys.* **77** 1303
[10] White R D, Robson R E, Dujko S, Nicoletopoulos P and Li B 2009 *J. Phys. D: Appl. Phys.* **42** 194001
[11] Donkó Z, Hartmann P and Kutasi K 2006 *Plasma Sources Sci. Technol.* **15** 178
[12] Donkó Z 2011 *Plasma sources Sci. Technol.* **20** 024001
[13] Fiala A, Pitchford L C and Boeuf L C 1994 *Phys. Rev.* E **49** 5607
[14] Boeuf J P and Marode E 1982 *J. Phys. D: Appl. Phys.* **15** 2169
[15] Surenda M, Graves D B and Jellum G M, 1990 *Phys. Rev.* A **41** 1112
[16] Kushner M J 2009 *J. Phys. D: Appl. Phys.* **42** 194013
[17] Dijk J, Kroesen G M W and Bogaerts A 2009 *J. Phys. D: Appl. Phys.* **42** 190301
[18] Wilhelm J and Kind W 1965 *Beitr. Plasmaphys.* **5** 295
[19] Kolobov V I and Tsendin L D 1992 *Phys. Rev.* A **46** 7837
[20] Peres I, Ouadoudi N, Pitchford L C and Boeuf J P 1992 *J. Appl. Phys* **72** 4533
[21] Phelps A V and Petrović Z Lj 1999 *Plasma Sources Sci. Technol.* **8** R21
[22] Raizer Yu P 1991 *Gas Discharge Physics* (Berlin: Springer-Verlag)
[23] Blagoev A B, Kolokolov N B and Kudrjavtsev A A 1994 *Physica Scripta* **50** 371
[24] Bedenkov N V, Kudryavtsev A A, Romanenko V A and Solodkiy S A 1996 *Physica Scripta* **53** 490
[25] Bogdanov E A, Kapustin K D, Kudryavtsev A A and Chirtsov A S 2010 *Tech. Phys.* **55** 1430
[26] Bano G, Hartmann P, Kutasi K, Horvath P, Plasil R J, Hlavenko P, Glosik J and Donko Z 2007 *Plasma Sources Sci. Technol.* **16** 492
[27] Donkó Z 2001 *Phys. Rev.* E **64** 026401 *Phys. Rev.* E **62** 1452
[28] Rozsa K, Gallagher A and Donkó Z 1995 *Phys. Rev.* E **52** 913
[29] Stefanović I and Petrović Z Lj 1997 *Japan. J. Appl. Phys.* **36** 4728

Journal of Physics: Conference Series **406** (2012) 012033 doi:10.1088/1742-6596/406/1/012033

Spectroscopic study of atmospheric pressure 915 MHz microwave plasma at high argon flow rate

R Miotk[1], B Hrycak[1], M Jasinski[1] and J Mizeraczyk[1,2]

[1] *Centre for Plasma and Laser Engineering, The Szewalski Institute of Fluid-Flow Machinery, Polish Academy of Sciences, Fiszera 14, 80-952 Gdansk, Poland*
[2] *Department of Marine Electronics, Gdynia Maritime University, Morska 81-87, 81-225 Gdynia, Poland*

E-mail: rmiotk@imp.gda.pl

Abstract In this paper results of optical emission spectroscopic (OES) study of atmospheric pressure microwave 915 MHz argon plasma are presented. The plasma was generated in microwave plasma source (MPS) cavity- resonant type. The aim of research was determination of electron excitation temperature T_{exc} gas temperature T_g and electron number density n_e. All experimental tests were performed with a gas flow rate of 100 and 200 l/min and absorbed microwave power P_A from 0.25 to 0.9 kW. The emission spectra at the range of 300 - 600 nm were recorded. Boltzmann plot method for argon 5p - 4s and 5d – 4p transition lines allowed to determine T_{exc} at level of 7000 K. Gas temperature was determined by comparing the measured and simulated spectra using LIFBASE program and by analyzing intensities of two groups of unresolved rotational lines of the OH band. Gas temperature ranged 600 - 800 K. The electron number density was determined using the method based on the Stark broadening of hydrogen H_β line. The measured n_e ranged $2\times10^{15} – 3.5\times10^{15}$ cm^{-3}, depending on the absorbed microwave power. The described MPS works very stable with various working gases at high flow rates, that makes it an attractive tool for different gas processing.

keywords: OES, electron excitation temperature, electron number density, gas temperature

1. Introduction

In plasma sources plasma is excited by applying high electric field to a gas. Tendero et al. give an overview of the numerous atmospheric pressure plasma sources which can be classified as DC, RF and microwave sources [1]. Atmospheric plasma sources are important tools for technological applications, e.g. the treatment of critical exhaust gases, plasma chemistry and activation and cleaning of materials [2-6]. Since there is no need for vacuum components, they are mostly easy to integrate in the process line.

The microwave plasma sources (MPSs) found applications in the processing of decomposition of gaseous pollutants like freon [7-11] and also production of hydrogen via methane conversion [12, 13].

Since thermal and electron reactions play an important role in the gas pollutants decomposition, the knowledge of the neutral gas temperature and electron number density in such plasmas is crucial for understanding the chemical kinetics of the gas decomposition process and its optimization.

In our previous spectroscopic investigations, the microwave plasma devices operated at 2.45 GHz and at low gas flow rates (0.5 - 2 l/min) were used [14-16]. The gas temperature and electron number density were about 4000 K and 10^{15} cm^{-3}, respectively.

Journal of Physics: Conference Series **406** (2012) 012033 doi:10.1088/1742-6596/406/1/012033

In this paper, results of spectroscopic study of microwave plasma at high argon flow rates are presented. The plasma was generated in waveguide-supplied MPS cavity-resonant type operated at 915 MHz. At this frequency the skin effect is weaker, so electromagnetic wave more deeply penetrates the plasma. It can be helpful in the gas processings.

The use of high gas flow rate in the presented device is expected to be of low cost and effective, and thus promising for applications in the treatment of gases. Moreover, the high gas flow rate cools device and protects the electrodes from erosion.

2. Theory

In this paper optical emission spectroscopic (OES) was used for diagnostics of 915 MHz microwave argon plasma. Plasma parameters were determined using the method based on the measurements of intensity and spectral line broadening.

2.1. Electron excitation temperature

Electron excitation temperature is a parameter describing a population of excited atomic levels. The excitation temperature may be determined from argon emission lines intensities. Atmospheric pressure microwave plasma are generally in partial local thermodynamic equilibrium (PLTE) [17, 18]. This suggests to use a Boltzmann plot method to determine the T_{exc}. Thus an observed light intensity I_{nm} of radiative transition between levels n and m may be given by [17-18]:

$$I_{nm} = A_{nm} \frac{hc}{\lambda_{nm}} N \frac{g_n \exp(-E_n / kT_{exc})}{Z} \quad (1)$$

where: A_{nm} is the Einstein coefficient for transition $n \rightarrow m$, h is the Planck constant, c is the light speed constant, λ_{nm} is wavelength for transition $n \rightarrow m$, g_n is the statistical weight of the upper level n, Z is the canonical partition function, N is the population of the ground state level, E_n is the energy of the upper level n and k is the Boltzmann constant.

From the plot $\ln(I_{nm}\lambda_{nm} / g_n A_{nm})$ versus the energy E_u, we can estimate the excitation temperature from the slope of the best line that fits the experimental points [17-18].

2.2. Gas temperature

Gas temperature in microwave plasma is close to the rotational temperature T_{rot} of the OH radicals [19-21]. The OH ($A^2\Sigma \rightarrow X^2\Sigma$) transition is one of most intense systems emitted by low temperature plasmas containing even a small amount of H_2O. This band is widely used for gas temperature estimations.

The rotational temperature can be obtained by fitting the entire band in LIFBASE [22] program or more simply from the relative intensities of groups of unresolved rotational lines corresponding G_0, G_1 and G_{ref} branches of the OH ($A^2\Sigma \rightarrow X^2\Sigma$) band. These branches form distinct peaks at about 306.4 nm, 306.8 and 309.1 nm, respectively. This OH band has been studied thoroughly by Ch. de Izarra [21]. In order to avoid the difficulties encountered in experimental measurements of group line intensities it has been decided to only consider the maximum amplitude of the groups of lines G_0 and G_1. His research has shown that the intensity of G_0 and G_1 strongly depends on the rotational temperature. A more detailed analysis of the groups G_0 and G_1 shows that the wavelength of their maximum amplitude varies as a function of the T_{rot}. It is obvious that this amplitudes are dependent on the instrumental broadening $\Delta\lambda_I$ that must be precisely determined. Izarra publication contains a set of relations G_0/G_{ref} and G_1/G_{ref} as a function of the gas temperatures varying from 600 – 9000 K, in 200 K steps, and for $\Delta\lambda_I$ varying from 0.02 – 0.98 nm, in 0.02 nm steps. The similar method was used by Jasinski et al. in [14].

2.3. Electron number density

In plasmas with the electron number densities greater than ~ $5 \cdot 10^{13}$ cm^{-3}, spatially and temporally resolved n_e can be determined using the OES of the line shape of the Balmer transition (4-2) of atomic hydrogen at 486.13 nm (H_β line) [15, 18, 23-26].

The lineshape of the H_β emission is determined by Lorentzian (Stark, van der Waals, resonance, natural) and Gaussian (Doppler, instrumental) broadening mechanisms that result in a Voigt profile. The convolution of the Gaussian and Lorentzian components results in a Voigt profile. The relation between Gaussian, Lorentzian and the resulted Voigt full-widths at half-maximum (FWHM) profiles is given by [18, 26]:

$$\begin{aligned} \Delta\lambda_V &= \frac{\Delta\lambda_L}{2} + \sqrt{\frac{\Delta\lambda_L^2}{4} + \Delta\lambda_G^2} \\ \Delta\lambda_G^2 &= \Delta\lambda_D^2 + \Delta\lambda_I^2 \\ \Delta\lambda_L &= \Delta\lambda_W + \Delta\lambda_S + \Delta\lambda_N + \Delta\lambda_R \end{aligned} \quad (2)$$

where the subscripts represent: V - Voigt , L - Lorentzian, G - Gaussian, D - Doppler, I - instrumental, W - Van der Waals, S - Stark, N - natural, R - resonance. Resonance and natural broadenings are usually negligible due to very low FWHM values in comparison to the Stark, the Doppler, van der Waals and instrumental broadenings [18, 26].

The random motion of the atoms in the plasma causes the Doppler broadening of the spectral lines. The FWHM of the Doppler profile is [24, 26]:

$$\Delta\lambda_D = 7{,}17\times10^{-7}\lambda_0\sqrt{\frac{T_g}{m}} \quad [nm] \quad (3)$$

where T_g is the gas temperature in Kelvin, m is the atomic mass in a.m.u. and λ_0 is the wavelength.

Van der Waals broadening results from the interaction of an excited atom with neutral ground state atom of another species and is an important broadening mechanism in high-pressure plasmas. The FWHM of the van der Waals profile is [24]:

$$\Delta\lambda_W = 6.48\times10^{-22}\frac{p}{kT_g^{0.7}} \quad [nm] \quad (4)$$

where p is the pressure. One can see that the van der Waals broadening is a gas temperature dependence: if temperature is higher, the broadening is lower.

Perturbation in the wave function on the excited states of hydrogen by the electric field induced by the charged particles of the plasma causes the Stark broadening of the spectral lines. The n_e can be calculated using the theory GKS [27], where Stark broadening is estimated in a quasi-static approximation using the classic Holtsmark field. The more recent Gig-Card theory is suitable to be used to determine the electron density in atmospheric pressure microware discharges and give best results that correlates $\Delta\lambda_S$ and n_e [23-26, 28]. This theory incorporates the ion dynamics to evaluate the Stark broadening of lines spontaneously emitted by the plasma.

3. Experiments

The experimental setup consisted of: microwave generator (915 MHz), isolator, rectangular waveguide (WR 975) as a feeding line, directional coupler, power meters, MPS cavity-resonant type with movable plunger, gas supply and flow control system (see figures 1, 2).To measure the plasma spectrum we used the spectrograph (DK-480 CVI with 1200 and 3600 grooves/mm) equipped with CCD sensitivity calibrated camera (SBIG ST-6 750*242). A PC computer was used to control the spectrograph and acquire the data.

In the resonant cavity two opposite electrodes were placed coaxially in the cavity axis. A gas to be treated at atmospheric pressure flows into the cavity through a gas inlet forming swirl flow round electrodes and then passes between them, so that a microwave plasma is initiated between the two electrodes. One of the electrodes has an axial hole to provide an outlet from the resonant cavity for the plasma and operating gas (based on patent rights [29, 30]).The absorbed microwave power P_A by the discharge was calculated as $P_I - P_R$, where P_I and P_R are the incident and reflected microwave powers, respectively. The incident and reflected microwave powers P_I and P_R were directly measured using directional coupler equipped with bolometric heads and HP power meters.

Window in the wall of the generator allows for observation of the microwave discharge. To measure the spectra, the light emitted by the plasma was focused with a quartz lens (50 mm in diameter, focal length – 75 mm) onto double diaphragm with pinholes of a 1 mm diameter

Journal of Physics: Conference Series **406** (2012) 012033 doi:10.1088/1742-6596/406/1/012033

at the entrance slit of the spectrometer (see figure 3). The width of the spectrometer entrance slit was 50 μm (20 mm height). The double diaphragm were placed to collimate the light. Photos of microwave argon plasmas at different values of absorbed microwave power are shown in figure 4. The figure shows, that the light intensity increases with increasing the value of absorbed microwave power.

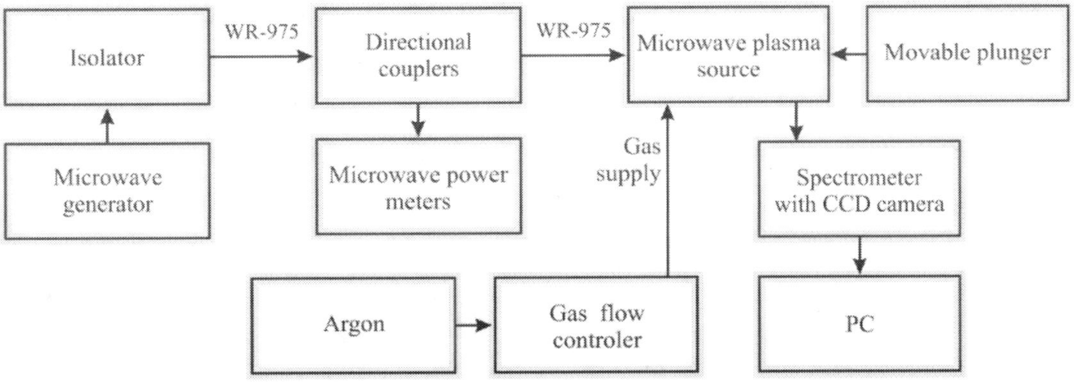

Figure 1. Diagram of the experimental setup.

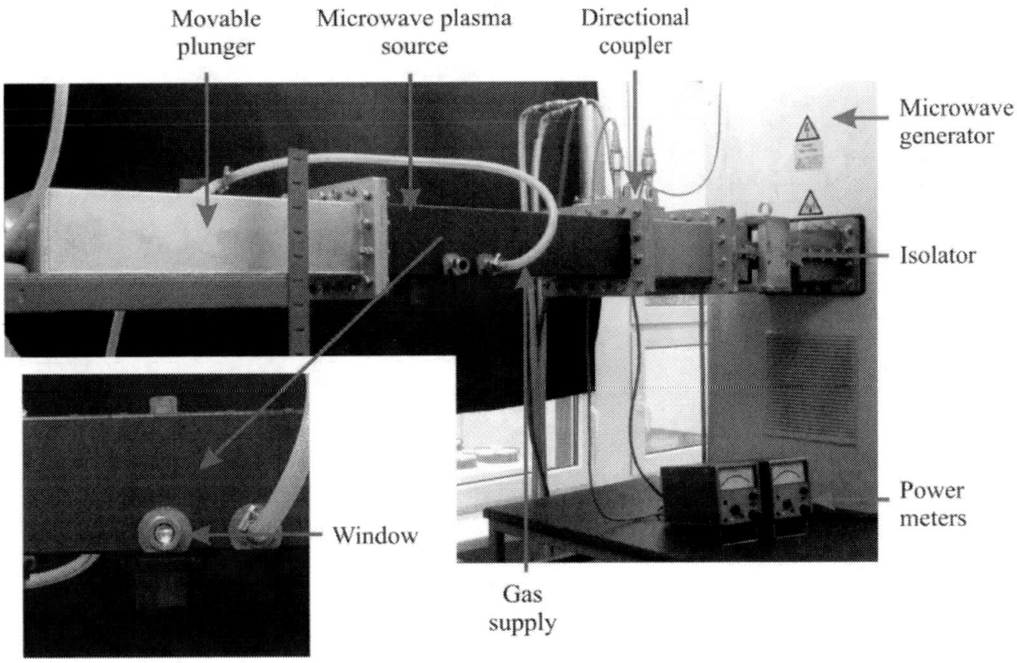

Figure 2. Photo of the experimental setup.

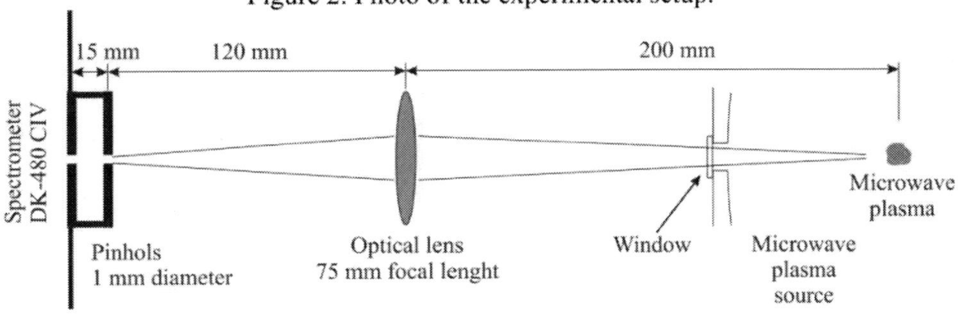

Figure. 3. The experimental setup used for the spectroscopic investigation.

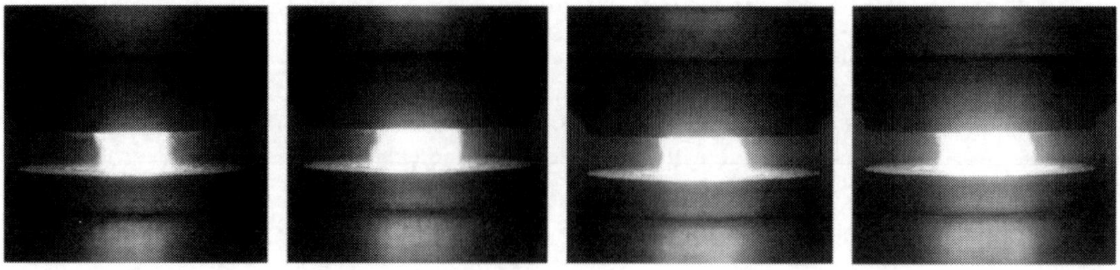

Figure 4. Photos of the microwave argon plasma at P_A: a) 0.3 kW, b) 0.4 kW, c) 0.5 kW, d) 0.6 kW. Flow rate - 200 l/min, distance between electrodes - 4.6 mm.

4 Results and discussion

In this section we present OES measurements of electron excitation, gas temperature and electron number density. Spectra were recorded and then corrected according to the wavelength sensitivity of CCD camera. The wavelength sensitivity of CCD camera was determined using a tungsten halogen calibration lamp. Using a Hg-Ne low-pressure calibration lamp (λ = 365.02 nm, 435.84 nm and 546.07 nm, Hg I) we measured that the Gaussian instrumental line profile FWHM was about 0.134 nm and 0.06 nm for 1200 and 3600 grooves/mm grating, respectively. Typically, spectra were measured with 1200 grooves/mm grating. Only for OH ($A^2\Sigma \rightarrow X^2\Sigma$) rotational band measurements the 3600 grooves/mm grating was used. All experimental tests were performed with the argon flow rate of 100 and 200 l/min.

4.1. Electron excitation temperature measurements

The electron excitation temperature can be determined using the Boltzmann-plot method from the selected argon $4p$–$4s$ and $5p$–$4s$ transition lines (figure 5). The list of selected argon lines with spectroscopic parameters: g_n, A_{nm}, λ_{nm} and E_n are shown in table 1. Electron excitation temperature can be estimated from the best linear fit of plot $\ln(I_{nm}\lambda_{nm}/g_nA_{nm})$ versus the energy E_u [17-18] (figure 6a). Figure 6b shows electron excitation temperature as a function of absorbed microwave power. Obtained T_{exc} at level 7000 K is independent of absorbed microwave power and flow rate.

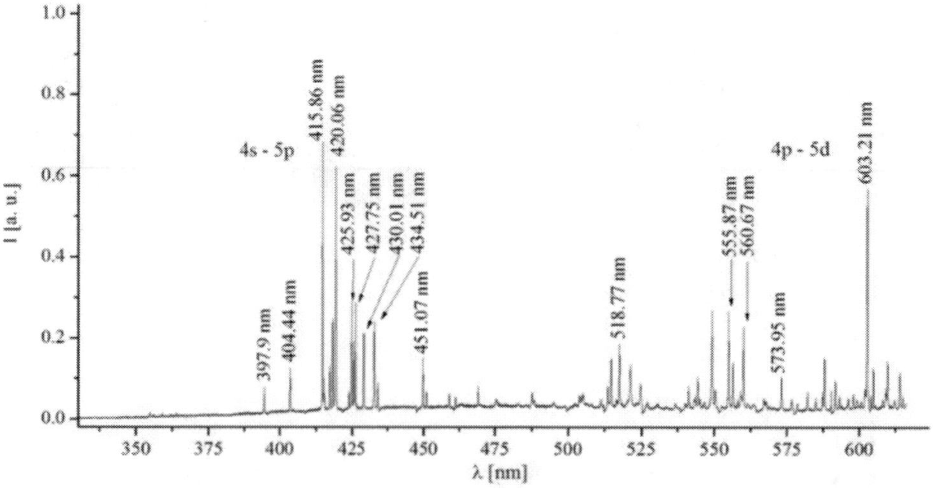

Figure 5. Emission spectrum of argon plasma with selected argon $4p$–$4s$ and $5p$–$4s$ transition lines for Boltzmann-plot method. Absorbed microwave power - 0.4 kW, flow rate - 200 l/min.

Table 1. Spectroscopic parameters of argon lines for Boltzmann plot method [31].

λ_{nm} [nm]	Radiative transition	A_{nm} [10^7 s^{-1}]	g_n	E_n [cm^{-1}]
394.9		0.0333	3	118459
404.44		0.0455	2	118469
415.86		0.1400	5	117183
420.06		0.0967	7	116942
425.93	4s - 5p	0.3980	1	118870
427.75		0.0797	3	117151
430.01		0.0377	5	116999
434.51		0.0297	3	118407
451.07		0.1180	1	117562
518.77		0.1380	5	123372
555.87		0.1420	5	122086
560.67	4s - 5d	0.0333	3	121932
573.95		0.0455	2	123505
603.21		0.1400	5	122036

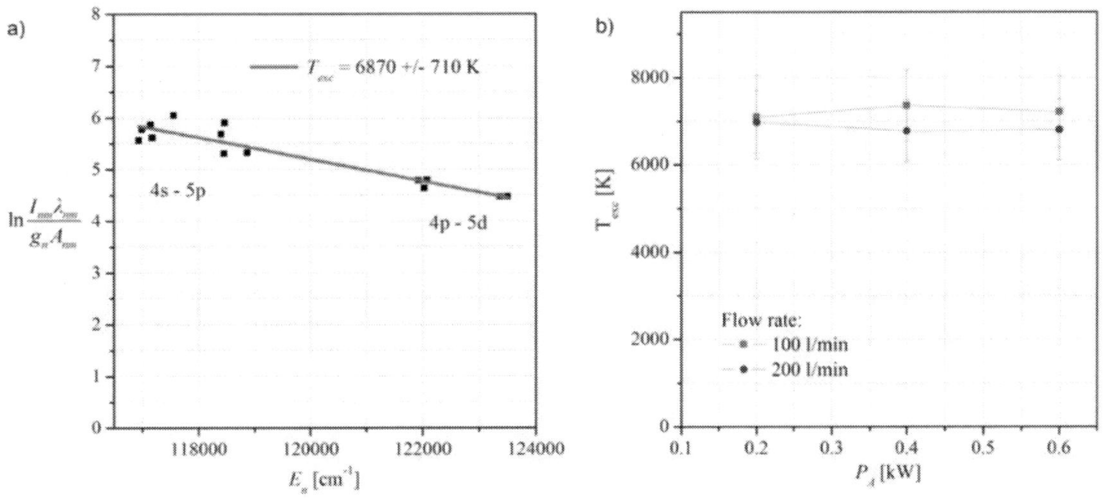

Figure 6. a) Boltzmann plot for argon $4p$–$4s$ and $5p$–$4s$ transition lines (P_A = 0.4 kW, Q = 200 l/min), b) Electron excitation temperature as a function of absorbed microwave power P_A for two values of argon flow rate 100 and 200 l/min.

4.2. Gas temperature measurements

To determine rotational temperature the emission spectra at the range of 305.5 - 309.5 nm were recorded (figure 7). The OH radical emissions are results of water vapor naturally present as impurity in the gas, so the intensity of the ($A^2\Sigma \rightarrow X^2\Sigma$) rotational band was very low. According to this in measurements the width of the entrance slit of the spectrograph was set at 100 μm. Due to the low light intensity, the double diaphragm (at the entrance slit) was not used.

After having the experimental spectrum of OH, a simulation of this band was carried out using LIFBASE program. The simulated OH ($A^2\Sigma \rightarrow X^2\Sigma$) band was fitted to the experimental one by variation of the rotational temperature. Figure 7 shows the comparison of the both spectra: experimental and simulated fitted to the experimental one. Figure 8 presents obtained T_{rot} as a function of absorbed microwave power. The temperature slowly increases with increasing absorbed microwave power and with decreasing the working gas flow rate (figure 8). The second way to determine T_{rot} was maximum amplitude relations G_0/G_{ref}, G_1/G_{ref} from the measured spectrums.

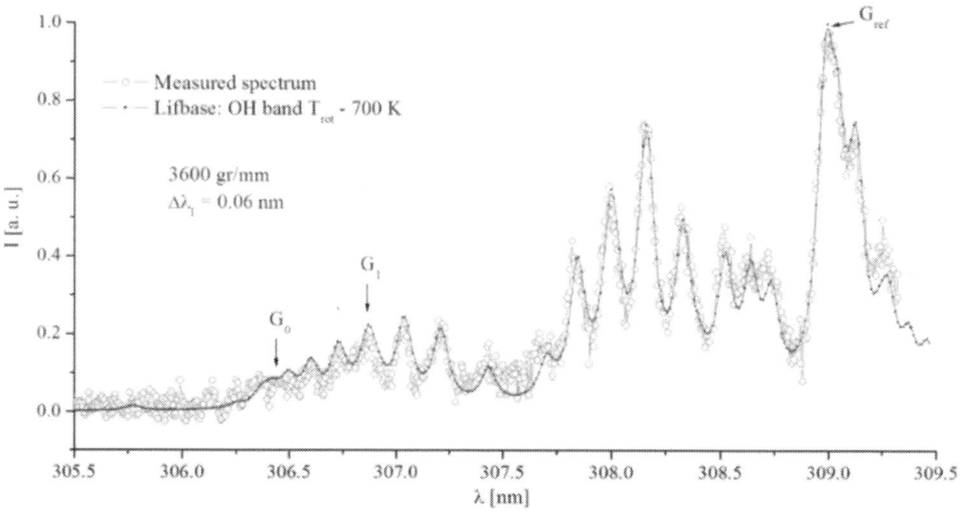

Figure 7. Comparison of the measured and simulated in LIFBASE emission spectra of OH rotational band in argon plasma (P_A = 0.6 kW, Q = 200 l/min).

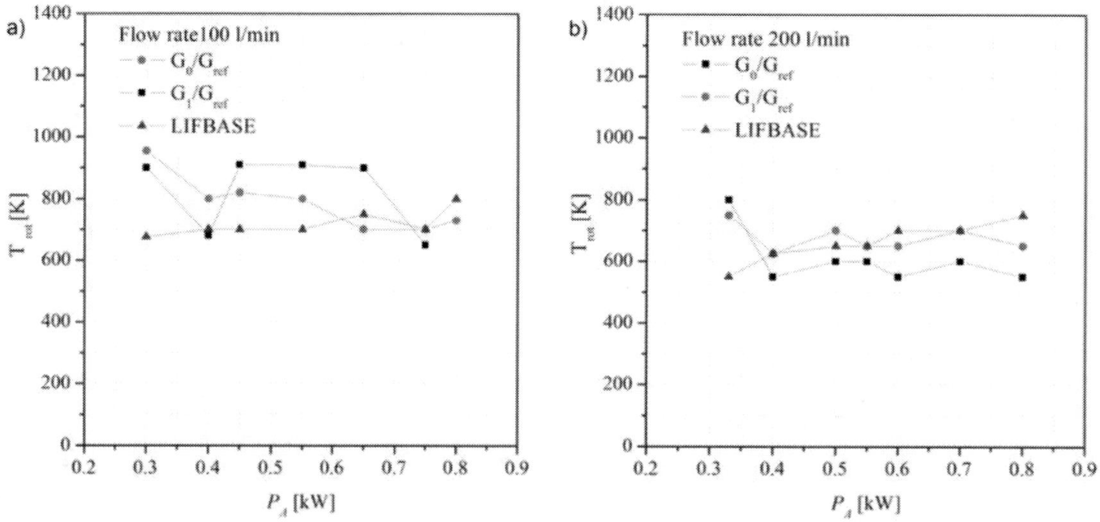

Figure 8. Measured rotational temperatures of OH radicals as a function of absorbed microwave power for two values of argon flow rate: a) 100 l/min, b) 200 l/min.

4.3. Electron number density measurements

To obtain a detectable intensity of H_β emission spectral line a small addition of H_2 (0.5%) was introduced to argon flow. Assuming that the gas temperature in discharge is constants (absence dependence of the absorbed microwave power, figure 8) and is at the level of 700 K for Q = 100 l/min and 650 K for Q = 200 l/min, respectively. For this values, dependence (3) allows to estimate the value of the Doppler broadening for the H_β line. From equation (2) the FWHM of Gaussian profile can be determine. The FWHM values of the Gaussian profile of H_β line is given in table 2. The parameters presented in table 2 are independent of the absorbed microwave power.

Table 2. H_β line Gaussian broadening effect FWHM.

Q [l/min]	$\Delta\lambda_I$ [nm]	$\Delta\lambda_D$ [nm]	$\Delta\lambda_G$ [nm]
100	0.134	0.009	0.134
200		0.008	

Figure 9 shows an example of H_β spectral line profile measured in microwave argon plasma and the Voigt profile fitted to the experimental points.

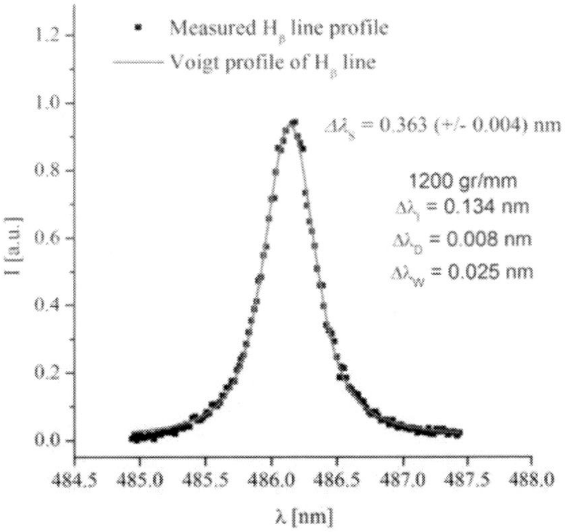

Figure 9. Measured H_β line profile and the Voigt function fitted to the experimental points for argon plasma. Absorbed microwave power P_A - 0.4 kW, flow rate - 200 l/min.

The fitting was made using the Voigt function with a Gaussian line width of 0.134 nm (table 2) in Origin software. Deconvolution the Voigt profile return value of the FWHM Lorentzian profile (Stark and van der Waals). The van der Waals broadening can be calculated from equation 4. Obtained $\Delta\lambda_W$ at the level of 0.051 nm for 650 K and 0.046 nm for 700 K, respectively. The difference between the measured FWHM of Lorentzian profile and calculated the van der Waals broadening gives the Stark broadening (2).

The electron number density calculated from Gig-Card and GKS theory is shown in figure 10. Obtained electron number density n_e ranged 2×10^{15} – 3.5×10^{15} cm^{-3} depending on the absorbed microwave power and flow rate. Generally, Gig-Card theory always gives lower values of the electron number density than GKS theory.

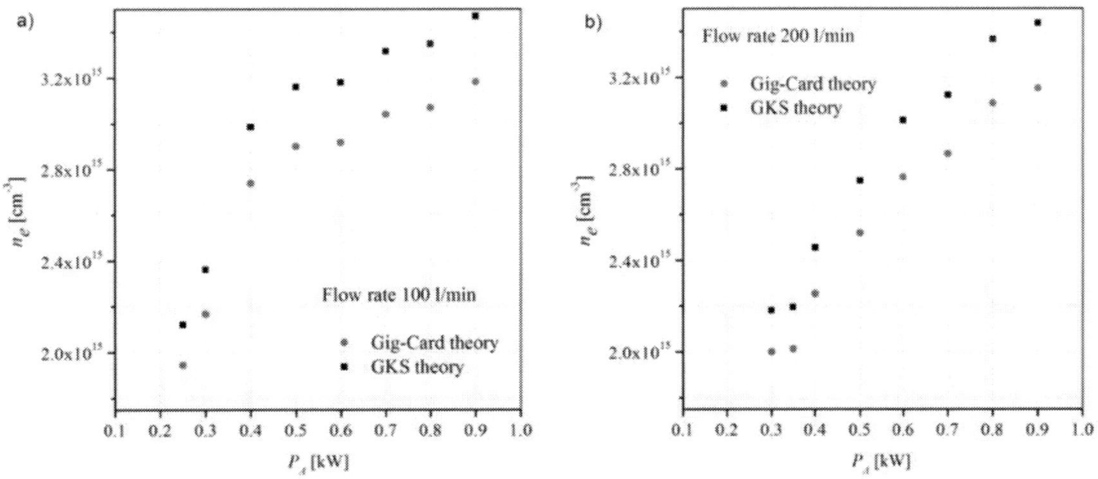

Figure 10 Electron density as a function of absorbed microwave power for two values of argon flow rate: a) 100 l/min, b) 200 l/min.

5. Conclusions

Spectroscopic study of microwave 915 MHz argon plasma at atmospheric pressure and high flow rate was presented in this work. The study concerns to determine electron excitation, gas temperature t and electron number density. The results showed that:

- Electron excitation temperature obtained at level 7000 K is independent of absorbed microwave power and argon flow rate,
- The rotational temperatures of OH radical ranged from 600 up to 800 K, depending on absorbed microwave power and argon flow rate,
- Electron number density was from 2×10^{15} – 3.5×10^{15} cm^{-3}, depending on the operating parameters (absorbed microwave power, argon flow rate) and theory that was used (GKS or Gig-Card theory).

The investigated microwave plasma source cavity-resonant type works very stable with various working gases. The parameters (flow rate, absorbed microwave power) can be changed in wide range that makes MPS an attractive tool for different gas processing at high flow rates.

References

[1] Tendero C Tixier C Tristant P Desmaison J and Leprince P 2006 Atmospheric pressure plasmas: a review *Spectrochim. Acta* B **61** 2-30

[2] Leins M Alberts L Kaiser M Walker M Schulz A Schumacher U and Stroth U 2009 Development and characterisation of a microwave-heated atmospheric plasma torch *Plasma Process. Polym.* **6** 227-232

[3] Baeva M Pott A and Uhlenbusch J 2002 Modelling of NOx removal by a pulsed microwave discharge *Plasma Sources Sci. Technol.* **11** 135–141

[4] Uhm H S Hong Y C and Shin D H 2006 A microwave plasma torch and its applications *Plasma Sources Sci. Technol.* **15** 26-34

[5] Mizeraczyk J Jasinski M and Zakrzewski Z 2005 Hazardous gas treatment using atmospheric pressure microwave discharges *Plasma Phys. Control. Fusion* **47** 589-602

[6] Kabouzi Y Moisan M Rostaing J C Trassy C Guerin D Keroack D and Zakrzewski Z 2003 Abatement of perfluorinated compounds using microwave plasmas at atmospheric pressure *J. Appl. Phys.* **93** 9483- 9497

[7] Jasinski M Mizeraczyk J Zakrzewski Z Ohkubo T and Chang J 2002 CFC-11 destruction by microwave torch generated atmospheric-pressure nitrogen discharge *J. Phys. D: Appl. Phys.* **35** 2274–2280

[8] Jasinski M Mizeraczyk J and Zakrzewski Z 2004 Microwave torch plasmas for decomposition of gaseous pollutants *J. Adv. Oxid. Technol.* **40** 51-58

[9] Jasinski M Mizeraczyk J Zakrzewski Z and Chang J 2002 Decomposition of C2F6 in atmospheric-pressure nitrogen microwave torch discharge *Czech. J. Phys.* D **52** 743-749

[10] Jasinski M Mizeraczyk J and Zakrzewski Z 2002 Decomposition of freons in atmospheric-pressure air using coaxial-line-based low-power microwave torch plasma *Journal of High Temperature Material Processes* **6** 317-320

[11] Jasinski M Dors M and Mizeraczyk J 2009 Destruction of freon HFC-134a using a nozzleless microwave plasma source *Plasma Chem. Plasma Process.* **29** 363-372

[12] Jasinski M Dors M and Mizeraczyk J 2009 *Przeglad Elektrotechniczny* **85** 121-123

[13] Jasinski M Dors M and Mizeraczyk J 2009 *Eur. Phys. J.* D **54** 179-183

[14] Jasinski M Mizeraczyk J and Zakrzewski 2002 Measurements of neutral gas temperature in microwave torch plasmas at atmospheric pressure *Czech. J. of Phys. D* **52** 421-426

[15] Jasinski M Mizeraczyk J and Zakrzewski 2006 Spectroscopic measurements of electron density in atmospheric-pressure surface wave sustained discharge in argon *Czech. J. of Phys.* D **56** 787-794

[16] Czylkowski D Jasinski M and Mizeraczyk J 2006 Argon and neon plasma columns in continuous surface wave microwave discharge at atmospheric pressure *Czech. J. of Phys.* B **56** 684-689

[17] Griem H R 1964 *Plasma Spectroscopy* (New York, McGraw-Hill)

[18] Sismanoglu B N Grigorov K G Santos R A Caetano R Rezende M V O Hoyer Y D and Ribas V W 2010 Spectroscopic diagnostics and electric field measurements in the near-cathode region of an atmospheric pressure microplasma jet *Euro. Phys. J.* D **60** 505-16

[19] Moon S and Choe W 2003 A comparative study of rotational temperatures using diatomic OH, O_2 and N_2+ molecular spectra emitted from atmospheric plasmas *Spectrochim. Acta Part* B **58** 249-257

[20] Kopecki J Kiesler D Leins M Schulz A Walker M Kaiser M Muegge H and Stroth U 2010 Investigations of a high volume atmospheric plasma torch at 915 MHz *Surface & Coatings Technology* **205** 342-346

[21] Izarra Ch 2000 UV OH spectrum used as a molecular pyrometer *J. Phys. D: Appl. Phys.* **33** 1697-1704

[22] Luque J and Crosley D R 1999 LIFBASE: Database and spectral simulation program (Version 1.5) *SRI International Report* **MP** 99-009

[23] Hrycak B Jasinski and M Mizeraczyk J 2010 Spectroscopic investigations of microwave microplasmas in various gases at atmospheric pressure, *Euro. Phys. J.* D **60** 610 - 619

[24] Lazzaroni C Chabert P Rousseau A and Sadeghi N 2010 Sheath and electron density dynamics in normal and self-pulsing regime of a micro hallow cathode discharge in argon gas *Euro. Phys. J.* D **60** 555-563

[25] Jasinski M Mizeraczyk J and Zakrzewski Z 2004 *in Proceedings of the XVth Int. Conf. on Gas Discharges and their Applications,* Toulouse

[26] Sismanoglu B N Grigorov K G Caetano R Rezende M V O and Hoyer Y D 2010 Spectroscopic measurements and electrical diagnostic of microhallow cathode discharges in argon flow at atmospheric pressure *Euro. Phys. J.* D **60** 505-516

[27] Griem H R 1974 Spectral line broadening by plasmas *Academic Press New York* 316

[28] Gigosos M A and Cardenosos V 1996 *J. Phys. B: AT. Mol. Opt. Phys.* **29** 4795

[29] Gillespie R F Hall S I and Raybone D and Winterbottom F 2000 Plasma gas processing U.S. patent no. US 6126779

[30] Bayliss K H 1995 Plasma generator with field-enhancing electrodes U.S. patent no. US 5418430

[31] http://physics.nist.gov/PhysRefData

Journal of Physics: Conference Series **406** (2012) 012034 doi:10.1088/1742-6596/406/1/012034

Shock waves in water at low energy pulsed electric discharges

M.E. Pinchuk, V.A. Kolikov, Ph.G. Rutberg, A.G. Leks, R.V. Dolinovskaya, V.N. Snetov, A.Yu. Stogov

Institute for Electrophysics and Electric Power of Russian Academy of Sciences, Dvortsovaya nab. 18, St.-Petersburg, 191186, Russia

E-mail: rc@iperas.nw.ru, pinchme@mail.ru

Abstract. Experimental results of shock wave formation and propagation in water at low energy pulsed electric discharges are presented. To study the hydrodynamic structure of the shock waves, the direct shadow optical diagnostic device with time resolution of 5 ns and spatial resolution of 0.1 mm was designed and developed. Synchronization of the diagnostic and electrodischarge units by the fast optocouplers was carried out. The dependences of shock wave velocities after breakdown of interelectrode gap for various energy inputs (at range of $\leq$1 J) into discharge were obtained. Based on the experimental results the recommendations for the adjustment parameters of the power supply and load were suggested.

1. Introduction

Nowadays a low energy pulsed electric discharges have wide prospects in various applications such as materials processing, plasma-chemical technologies and so on. Widespread use is also found in medical or biological applications [1-4]. One of widely spread practice is using the discharge for water purification [5]. In each case advantages of pulsed discharge is caused by many physical, technical, and economical factors. That is concerned the last mentioned application areas, this great perspectives are connected with prolonged microbial resistance of water [6,7], high efficiency of conversion or purification from organic substance [8] and from microorganisms during the discharge[9].
Apparently, UV-radiation and shock waves play main role for disinfection of water by pulsed electric discharges among the factors acting directly on pathogenic bacteria during the discharge [7].
To kill the bacteria by the shock waves it is necessary to break the bacteria membrane. In this case the pressure front thickness must be comparable with size of microorganisms, and the amplitude of pressure in the front of shock waves must be >2 MPa.
The energy transfer from the discharge channel to the shock wave is not fully studied so far. The simple increasing of energy input in the discharge does not lead to significant increasing of the pressure steepness and amplitude pressure of shock wave [6,7]. Thus the electro acoustic coefficient of energy transfer from discharges to water is low, the energy released in the channel of leaders, is small due to high ballast resistance of water [10]. Main channel, formed in a subsequent step of discharges, also has low resistance and energy-release is redistributed in the power supply system. Discharge channel is practically no longer expands and not transmits the mechanical energy to the surrounding water. From an economic and technological points of view it is advantageously to use the low power installations.
Thus the purpose of this work was investigations of shock waves formation, research and development industrial installation with controllable level of energy transfer to shock wave at low energy in pulse.

2. Experimental setup

Electric discharge under investigation were initiated in discharge chamber with volume 8 cm^3 made of an insulating material by electric pulse generator. Parameters of installation are: output water flow from 0 (without water flowing) to 2 l/min, electric pulses with a duration of 1–20 µs, energy in pulse of ≤1 J, and a repetition rate of 1–150 Hz. Water inlet was diameter of 10 mm. In the chamber quartz windows were installed for optical diagnostic. Diameter of copper electrode was 0.5 mm and interelectrode gap of axisymmetric electrode system "wire to wire" from 1 to 10 mm was varied. A scheme of the experimental setup is shown on figure. 1.

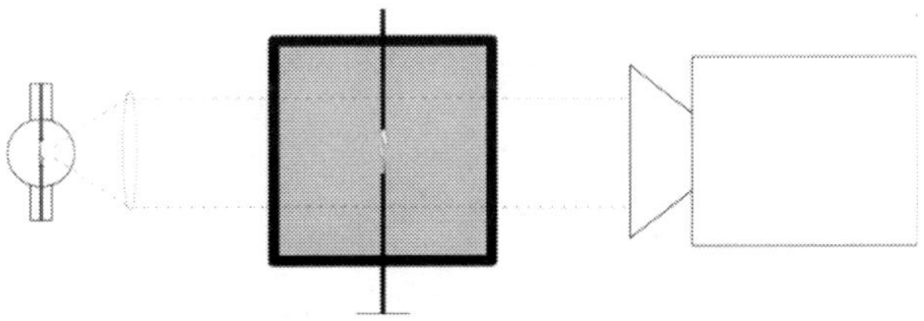

Figure 1. Experimental setup

To study the hydrodynamic structure of the shock waves, the optical diagnostic installation using direct shadow method with a time resolution of 5 ns and a spatial resolution of 0.1 mm was designed and developed. Gas-discharge xenon lamp was used as backlight source for shadow method. Shadow image registration was made by 16-frames high-speed CCD camera Cordin-222-16 with exposure times down to 5 ns. In condition of high electromagnetic level of interferences from electrodischarge installation we were compelled to make original system of synchronisation, based on fast optocouplers, with diagnostic devices. Diagnostic system was started by synchrosignal from special digital filter considering rather small reduction of voltage on discharge gap, signalling about the breakdown beginning. The dead time from synchrosignal to image exposition start was about 50 ns. So this system allowed to record series of images for subsequent independent discharges just from its beginning breakdowns.

Power supply with energy recuperation was specially designed and made for intelligent regulation of input energy level into the discharge gap [11]. Turning on of the transformer with saturating core in the circuit of the load (with water gap) allows to limit a current in the discharge gap at the main channel stage of the discharge and to return energy on a storage capacitor in the source. And regulation of input energy into discharge channel can be make by change of feedback factor in recuperation energy level. Current and voltage oscillograms at input energy into interelectrode gap of 0.15 J and 0.35 J are presented on figure 2. And respectively, for example, it is possible to provide optimal energy input for generation of a shock wave.

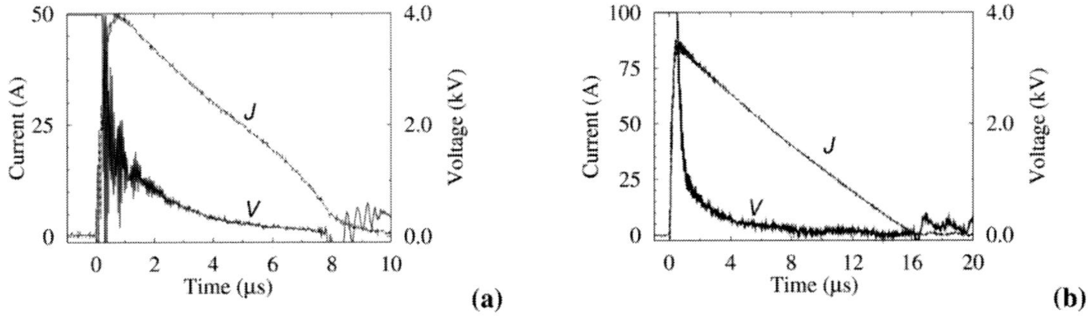

Figure 2. Current and voltage curves. Energy in pulse: (a) – 0.15 J, (b) – 0.35 J.

3. Experimental results

Shock wave velocities were measured for different energy input into discharge and for different interelectrode gaps. Typical shadow image is shown on figure 3. A shock wave place changing was measured from frame to frame and the shock wave velocity was calculated at different time from water breakdown by this measurements.

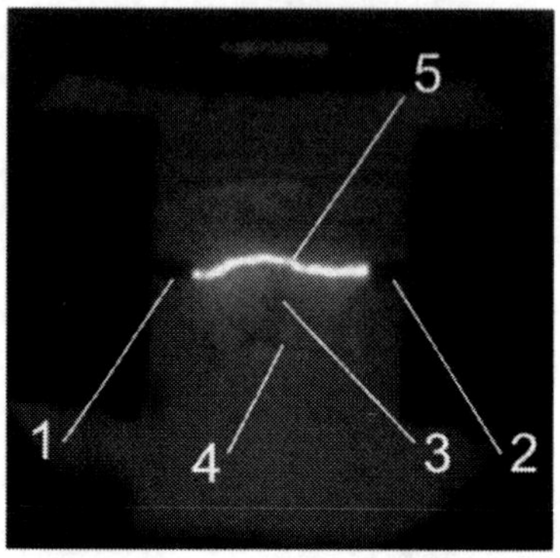

Figure 3. Shadow image of discharge at 2 μs after breakdown (interelectrode gap of 7 mm): 1 — anode, 2 — cathode, 3 — shock wave from anode, 4 - shock wave from cathode, 5 — discharge channel.

The pictures of streamers with different structures of three varieties: from the anode, from the cathode, and from both electrodes simultaneously (figure 4a) were obtained at initial stage of the gap breakdown. In most of the cases, the shock waves were formed near to the ends of electrodes and have similar velocities. But schock waves were formed by forming main discharge channel (figure 4c), not streamers.

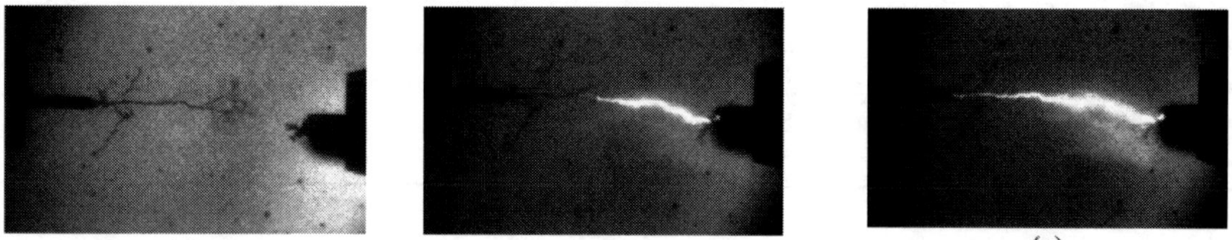

(a) **(b)** **(c)**

Figure 4. Shadow image of discharge gap at stage of formation main discharge channel. Time between frames is 100 ns.

Images on figure 3 showed two spherical shock waves forming near the ends of the electrodes. And at an very early stage their velosity was ~1×10^4 m/s. The initial stage of shock wave formation is characterized by local extreme parameters of substance. Pressure behind the shock wave front at speed of ~10 km/s according to [12] corresponds to pressure 1.2×10^{11}GPa. After ~500 ns in the far Fraunhofer zone both waves merged in visually one quasispherical wave with velocity of ~1.5×10^3 m/s. Dependencies of shock wave velocity versus time are presented on figure 5 for other

different parameters. This data correspond to the results obtained previously under similar conditions [13].
At an initial voltage rising before breakdown (that corresponds to increasing of an interelectrode gap) corresponds increasing in number of streamers and also embeded energy into an interelectrode gap is increased at stage of forming main discharge channel. So shock wave intensity is increased too. An increasing of power at constant energy in an pulse leads to growing in an amplitude of pulse pressure [7]. In this system the current steepness practically doesn't depend from the storage (and input) energy. And energy in a shock wave at the less energy in an pulse is same as one at more energy (figure 5b). Shock wave intensity was increased by growing of interelectrode gap (figure 5a).

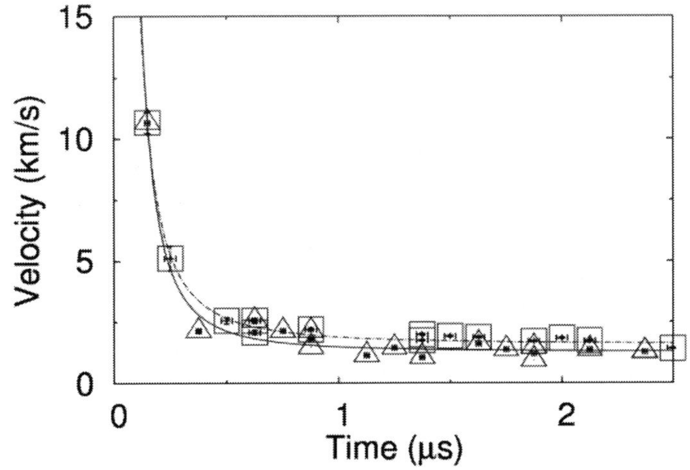

(a)

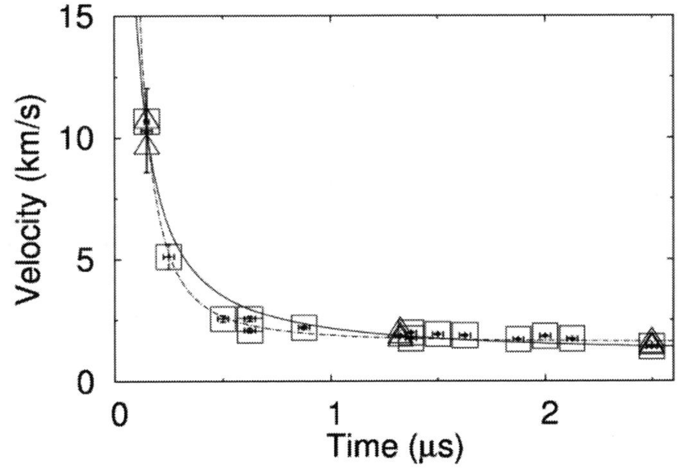

(b)

Figure 5. Shock wave velocity vs time after breakdown: (a) — Energy in pulse of 0.35 J, interelectrode gap: Δ - 1.7 мм, □ - 7 мм; (b) — Interelectrode gap of 7 mm. Energy in pulse: Δ - 0.15 J, □ - 0.35 J.

4. Conclusion

Thus, although a shock wave intensity is grown at increasing of a pulse steepness and a general pulse energy. Nevertheless, general energy reduction in a pulse doesn't lead to considerable reduction of shock wave intensity. On length of some millimetres from the discharge channel intensities of shock waves for different energy in pulse are became approximately equal and shock waves degenerated into weak wave. The formation of shock waves occurs at the stage of formation main discharge channel, and there is insignificant influence of a further energy input into channel at the main discharge stage.

Acknowledgments
This work was partially supported by the Russian Foundation for Basic Research, project nos. 11-08-00674.

5. References

[1] Ph.G.Rutberg, V.L.Gorjachev, V.A.Kolikov, V.N.Snetov, A.Yu.Stogov. Main results of investigations of pulsed electric discharges in water carried out at IEE RAS. *Journal of High Temperature Material Processes* **14** n1 (2010) 167-174

[2] Ph. Rutberg, V. Kolikov, V. Snetov, A. Stogov, A. Moshkin, and M. Khalilov. Application of aqueous dispersions of silver nanostructures for treatment of pyoinflammatory diseases with a chronic component. *J. Phys.: Conf. Ser.*. **275** (2011) 012010 6pp

[3] Ph.Rutberg, V.Kolikov, V.Snetov, A.Stogov, L.Noskin, S.Landa, A.Arutjuman, V.Egorov, A.Sirotkin. Uppermolecule complexes of oxide nanostructures and albumins formation. *Journal of High Temperature Material Processes.* **13** n3 (2009) 277-284

[4] Ph.G.Rutberg, M.V.Dubina, V.A.Kolikov, F.V.Moiseenko, E.V.Ignat'eva, N.M.Volkov, V.N.Snetov, and A.Yu.Stogov. Effect of Silver Oxide Nanoparticles on Tumor Growth in Vivo. *Doklady Biochemistry and Biophysics.* **421** (2008) 191-193.

[5] V.L. Goryachev, F.G. Rutberg and A.A. Ufimtsev. Photolytic properties of a pulsed discharge in water. *Technical Physics Letters.* **24** n2(1998) 122-123

[6] Ph.G.Rutberg, V.A.Kolikov, V.E.Kurochkin, L.K.Panina and A.Ph.Rutberg, Electric Discharges and the Prolonged Microbial Resistance of Water. *IEEE Transactions on Plasma Science.* **35** n4 August (2007) 1111-1118

[7] V.A. Kolikov, V.E. Kurochkin, L.K. Panina, A.F. Rutberg, F.G. Rutberg, V.N. Snetov and A.Yu. Stogov. Prolonged microbial resistance of water treated by a pulsed electrical discharge *Technical Physics.* **52** n2 (2007) 263-270

[8] L.A. Yutkin. Elektrogidravlicheskiy effekt i ego primenenie v promyshlennosti. (Electrohydraulic Effect and Its Industrial Application) Mashinostroenie, Leningrad, 1986 [in Russian].

[9] A.A. Bogomaz, V.L. Goriachev, A.S. Remennui, Ph.G. Rutberg. About of pulse electric discharge efficiency for water disinfection. *Technical Physics Letters.* **17** n12 (1991) 65-68

[10] K.A. Naugol'nykh and N.A. Roy. Elektricheskie razryady v vode. (Electrical Discharge in Water). Nauka, Moscow, 1971 [in Russian].

[11] S. V. Korotkov, Yu. V. Aristov, A. K. Kozlov, D. A. Korotkov, and I. A. Rol'nik. A generator of electrical discharges in water. *Instruments and Experimental Techniques.* **54** n2 (2011) 190-193.

[12] Shock wave database. http://ihed.ras.ru/rusbank/

[13] V.A. Kolikov, M.E. Pinchuk, A.G. Leks, Ph.G. Rutberg. High-speed diagnostic of pulsewise-periodic electric discharge in water. *Proceeding of SPIE.* **6279** paper192 (2007) 62795C1-62795C7

Dynamics of bubble generated by low energy pulsed electric discharge in water

M.E. Pinchuk, V.A. Kolikov, Ph.G. Rutberg, A.G. Leks, R.V. Dolinovskaya, V.N. Snetov, A.Yu. Stogov

Institute for Electrophysics and Electric Power of Russian Academy of Sciences, Dvortsovaya nab. 18, St.-Petersburg, 191186, Russia

E-mail: rc@iperas.nw.ru, pinchme@mail.ru

Abstract. Results of investigations of bubble formation and dynamics for discharge in water are presented. Experiments were carried out in discharge chamber with axisymmetric electrode system "wire to wire". Interelectrode gap was varied from 1 to 10 mm. Energy in a pulse was <1 J. Velocity of bubble expantion and collapse is about several hundreds meter per second at early stage of discharge. Bubble pulsation period is 0.5 - 1 ms. Increasing of energy released in the discharge gap will increase bubble pulsation period. Little bubble was formed by reducing energy input into discharge. But the main stage of discharge always followed by bubble formation. Specific erosion is measured for different energy in pulse and matched up with bubble collapse.

1. Introduction

Pulsed electrical discharge in water is an object of intense researches now. Pulsed electrical discharge with a low pulse energy has good prospects for use in various applications [1] such as materials processing, plasma-chemical technologies and so on. One of the most promising area for its practice using is metal oxide nanoparticles manufacture [2]. Its production based on water pulsed discharge methode has high productivity [3]. This nanoparticles (and substance with it) start to play a great role in many biomedical technology [4-7].

For this it is very important to know the sizes and mass of nanoparticles manufactured in dispersions and how their sizes and mass depend on characteristics of discharge and other conditions. But the electrical discharge in water is very complex physical phenomena. And there are many factors affecting on relations between the distribution of nanoparticles subfractions in size, specific erosion of electrodes and discharge parameters.

Bubble is always formed in underwater spark discharge. And electrode erosion depends from its appearance and dynamics [8]. Thus the purpose of this work was observation of bubble dynamics matching up electrode erosion.

2. Experimental setup

Electric discharge under investigation were initiated in discharge chamber with volume 8 cm^3 made of an insulating material by electric pulse generator. Parameters of installation are: output water flow from 0 (without water flowing) to 2 l/min, electric pulses with a duration of 1–20 μs, energy in pulse of ≤1 J, and a repetition rate of 1–150 Hz. Water inlet was diameter of 10 mm. In the chamber quartz windows were installed for optical diagnostic. Diameter of copper electrode was 0.5 mm and interelectrode gap of axisymmetric electrode system "wire to wire" from 1 to 10 mm was varied. A scheme of the experimental setup is shown on figure. 1.

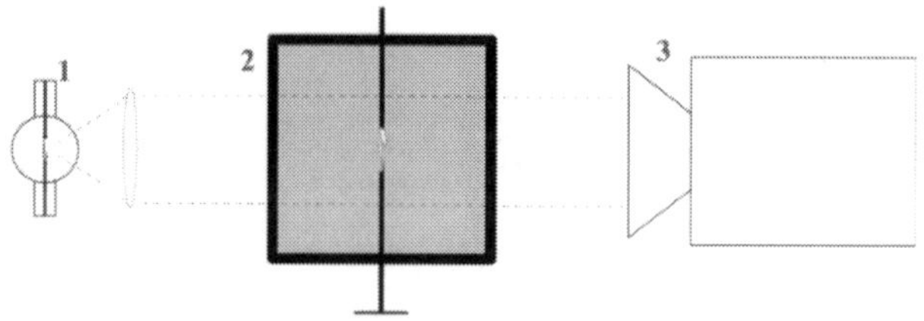

Figure 1. Experimental setup. 1 -backlight source, 2 – discharge chamber, 3 – high-speed CCD camera.

To study the hydrodynamic structure of the discharge, the optical diagnostic installation using direct shadow method with a time resolution of 5 ns and a spatial resolution of 0.1 mm was designed and developed. Gas-discharge xenon lamp was used as backlight source for shadow method. Shadow image registration was made by 16-frames high-speed CCD camera Cordin-222-16 with exposure times down to 5 ns. In condition of high electromagnetic level of interferences from electrodischarge installation we were compelled to make original system of syncronisation, based on fast optocouplers, with diagnostic devices. Diagnostic system was started by synchrosignal from special digital filter considering rather small reduction of voltage on discharge gap, signalling about the breakdown beginning. The dead time from synchrosignal to image exposition start was about 50 ns. So this system allowed to record series of images for subsequent independent discharges just from its beginning breakdowns.

Power supply with energy recuperation was specially designed and made for intelligent regulation of input energy level into the discharge gap [9]. Turning on of the transformer with saturating core in the circuit of the load (with water gap) allows to limit a current in the discharge gap at the main channel stage of the discharge and to return energy on a storage capacitor in the source. And regulation of input energy into discharge channel can be make by change of feedback factor in recuperation energy level. Current and voltage oscillograms at input energy into interelectrode gap of 0.15 J and 0.35 J are presented on figure 2.

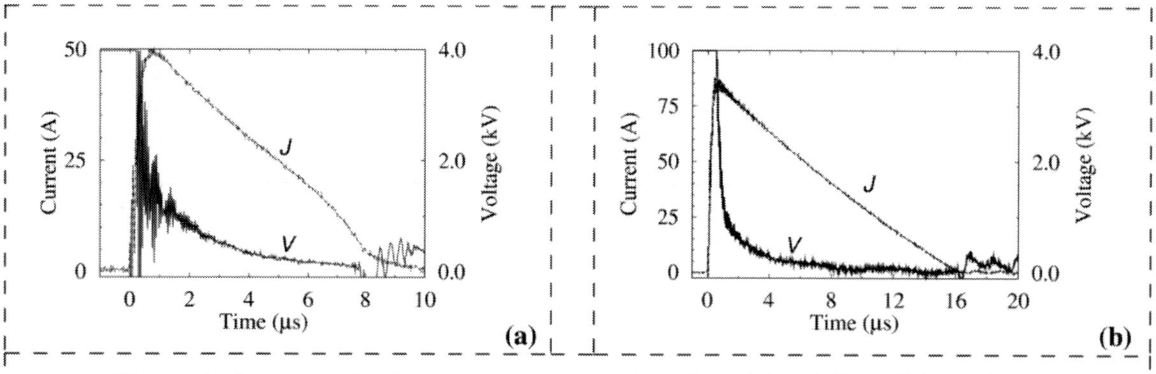

Figure 2. Current and voltage curves. Energy in pulse: (a) – 0.15 J, (b) – 0.35 J.

3. Experimental results

Bubble expansion and contraction are presented on figure 3. Usually bubble became hemispherical near its maximum radius. But its initial shape is depended from breakdown channel form. Bubble lifetime and radius are defined by input energy into discharge gap. Dependencies of bubble radius versus time are presented on figure 4. After expansion bubbles contracted and disappeared, often collapsed near electrode surface (figure 5).

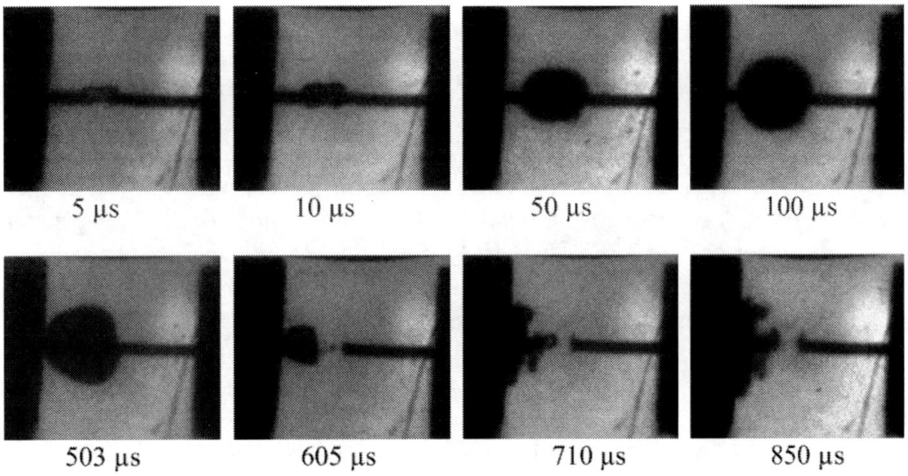

Figure 3. Bubble evolution.

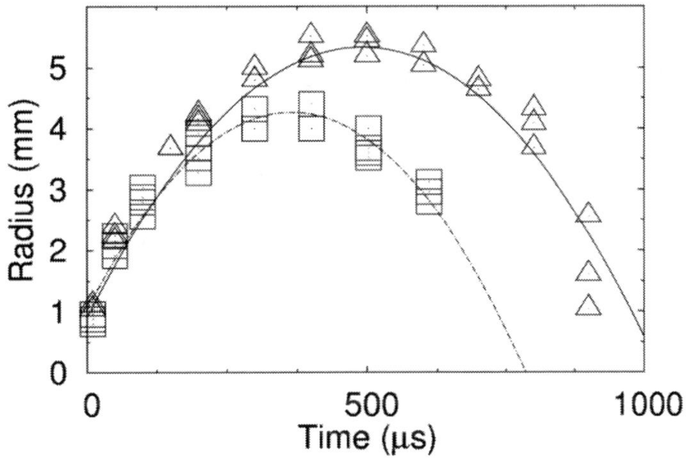

Figure 4. Bubble radius vs time. Interelectrode gap of 7 mm.
Energy in pulse: Δ - 0.35 J, □ - 0.15 J.

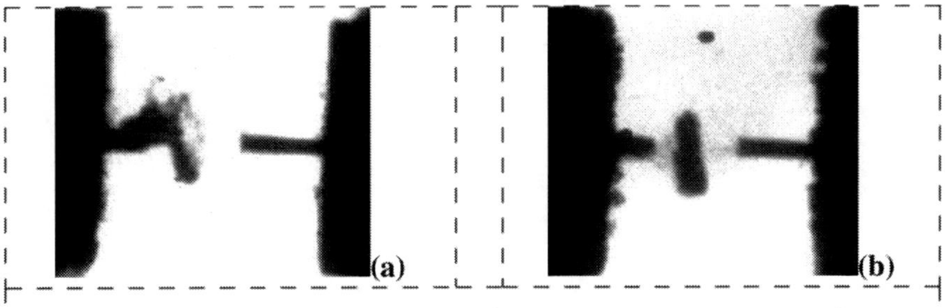

Figure 5. Bubble collapse: (a) – on electrode, (b) – between electrodes.

Channel discharge stage are occured in water vapour media of expanding bubble at short interelectrode gaps. At early phase of discharge the pressure value is of hundreds of atmospheres. Pressure in bubble is decraesed down to 1 MPa by the end of discharge current. So the processes at the stage are similar to discharge in high pressure gas media. Calculation of heat flux on electrode tips shown that liquid phase appeared on electrodes surface after several mikroseconds. And collapsed bubble sprinkles the liquid film on electrodes.

Specific electrode erosion really more higher at short interelectrode gap then long gap. And specific erosion has similar value for different charge throgh gap at current pulse (different energy in pulse). Specific electrode erosion for anode is of 2.2×10^{-4} g/C and for cathode of 2.2×10^{-4} g/C correspondingly in gap of ~2 mm at energy of 0.35 J in pulse. So erosion of both electrodes are very close. And specific erosion is of 3.8×10^{-4} g/C for both electrodes in gap of ~2 mm at energy of 0.15 J in pulse that close to previous value for both electrodes. But specific erosion for anode is of 1.7×10^{-4} g/C and for cathode of 0.9×10^{-4} g/C correspondingly for gap of ~7 mm at energy of 0.35 J in pulse.

Acknowledgments

This work was partially supported by the Russian Foundation for Basic Research, project nos. 11-08-00674.

4. References

[1] *Plasma Sources Sci. Technol.* Special Issue on Plasmas with Liquids. **20** n 3 June (2011) 030201-035024

[2] Ph. Rutberg, V. Kolikov, V. Snetov, A. Stogov, L. Noskin, S. Landa, A. Arutjunan. Pulsed electric discharges in water and oxide nanoparticles. *Proceeding of 16th IEEE International Pulsed Power Conference.* **V2** #1528 (2007) 1244-1247

[3] Ph.G. Rutberg, V.A. Kolikov, V.E. Kurochkin, V.G. Mal'tsev. Method to manufacture nanoparticles. Russian Federation patent number 2 272 697 C1, patent date Sept. 7 (2004).

[4] V.A. Kolikov, V.E. Kurochkin, L.K. Panina, A.F. Rutberg, F.G. Rutberg, V.N. Snetov and A.Yu. Stogov. Prolonged microbial resistance of water treated by a pulsed electrical discharge *Technical Physics.* **52** n2 (2007) 263-270

[5] Ph.G.Rutberg, V.A.Kolikov, V.E.Kurochkin, L.K.Panina and A.Ph.Rutberg, Electric Discharges and the Prolonged Microbial Resistance of Water. *IEEE Transactions on Plasma Science.* **35** n4 August (2007) 1111-1118

[6] Ph.G.Rutberg, M.V.Dubina, V.A.Kolikov, F.V.Moiseenko, E.V.Ignat'eva, N.M.Volkov, V.N.Snetov, and A.Yu.Stogov. Effect of Silver Oxide Nanoparticles on Tumor Growth in Vivo. *Doklady Biochemistry and Biophysics.* **421** (2008) 191-193.

[7] Ph. Rutberg, V. Kolikov, V. Snetov, A. Stogov, A. Moshkin, and M. Khalilov. Application of aqueous dispersions of silver nanostructures for treatment of pyoinflammatory diseases with a chronic component . *J. Phys.: Conf. Ser.*. **275** (2011) 012010

[8] M. Kanemaru, S. Sorimachi, S. Ibuka and S. Ishii. Single bubble generated by a pulsed discharge in liquids as a plasma microreactor. *Plasma Sources Sci. Technol.* Special Issue on Plasmas with Liquids. **20** n3 June (2011) 034007 7pp

[9] S. V. Korotkov, Yu. V. Aristov, A. K. Kozlov, D. A. Korotkov, and I. A. Rol'nik. A generator of electrical discharges in water. *Instruments and Experimental Techniques.* **54** n2 (2011) 190-193

Arc Motion in an Obliquely Imposed Alternating Magnetic Field

R Akiho[1], K Takeda[1] and M Sugimoto[1]

[1] Department of Machine Intelligence and Systems Engineering, Akita Prefectural University, 84-4 Tsuchiya-Ebinokuchi, Yurihonjo, Akita 015-0055, Japan

E-mail: M14A001@akita-pu.ac.jp

Abstract. The arc motion is theoretically investigated under an alternating magnetic field imposed obliquely to the arc. The arc is known to oscillate on a 2-D plane when the alternating magnetic field is imposed perpendicularly to the arc. If the alternating magnetic field is imposed obliquely to the arc, then it is expected that the arc oscillates not on the 2-D plane but in a 3-D space. For this study, 3-D simulation was performed on the motion of the plasma gas under an alternating magnetic field crossing obliquely to the arc. It was also assumed that a stream line of the plasma gas represented the arc profile. The momentum equation for the plasma gas was solved together with the continuity equation. Governing parameters for the gas motion are θ (crossing angle), v_0 (initial velocity of the plasma gas), and λ. Parameter λ is defined as $\lambda = (I_a B_0)/Q_0$. Numerical results are reported under different operating conditions such as magnetic flux densities and the angles between the arc and the magnetic flux. If the crossing angle is larger than $4/\pi$, the arc might be extinguished because of the drastic increase of the arc length.

Keywords. Transferred arc, Alternating magnetic field, Momentum equation, 3-D simulation

1. Introduction

Arcs, which are widely used as high-intensity heat sources in various industrial fields, have power output that is easily controllable by adjusting the arc current. Various chemical atmospheres can be provided using different plasma gases. Among different modes of arc operation, a transferred arc is often used in many material processes such as welding, cutting, melting, and refining metals because of its high-efficiency heat transfer. Thermal and electromagnetic pinch effects stabilize the arc, which concentrates the heat flow into a small area. This property is desirable to obtain high temperatures in a narrow zone of the material, but it is not convenient for heating wide areas of materials.

Several trials were attempted to expand the arc heat flow [1–3]. The authors obtained a wide heat source by imposing the alternating magnetic field perpendicularly to a transferred arc [4–6]. As shown in Figure 1, the arc is driven by the electromagnetic force induced from the interaction between the magnetic field and arc current. The high-speed oscillating arc can be regarded as an expanded heat source with a width of the oscillation amplitude. In addition to conventional arc characteristics, a magnetically driven arc presents the following advantages: (1) the arc width can be varied easily by adjusting the imposed magnetic field, and (2) the density distribution of heat flux can be controlled by varying the waveform of the imposed magnetic field. For example, if the waveform of the magnetic field is rectangular, then the heat flow is concentrated at both ends of the oscillation.

In previous work on a magnetically driven arc, theoretical, and experimental investigation was restricted to the magnetic field perpendicular to the arc, where the arc moved two-dimensionally in a single plane. In the present work, the authors consider the arc motion under the magnetic field imposed obliquely at an arbitrary crossing angle to the arc. In such an arrangement, the arc is expected to move in a three-dimensional space.

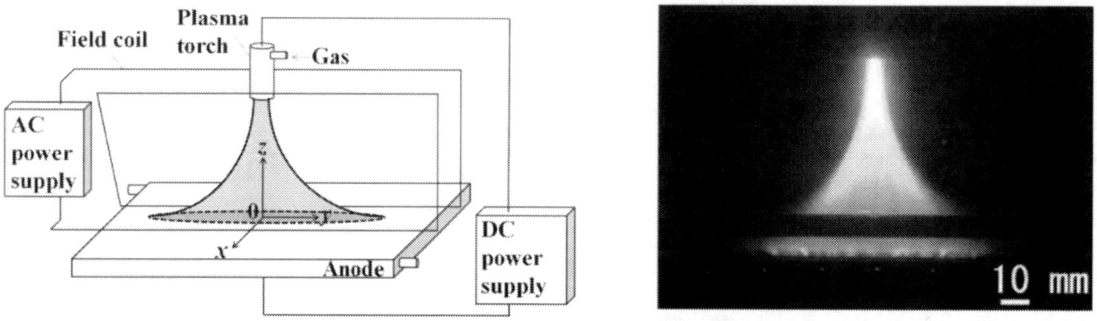

Figure 1. (a) Schematic illustration of the arrangement used to produce the magnetically driven arc, and (b) a typical expanded arc [5].

2. Modeling of an arc motion under the external magnetic field at an arbitrary crossing angle

2.1. Governing equation

A theoretical model is considered under the arrangement shown in Figure 2. A DC transferred arc is generated between cathode in a plasma torch and a metallic plate that serves as an anode. Plasma gas is supplied to the torch and is ejected from the torch orifice. The origin of the coordinate system used in the theoretical modeling is located at the torch orifice exit. The direction and strength of the imposed magnetic field change periodically with time, but they are uniform in space. Physical parameters and their nomenclature are presented in Table 1.

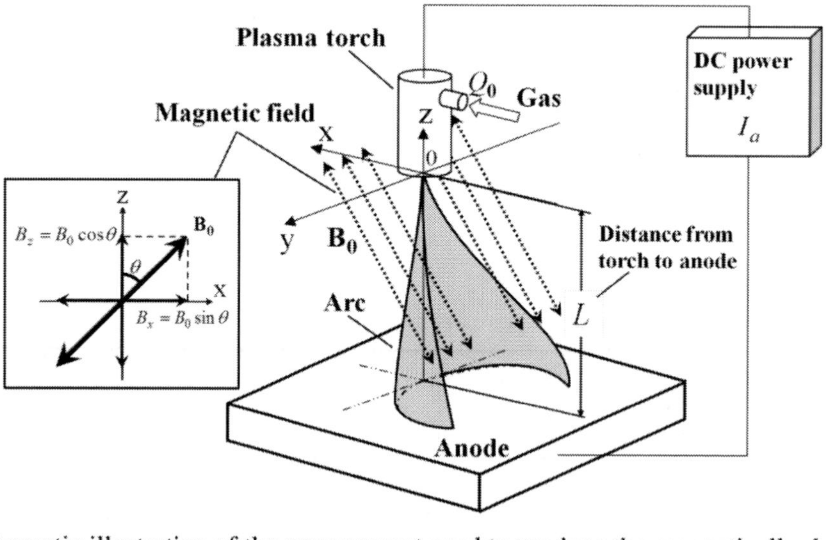

Figure 2. Schematic illustration of the arrangement used to produce the magnetically driven arc. The magnetic field is imposed at the crossing angle of θ.

Table 1. Parameters used in theoretical consideration

symbol	Nomenclature
ρ	Density of plasma gas [kg/m^3]
P	Pressure of plasma gas [Pa]
μ	Coefficient of viscosity [Pa/s]
$\boldsymbol{v}$	Velocity of plasma gas [m/s]
$\boldsymbol{j}$	Arc current density [A/m^2]
$\boldsymbol{F}$	External force (Electromagnetic force) [N]

$\boldsymbol{B}$	Magnetic flux density [T]
L	Distance between torch and anode [m]
F	Frequency of alternating magnetic field [Hz]
θ	Crossing angle of magnetic flux density [rad]
v_0	Initial velocity plasma gas [m/s]
v_x	x-component of plasma gas velocity [m/s]
v_y	y-component of plasma gas velocity [m/s]
v_z	z-component of plasma gas velocity [m/s]
I_a	Arc current [A]
Q_0	Flow rate of plasma gas [kg/s]
B_0	Magnetic flux density [T]
B_x	x-component of magnetic flux density [T]
B_z	z-component of magnetic flux density [T]
t	Time [s]

* Vector quantities are shown in bold.

The fundamental concept of the present modeling is based on the idea that the stream line of the plasma gas after the torch orifice represents the arc profile because the arc current is able to pass only through the conductive plasma gas. The relations governing the plasma gas motion are a continuity equation and a momentum equation. Their general forms are written respectively as shown below.

$$\frac{\partial \rho}{\partial t} + \nabla \cdot (\rho \boldsymbol{v}) = 0 \tag{1}$$

$$\rho \frac{d\boldsymbol{v}}{dt} = \boldsymbol{F} - \nabla p + \mu \Delta \boldsymbol{v} + \frac{\mu}{3} \nabla (\nabla \cdot \boldsymbol{v}) \tag{2}$$

To simplify the consideration, the following assumptions are used.

(1) Pressure is constant
(2) Plasma gas is incompressible outside the torch
(3) Viscous effects are negligible
(4) External body force is an electromagnetic one written as $\boldsymbol{F} = \boldsymbol{j} \times \boldsymbol{B}$
(5) The alternating magnetic field frequency is so low that the imposed magnetic field can be assumed as steady.

Consequently, eq. (2) takes the following form.

$$\rho \frac{d\boldsymbol{v}}{dt} = \boldsymbol{j} \times \boldsymbol{B} \tag{3}$$

According to the fundamental concept by which the arc current passes along the channel of the plasma gas flow, the vector of plasma gas velocity and that of arc current are denoted as

$$\boldsymbol{j} = -\alpha \cdot \boldsymbol{v}, \tag{4}$$

where α is a positive constant. A minus sign in eq. (4) results from the anti-parallel relation between the plasma gas motion and the arc current. Consequently, the following relation is obtained.

$$\rho \frac{d\boldsymbol{v}}{dt} = -\alpha (\boldsymbol{v} \times \boldsymbol{B}) \tag{5}$$

The validity of the relation simplified in eq. (5) was confirmed in an earlier experiment conducted under the imposed magnetic field perpendicular to the arc [4]. In the present study, the magnetic flux density has not only an x-component but also a z-component, as shown in Figure 2.

$$\boldsymbol{B} = (B_x, 0, B_z), \quad B_x^2 + B_z^2 = B_0^2$$
$$B_x = B_0 \sin\theta, \quad B_z = B_0 \cos\theta \tag{6}$$

Then, eq. (5) expressed in vector form is decomposed into three scalar equations as shown below.

$$\rho \frac{dv_x}{dt} = -\alpha v_y B_z \tag{7}$$

$$\rho \frac{dv_y}{dt} = -\alpha (v_z B_x - v_x B_z) \tag{8}$$

$$\rho \frac{dv_z}{dt} = \alpha v_y B_x \tag{9}$$

Differentiating eq. (8) with respect to t and substituting eqs. (7) and (9), following eq. (10) is obtained.

$$\frac{d^2 v_y}{dt^2} + \lambda^2 v_y = 0 \tag{10}$$

Therein, λ is defined as shown below.

$$\lambda = \frac{\alpha}{\rho} B_0 \tag{11}$$

Similarly, eq. (7) and eq. (9) can be transformed respectively into eq. (12) and eq. (13).

$$\frac{d^2 v_x}{dt^2} = -\frac{\alpha}{\rho} B_z \left\{ -\frac{\alpha}{\rho} \left(v_z B_x - v_x B_z \right) \right\} \tag{12}$$

$$\frac{d^2 v_z}{dt^2} = \frac{\alpha}{\rho} B_x \left\{ -\frac{\alpha}{\rho} \left(v_z B_x - v_x B_z \right) \right\} . \tag{13}$$

Rearranging both eqs. (12) and (13), the following eqs. (14) and (15) are obtained, each of which corresponds respectively to the equation for v_x and that for v_z.

$$\frac{d^2}{dt^2} \left\{ \left(\frac{\rho}{\alpha} \right)^2 \frac{1}{B_x B_z} \frac{d^2 v_x}{dt^2} + \frac{B_0^2}{B_x B_z} v_x \right\} = 0 \tag{14}$$

$$\frac{d^2}{dt^2} \left\{ \left(\frac{\rho}{\alpha} \right)^2 \frac{1}{B_x B_z} \frac{d^2 v_z}{dt^2} + \frac{B_0^2}{B_x B_z} v_z \right\} = 0 . \tag{15}$$

Integrating the equations above with respect to t,

$$\frac{d^3 v_x}{dt^3} + \lambda^2 \frac{dv_x}{dt} + C_1 = 0 \tag{16}$$

and

$$\frac{d^3 v_z}{dt^3} + \lambda^2 \frac{dv_z}{dt} + C_2 = 0 \tag{17}$$

further integration of eq. (16) and eq. (17) result in the following equations.

$$\frac{d^2 v_x}{dt^2} + \lambda^2 v_x + C_1 t + C_3 = 0 \tag{18}$$

$$\frac{d^2 v_z}{dt^2} + \lambda^2 v_z + C_2 t + C_4 = 0 . \tag{19}$$

The constants of integration, C_1, C_2, C_3, and C_4 are determined respectively by considering the initial conditions for the plasma gas velocity as

$$\left(v_x, v_y, v_z \right) = \left(0, 0, v_0 \right) , \text{ at } (x, y, z, t) = (0,0,0,0) . \tag{20}$$

Using the relations of eqs. (7), (8), and (9), further initial conditions are obtained as eqs. (21), (22), and (23).

$$\left(\frac{dv_x}{dt}, \frac{dv_y}{dt}, \frac{dv_z}{dt} \right)_{t=0} = \left(0, -\lambda v_0 \sin\theta, 0 \right) , \text{ at } \left(x, y, z, t \right) = \left(0,0,0,0 \right) \tag{21}$$

Journal of Physics: Conference Series **406** (2012) 012036 doi:10.1088/1742-6596/406/1/012036

$$\left(\frac{d^2v_x}{dt^2},\frac{d^2v_y}{dt^2},\frac{d^2v_z}{dt^2}\right)_{t=0} = \left(-\lambda^2 v_0 \sin\theta\cos\theta, 0, -\lambda^2 v_0 \sin^2\theta\right),\ \text{at}\ (x,y,z,t)=(0,0,0,0) \tag{22}$$

$$\left(\frac{d^3v_x}{dt^3},\frac{d^3v_y}{dt^3},\frac{d^3v_z}{dt^3}\right)_{t=0} = \left(0, -\lambda^3 v_0 \sin\theta, 0\right),\ \text{at}\ (x,y,z,t)=(0,0,0,0). \tag{23}$$

Considering these initial conditions of eqs. (20), (21), (22), and (23), eq. (18) and eq. (19) are expressed as

$$\frac{d^2v_x}{dt^2} + \lambda^2 v_x + \lambda^2 v_0 \sin\theta\cos\theta = 0 \tag{24}$$

and

$$\frac{d^2v_z}{dt^2} + \lambda^2 v_z - \lambda^2 v_0 \cos^2\theta = 0 \tag{25}$$

Then, substituting the following relations eq. (26) to eq. (10), eq. (24) and eq. (25),

$$v_x = \frac{dx}{dt},\quad v_y = \frac{dy}{dt},\quad v_z = \frac{dz}{dt}, \tag{26}$$

the equations which govern the displacement of the plasma gas are finally rewritten in the following forms:

$$\frac{d^3x}{dt^3} + \lambda^2\frac{dx}{dt} + \lambda^2 v_0 \sin\theta\cos\theta = 0 \tag{27}$$

$$\frac{d^3y}{dt^3} + \lambda^2\frac{dy}{dt} = 0 \tag{28}$$

$$\frac{d^3z}{dt^3} + \lambda^2\frac{dz}{dt} - \lambda^2 v_0 \cos^2\theta = 0 \tag{29}$$

Solving these differential equations, the position of the plasma gas with t, $\{x(t), y(t), z(t)\}$ is obtained. Initial conditions at $t = 0$ for eqs. (27), (28), and (29) are as shown below:

$$\begin{gathered}(x,y,z)_{t=0} = (0,0,0)\\ \left(\frac{dx}{dt},\frac{dy}{dt},\frac{dz}{dt}\right)_{t=0} = (0,0,v_0)\\ \left(\frac{d^2x}{dt^2},\frac{d^2y}{dt^2},\frac{d^2z}{dt^2}\right)_{t=0} = (0,\lambda v_0 \sin\theta, 0)\end{gathered} \tag{30}$$

2.2. *Conservation of kinetic energy for the plasma gas*

The equation shown as eq. (5) induces the conservation of kinetic energy for the plasma gas easily. The scalar products of vector $\boldsymbol{v}$ to both sides of eq. (5) results in

$$\rho\left(\frac{d\boldsymbol{v}}{dt}\right)\cdot\boldsymbol{v} = -\alpha(\boldsymbol{v}\times\boldsymbol{B})\cdot\boldsymbol{v}. \tag{31}$$

The left-hand-side of eq. (31) is written as

$$\rho\left(\frac{d\boldsymbol{v}}{dt}\right)\cdot\boldsymbol{v} = \frac{d}{dt}\left(\frac{1}{2}\rho\boldsymbol{v}^2\right), \tag{32}$$

and its right-hand-side is zero. Therefore,

$$\left(\frac{1}{2}\rho\boldsymbol{v}^2\right) = \text{constant}. \tag{33}$$

This relation is that of the kinetic energy conservation for the plasma gas during the flight. The velocity of the plasma gas might change its direction, but v^2 must always equal v_0^2 where v_0 represents the initial velocity of the plasma gas. Because the plasma gas is ejected in the downward direction, the initial velocity v_0 is negative.

Then we consider the assumption (5) described in the previous section. For the plasma gas, the travelling time from the torch orifice to the anode, τ_{travel}, is estimated as

$$\tau_{travel} \approx \frac{L}{|v_0|}, \tag{34}$$

whereas the characteristic value τ_{field} which expresses the time to change the magnetic field is evaluated as

$$\tau_{field} \approx \frac{1}{f}. \tag{35}$$

Therein, f represents the frequency of the imposed alternating magnetic field. If the following condition is satisfied, it is reasonable that the imposed magnetic field is assumed to be steady during the flight of the plasma gas, as

$$\tau_{field} >> \tau_{travel}, \tag{36}$$

or

$$f << \frac{|v_0|}{L}. \tag{37}$$

We must bear in mind that our consideration is restricted to the condition written in eq. (37).

2.3. *Determination of λ.*

We consider the relation between λ and the operating parameters such as the arc current, the imposed magnetic flux density, and the plasma gas flow rate. The governing equations (27), (28), (29) under the initial conditions (30) show that the displacement of the plasma gas with time (or the arc profile) is determined by "λ", "v_0" and "θ". Using the relation of eq. (4), eq. (11) is rewritten as

$$\lambda = -\frac{j}{\rho v}B_0 = \frac{j}{\rho v_0}B_0. \tag{38}$$

If the cross sectional area of the plasma gas stream tube is represented as S, then the flow rate of the plasma gas Q_0 is expressed as $\rho v S$, whereas the arc current I_a is equal to jS according to the assumption that the arc current passes along the stream line of the plasma gas. Consequently, the following relation is obtained.

$$\lambda = \frac{I_a B_0}{Q_0} \tag{39}$$

3. Numerical Results and Discussion

Numerical calculations were performed. Commercial software (Mathematica;)[9] was used to solve the differential equations shown as (27), (28), and (29) under the initial condition of eq. (30). The values of parameters used in the present calculation are presented in Table 2. Except for crossing angle, numerical values shown in Table 2 are the same as those used in the previous work [4]. The numerical value of λ was determined easily by operating parameters such as arc current I_a, imposed magnetic flux density B_0, and mass flow rate of plasma gas Q_0, according to the relation of eq. (39). Because the imposed magnetic field is alternating, the magnetic flux density, B_0 varies between B_{max} to $-B_{max}$.

Table 2. Numerical values used in the calculation

Parameters	Symbol	Value
Arc current [A]	I_a	90
Flow rate of plasma gas [kg/s]	Q_0	3.0×10^{-4}
Initial velocity of plasma gas [m/s]	v_0	-80
Magnetic flux density [T]	B_0	20×10^{-3}

Crossing angle [rad]	θ	$0-\pi/2$
Standoff distance between torch and anode [m]	L	0.05
Frequency of the alternating magnetic field [Hz]	f	50

Under the conditions presented above, τ_{travel} was estimated at 0.6×10^{-3} [s] and τ_{field} was 20.0×10^{-3} [s]. The condition expressed in eq. (36) was satisfied in the present calculation.

For different crossing angles of θ, variations of arc profiles with magnetic flux density are depicted in Figure 3. As shown in Figure 3(a) in which $\theta=\pi/2$, the curvature of the profiles increases with increased B_0. Furthermore, the arc root changes its position from A, B, C, and D along the y axis. Beyond a critical strength of the magnetic flux density corresponding to point D, arc has no arc root on the anode plate, which means that the arc is not able to exist beyond a certain magnetic flux density. Figures 3(b)–3(d) show variations of arc profiles with λ, respectively, for $\theta=\pi/3$, $\theta=\pi/4$, $\theta=\pi/6$.

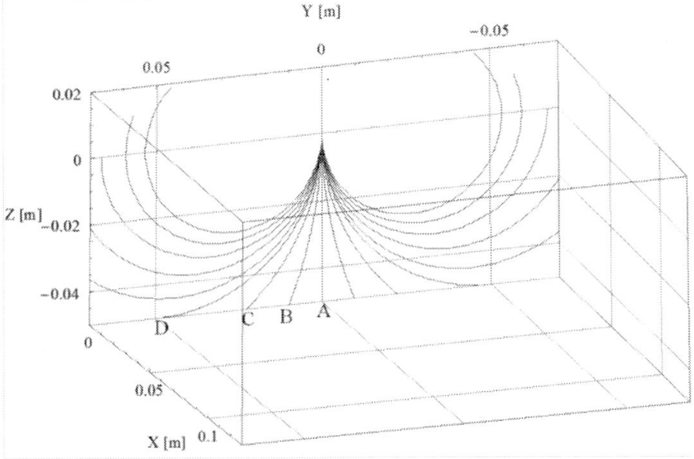

Figure 3(a). Arc profile variation with the increase of the magnetic field at $\theta=\pi/2$.

In Figure 3(b), arc root movements are shown between the positive magnetic flux density (or the positive λ) to clarify the movement of the arc root on the anode. Because the arc movement is symmetric with respect to the x-z plane for the negative magnetic flux density, the total movement of the arc root can be imagined in the whole range of $-B_{max}$ to B_{max}.

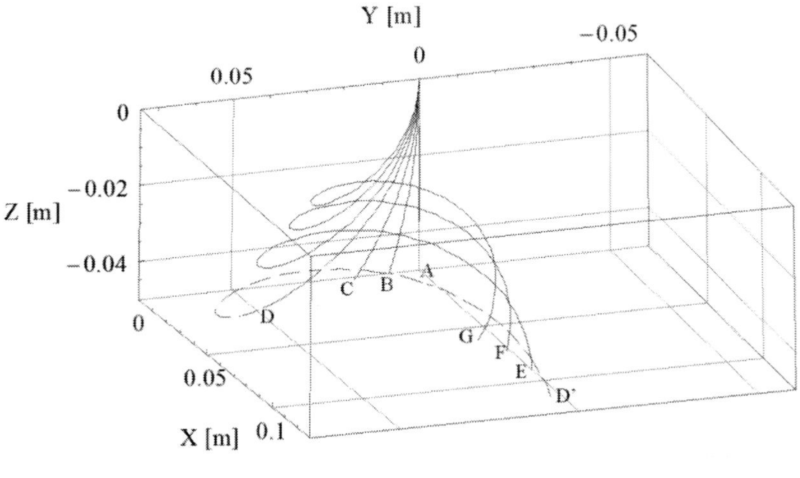

Figure 3(b). Arc profile variation with the increase of the magnetic field at $\theta=\pi/3$.

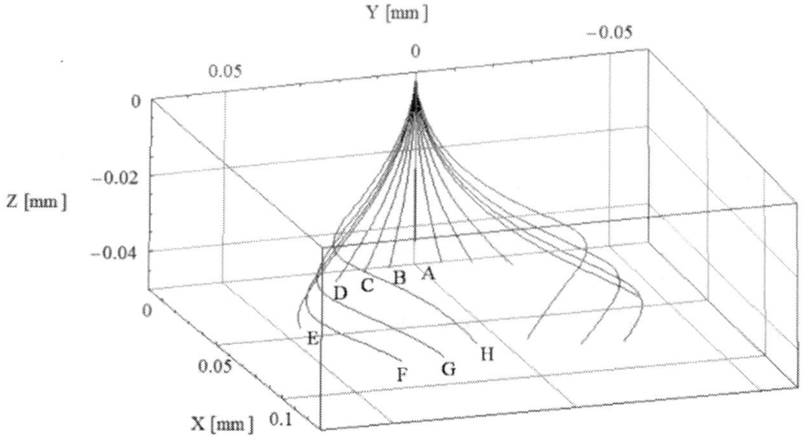

Figure 3(c). Arc profile variation with the increase of the magnetic field at $\theta = \pi/4$.

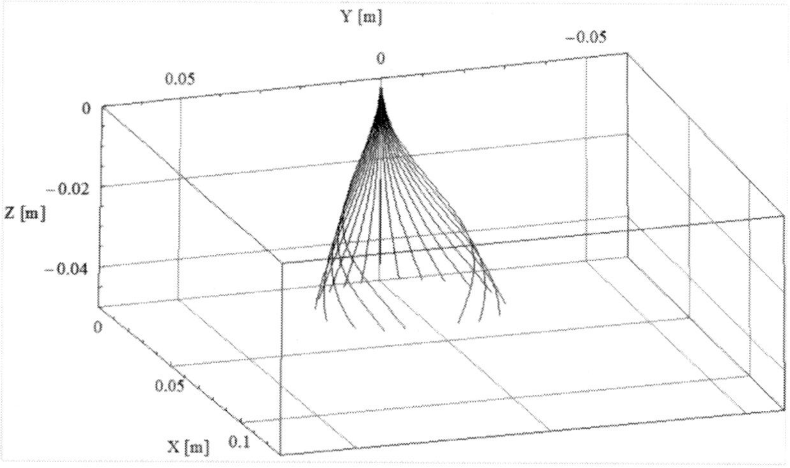

Figure 3(d). Variation of arc profiles with increased magnetic field at $\theta = \pi/6$.

As shown in Figure 3(b), the arc root shifts with the increase of the magnetic flux density from points A to B, C, and D; then it jumps to D' and with the further increase of the magnetic flux density, it moves to the points E, F, and G. Under the numerical condition presented in Table 2, a discrete jump occurs at $B_0 = 6.4\times10^{-3}$ [T]. In Figure 3(c), the arc root moves with the increase of the magnetic field from A to B, C, D and E. At point E, the arc is tangential to the anode plane. Then it goes to F, G, and H. No discrete jump appears at $\theta = \pi/6$, as shown in Figure 3(d).

The arc root movements on the anode for different values of θ are depicted in Figure 4. Further detailed analyses for various crossing angles reveal that the discrete jump occurs in the case of $\theta > \pi/4$. The arc length, H is evaluated as

$$H = \int v \cdot dt = v_0 \tau_{travel} \; . \qquad (40)$$

Its flight time of the plasma gas from the torch orifice to the anode, τ_{travel}, satisfies the following relation:

$$z(t = \tau_{travel}) = L \; . \qquad (41)$$

Therein, $z(t)$ is obtained by solving eq. (29). The arc length variation with λ is shown for various crossing angles in Figure 5, where the ratio of H/L, as the ordinate, is shown with λ as the abscissa.

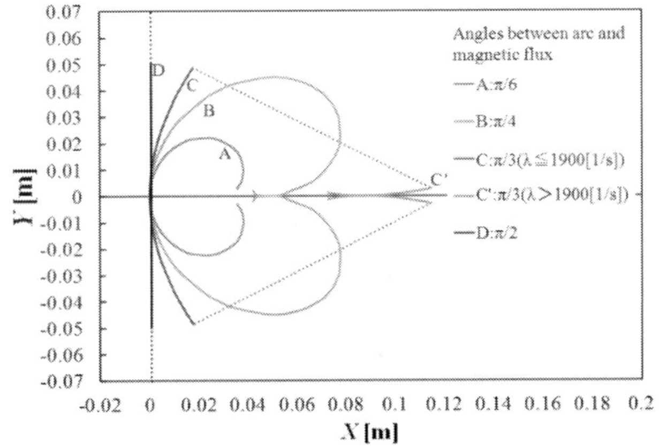

Figure 4. Arc root movements on the anode plate for different values of crossing angle.

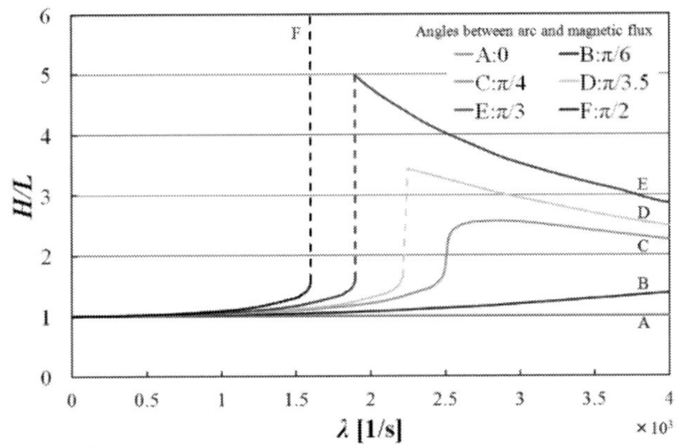

Figure 5. Arc length variation with the increase of λ for various θ.

For $\theta = \pi/2$, the maximum arc length is $\pi/2$ times longer than the stand-off distance at $\lambda = 1600$. When the crossing angle is between $\pi/4 < \theta < \pi/2$, a jump in the arc length occurs. When $\theta = \pi/3$, the arc elongates to five times larger than the stand-off distance. Such an arc length increase results in the drastic increase of the arc voltage. Theoretically, it is possible for the arc to exist, but the arc will practically disappear. If $\theta = \pi/4$, then no jump occurs. The maximum elongation *H*/*L* is 2.54 at $\theta = \pi/4$. It becomes 1 when θ approaches 0.

4. Conclusions

Theoretical consideration was performed to investigate the arc motion under the alternating magnetic field imposed obliquely to the arc. The fundamental concept of the modeling was based on the idea that the stream line of the plasma gas represents the arc profile.

- Theoretical analyses reveal the following results:

(1) Displacement of the plasma gas with time is determined solving a set of three third-order linear ordinary differential equations.

(2) Governing parameters for the gas motion are θ (crossing angle), v_0 (initial velocity of the plasma gas), and λ.

(3) Parameter λ is defined as $\lambda = (I_a B_0)/Q_0$.

- Numerical calculations reveal the following results:

(4) The arc under the obliquely imposed alternating magnetic field moves three-dimensionally in space, different from the two-dimensional movement found under the perpendicular magnetic field.

(5) If a crossing angle is larger than $\pi/4$, then the arc root on the anode moves discretely at a certain value of λ. At this critical λ, the arc might be extinguished because of the drastic increase of the arc length.

References

[1] Y. Arata and H. Maruo, 1972, magnetic control of arc plasma and its application for welding, *Technology Reports of the Osaka University*, **22**, 135-145

[2] J. E. Harry and D. Goodwin, 1978, Surface heat treatment using a plasma torch with a magnetically traversed arc, *Proceedings of Fourth Intl. Conf. on Advances in Welding Processes*, England, 181.

[3] K. Takeda, 1990, Oscillating plasma arc by external AC magnetic field, *Proceedings Int. Workshop on Plasma Jet in the Development of New Materials Technology*, Frunze, 485-492.

[4] T. Yamamoto, 2010, Fundamental study of the arc control for heating over large surface, *Doctor Thesis, Akita Prefectural University*, Japan

[5] T. Yamamoto, K. Takeda, K. Toh, and J. Tanaka, 2007, Distribution of heat flux transported by a magnetically driven arc, *Thin Solid Films*, **515**, 947-953

[6] T. Toh, J. Tanaka, J. Muraki, Y. Yamamoto and K. Takeda, 2005, Magneto-hydrodynamic Simulation of DC Arc Plasma under AC Magnetic field, *ISIJ International*, **45**, No. 7

[7] T. Yamamoto, N. Matsumoto, M. Sugimoto, and K. Takeda, 2009, Theoretical consideration of heat flux distribution of arc driven by ac magnetic field, *ISIJ International*, **49**, 1845-1849

[8] N. Matsumoto, I. Kuno, T. Yamamoto, M. Sugimoto and K. Takeda, 2011, Arc Behavior in non-uniform ac magnetic field, *ISIJ International*, **52**, 488-492

[9] S. Wolfram, 1998, *The Mathematic Book Third Edition*, Wolfram Research, Inc., USA

Spectroscopic characterization of nitrogen plasma generated by waveguide-supplied coaxial-line-based nozzleless microwave source

B Hrycak[1], **M Jasiński**[1] **and J Mizeraczyk**[1,2]
[1]Centre for Plasma and Laser Engineering, The Szewalski Institute of Fluid Flow Machinery, Poland
[2]Faculty of Marine Electrical Engineering, Gdynia Maritime University, Gdynia, Poland

e-mail: bhrycak@imp.gda.pl

Abstract. In this paper, results of spectroscopic study of microwave (2.45 GHz) plasma at atmospheric pressure and high flow rate are presented. The plasma is generated by waveguide-supplied coaxial-line-based microwave plasma source (MPS). Pure nitrogen was used as working gas. An additional nitrogen flow rate of 50l/min was introduced to the plasma by four gas ducts which formed a swirl flow inside the quartz tube. The swirl concentrated near the quartz cylinder wall, stabilized plasma generation and protects the quartz tube wall from overheating. Working gas flow rate and microwave absorbed power varied from 50 up to 200 l/min and from 600 up to 5500 W, respectively. The emission spectra in the range of 300 - 600 nm were recorded. The rotational and vibrational temperatures of N_2+ ions and N_2 molecules, as well as the rotational temperature of OH radicals were determined by comparing the measured and simulated spectra. The plasma gas temperature inferred from rotational temperature of heavy species ranged 4000 to 6200 K. It depended on location in plasma, microwave absorbed power and working gas flow rate. The MPS can be used in various gas processing applications.

Keywords—microwave plasma, OES, rotational temperature, gas processing

1. Introduction

Recently, microwave plasma sources (MPSs) operated at atmospheric pressure have been developed [1, 2]. MPSs were used as a light sources [3], in sterilization [4] and technological processes like surface treatment. Various materials could be treated with microwave plasma: metals (steel [5] and aluminium [6]), ceramic [7] and polymers [8]. Plasma can be used for deposition of thin films [9], as well as etching [10]. MPSs also found applications in the processing of various gases [11, 12]. Destruction of Freon CFC-11 [13] and HFC-134a [14] and production of hydrogen via methane conversion [15] in microwave atmospheric pressure plasmas were reported by us elsewhere. In papers [14] and [15] a waveguide-supplied nozzleless MPS was used.

The plasma gas temperature is one of the most important parameter in plasma technology and applications. This temperature can often be inferred from the rotational temperature of the heavy species of the gas [16-23]. The rotational and vibrational temperatures in plasmas are often determined comparing experimental and simulated optical spectra.

In this paper, results of spectroscopic study of microwave plasma at high nitrogen flow rates are presented. The plasma was generated in waveguide-supplied coaxial-line-based nozzleless MPS operated at 2.45 GHz.

2. Optical emission spectroscopy (OES)

Measurements of the plasma parameters in high-pressure plasma environment pose requirements that are different compared to low-pressure plasmas (<10 mbar). The highly collisional nature of atmospheric pressure plasma can significantly modify the data analysis procedure and sometimes questions the applicability of methods used in the low-pressure plasmas. However, as many research showed, well developed low-pressure plasma diagnostics methods for both partially ionized [24, 25] and highly ionized plasmas [26, 27] can be adopted for collisionally dominated plasmas.

In the thermal equilibrium plasmas the temperature measurements using OES are usually based on the absolute or relative intensities of various atomic lines of oxygen and nitrogen. In the case of non-equilibrium plasmas at atmospheric pressure, the techniques used for the temperature measurements in the equilibrium plasmas may not provide reliable information about the gas temperature because the population distribution of internal energy states tends to depart from the Boltzmann distribution. This concerns mainly the case of the electronic and vibrational population distributions, but the rotational populations tend to follow the Boltzmann distribution owing to fast rotational relaxation at atmospheric pressure. Thus, the gas temperature can often be inferred from the rotational temperature of the heavy species of the gas [16-23]. Various transitions of O_2, N_2, N_2+, and NO (in dry air), OH (in humid air) and other atoms and molecules present in the plasma can be used for the temperature measurements, depending on the level of plasma excitation and interference of the other transitions.

In case of resolved spectra the rotational and vibrational temperatures can be obtain from the Boltzmann plot [21, 22]. But in most experiments obtaining of resolved spectra is impossible due to high spectral resolution requirements. Fortunately, also in case of unresolved spectra some methods were developed to determine temperatures [16, 17]. Rotational and vibrational temperatures in plasmas are often determined comparing relative intensities of experimental and simulated optical spectra [20]. In this work for simulation of the optical spectra the Specair [28, 29] and Lifbase [30, 31] programs were used. The codes of these programs calculate the spectral radiation emitted by gases and plasmas of various compositions at given individually vibrational and rotational temperatures.

3. Microwave Plasma Source (MPS)

The sketch of the nozzleless waveguide-supplied coaxial-line-based MPS is shown in figure 1. The source was based on a standard WR 430 rectangular waveguide with a section of reduced-height preceded and followed by tapered sections. The tapered section assured smooth transition from the standard waveguide dimensions to the waveguide of reduced height. There was an inner cylinder electrode mounted perpendicular to the waveguide in the center of the circular openings on the axis of the waveguide wide wall. The inner cylindrical electrode was surrounded by a quartz tube. Both the inner cylindrical electrode and the quartz tube were surrounded by outer cylindrical shield on the outside of the waveguide. The inner cylindrical electrode and the outer cylindrical shield formed a coaxial line. The working gas was injected to the plasma by the inner cylindrical electrode. An additional gas was introduced to the plasma by four gas ducts which formed a swirl flow inside the quartz tube. The swirl concentrated near the quartz cylinder wall, stabilized plasma generation and protects the quartz tube wall from overheating. The plasma in the form of a flame was generated inside a quartz tube above the end part of the inner cylindrical electrode. The plasma could be observed through visualization slit in outer cylindrical shield (see figure 2 and 4). Figure 2 shows the photos of nitrogen plasma at different discharge conditions and figure 3 shows the length of nitrogen plasma (measured from electrode) as a function of microwave absorbed power for two different axial flow rates. Operation of the MPS was described in details in [32].

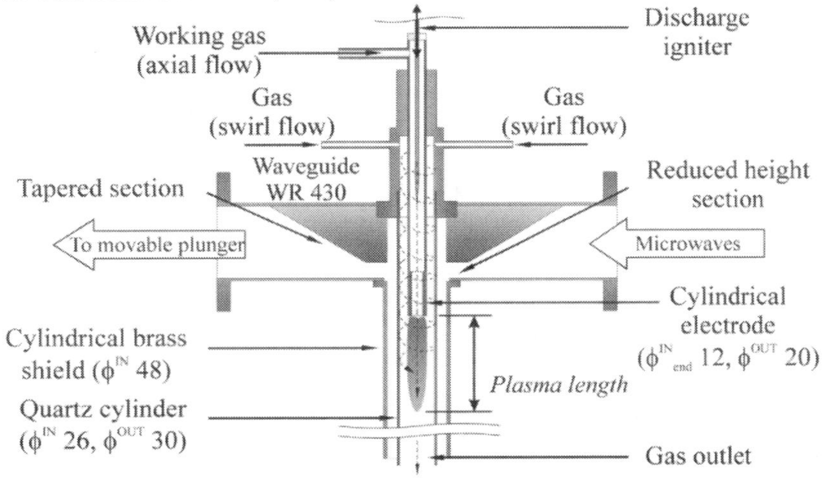

Fig. 1. The sketch of waveguide-supplied coaxial-line-based nozzleless microwave plasma source.

Fig. 2. Microwave nitrogen plasmas for different conditions.

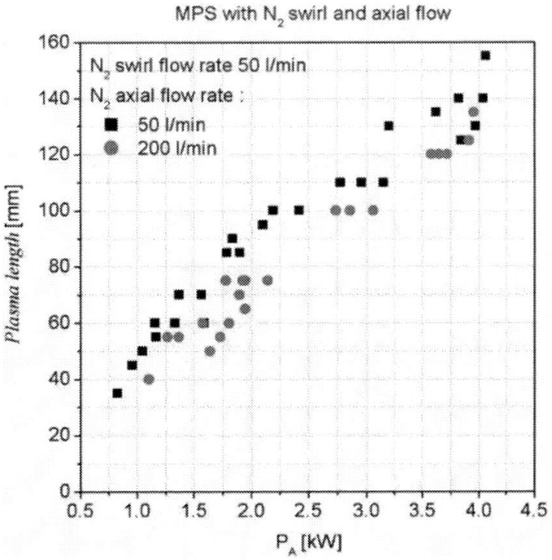

Fig. 3. The length of nitrogen plasma (measured from electrode) as a function of microwave absorbed power P_A for different axial flow rates.

Journal of Physics: Conference Series **406** (2012) 012037 doi:10.1088/1742-6596/406/1/012037

4. Experiments

The experimental setup is shown in figure 4. It consisted of a 2.45 GHz magnetron generator, microwave power supplying and measuring system, MPS, movable plunger for impedance matching, gas supplying and flow control system, spectrometer [DK-480 (CVI), 1200 and 3600 grooves/mm grating] for spectral emission measurements, equipped with a CCD camera [SBIG ST-6, 750 × 242], and a PC computer. The microwave generator is composed of a high voltage power supply, a control unit and a magnetron head. The magnetron head is equipped with a water cooled circulator which protects it against damages caused by the reflected microwave power. The microwave power measuring system includes a directional coupler, two power meter heads and a digital dual-channel power meter. This system enables direct measurements of an incident P_I and reflected P_R microwave powers. An absorbed power P_A was obtained from the subtraction of P_I - P_R.

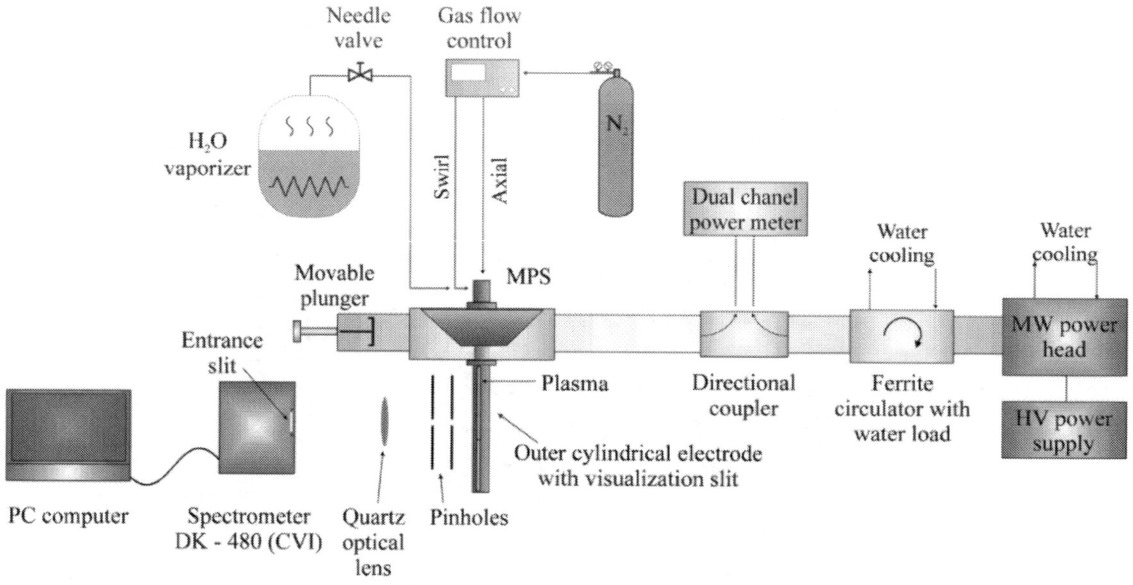

Fig. 4. Experimental setup.

Pure nitrogen was used as working gas. An additional nitrogen flow rate of 50 l/min was introduced to the plasma by four gas ducts which formed a swirl flow inside the quartz tube. A small amount of water vapor was added optionally to swirl gas flow in order to achieve detectable intensity of OH spectra. Working gas flow rate and microwave absorbed power varied from 50 up to 200 l/min and from 600 up to 5000 W, respectively.

To measure the spectra, the light emitted by the plasma was focused with a quartz lens (50 mm in diameter, focal length – 75 mm) onto the entrance slit of the spectrometer (see figure 4). The width of the spectrometer entrance slit was 50 μm (20 mm height). Two opaque screens with pinholes of a 1 mm diameter were placed near the plasma. These pinholes, together with the lens and the pinhole at the spectrometer entrance as well as the entrance slit formed a spatial resolution unit for selection of an area of the plasma, emission of which was recorded by the spectrometer. We estimate that the measured area was about 8 mm diameter. Spectra were recorded and then corrected according to the wavelength sensitivity of CCD camera. The wavelength sensitivity of CCD camera was determinated using a tungsten halogen calibration lamp. Using a Hg-Ne low-pressure calibration lamp (λ = 365.02 nm, 435.84 nm and 546.07 nm, Hg I) we measured that the Gaussian instrumental line profile FWHM was about 0.12 nm and 0.05 nm for 1200 and 3600 grooves/mm grating, respectively. All spectra were measured with 1200 grooves/mm grating. Only for OH (A-X) rotational band measurements the 3600 grooves/mm grating was used.

In this experiment, for the temperatures determination we used the transitions as follows:

- OH ($A^2\Sigma \rightarrow X^2\Pi$, 306-310 nm band),
- N_2 second positive system ($C^3\Pi \rightarrow B^3\Pi$, 308-317 nm band),
- N_2+ first negative system ($B^2\Sigma \rightarrow X^2\Sigma$, 380-392 nm and 463-472 nm band).

5. Results

Figure 5 shows spectrum emitted by the nitrogen plasma. As seen, the dominant spectrum is N_2+ first negative system ($B^2\Sigma \rightarrow X^2\Sigma$). The spectrum contained also: N_2 second positive system ($C^3\Pi \rightarrow B^3\Pi$) and N_2 first positive ($B^3\Pi \rightarrow A^3\Sigma$). The Zn I lines (identified with NIST database [33]) observed in spectrum indicates degradation of the brass inner electrode end. The intensity of all measured systems increased linearly with increasing microwave absorbed power.

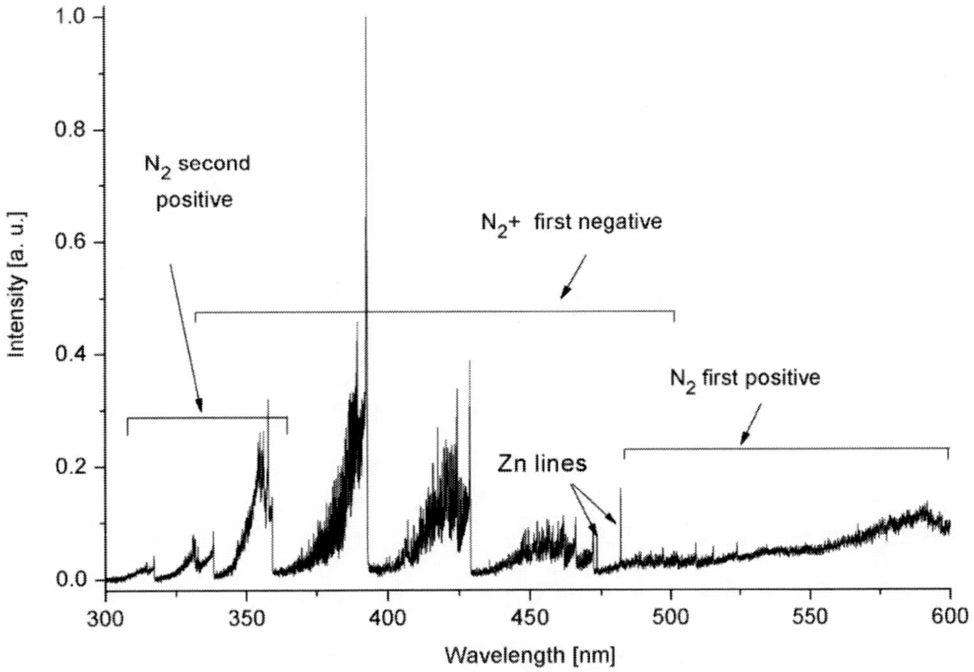

Fig. 5. Measured emission spectrum of nitrogen plasma (P_A – 4 kW, nitrogen flow rate - 50 l/min, 25 mm below the electrode end)

A small amount of water vapor was added to the swirl nitrogen flow. This addition of water vapor did not changed the plasma significantly: the microwave incident and reflected powers, as well as the plasma length remained unchanged. The benefit of such addition was occurring of OH (A-X) rotational band. This band is widely use for gas temperature estimations, e.g. [17, 18, 23].

After having the experimental spectra, a simulations of this spectra was carried out using Specair or Lifbase program. The simulated spectrum was fitted to the experimental one by variation of the rotational T_{rot} temperature. Figures 6 and 7 show the comparisons of the both spectra: experimental and simulated fitted to the experimental one. The comparasion of measured and simulated spectra of OH (A-X) (figure 6a), N_2 second positive system (figure 6b) N_2+ first negative system for two different bands (figure 7), were performed and rotational T_{rot} and vibrational T_{vib} temperatures were determineted.

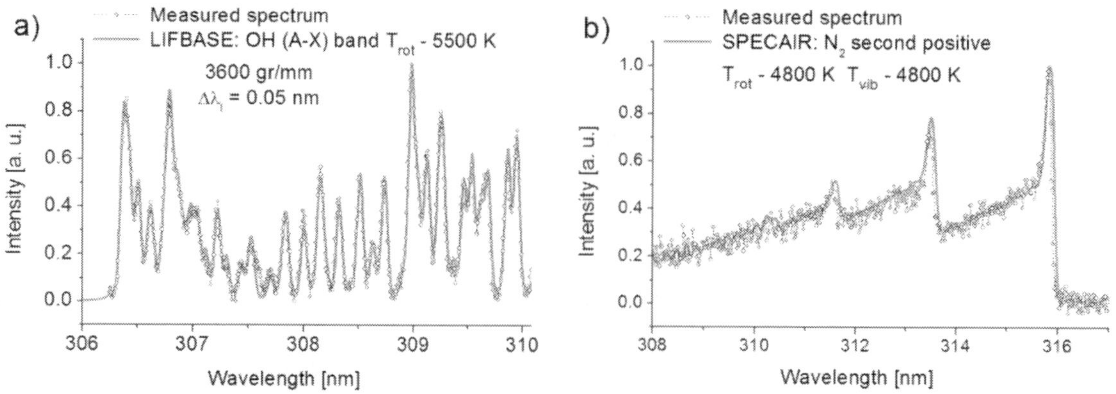

Fig. 6. Comparison of the measured and simulated emission spectra of OH (A-X) rotational band (a) and N_2 second positive system (b) in nitrogen plasma (P_A – 2 kW, nitrogen flow rate - 50 l/min, 25 mm below the electrode end)

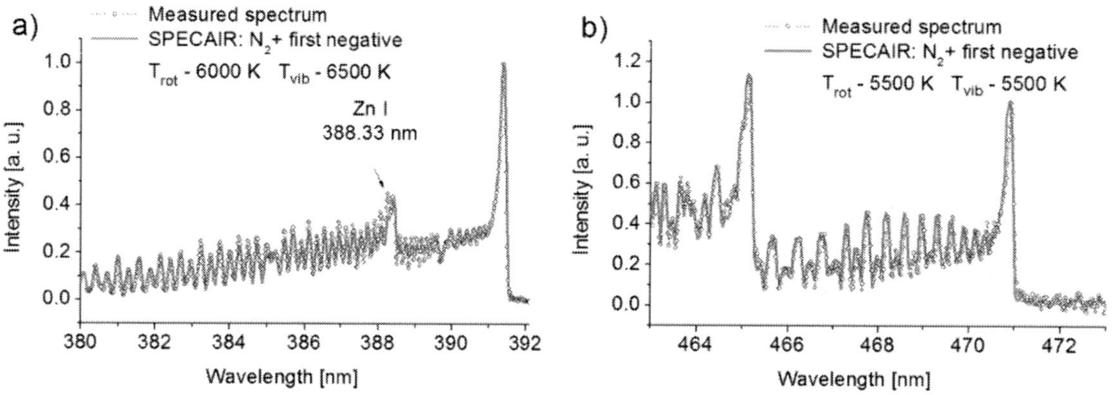

Fig. 7. Comparison (two different bands) of the measured and simulated emission spectra of N_2+ first negative system in nitrogen plasma (P_A – 2 kW, nitrogen flow rate - 50 l/min, 25 mm below the electrode end)

The rotational temperatures of OH radicals, N_2 molecules and N_2+ ions as a function of distance below the inner electrode end top (distance BIEE) in nitrogen plasma (P_A - 2kW, nitrogen flow rate - 50 l/min) are presented in figure 8. At this condition the plasma length was about 85 mm (see figure 3). Regardless the plasma length it was impossible to determine the temperature of N_2+ ions and N_2 molecules at the further area of plasma. The reason of this was low intensities of band of N_2+ first negative system and N_2 second positive system used for temperature determinations. As seen in picture we could observe effect of cooling the gas by the inner electrode. The highest temperature occurred at distance of 15 - 25 mm. Further from the inner electrode temperature decreased. Interesting phenomenon could be observed for N_2+ ions temperatures determined from 380-392 nm band. Close to the inner electrode, at excitation region the vibrational temperature was 1000 K higher than the rotational temperature. At the end of plasma flame, due to thermal relaxation both temperatures were in equilibrium [23].

The rotational and vibrational temperatures increased with microwave absorbed power (figure 9). This increse was not significant. As it could been seen in figure 3 the increase of the microwave power coused almost proportional increse of the pasma length and thus the plasma volume. It could be concluded that the microwave power influence the plasma volume and much less the plasma gas temperature. The influence of axial flow was more apparent for higher microwave absorbed powers. Increasing the flow caused slight decrease of temperature (figure 10). For lower absorbed powers this effect was less apparent. At location in the plasma placed 25 mm below the inner electrode determination of the rotational and vibrational temperatures of N_2+ ions temperatures from 380-392 nm band provided the constant result (figures 8-10). The initial distribution (i.e. distribution just after the excitation) of rotational and vibrational states in this case was non-thermalized, but was the effect of excitation mechanisms [23]. The effect of thermal relaxation was observed at the end of the plasma flame (see figure 8).

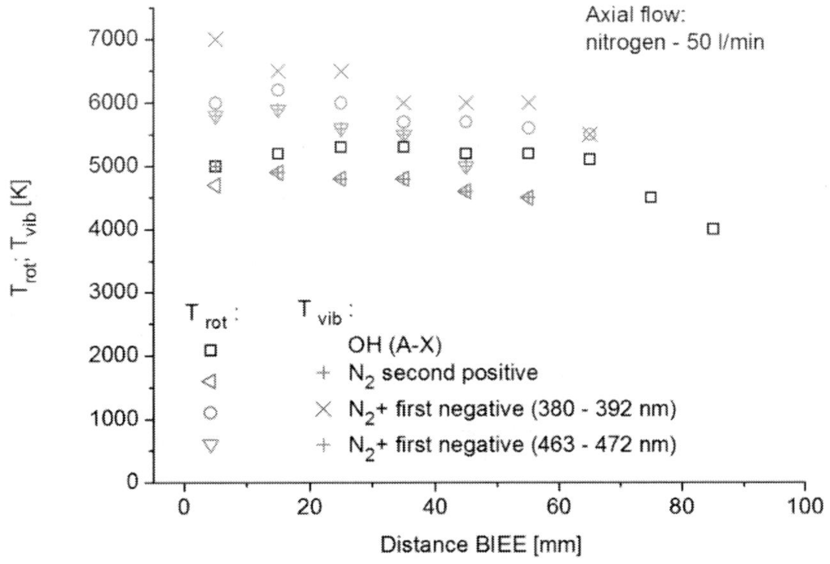

Fig. 8. Measured rotational and vibrational temperatures of OH radicals, N_2 molecules and N_2+ ions as a function of distance below inner electrode end (Distance BIEE) (P_A - 2 kW, nitrogen flow rate - 50 l/min)

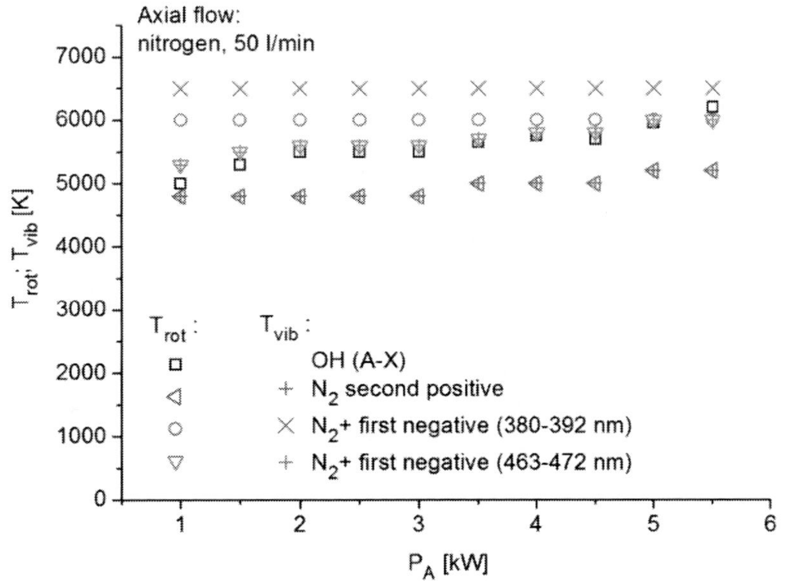

Fig. 9. Measured rotational and vibrational temperatures of OH radicals, N_2 molecules and N_2+ ions as a function of microwave absorbed power (nitrogen flow rate - 50 l/min, 25 mm below the electrode end)

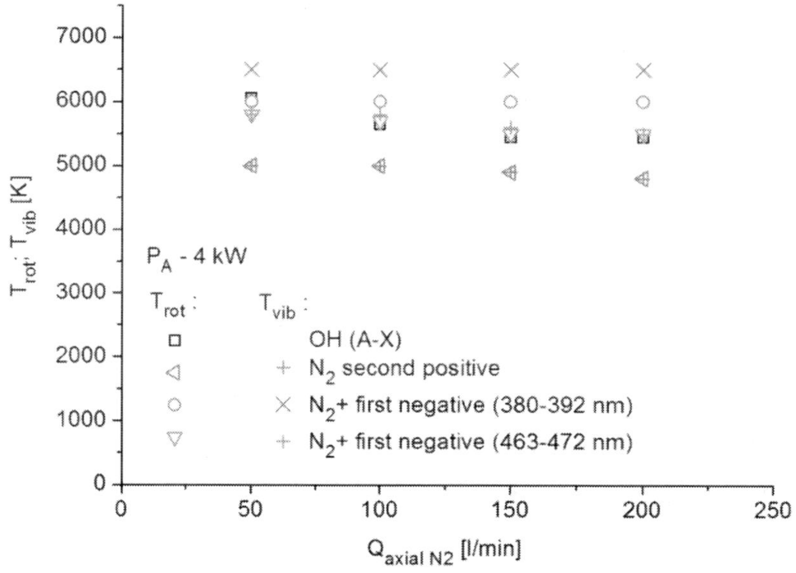

Fig. 10. Measured rotational and vibrational temperatures of OH radicals, N_2 molecules and N_2+ ions as a function of axial nitrogen flow rate Q_{axial} N_2 (P_A - 4kW, 25 mm below the electrode end)

As seen in figures 8, 9 and 10 the results for the each selected specie were slightly different for the same discharge conditions. Obtained rotational and vibrational temperatures ranged from 4000 to 6200 K and from 4500 to 7000 K respectively, depending on the location in the plasma, the microwave absorbed power and axial nitrogen flow rate. OH radicals and N2+ ions from 463-472 nm band provided comparable results. N2 molecules in all cases provided slightly lower temperatures. In all conditions T_{vib} were greater or equal than T_{rot}. It is in good agreement with theory [20]. The rotational and the vibrational temperatures, except these obtained from N2+ ions determined from 463-472 nm band, were in equilibrium in nitrogen microwave plasma. The results gave comparable temperatures to that presented by us [34] measured in similar device: nozzleless waveguide-supplied metal-cylinder-based MPS. Rotational temperature of OH radicals was assumed to be the best estimation of the plasma gas temperature in nitrogen microwave plasma.

6. Conclusion

Spectroscopic study of nitrogen microwave (2.45 GHz) plasma at atmospheric pressure and high flow rates was presented in this work. The study concerns the rotational and vibrational temperatures of selected heavy species present in microwave plasma in order to estimate gas temperature. The plasma gas temperature inferred from rotational temperature of OH radicals ranged 4000 to 6200 K. It depended on location in plasma, microwave absorbed power and working gas flow rate.

The investigated nozzleless, waveguide-supplied, coaxial-line-based MPS works very stable with various working gases. The parameters of the plasma can be changed in wide range. The high gas temperature makes it attractive tool for different gas processing at high flow rates. Owing to high gas temperature MPS was successfully used for Freon destruction [14], as well as for hydrogen production via hydrocarbon conversion [15].

Acknowledgments

This research was supported by The National Centre for Research and Development (NCBiR) under the programme NR14-0091-10/2010.

Journal of Physics: Conference Series **406** (2012) 012037 doi:10.1088/1742-6596/406/1/012037

References:

[1] Tendero C, Tixier C, Tristant P, Desmaison J and Leprince P 2006 Atmospheric pressure plasmas: A review *Spectrochim. Acta Part B* **61** 2-30

[2] Bogaerts A, Neyts E, Gijbelsa R and van der Mullen J 2002 Gas discharge plasmas and their applications *Spectrochim. Acta Part B* **57** 609-58

[3] Kono A, Wang J and Aramaki M 2006 Production and characterization of high-pressure microwave glow discharge in a microgap aiming at VUV light source *Thin Solid Films* **506-507** 444-8

[4] Lee K-Y, Park B J, Lee D H, Lee I-S, Hyun S O, Chung K-H and Park J-C 2005 Sterilization of Escherichia coli and MRSA using microwave-induced argon plasma at atmospheric pressure *Surf. & Coat. Technol.* **193** 35-8

[5] Peelamedu R, Kumar D and Kumar S 2006 Microwave atmospheric pressure plasma for surface treatment and reactive coating on steel surfaces *Surf. & Coat. Technol.* **201** 4008–13

[6] Cho S C, Hong Y C and Uhm H S 2007 Surface Treatment of Aluminum Sheets by Pulsed Microwave Plasma Discharge at Atmospheric Pressure *Jpn. J. Appl. Phys.* **46** 3583-8

[7] Wylie S R, Al-Shamma'a A I, Lucas J and Stuart R A 2004 An atmospheric microwave plasma jet for ceramic material processing *J. Mater. Process. Technol.* **153-4** 288-93

[8] Yuji T, Urayama T, Fujii S, Mungkung N and Akatsuka H 2008 Temperature behavior of atmospheric-pressure non-equilibrium microwave discharge plasma jets for poly(ethylene naptharate)-surface processing *Surf. & Coat. Technol.* **202** 5289–92

[9] Pfuch A and Cihar R 2004 Deposition of SiO thin films by microwave induced x plasma CVD at atmospheric pressure *Surf. & Coat. Technol.* **183** 134–40

[10] Huet S, Belmonte T, Thiebaut J M, S. Bockel S and Michel H 2005 Reduction of TiO_2 assisted by a microwave plasma at atmospheric pressure *Thin Solid Films* **475** 63–7

[11] Baeva M, Gier H, Pott A, Uhlenbusch J, Hoschele J and Steinwandel J 2002 Pulsed microwave discharge at atmospheric pressure for NOx decomposition *Plasma Sources Sci. Technol.* **11** 1-9

[12] Henriques J, Bundaleska N, Tatarova E, Dias F M and Ferreira C M 2011 Microwave plasma torches driven by surface wave applied for hydrogen production Int. *J. Hydrogen Energy* **36** 345-54

[13] Jasiński M, Mizeraczyk J, Zakrzewski Z , Ohkubo T and Chang J S 2002 CFC-11 destruction by microwave plasma torch generated atmospheric-pressure nitrogen discharge *J. Phys. D: Appl. Phys.* **35** 2274-80

[14] Jasiński M, Dors M and Mizeraczyk J 2009 Destruction of Freon HFC-134a Using a Nozzleless Microwave Plasma Source *Plasma Chem. Plasma Process.* **29** 363- 72

[15] Jasiński M, Dors M and Mizeraczyk J Production of hydrogen via methane reforming using atmospheric pressure microwave plasma 2008 *J. Power Sources* **181** 41-5

[16] Pawelec E, Simek M, Nassar H, Czernichowski A, Musioł K and Dittrichova L 1996 Temperature measurements in non-equilibrium "ferroelectric" plasma *Acta Physica Polonica A* **89** 503–7

[17] Izarra Ch 2000 UV OH spectrum used as a molecular pyrometer *J. Phys. D: Appl. Phys.* **33** 1697-704

[18] Moon S Y and Choe W 2003 A comparative study of rotational temperatures using diatomic OH, O_2 and N_2+ molecular spectra emitted from atmospheric plasmas *Spectrochim. Acta Part B* **58** 249-57

[19] Wang Q, Koleva I, Donnelly V M and Economou D J 2005 Spatially resolved diagnostics of an atmospheric pressure direct current helium microplasma *J. Phys. D: Appl. Phys.* **38** 1690-7

[20] Machala Z, Janda M, Hensel K, Jedlovsky I, Lestinska L, Foltin V, Martisovits V and Morvova M 2007 Emission spectroscopy of atmospheric pressure plasmas for bio-medical and environmental applications *J. Mol. Spectros.* **243** 194–201

[21] Pawelec E 2007 Estimation of the plasma temperature by using the resolved spectrum of N_2+ *Euro. Phys. J. Spec. Topics* **144** 227-31

[22] Janus H W and Musielok J 2010 Field strength and temperature distributions in N_2/H_2 DBD discharge determined from measured spectral distributions of H_α and N_2+ radiation *Proc. Int. Conf. on Plasma Diagnostics (Pont-á-Mousson, France, 12-16 April 2010)* P52

[23] Raud J, Laan M and Jogi I 2011 Rotational temperatures of N_2(C,0) and OH(A,0) as gas temperature estimates in the middle pressure Ar / O_2 discharge *J. Phys. D: Appl. Phys.* **44** 345201
[24] O. Aucillo O and D. L. Flamm D L 1989 *Plasma Diagnostics* vols 1, 2 (Academic Press, New York)
[25] Herman I P 1996 *Optical Diagnostics for Thin Film Processing* (Academic Press, New York)
[26] Fonck R J and Den Hartog D J 2003 *Rev. Sci. Instrum.* **74** 1409
[27] Hutchinson I H 2002 *Principles of Plasma Diagnostic* 2nd edn. (Cambridge University Press, Cambridge)
[28] http://www.specair-radiation.net (24.05.2012)
[29] Laux C O 2002 Radiation and Nonequilibrium Collisional-Radiative Models *Physico-Chemical Modeling of High Enthalpy and Plasma Flows* (von Karman Institute Lecture Series 2002-07) eds D Fletcher, J -M Charbonnier, G S R Sarma, and T Magin (Rhode-Saint-Gencse, Belgium)
[30] http://www.sri.com/psd/lifbase/ (24.05.2012)
[31] Luque J and Crosley D R 1999 LIFBASE: *Database and Spectral Simulation Program (Version 1.5)* SRI International Report MP 99-009
[32] Hrycak B, Czylkowski D, Jasiński M and Mizeraczyk J 2012 Tuning characteristics of coaxial microwave plasma source operated with argon, nitrogen and methane at atmospheric pressure *Electr. Rev.* in press
[33] http://www.nist.gov/pml/data/asd.cfm (31.05.2012)
[34] Hrycak B, Jasiński M, Mizeraczyk J, Naumov V and Chernyak V 2011 Spectroscopic study of atmospheric pressure microwave plasma at high nitrogen flow rate *Proc. Int. Conf. on Research and Applications of Plasmas (Warsaw, Poland, 12-16 September 2011)*

Comparing the effect of different atmospheric pressure non-equilibrium plasma sources on PLA oxygen permeability

M Boselli[1,3], V Colombo[1,3], M G De Angelis[2,3], E Ghedini[1,3], M Gherardi[1], R Laurita[1], A Liguori[1], M Minelli[2,3], F Rotundo[1,3], P Sanibondi[1] and A Stancampiano[1]

[1] Department of mechanical engineering (D.I.E.M), Alma Mater Studiorum – University of Bologna, 40123 Bologna, Italy
[2] Department of chemical engineering (D.I.C.M.A.), Alma Mater Studiorum – University of Bologna, 40123 Bologna, Italy
[3] Industrial Research Center for Advanced Mechanics and Materials (C.I.R.I.-M.A.M.), Alma Mater Studiorum – University of Bologna, 40123 Bologna, Italy

E-mail address: vittorio.colombo@unibo.it

Abstract. Plasma technology is widely adopted for polymer surface modification. In this work polylactide (PLA) samples have been exposed to the plasma region generated by three different plasma sources operating at atmospheric pressure: a floating electrode dielectric barrier discharge (FE-DBD), a novel linear corona discharge and a DBD roller. The sources have been supplied with a high voltage generator capable of producing pulses with a rise rate in the order of several kV/ns in order to obtain diffuse plasma and avoid local damage to the membrane; air and argon have been used as working gases. Pure oxygen permeation tests in PLA films have been carried out by means of a closed-volume manometric apparatus working at 35°C with a pressure difference of pure O_2 of about 1 bar applied across the membrane. Tests have been performed shortly after the plasma treatment and also replicated at different times in order to investigate the durability of surface modification. The effects of voltage, pulse repetition frequency (PRF) and exposure time on the membrane surface characteristics and barrier property have been studied.

1. Introduction

In past decades petroleum-based polymers such as polystyrene, polyamide and polyethylene have been widely used because of their availability in large quantities at low costs [1]. However, driven by growing environmental concerns, new polymers based on renewable materials are being designed and developed to replace conventional polymers in different applications like food packaging [2].

One of the most promising of these eco-friendly polymers is polylactide (known also as polylactic acid or PLA). Polylactide is an aliphatic biodegradable, thermoplastic polyester derived from vegetal sources such as sugarcane, corn starch and tapioca, which are fermented to lactic acid and then polymerized. PLA can be processed like common thermoplastics and possesses good mechanical properties and high transparency. For these reason PLA is widely used in the packaging field for bottles, trash bags or disposable tableware [3]. In the human body PLA undergoes scission into lactic acid, a natural intermediate in carbohydrate metabolism [4]. PLA is therefore biocompatible

Journal of Physics: Conference Series **406** (2012) 012038 doi:10.1088/1742-6596/406/1/012038

and is used for many medical applications such as drug delivery, resorbable sutures, vascular devices and orthopedic implants [5].

Despite its excellent properties PLA presents a number of drawbacks that limit its broad application, such as its brittleness and thermal instability. Moreover, PLA barrier properties are worse than those of some oil-based polymers such as PET, and even if production cost is constantly decreasing it is still higher than those of common plastics. In order to enhance PLA properties and broaden its application field, new treatments and solutions must be developed. Chemical and physical properties on the surface have great influence on the final characteristic of the material; plasma treatment is one of the new inexpensive technologies able to achieve surface modification of the top layer (1-10nm) of polymeric materials without altering their bulk properties [6]. Furthermore plasma treatment proves to be an environmental friendly process, solvent free and with good energy efficiency [3]. While many publications have been dedicated to plasma treatment of polymeric materials using low pressure RF plasmas [1,10,11], only few investigate PLA modification using atmospheric pressure plasma sources [3,12,13].

In this work we report on the treatment of PLA films using three different non-equilibrium atmospheric pressure plasma sources; the effect on surface wettability, morphology, O_2 barrier properties and treatment durability have been analyzed for various operative conditions.

2. Experimental setup

2.1. Plasma treatments of PLA sample

Three different plasma sources have been used for PLA surface modification. The three sources have been alimented by a pulse generator (FID GmbH – FPG 20-1NMK) producing high voltage pulses with a rise time of few kV/ns, a peak voltage of 7-20 kV into a 100-200 Ω load impedance and a maximum pulse repetition rate of 1000 Hz. The circular PLA treated sample had a 2.5 cm diameter and a thickness of 40μm.

2.1.1. Floating Electrode – Dielectric Barrier Discharge (FE-DBD): this plasma source is widely adopted for the treatment of thermosensitive and biological materials since it relies on a dieletric layer between the electrodes to prevent the formation of an arc discharge and consequently limit the operating temperature (figure 1) [14]. In this study we used a DBD plasma source composed of a cylindrical copper electrode (diameter 30 mm, length 45 mm), a dielectric plastic material that covers the curved face of the electrode and a thin quartz disk (diameter 33 mm, thickness 0.5 mm) that covers the electrode plane surface where the plasma is generated. As the copper electrode diameter is larger (3 cm) than the PLA film circular sample (2.5 cm) a static treatment has been realized.

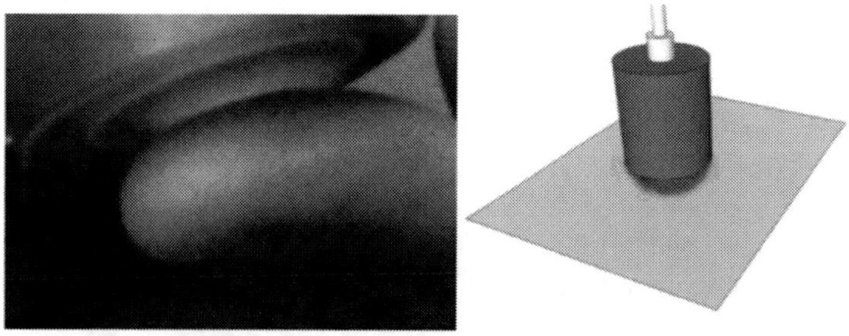

Figure 1. Floating Electrode Dielectric Barrier Discharge

2.1.2. Linear Corona: this source is based on the corona discharge principle, while instead of a pin the discharge is generated at the edge of a metal sheet with thickness lower than 1 mm (figure 2). The plasma region dimensions are limited by the electrode geometry, thus requiring handling for sample treatment. The distance between electrode and PLA samples has been controlled using two spacers fixed to the source.

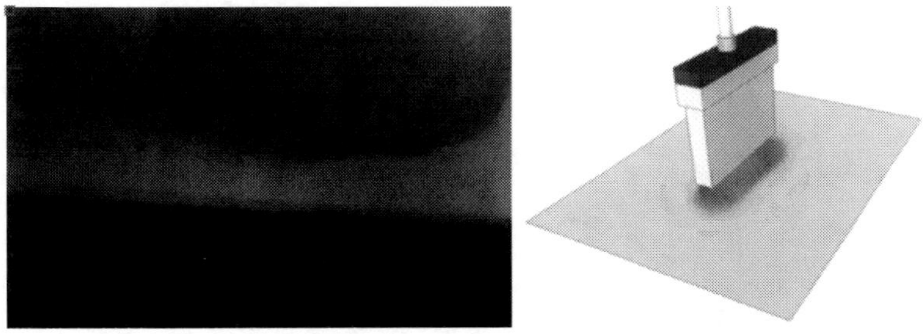

Figure 2. Linear Corona

2.1.3. DBD roller: this source has been specifically designed for the treatment of films. The plasma generated between the two rollers has a long and narrow geometry, suitable to treat large foils. Two different configurations have been investigated: a direct treatment (figure 3-b) where the PLA was directly in contact with the plasma and an indirect treatment (figure 3-c) where the PLA foil was posed under the plasma region and handled with a motorised slit.

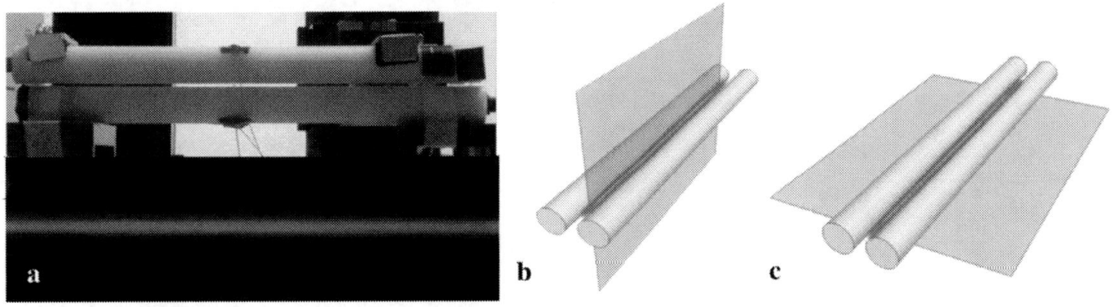

Figure 3. DBD roller: (a) DBD roller on, (b) direct PLA treatment (c) indirect PLA treatment

2.2. Diagnostic apparatus

In this work we investigated plasma modification of PLA samples in terms of oxygen barrier properties, wettability and surface morphology.

Change in wettability has been monitored with a drop shape analysis system (KRUSS DSA30) for the measurement of static water contact angle. PLA sample were placed on a horizontal plane with the plasma treated surface upwards. A distilled water drop of 5 μl was deposited on the sample surface. Measurements of contact angle were done immediately after the plasma treatment.

Pure oxygen permeation tests in PLA films were carried out at 35°C by means of a closed-volume manometric apparatus [7] which resembles the one described in ASTM D 1434. The oxygen permeability P of a sample of thickness l is the ratio between the steady state gas molar flux, J_{ss}, divided by the gas pressure gradient across the sample $\Delta p/l$. The flux is estimated as a function of the pressure p_d in the downstream volume V, the sample area A, the universal gas constant R, the temperature T, according to the ideal gas law:

$$P \equiv \frac{J_{SS}}{\Delta p / l} = \left(\frac{dp_d}{dt} \right)_{SS} \frac{V}{RTA} \frac{l}{\Delta p} \qquad (1)$$

Permeation experiments were carried out with a pressure difference across the membrane of about 1 bar, although the oxygen permeability does not depend on pressure in this range. In order to assess the repeatability of the permeability data, all the measurements were replicated at least twice (the resulting deviation is around 1%).

Morphological investigations have been carried out by means of Atomic Force Microscopy (AFM). Areas of 10x10 μm have been sampled in different positions on the surface for every PLA sample investigated.

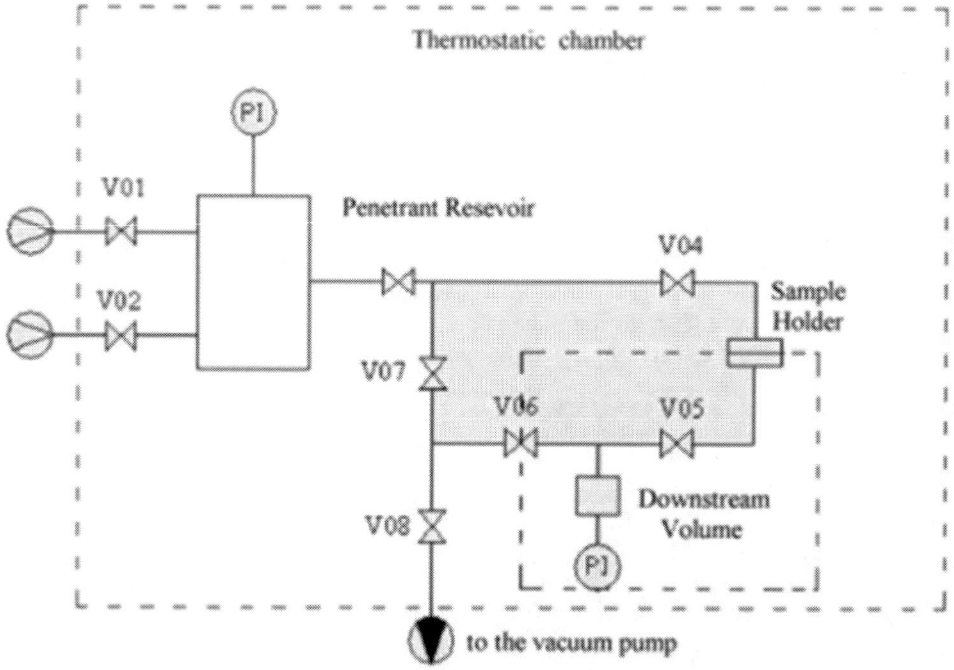

Figure 4. Manometric apparatus used for oxygen permeation measurements

3. Results

3.1. Contact angle

Contact angle measurements have been performed on PLA samples treated with both FE-DBD and the DBD roller sources. For each plasma source three different operating conditions were considered. Images and values of the contact angle measurements are shown in table 1 for the treated samples and an untreated PLA sample used as a reference for comparisons.

The contact angle of untreated PLA sample is 69.3°. After 1 minute of FE-DBD plasma treatment, static contact angle of PLA has shown a decrease to 48.2°. A continuous decrease in contact angle value has been observed for increasing treatment times, down to 33.7° after a 5 minutes treatment. Plasma generated by DBD roller has caused less straightforward effects on PLA samples; indeed, after 1 minute of plasma treatment static contact angle was still 64.1°. Instead, after 2 minutes, this parameter has increased up to 69.7°. Decreasing the peak voltage from 20kV to 18kV has produced a static contact angle of 63.7° for a 2 min treatment time.

Operative conditions	Drop image	Average value
UNTREATED		
		**69.3°**
FE-DBD		
Treatment time: 1 min PV: 16 kV, PRF : 125 Hz Discharge gap: 2 mm		**48.2°**
Treatment time: 3 min PV: 16 kV, PRF : 125 Hz Discharge gap: 2 mm		**39.8°**
Treatment time: 5 min PV: 16 kV, PRF : 125 Hz Discharge gap: 2 mm		**33.7°**
DBD ROLLER		
Treatment time: 1 min PV: 20 kV, PRF : 330 Hz Distance from the discharge: 11 mm		**64.1°**
Treatment time: 2 min PV: 20 kV, PRF : 330 Hz Distance from the discharge: 11 mm		**69.7°**
Treatment time: 2 min PV: 18 kV, PRF : 330 Hz Distance from the discharge: 11 mm		**63.7°**

Table 1. Static contact angle measurements

3.2. Pure oxygen permeability

The effect of treatment time over the variation of the oxygen permeability has been investigated for different plasma sources (FE-DBD, Linear Corona and DBD roller). In order to maximize discharge uniformity we have adopted different sets of operating condition depending on the plasma source:

- **FE-DBD**: voltage 16 kV, PRF 125 Hz, gap 2 mm
- **Linear Corona**: voltage 16 kV, PRF 1 kHz, gap 2 mm, flow rate (He) 5 slpm
- **DBD roller**: voltage 20 kV, PRF 330 Hz, distance from the discharge 11 mm

Results for this measurements are presented in figure 5. The effect of the plasma treatment depends strongly on the plasma source.

In the case of FE-DBD the oxygen permeability increases linearly with treatment times up to 180 s; for longer treatment times no further variation is observed.

PLA treated with Linear Corona shows a decrease in permeability for exposure times shorter or equal to 120 s; permeability values close to pristine PLA have been measured for longer exposure times.

The treatment with the DBD roller has achieved the most significant alteration in PLA oxygen permeability. A maximum increase in permeability (15%) has been obtained with 60 s of treatment time. A similar, but slightly lower oxygen permeability has been obtained for a 120 s treatment time.

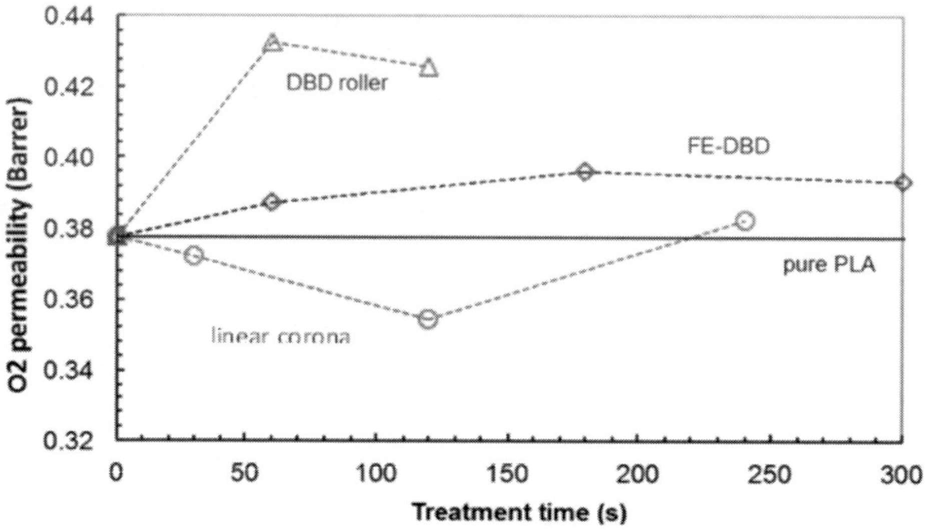

Figure 5. Pure oxygen permeability measurements

3.2.1. Ageing of PLA sample

After plasma treatment polymeric materials tend to recover their pristine properties in a process often referred to as ageing [2, 13]. In this work we investigate the effect of treatment time on the ageing of PLA permeability, measuring this property at different times after the treatment.

Results obtained for the Linear Corona source are reported in figure 6; for 30 s and 240 s treatment times no evident recovery effects has been appreciated up to 10 days after the treatment. Instead, samples treated for 120 s show a strong aging effect and measured oxygen permeability is very close to the original value after 8 days.

The samples treated with the FE-DBD and the DBD roller have also been investigated but show no apparent effect of recovery within 10 days.

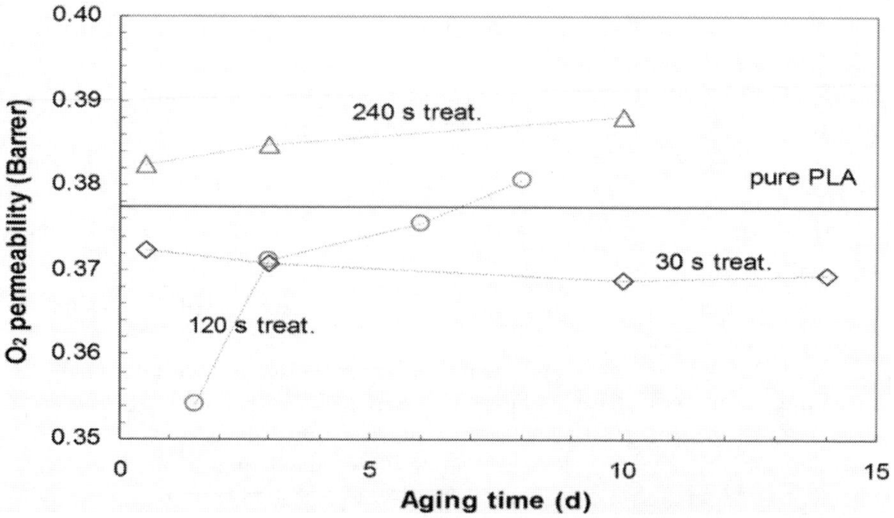

Figure 6. Aging of PLA sample treated with the Linear Corona plasma source

3.3. Morphological investigation with AFM

In this paragraph we present the morphological characterization of samples treated for 5 min with the DBD roller source (peak voltage 20 kV, pulse repetition frequency 125 Hz) in the direct configuration.

AFM images of PLA surface are presented in figure 7; images for the treated PLA refer to the surface exposed to the high voltage electrode during plasma modification. The untreated PLA surface (figure 7-a) appears smooth and characterized by a mean roughness (S_a) and a mean square roughness (S_q) of few nanometers. On the other side, plasma treated PLA surfaces (figure 7-b,c) show many morphological structures similar to bubbles, uniformly distributed on the sample area. These "bubble" structures, with a diameter in the range of 100-2000 nm, are visible in all the sampled areas of the treated PLA surface, proof that an uniform plasma treatment was achieved with the DBD roller. The measured mean roughness and mean square roughness of the modified PLA are an order of magnitude higher than for pristine PLA and uniform on the whole treated area.

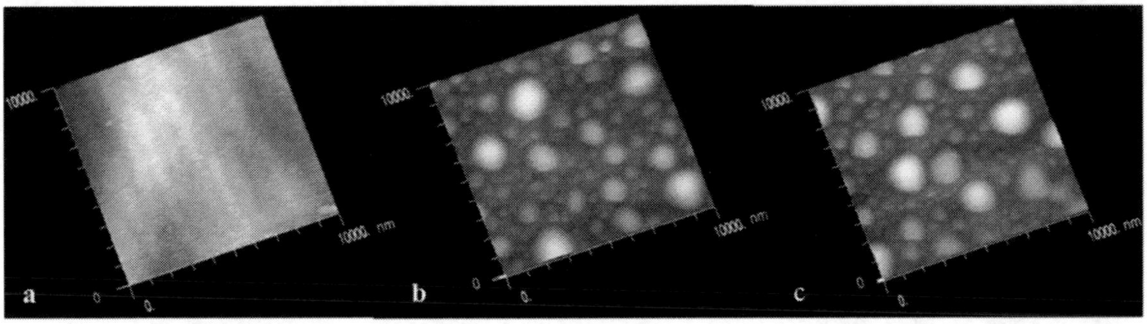

Untreated PLA	**Treated PLA (sampled area 1)**	**Treated PLA (sampled area 2)**
Sa: 2.43 nm	Sa: 28.17 nm	Sa: 28.47 nm
Sq: 3.08 nm	Sq: 36.81 nm	Sq: 37.61 nm

Figure 7. AFM images: (a) untreated PLA, (b) treated PLA surface sample 1, (c) treated PLA surface sample 2

4. Conclusions

In this paper we have reported on non-thermal atmospheric pressure plasma treatment of PLA, underlining the influence of the plasma source on the final surface modification.

Measurements performed on PLA samples treated with the FE-DBD have shown the highest increase, among the adopted plasma sources, in PLA wettability, without reducing significantly the oxygen barrier properties.

Very small alterations of the PLA oxygen permeability (below 2%) were noted for the samples exposed to the Linear Corona plasma source, and samples treated for 2 minutes tend to recover the original permeability after approximately 1 week.

DBD roller treatment results have shown that indirect treatment with this source was ineffective for the modification of the water contact angle and induced a considerable increase in the PLA oxygen permeability that indicates a deep deterioration of the surface. Direct DBD roller treatment has resulted in a uniform modification of PLA surface morphology; bubble like structures of variable size have been observed using AFM.

The nature of these bubble structures and their formation mechanism will be object of future investigations, together with the analysis of the apparent stability over time of samples treated with FE-DBD and DBD roller plasma. Since each of the tested plasma sources has shown the ability to differently modify PLA surface characteristics, future works will be directed towards their tailoring to specific applications and improvement of treatment performance.

References

[1] Chaiwong C *et al* 2010 *Surf. Coat. Technol.* **204** 2933-2939
[2] Guinault A *et al* 2010 *Int. J. Mater. Form vol.3* suppl **1**:603-606
[3] Vergne C, Bunchheit O, Eddoumy F, Sorrenti E, Di Martino J, Ruch D 2011 *J. Eng. Mater. Tech.* **133** 10.1115/1.4004156
[4] Drumright R E, Gruber P R, Henton D E 2000 *Adv. Mater.* **12**, 1814
[5] Slager J, Domb A J 2003 *Adv. Drug Delivery Rev.* **55,** 549
[6] Alves C M, Yang Y, Marton D, Carnes D L, Ong J L, Sylvia V L, Dean D D, Reis R L, Agrawal C M, 2008 *J. Biomed. Mater. Res., Part B:Appl. Biomater.*, **87**(1), pp. 59–66.
[7] Minelli M, De Angelis M G, Doghieri F, Marini M, Toselli M, Pilati F 2008 *Eur. Polym. J.* **44**(8), 2581-2588
[8] Forch R, Zhang Z, Knoll W 2005 *Plasma process Polym.* **2,** 351
[9] Renò F *et al* 2012 *Plasma Process. Polym.* **9,** 10.1002/ppap.201100139
[10] Dreux F *et al* 2003 *Mat. Res. Innovat.* **7**:183-190
[11] Inagaki N *et al* 2002 *J. Appl. Polym. Scie.* Vol. **89**, 96-103
[12] Morent R *et al* 2010 *Plasma Chem. Plasma Process.* **30**,525-536
[13] Liu C *et al* 2004 *Surf. Coating Tech.* **185**, 311-320
[14] Fridman A 2008 Plasma Chemistry *Cambridge Uni. Press.*

RF thermal plasma treatment of dredged sediments: vitrification and silicon extraction

V Colombo [1,2], E Ghedini [1,2], M Gherardi [1], V Mani [1], P Sanibondi [1], B Vazquez [1]

Alma Mater Studiorum – Università di Bologna, Via Saragozza 8, 40123 Bologna, Italy

[1] Department of Mechanical Engineering (D.I.E.M.)

[2] Industrial Research Center for Advanced Mechanics and Materials (C.I.R.I.-M.A.M.)

E-mail: emanuele.ghedini@unibo.it

Abstract. In the present work, dredged polluted sediments with high level of hydrocarbons and heavy metals have been treated by means of a laboratory scale radiofrequency (RF) thermal plasma source inside a graphite crucible. Two different experimental approaches have been utilized (1) to fully decontaminate and vitrify this material, and (2) to study the technical feasibility of metallurgical-grade silicon (MGS) smelting by carbothermal-reduction reactions of carbon with silica (SiO_2) content in these dredged sediments. A two-dimensional model of a commercial inductively coupled RF plasma torch has been used to investigate the effects of plasma flow and temperature distributions of the plasma discharge interacting with the material inside the crucible. Samples of both vitrification and carbothermal reduction processes were characterized by scanning electron microscopy analysis (SEM), energy dispersion spectroscopy (EDS), X-ray fluorescence (XRF) and leaching tests. Results obtained showed a fully decontamination of the dredged sediments, with release values of heavy metals in leachates well below the law limits. Moreover, SEM/EDS analyses suggested that separation of MGS by carbothermal-reduction process is possible.

1. Introduction

The annual volume of dredged sediments in the European Union is roughly estimated between 100M m^3-200 Mm^3 per year [1]. These sediments are generally contaminated with heavy metals and hydrophobic organic compounds such as Polychlorinated Biphenyl (PCB), Isopropyl Alcohol (IPA) and hydrocarbons, and can be considered as waste under the European Waste Framework Directive (Directive 2008/98/EC). In the past, sediment dredged operations were exclusively aimed at warranting a proper operational functionality of harbours, docks and waterways, and these polluted dredged sediments were usually drained into wastewater systems or sent to landfills. Fortunately, this tendency has been changing in recent years. Conventions for the protection of the marine environment [2-4], and some new European tight regulations concerning waste and water management [5-7], have been introduced (1) to set guidelines for a proper disposal of dredged material into the sea, (2) to establish a framework for water protection, (3) to avoid the traditional perception of these dredged pollutant sediments as waste, considering them as a commercially exploitable resource.

A large group of environmental projects are being currently carried out in the European Union to implement, update and develop the EU environmental policy and legislation, with the financial help of the EU's funding instrument for the environment, the LIFE+ programme. In particular, the "Recovery of dredged SEDIments of the PORT of Ravenna and SILicon extraction" project (SEDI.PORT.SIL), in which is included this work, represents a highly innovative alternative to the traditional treatment of the dredged polluted sediments. This project is aimed at laying down clear guidelines for the sustainable management of dredged sediments, establishing the basic pillars for the realization of a treatment plant in the port of Ravenna (Italy), and demonstrating the efficiency of some consolidated technologies for the treatment of sediments, e.g. landfarming for the microbial degradation of hazardous compounds and soil washing for scrubbing soils ex-situ to remove contaminants through particle size separation, coupled with innovative techniques to recycle and increase in value the contaminated sediments.

Among these new innovative techniques thermal plasma technology offers a great number of unique advantages to decontaminate these sediments and to obtain a final recyclable product: (1) the high temperatures ($T \geq 10^4$ K) and the high energy densities, ranging from 10 W/cm3 to above 10 kW/cm3 [8-9], help to decompose by combustion all organic contaminants presented in these dredged sediments, encapsulating heavy metals and mineral phases into a stable glassy matrix, (2) the use of electric energy to generate the plasma discharge allows decreasing the gas flow rate and the requirements to treat off-gases produced, facilitating control over the chemistry of the process and giving the possibility of generating saleable co-products [10]. One of the most extensively used methods to obtain thermal plasma is by means of radio-frequency (RF) induction discharges [11]. In an Inductively Coupled Plasma (ICP) the discharge is generated by an RF electric field coming from an external coil. This type of device iscommonly used as an effective mean to produce a plasma jet with high enthalpy content which can be employed in a wide range of applications and research activities, such as densification and spheroidization of powders, waste treatment, plasma spray deposition of materials, chemical synthesis of nanoparticles, and others [12-16].

In the present work, two different experimental approaches have been investigated by means of a laboratory scale radio-frequency (RF) thermal plasma source to completely decontaminate the dredged polluted sediments and to obtain a saleable end product which enable to defray the high cost of the plasma treatment: 1) carbothermal reduction approach, adding carbon to the untreated sediments, in order to analyse qualitatively the technical feasibility to separate silicon from the silica (SiO_2) content, and 2) vitrification process of the as-received polluted sediments. This plasma source allows reproducing the high temperatures and the high heat exchange conditions of the more widespread arc plasma furnaces. A customized 2-D version of the CFD commercial code FLUENT has been utilized to quickly optimize some of the main parameters of the plasma process and to have a deeper understanding of the different temperature fields inside the reaction chamber during the plasma discharge. Obtained samples of both carbothermal reduction and vitrification processes have been characterized by scanning electron microscopy analysis (SEM), energy dispersion spectroscopy (EDS), and X-ray fluorescence (XRF) to analyse their microstructure and their elemental composition, as well as to determine if the silicon separation is technically possible. Moreover, leaching tests have been carried out to evaluate toxic metals stabilization in the resulting material.

2. Methods and Materials

2.1. Simulation activities

As a consequence of the high temperatures and the aggressive environment inside the reaction chamber where the plasma discharge is generated, direct measurements are extremely complicated. Simulation activities represent an effective and widely used tool to have a deeper understanding of the physical and chemical processes of a plasma system [17, 18]. For the purpose of this study, a two-dimensional model of a commercial ICP torch already developed with commercial code FLUENT© has been used [19]. Numerical modelling results have allowed optimizing the torch operating conditions and understanding the influence on the plasma flow and temperature, as well as the behaviour of the plasma discharge interacting with the material inside the crucible. For the sake of simplicity, this material has been considered all silica (SiO_2) because it is the main compound of the dredged sediments, the effects of silica vapour on plasma have been neglected and the graphite crucible has been included in the heat transfer and fluid dynamics model domain.

2.1.1. Optimization of the distance between the exit nozzle of the plasma torch and the crucible position. The Optimal Zone where the crucible must be placed has been defined to avoid (a) high temperatures that partially evaporate the treated material, and (b) a low crucible position where heat is not enough to completely melt the material. After the first simulation results, two new distances have been taken into consideration for the laboratory experiments: Optimal_1 (85mm) and Optimal_2 (92mm).

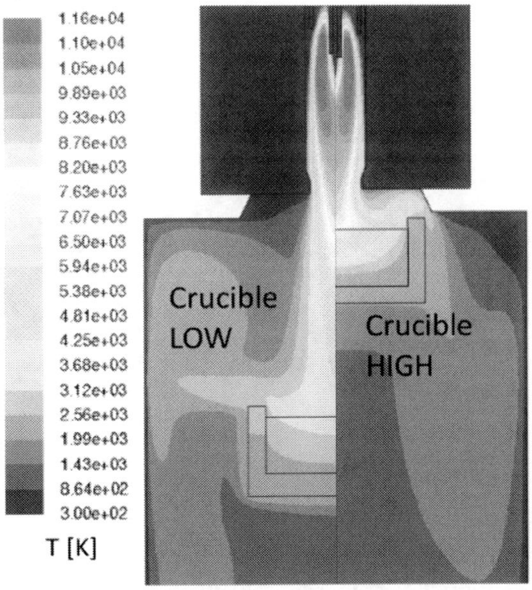

Fig.1 Detail of the temperature field inside the torch and in the highest part of the reaction chamber for two different positions of the crucible: LOW (distance 170mm) and HIGH (distance 70mm).

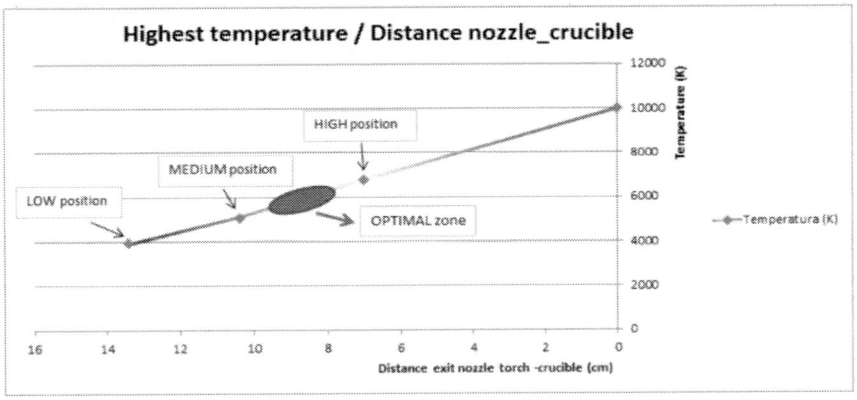

Fig.2 Evolution of the highest temperature reached inside the crucible depending on its position inside the reaction chamber; identification of the OPTIMAL ZONE where the crucible must be placed

2.1.2. Time-dependent simulations. These results help to optimize the minimum time necessary to melt the silica undergoing the plasma discharge inside the crucible. This time has been estimated in 5 minutes when a crucible of dimensions 110mm x 25mm (Diameter x Height) is placed in the highest position. On the other hand, when the crucible is in lower positions, the time needed increases drastically up to 13-14 minutes (figure 3).

2.1.3. Volume of material to treat inside the crucible. This volume has been reduced from 16 cm^3 to 12cm^3 because of the low thermal conductivity (1,3 W/m·K) and high specific heat (740 J/Kg·K) of the silica. As a consequence, a more homogeneous temperature profile has been achieved all along the treated material. This effect has been also confirmed in the first laboratory tests filling completely the crucible with the dredged sediments. After these trials, the lower part of the material inside the crucible was completely unreacted after the treatment.

Journal of Physics: Conference Series **406** (2012) 012039 doi:10.1088/1742-6596/406/1/012039

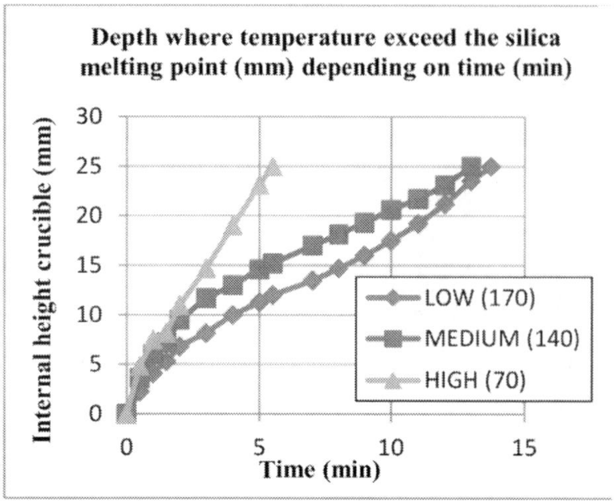

Fig. 3 Evolution of the silica melting depth for different crucible positions

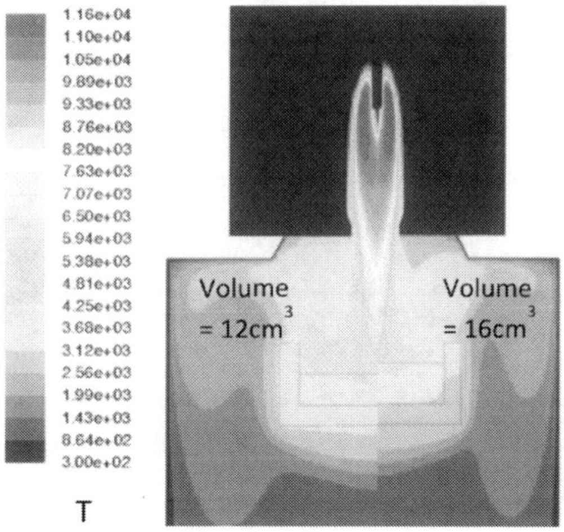

Fig. 4 Temperature field inside the crucible for two different volumes of material

2.2. Experimental set-up

In figure 5 the experimental setup used in this work is shown. The laboratory scale RF thermal plasma source consists of: RF plasma torch, model Tekna PL-35 (with ceramic confinement tube for operation up to 30kW); 35 kW LEVEL unit RF power supply working at 3 MHz; a control console Tekna Plasma Systems (Sherbrooke, Quebec, Canada) RF-CS-30 with gas flow meters for operation with Ar, Ar/H2 or any other auxiliary gas (Air, N2, O2); Tekna Plasma Systems external cooling system; adjustable water-cooled steel manipulator. The manipulator allows regulating the distance between the exit of the torch nozzle and the graphite crucible from 70mm down to 170mm. The importance of modifying this specific parameter has a large influence in the temperature field inside the graphite crucible, and more than 1500K of difference have been measured between the highest temperature in the upper crucible position with respect to the lower one [20].

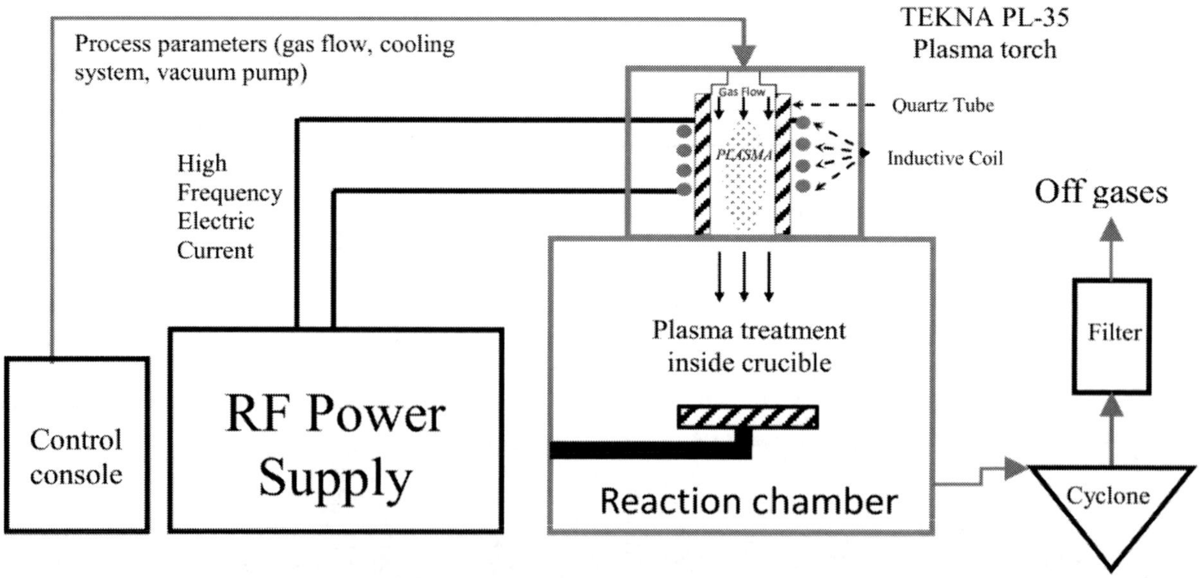

Fig. 5. Schematic illustration of the inductive plasma reactor system

The off gases produced during the plasma treatment pass through a cyclone to separate the largest particles, and through a dry collector which has a filter bag membrane with minute pores to retain the smallest particles near the filter surface.

2.3. Dredged sediments characterization

Sediments from the Port of Ravenna have been fully characterized, in accordance with the existing Ministerial Decree DM 471/99, to establish an accurate quantification of their different levels of pollution. As a consequence, sediments have been subdivided in three different categories depending on their degree of contamination: Green (no polluted sediments), Yellow (mean polluted sediments) and Red (highly polluted sediments, mainly due to the high content of hydrocarbons, HC). Composition data of heavy metals and hydrocarbons for each category measured in mg/kg on a dry basis is shown in table 1.

Both yellow and red sediments represent the most polluted fractions, especially because of the high hydrocarbons concentration. Nevertheless, thermal plasma technology is a well-established and efficient technology for destruction of organic contaminants [21], and it can be assumed that slight differences in the resulting material for each of these fractions of sediments will be found after plasma treatment. Also common to the main purpose of the project, these categories of sediments have been further subdivided after a sorting and soil washing process into two additional categories, depending on their grain size: sandy fraction (particle size > 75μm) and silty fraction (particle size < 75μm). Both fractions present marked differences, especially related to their chemical composition.

From the first XRF analyses to characterize the different categories of sand and silt fractions, it was determined that the percentage of silica was about 43%-53% depending on the polluted grade (green, yellow and red) and grain size (sandy and silty fractions) of the sediments (table 2). To determine the volatile substances in the each sample, a Loss On Ignition (LOI) test have been carried out: sediments have undergone a temperature of 1223K for 12 hours, losing between 10%-15% in weight after the test.

Table 1. Metal and hydrocarbon concentration in dredged sediments with respect to maximum acceptable limits (APAT, Agenzia per la Protezione dell'Ambiente e per i Servizi Tecnici)

	Cd	Ni	Pb	Cu	Zn	Hg	As	V	Cr	Se	HC>12	HC<12
VALUE LIMIT REUSE IN INDUSTRIAL AREAS	15	500	1000	600	1500	5	50	250	800	15	750	250
VALUE LIMIT REUSE IN PUBLIC PARKS	2	120	100	120	150	1	20	90	150	3	50	10
GREEN	0,1	71,5	21,5	31,5	71,5	0,2	6,6	25,9	41,0	0,4	1,0	0,2
YELLOW	0,2	59,5	30,0	41,5	280	3,1	7,7	24,0	47,5	0	190,4	0,6
RED	0,1	73,5	64,0	34,0	66,5	0,7	38,7	20,2	43,5	0,2	764,0	8,7

[a] Values which exceed the limits of law for the reuse of the material are underlined

Table 2. XRF analyses for the various categories of sediments

SAMPLE DESCRIPTION	SiO2	TiO2	Al2O3	Fe2O3	MgO	CaO	Na2O	K2O	LOI [a]	Tot.
Sand - GREEN	49,5	0,5	11,1	3,9	3,8	17,1	1,7	2,1	10,2	99,7
Sand - YELLOW	49,9	0,5	11,0	3,6	4,1	16,0	1,8	1,9	11,3	100,0
Sand - RED	53,0	0,5	10,7	3,8	4,0	14,2	1,8	1,8	10,0	99,8
Silt - GREEN	44,6	0,6	14,0	5,5	4,1	11,3	1,2	2,3	16,2	99,7
Silt - YELLOW	45,2	0,6	14,0	5,5	4,3	11,3	1,3	2,2	15,5	99,8
Silt - RED	43,6	0,6	13,6	6,0	4,4	12,6	1,2	2,1	15,5	99,7

[a] Loss On Ignition

Journal of Physics: Conference Series **406** (2012) 012039 doi:10.1088/1742-6596/406/1/012039

3. Results and Discussion

Material treatment in an ICP torch is commonly performed with a powder feeder equipment, which injects powders from the top of the plasma torch, along its axis and into the plasma region by means of a water cooled injection probe. However, because of the unexpected low flowability of the dredged sediments, constant blockages in the feed pipe were experimented with this technique. Therefore, in all these experimental activities, sediments have been treated directly in a graphite crucible inside the reaction chamber, where they have been exposed to the high temperatures and the high heat flux of plasma discharge in controlled atmosphere, resulting in a completely melted batch after the treatment. For the purpose of this work, two different experimental tests have been performed depending on whether carbon was added or not to the dredged sediments: 1) carbothermal reduction process for the Metallurgical-Grade Silicon (MG-Si) smelting, and 2) vitrification of the as-received sediments.

3.1. Carbothermal reduction process

This process has been aimed at separating silicon from the silica content presented in the sediments. It is well-known that this process proceeds through more steps [22, 23], especially because of the different temperature distribution along the reaction chamber. Only considering a Si-C-O system, the intermediate reactions of this process with their respective temperature would be:

- ***T ≈ 1500 - 1800 K*** $SiO_2\ (s) + C\ (s) \rightarrow SiO\ (g) + CO\ (g)$ (1.1)
- ***T > 1800 K*** $SiO\ (g) + 2C\ (s) \rightarrow SiC\ (s) + CO\ (g)$ (1.2)
  $2SiO_2\ (g) + SiC\ (s) \rightarrow 3SiO\ (g) + CO\ (g)$ (1.3)
- ***T ≈ 2100 K*** $SiO\ (g) + SiC\ (s) \rightarrow 2Si\ (s,l) + CO\ (g)$ (1.4)
  $SiO_2\ (g) + Si\ (l) \rightarrow 2SiO\ (g)$ (1.5)

Moreover, all the various types of dredged sediments contain a great number of other compounds, such as Hematite (Fe_2O_3), Aluminium Oxide (Al_2O_3), Calcium Oxide (CaO) or Magnesia (MgO), which will undoubtedly modify the progress of the global reaction for the silicon separation. Consequently, the overall reduction reaction can be simplified as follows in order to estimate the amount of carbon necessary for the total reaction of silica:

$$SiO_2\ (s) + 2C\ (s) \rightarrow Si\ (s,\ l) + 2CO\ (g) \quad (1.6)$$

Operating conditions of an optimized carbothermal reduction test are shown in table 3. Argon has been used as plasma and carrier gas, whereas a proportional mixture of argon and hydrogen has been utilized as sheath gas because the high thermal conductivity and specific heat of the hydrogen results in a higher plasma enthalpy and in a higher heat flux to the particles. The crucible has been placed in the Optimal_2 position, included in the Optimal Zone previously determined with the simulation activities. A high concentration of minute metallic spheres has been visibly observed, and later analysed with SEM/EDS microscopy. The remaining material is roughly 60% of the total, mainly because of a partial evaporation of some volatile elements and other compounds formed as a consequence of the high temperatures of the plasma discharge.

3.2. Vitrification process

The other group of tests have been aimed at treating directly with the thermal plasma source the as-received sediments in order to vitrify them, analysing the resulting material to check out the leaching properties. This is a well-known technology and just few tests were enough to prove the feasibility of this technology to obtain a homogeneous glass, with a partial separation of metallic spheres which are formed in the surface of the final material. As it was expected, the percentage of the resulting material after the thermal treatment is higher with respect to the carbothermal reduction tests, approximately 74%.

Table 3. Operating conditions of a carbothermal reduction test

	CARBOTHERMAL REDUCTION	*FINAL RESULT*
Plasma gas	Ar 15 slpm	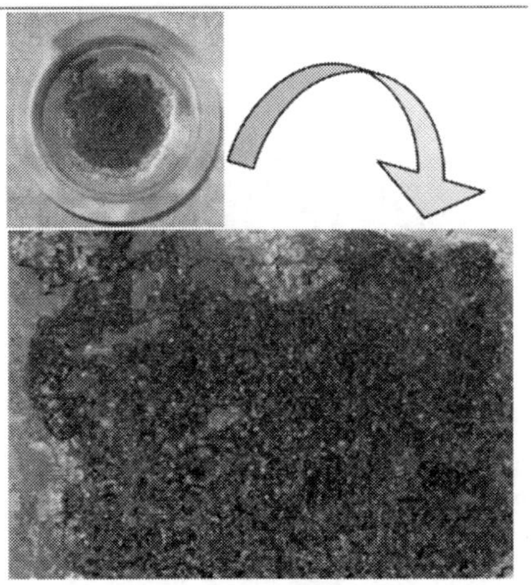
Sheath gas	Ar 30 slpm + H_2 30 slpm	
Carrier gas	Ar 3 slpm	
Crucible position	Optimal_2 (92mm)	
Dredged sediment	Red sediments	
Sediments/carbon	20g / 5g = 25g	
Resulting material	15,2g	
Percentage	60,8%	
Plate power	27,30 kW	
Chamber pressure	14 kPa	
Exposure time	5 minutes	

3.3. Characterization of plasma treated sediments

Material resulting from plasma treatment of carbothermal reduction and vitrification processes have been analysed with different characterization techniques: (1) X-Ray Fluorescence (XRF) to determine the main compounds of the sediments after the thermal treatment, (2) SEM/EDS to check out the silicon extraction from all the other compounds, and (3) leaching tests to predict whether these treated sediments are likely to leach contaminants, generally metals, at levels of concern for their possible re-uses. The most relevant results are presented.

Table 4. Operating conditions of a vitrification test

	VITRIFICATION	*FINAL RESULT*
Plasma gas	Ar 15 slpm	
Sheath gas	Ar 30 slpm + H_2 30 slpm	
Carrier gas	Ar 3 slpm	
Crucible position	Optimal_2 (92mm)	
Dredged sediment	Silt-red fraction	
Sediments	40g	
Resulting material	29,6g	
Percentage	74%	
Plate power	29,23 kW	
Chamber pressure	14 kPa	
Exposure time	5 minutes	

3.3.1. XRF. Plasma treated powders of 10 tests (8 carbothermal reductions and 2 vitrifications) carried out with similar operating conditions have been analysed by wave-length dispersive XRF to characterize every single sample after plasma treatment. The SiO_2 percentage varies from 42% up to 67%, slightly higher with respect to the untreated sediments. In particular, in sand-red fraction tests, the %SiO_2 has increased from 53% (table 2) up to 66%-67% (table 5). Moreover, in both silt-fraction tests the silica percentage has increased of about 9%-11%. These significant results are probably a result of a partial evaporation of some volatile elements and other compounds formed as a consequence of the high temperatures of the plasma discharge.

Journal of Physics: Conference Series **406** (2012) 012039		doi:10.1088/1742-6596/406/1/012039

Table 5. Percentage of the different compounds measured in the final samples

Sample	SiO_2	TiO_2	Al_2O_3	Fe_2O_3	MgO	CaO	Na_2O	K_2O	Sample description
1	52,7	0,8	16,5	5,7	4,2	15,9	2,1	1,9	Yellow sediment
2	52,7	0,9	15,4	5,8	4,9	15,5	2,3	2,4	Yellow sediment
3	54,9	0,8	14,3	4,6	3,7	17,6	2,1	1,8	Red sediment
4	42,1	1,1	20,7	6,3	3,0	24,7	1,0	0,8	Red sediment
5	66,6	0,5	10,0	3,4	3,3	12,6	1,8	1,7	Sand-red fraction
6	65,4	0,5	10,6	3,4	3,3	12,7	2,0	1,8	Sand-red fraction
7	53,5	0,8	15,8	6,0	5,1	14,4	1,5	2,5	Silt-yellow fraction
8	53,9	0,8	15,3	4,9	5,2	16,1	1,3	2,1	Silt-red fraction
9	53,8	0,6	13,1	5,1	4,6	11,3	7,9	3,4	Vitrification
10	55,5	0,8	15,0	5,1	5,1	15,0	1,3	2,0	Vitrification

3.3.2. SEM/EDS. Chemical analysis and imaging on a variety of zones of the different analysed tests have been performed using a Scanning Electron Microscope (SEM). The Scanning Electron Microscope is equipped with an Energy Dispersive Spectrometer (EDS), which provides chemical analysis of the field of view or spot analyses of minute particles. This kind of micro-chemical analysis is a non-destructive test. A couple of micrographs from two different carbothermal reductions tests are shown in figure 6. Results of SEM/EDS analyses evidence that, despite the significant percentage of carbon which remains probably unreacted within the sediments, a significant percentage of silicon is still present in the material. Moreover, some metallic spheres have been separated from the inert matrix .Most of them are ferrosilicon alloys with different percentages of silicon, from 10% up to 35%. In addition, some EDS analyses showed that some of these metallic particles have a very high percentage of silicon (85-90%), very close to the MG-Si.

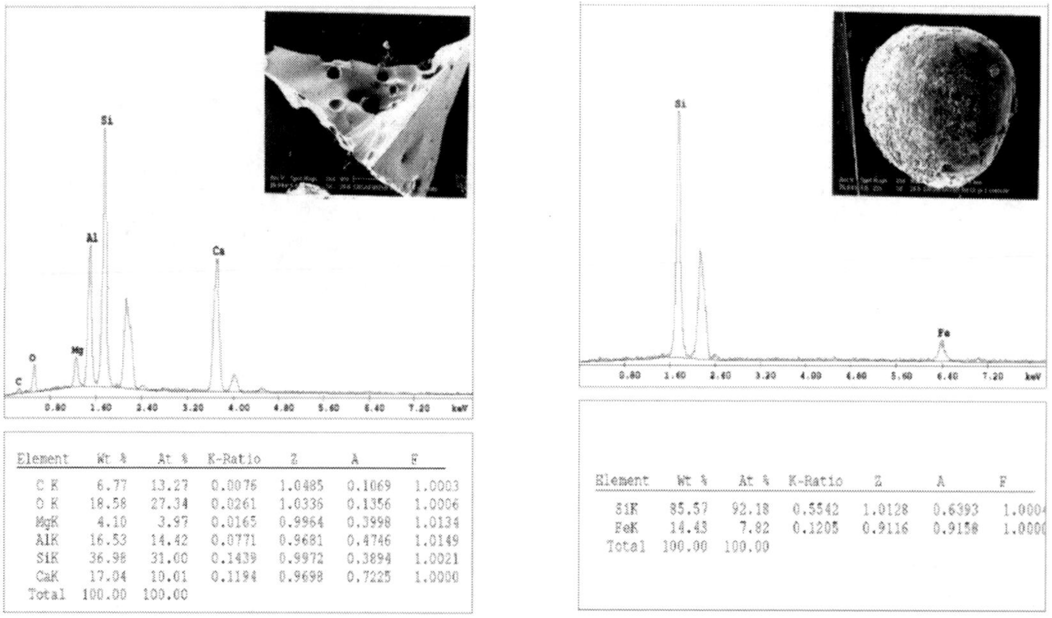

Fig. 6 SEM images and EDS spectra of two sections of different carbothermal reduction tests

3.3.3 Leaching test. The same tests analysed with the XRF have been further characterized to measure the amount of metals released into the leachate, simulating the environmental conditions of which the waste would become part. Table 6 shows that concentrations of all the metals analysed were well below the law limits for every sample. In addition, a Gas Chromatography-Mass Spectrometry (GC-MS) technique has been used to detect possible organic pollutants in the samples. The screening with the GC-MS showed that no hydrocarbons were found in every single test ("Organics screening"). Therefore, decontamination of dredged sediments with a thermal plasma source, of both carbothermal reduction and vitrification processes has been demonstrated, and these materials can be all considered inert.

Table 6. Compliance leaching test for different carbothermal reduction and vitrification tests

LEACHING TEST	Limits DM 27/09/2010 Tab. 5 Non-hazardous Waste	Limits DM 27/09/2010 Tab. 2 Inert Waste	**Test 1**	**Test 2**	**Test 3**	**Test 4**	**Test 5**	**Test 6**	**Test 7**	**Test 8**	**Test 9**	**Test 10**
Al [b]	-	-	0,678	0,006	0,032	0,044	0,018	0,024	0,030	0,040	0,024	0,016
As [b]	**0,2**	**0,05**	0,002	0,025	0,033	0,028	0,001	0,001	0,002	0,004	0,021	0,035
Ba [b]	**10**	**2**	1,355	0,012	0,064	1,000	0,036	0,048	0,060	0,080	0,047	0,032
B [b]	[b] -	-	<0.001	<0.001	<0.001	<0.001	<0.001	<0.001	<0.001	<0.001	<0.001	<0.001
Cd [b]	**0.1**	**0,004**	<0.001	<0.001	<0.001	<0.001	<0.001	<0.001	<0.001	<0.001	<0.001	<0.001
Cr tot [b]	**1**	**0,05**	0,0003	0,0000	0,0013	0,0008	0,0002	0,0003	0,0000	0,0000	0,0004	0,0002
Fe [b]	-	-	0,0109	0,0083	0,0025	0,0067	0,0065	0,0059	0,0037	0,0024	0,0054	0,0083
Mn [b]	-	-	0,0004	0,0003	0,0007	0,0007	0,0006	0,0009	0,0003	0,0004	0,0003	0,0010
Hg [b]	**0,02**	**0,001**	0,0004	0,0002	0,0001	0,0002	0,0001	0,0001	0,0005	0,0002	0,0004	0,0001
Ni [b]	**1**	**0,04**	0,0001	0,0005	<0.0001	0,0001	0,0001	<0.0001	0,0001	0,0002	0,0001	0,0001
Pb [b]	**1**	**0,05**	0,0000	0,0000	0,0000	0,0001	0,0002	0,0001	0,0001	0,0002	0,0000	0,0003
Cu [b]	**5**	**0,2**	0,0007	0,0004	0,0005	0,0005	0,0004	0,0004	0,0008	0,0012	0,0003	0,0002
Se [b]	**0,05**	**0,01**	0,0007	0,0012	0,0037	0,0044	0,0003	0,0003	0,0026	0,0030	0,0012	0,0030
Sn [b]	-	-	0,0001	<0.0001	0,0003	0,0003	0,0010	<0.0001	<0.0001	<0.0001	<0.0001	0,0004
Zn [b]	**5**	**0,4**	0,0017	<0.0001	0,0087	0,0027	0,0002	0,0016	0,0001	0,0018	0,0012	0,0007
Total HC [a] (C>12)	-	-	<10	<10	<10	<10	<10	<10	<10	<10	<10	<10
Organics screening	-	-	<10	<10	<10	<10	<10	<10	<10	<10	<10	<10

[a] Hydrocarbons
[b] Unit of measurement mg/L

4. Conclusions

A new and innovative approach has been accomplished to treat polluted dredged sediments with a thermal plasma source. A two-dimensional model of a commercial inductively coupled RF plasma torch with customized CFD commercial code FLUENT© has been used to investigate in the early stages of the project the effects of plasma flow and temperature distributions of the plasma discharge interacting with the material inside the crucible.

Despite the few scientific publications available related to this topic, some important obtained results highlight the great potential of thermal plasmas to decontaminate and vitrify dredged sediments. All the different categories of sediments have been successfully decontaminated, and from the leaching test it was concluded that the resulting material of both carbothermal reduction and vitrification processes were completely inert. Moreover, the first positive results to estimate qualitatively the possibility to extract MGS from marine sediments could represent a major opportunity to the SEDI.PORT.SIL project to become economically sustainable.

Acknowledgements
This study was completed as part of the project "Recovery of dredged SEDIments of the PORT of Ravenna and SILicon extraction" funded by the European Life+ Programme. LIFE+ Environment Policy and Governance is one of the strands of the European Union's main funding programme for the environment. It supports technological projects that offer significant environmental benefits. The authors would like to acknowledge the collaboration between all the partners involved in this project.

References

[1] SedNet Consortium 2004 *Contaminated Sediments In European River Basins*, SedNet booklet as final report for the EC FT5 Thematic Network Project SedNet (EVK1 –CT-2001-20002), ed W Salomons and J Brils
[2] The London Convention LC 1972 *Prevention of Marine Pollution by Dumping of Wastes and Other Matter*
[3] The OSLO-PARIS (OSPAR) Convention 1992 *Protection of the Marine Environment of the North-East Atlantic*
[4] The Helsinki Convention (HELCOM) 1992 *Protection of the Marine Environment of the Baltic Sea Area*
[5] Council Directive 91/692/EEC 1991 Standardizing and Rationalizing Reports on the Implementation of Certain Directives Relating to the Environment, *OJ No L* 377, 31.12.1991, 48
[6] Council Directive 1999/31/EC 1999 Landfill of Waste, *OJ No L 182/1*, 16.7.1999
[7] Council and European Parliament Directive 2008/98/EC 2008 Waste and Repealing Certain Directives, *OJ No L 312/3*, 22.11.2008
[8] Gleizes, Gonzalez J J and Freton P 2005 Thermal Plasma Modelling *J Phys. D: Appli.Phys*. **38** 153-183
[9] Roth J R 2001 *Industrial Plasma Engineering: Applications to Nonthermal Plasma Processing* vol 2 (Taylor & Francis) 658
[10] Boulos M I, Fauchais P and Pfender E 1994 *Thermal Plasmas:Fundamentals and Applications* vol 1 (New York: Plenum) 452
[11] Chen F F and Chang J P 2002 *Lecture Notes on Principles of Plasma Processing* (Los Angeles: Springer) 208
[12] Deegan D E, Chapman C D, Ismail S A, Wise M L H and Ly H 2006 The Thermal Treatment of Hazardous Waste Materials Using Plasma Arc Technology, *IMechE Conf. Delivering Waste Solutions—Balancing Targets, Incentives and Infrastructure*, vol. 9 (London)
[13] Barton T G and Mordy J A (1984) The Destruction of Halogenated Organic Chemicals by Plasma Pyrolisis, *Can. J. of Physiol. and Pharmacol.* **62** 976-978
[14] Drouet M G, Meunier J, Laflamme C B, Handfield M D, Biscaro A and Lemire C 1995 A rotary arc furnace for aluminum dross processing *Proc. Int. Symp. On Environmental Technologies: Plasma Systems and Applications* (*Atlanta, EEUU, 8-11 October 1995, Georgia Tech Research Corp*) vol 1,2 ed P W Maynl and J A Mulholland, 539-548
[15] Boulos M I 1997 The Inductively Coupled Radio Frequency Plasma, *J. High Temp. Mat. Proces.* **1** 17-39
[16] Colombo V, Ghedini E, Masini G, Russo D, Andreola, F, Barbieri L, Belviso D, Lancelloti I and Pozzi P (2006) RF Thermal Plasma Treatment of Waste Glass and its Reutilization in Composite Materials *High Temp. Mater. Proc* **10** 207-218
[17] Bernardi D, Colombo V, Ghedini E and Mentrelli A 2005 Three-Dimensional Modelling of Inductively Coupled Plasma Torches *Pure Appl. Chem.* **77** 359-372
[18] Colombo V, Ghedini E and Sanibondi P 2010 3-D Static and Time-Depending Modelling of DC and RF Thermal Plasmas for Industrial Applications *Advances Sci. Technol.* **66** 1-10
[19] Bernardi D, Colombo V, Coppa G G M and D'Angola A 2001 Simulation of the Ignition Transient in RF Inductively-Coupled Plasma Torches *Eur. Phys. J. D* **14** 337-348
[20] Colombo V, Ghedini E and Sanibondi P 2010 A Three-Dimensional Investigation of the Effects of Excitation, Frequency and Sheat Gas Mixing in an Atmospheric-Pressure Inductively Coupled Plasma System, *J. Phys.D: Appl.Phys* **43** 105202 – 105214
[21] Gomez E, Rani D A, Cheeseman C R, Deegan D, Wise M and Boccaccini A R 2008 Thermal Plasma Technology for the Treatment of Wastes: A Critical Review *J. Hazard. Mater.* **161** 614-626
[22] Enger R, Piene K, Skreien N, Saethre A and Throslund J E (1975) Reduction of Silicon Dioxide by Means of a Carbon in Electric Furnace, United States Patent Number 3,877,359
[23] Abdellatif M 2011 DC Arc Smelting of Silicon: is it Technically Feasible? *Proc. Inter. Conf. Southern African Pyrometallurgy (Johannesburg, SouthAfrica 6-9 March 2011)* ed R T Jones and P. Den Hoed

AUTHOR INDEX

AUTHOR INDEX

Institute of Physics
Dirac House, Temple Back
Bristol BS1 6BE UK

CURRAN ASSOCIATES INC
proceedings
.com